TEXTILE SCIENCE AND TECHNOLOGY 13

ABSORBENT TECHNOLOGY

TEXTILE SCIENCE AND TECHNOLOGY

Volume 1	Open-end Spinning *by* V.Rohlena *et al.*
Volume 2	Processing of Polyester Fibres *by* O. Pajgrt *and* B. Reichstädter
Volume 3	Shuttleless Weaving Machines *by* O. Talavášek *and* V. Svatý
Volume 4	Fluorescent Brightening Agents *by* R. Williamson
Volume 5	Polypropylene Fibres - Science and Technology *by* M. Ahmed
Volume 6	Production and Applications of Polypropylene Textiles *by* O. Pajgrt *et al.*
Volume 7	Absorbency *edited by* P.K. Chatterjee
Volume 8	Needle Punching Textile Technology *by* V. Mrština *and* F. Fejgl
Volume 9	Industrial Textiles *edited by* J. Švédová
Volume 10	Modified Polyester Fibres *by* J. Militký *et al.*
Volume 11	Textile Processing and Properties: Preparation, Dyeing, Finishing and Performance *by* T.L. Vigo
Volume 12	Chemical Technology in the Pre-treatment Processes of Textiles *by* S.R. Karmakar
Volume 13	Absorbent Technology *edited by* P.K. Chatterjee *and* B.S. Gupta

TEXTILE SCIENCE AND TECHNOLOGY 13

ABSORBENT TECHNOLOGY

EDITED BY

P.K. CHATTERJEE
Nutech International Co.
East Brunswick, NJ 08816, USA

and

B.S. GUPTA
North Carolina State University
Raleigh, NC 27695, USA

2002

ELSEVIER
Amsterdam - Boston - London - New York - Oxford - Paris
San Diego - San Francisco - Singapore - Sydney - Tokyo

ELSEVIER SCIENCE B.V.
Sara Burgerhartstraat 25
P.O. Box 211, 1000 AE Amsterdam, The Netherlands

First edition 2002

Library of Congress Cataloging in Publication Data
A catalog record from the Library of Congress has been applied for.

ISBN: 0-444-50000-6

♾ The paper used in this publication meets the requirements of ANSI/NISO Z39.48-1992 (Permanence of Paper).
Printed in The Netherlands.

Preface

During the past two decades the disposable absorbent market has grown considerably and become highly competitive. New products appeared in the market with newly developed materials and more sophisticated structural designs. The challenge to remain competitive in this business primarily depends on how well the fundamental science and technology in this field are implemented to attain the product and process innovations. This book will serve the function of providing fundamental scientific and engineering information that are needed to develop and bring into the market place new and innovative absorbent materials and structures. It has all the aspects of absorbency that are essential to understanding the attributes of any absorbent materials or composites and to designing new products with unique characteristics.

The book contains some chapters, thoroughly revised, taken from an earlier edition entitled "Absorbency" published in 1985 by Elsevier Science Publisher and some new chapters that discuss more recent developments on the subject. The chapters are arranged in a sequence that facilitates a reader to advance progressively through fundamental theories of absorbency to more practical aspects of the technology. Each chapter provides the current status of the technology as well as the future prospects that would stimulate further research in the subject area. The book is intended for both the academic and the industrial scientists and engineers engaged in research and development on absorbency and absorbent products.

Our special thanks go to individual authors of the chapters of the current book as well as those who contributed in the previous edition. We would like to express our appreciation to Elsevier Science Publisher for providing us with the opportunity to participate in this publication. One of us (BSG) would like to extend his appreciation to the management of the Department of Textiles Engineering, Chemistry and Science, College of Textiles, North Carolina State University, for its encouragement and support.

Our acknowledgement and appreciation are also due to Dr. Sumedha Gupta Ariely for many helpful suggestions during the review of chapters and the cooperation of Ajit Moghe and Chad Wade in preparing many drawings and photographs and providing assistance all along, as needed. We would like to express our sincere thanks to Mrs. Susan Olsen who assisted us with secretarial help all through the preparation of the present manuscript.

And, finally, we take this opportunity to express our deep appreciation to our families for their understanding and support during the course of this project that required many long hours of activities during evenings and weekends.

East Brunswick, New Jersey — *Pronoy K. Chatterjee*
Raleigh, North Carolina — *Bhupender S. Gupta*

October 2001

List of Contributors

JOHN C. BERG
Department of Chemical Engineering, Box 351750, University of Washington, Seattle, WA 98195, USA

PRONOY K. CHATTERJEE
Nutech International Co., 331 McDowell Drive, East Brunswick, NJ 08816, USA

W. M. DOANE
National Center for Agricultural Utilization Research, USDA-ARS, Peoria, IL 61604, USA

GEORGE. F. FANTA
National Center for Agricultural Utilization Research, USDA-ARS, Peoria, IL 61604, USA

BHUPENDER S. GUPTA
Department of Textile Engineering, Chemistry and Science, College of Textiles, North Carolina State University, Raleigh, NC 27695-8301, USA

SACH KANGOVI
Simutel Associates, P. O. Box 252, Princeton Junction, NJ 08550, USA

LUDWIG REBENFELD
TRI/Princeton, P. O. Box 625, Princeton, NJ 08542, USA

ANTHONY M. SCHWARTZ
Nutech International Co., 331 McDowell Drive, East Brunswick, NJ 08816, USA

D. K. SMITH
Smith, Johnson & Associates, 2709 Edgewood, Provo, UT 84604, USA

VIVIAN T. STANNETT
Chemical Engineering Department, North Carolina State University, Raleigh, NC 27695, USA

THOMAS L. STAPLES
Superabsorbents R & D, The Dow Chemical Company, Midland, MI 48674, USA

RAYMOND A. YOUNG
Department of Forest Ecology and Management, University of Wisconsin, Madison, WI 53706, USA

Contents

Introduction

The current monograph, "Absorbent Technology," is a thoroughly revised and expanded edition of "Absorbency" published in 1985. As mentioned in that publication, the term "absorbency" is used in many different fields with different connotations. In the biomedical field, for example, absorbency refers to a phenomenon related to the consumption (degradation or decomposition) of a material in a biological or physiological environment. In skin biology, the diffusion of ionic species, gases, vapor or oil, through the epidermis is frequently referred to as absorbency. In this book, the term "absorbency" is described as an interdisciplinary scientific phenomenon that deals with the absorption of aqueous fluids by porous media, or more specifically, by fibrous and polymeric systems. The theoretical aspects of absorbency, as well as the structure, properties and performance of materials, currently being used or considered for use in absorbent structures, have been illustrated in various chapters. The technology related to manufacturing and designing of absorbent products has not been the primary objective of this book, neither has been the business aspect of these materials.

Absorbency criteria are characterized by the mode and the extent of the transport of liquid into an absorbing material. Numerous attempts have been made to define and predict absorbency using many classical theories, dating back to the nineteenth century, as well as those developed more recently, i.e. within the last few decades. Unfortunately, none of those theories can be utilized to effectively elucidate the intricate mechanism that governs fluid absorption in products that are available in the marketplace. None-the-less, many unique absorbent products have been developed over the years, which at least partially fulfill the need of consumers. The development of the majority of those products was not based on a scientific or mechanistic approach, but on trial and error techniques and on the intuitive imagination of innovative technologists. Since the absorbent products field has become more and more competitive, the necessity of more in-depth scientific study has become vitally important.

Product innovation comes through an intimate relationship between the development of fundamental science and technologies related to material characterization and the manufacturing processes. Industries, constrained by high labor costs and low profit margins, are inherently slow in investing in research on fundamental science where the payoff is not immediate but long term. Technology feeds on science and it is the lack of scientific study that has shortchanged the field of fluid absorbency for many decades. This edition of the book, it is hoped, will stimulate industry as well as academia to undertake research activities on more fundamental aspects of absorbency that will lead to technological innovations of the future.

Regarding the materials for absorbent products, cotton cellulose has been utilized since the time of the Egyptian civilization. Not until the end of the nineteenth century was wood pulp, in the form of fluff and then tissue, used as an absorbent material. Shortly thereafter rayon was introduced into the field. The growth of new materials for absorbent products remained status quo until late 1960's when a new class of materials called "superabsorbents" appeared and made a great impact in the field. Its absorption characteristic, particularly liquid retention capacity under pressure, is unique and helped the industry to come up with new and innovative, and more comfortable and hygienic, products. The development of superabsorbent materials goes back to 1960's and since then many different kinds with more efficient absorption properties have been developed and are still being developed.

The availability of such polymers with different degrees and distribution of cross-linking, and particle sizes and shapes, contributed to the advancement of absorbent products, in some cases radically and in others incrementally. The development of the textile fiber form of superabsorbent also presents substantial promise for disposable absorbent products; however, this could be realized in practice only if the cost is reduced or new applications technology, specific to superabsorbent long fiber, is discovered.

The incorporation of superabsorbent polymers in airlaid structures has advanced the technology further. There had been many attempts in the past to introduce tissue-superabsorbent composites into the absorbent products but none indicated as much promise as did the airlaid superabsorbent composites. This technology is expected to grow and will continue to have an impact on all types of absorbent products, including personal care, wound care, wipes, as well as those used in agricultural applications.

Among all the recent developments in the absorbent core technology area, the most prominent one has been the application of the preformed structures of one type or another. In sanitary napkin products, the use of preformed airlaid materials with short fibers has grown rapidly, but in baby diapers and adult incontinence products their adoption has been much slower than generally expected. We believe that the preformed airlaid material, with or without superabsorbent polymers, has tremendous potential for application in all types of absorbent products but its production technology needs to be advanced further in order to reduce the cost and improve the ease of handling on the converting line.

New polymers through biotechnology are beginning to proliferate. It is conceivable that tailor-made absorbent materials could be produced by identifying and isolating appropriate bacterial species or through modern genetic engineering. Possibilities of adapting hitherto unusable materials for improving absorbent products through effective utilization of plasma treatment, electron beam irradiation, acoustical treatment, or laser technology, are unlimited. Even though the prospects are excellent, the research activities on new materials are few compared to the overall activities directed at reducing the cost of the current raw materials.

Absorbent technology refers to managing fluid with a compatible porous medium. There are two different ways to approach the challenge involved in improving any absorbent product. The first, which is most widely followed, is to modify the porous structure through innovative composite constructions and/or incorporation of improved materials to control the fluid leakage. The second, which has been given less attention by far, is to change the fluid characteristic as it enters the product so that it can be more effectively managed within a given structure. The former requires more physics and engineering skills, not accounting for the material development part, and the latter requires more chemistry and biochemistry skills. The patent literature clearly indicates that the emphasis is towards composite design criteria that would produce incremental benefit in leakage reduction, comfort and/or hygiene. The efforts towards any material development, targeted to come up with altogether new material characteristics, which would generally require a long-term commitment, have been drastically reduced during the past two decades. There were also sporadic efforts in changing the fluid characteristics, as revealed in the past patents, but those were never pursued to perfection. Two schemes were revealed: one, to liquefy the proteinaceous thick body discharge like menstrual fluid, and other, to thicken the watery body fluid, such as urine. Any practical benefit from these approaches would also require longer-term research commitment.

"Interactive Computer Graphics for Communication of Chemical Structure", an idea originating from Harvard University, is being transformed into a powerful technology for chemical synthesis. Recent work in many noted universities around the world is expected to bring the concept to the real world of the laboratory chemist. Designing new polymers for absorbent materials will become much simpler once this technology is firmly established. Among other computer technologies, computer-aided mathematical modeling is expected to have unique roles in designing new and improved absorbent structures.

The current trend concerning products in the marketplace is towards thinner and/or smaller products with maximum efficiency. Obviously, for the development of such an article a material that absorbs large quantity of fluid in a small volume would be desirable as absorbent core structure. The development of such structures are being pursued by various means, including the modification of cellulose and the incorporation of a large quantity of high gel strength superabsorbent in a fibrous capillary network. Superabsorbent holds the fluid by a different mechanism than does the fibrous capillary material and, therefore, the combination, if properly designed, would present a superior absorbent structure. However, as we approach a thinner and thinner product, we reach a thickness limit, because the product must possess a "critical minimum volume" to absorb a specified amount of fluid. Accordingly, if the thickness is less than the critical value, assuming that all other parameters remain the same, the product would fail to hold the amount of fluid that is desired to be held under a given circumstance. Obviously, a question may arise, can the product be made thinner than this "critical thickness" and the answer is, yes. A solution is to develop an absorbent core that will grow only when it interacts with fluid and thus provides localized critical volume to absorb and retain the fluid locally. Further incoming fluid may be held at the same place and when the latter reaches its expansion limit the fluid will be migrated to the adjacent region that will grow similarly. This way, one can create a super thin product, which to start with may not have adequate pore volume to accommodate the total liquid but it will spontaneously generate the space on demand. Development of new materials or composites along this direction was disclosed in several patents in the 1960s through the 80s, however, the concept was not pursued diligently in the years that followed. Recently, a few patents have emerged on superabsorbent composites that expand on wetting. Such a renewed interest on an important development in the field would be welcomed by the personal hygiene industry.

All absorbent products are composites; it is not the individual materials that determine the final performance of the product but their interactions with each other that influence the characteristics of a composite. Due to the lack of hard data on composite properties, it is difficult to predict the ultimate behavior of the product. The science and technology of composites in general have been extensively studied by automotive and aerospace industries. The technology already developed by those industries could be used as a basis for the development of absorbent structures. To achieve this goal more effectively, the absorbent product industry, jointly with academic institutions, should make a concentrated effort to uncover first the mystery of the natural absorbent polymer composites, and then design a synthetic composite that simulates the natural polymer.

As stated earlier, the main objective of this book is to present in detail the current state of the art with a brief perspective on the technological prospect. The chapters have been organized in a manner such that readers can get a coherent picture of the interrelated concepts that define the science and technology of absorbency.

The mechanism of liquid flow in porous structure has been illustrated in Chapter I. This is a subject that has been extensively studied in soil physics but has not been seriously pursued in fiber science. Many classical theories on liquid flow in porous materials have been applied to fibrous assemblies in order to predict absorbency, but with limited success. Absorbent articles are usually composites of different types of polymeric materials with highly complex design and intricate pore geometry. The problem of defining the pore structure in sufficient detail complicates the application of standard mathematical treatments. The chemistry and biochemistry of absorbing fluid and its interaction with absorbent elements in a composite structure further add to the complexity of the problem. In this chapter, classical concepts are given a detailed review along with a discussion of their limitations when applied to fibrous assemblies. Structural models are presented that could be used to characterize pore geometry and fluid flow behavior in fibrous masses.

The pores of inter-particulate spaces are occupied by gas or vapor at atmospheric pressure. Contrary to the assumption in many classical theories, the liquid enters the pores by bulk convection that cannot be accounted for by pressure external to the system. In reality, the driving forces for wicking in an absorbent medium arise from the free energies of the absorbent elements, which is the main theme of discussion in Chapter II. The fundamentals of force and energy applicable to different phase interfaces, with practical applications and measuring techniques, are discussed.

For textile fiber assemblies, e.g. woven fabrics, needlepunched and other high bulk structures, the studies have indicated that the ability to wick and hold fluid is greatly influenced by the physical and mechanical properties of fibers and the structural and compressional characteristics of fabrics. Many factors affect the latter, including the technology used in bonding, the weight of the fabric, and the nature and the composition of the blend. An understanding of the role these and other factors play in absorbency will be important in designing new and functionally more optimum products of the future. These informations as well as a theoretical model that can be used to predict the absorbency performance of textile fiber structures are presented in Chapter III.

Chapter IV describes the application of computational modeling to certain transport related phenomenons in absorbent technology, including penetration absorbency and pneumatic transport of fibers in a web-forming machine. Such computation modeling, when applicable would be a cost-effective tool to study and solve a problem efficiently and economically.

Among many additives, which influence absorbency behavior, the most significant class of materials used is known in the art as "surfactants". No treatment of absorbency would be complete without a discussion of the role of surfactants. The two major wetting parameters of absorbency, the surface energy of fibers and the surface tension of liquid, can be modified by this compound. Chapter V deals with the manner in which interfacial property influence absorbency and the manner in which surfactants influence the properties.

The next four chapters deal with the materials for various functional layers, i.e. the absorbent core and the facing of an absorbent product. Chapter VI, for example, describes the properties and morphology of natural and synthetic fibers used in various types of absorbent articles. While some of these fibers, particularly those which are hydrophobic in nature, may not have a direct relationship to absorbency, they nevertheless possess the potential for improving the functioning of the network that imbibes and holds fluid. An understanding of their structure-property relationship will aid in achieving a deeper perspective of the behavior of a product containing the material. Several specialty fibers, mentioned in this chapter, which were originally designed for different types of applications, are finding important roles in designing improved and more comfortable absorbent products.

Chapters VII, VIII and IX deal with the science and technology of superabsorbent materials developed within the past three decades. The subject is divided into three chapters as follows: cross-linked cellulose and cellulose derivatives, synthetic superabsorbents, and polymer-grafted cellulose and starch. A superabsorbent absorbs a considerable amount of fluid and, because of its unique structure, is able to retain a substantial quantity of liquid in its internal network, significantly more than a conventional absorbent fiber such as cotton, rayon or wood pulp, is able to hold.

Certain critical functional aspects of absorbent products, including the application of new and innovative composites and nonwovens, are the topics of discussion in Chapter X. With the advent of new types of fibers and new and unique fabric formation technologies, more cost and functionally effective nonwovens are developed and used in today's absorbent products.

Absorbent products being highly engineered structures and existing in a very competitive market, the survival of a specific brand depends on the scrutiny of and acceptance by consumers. A sound scheme of characterization of materials, at different levels of operation, that supports selection of materials, design and optimization of intermediate components, and field evaluation of final assembly, must be incorporated. A detailed discussion of various test methods, some used routinely by industry and others employed as research tools, is given in Chapter XI.

The book concludes with Chapter XII, comprising specific comments on absorbent composite structures, fluid characteristics, product design, fiber processing and technology forecasting. The chapter also includes general comments on absorbent products and research and potential future developments.

East Brunswick, New Jersey
Raleigh, North Carolina

Pronoy K. Chatterjee
Bhupender S. Gupta

October 2001

Absorbent Technology.
P.K. Chatterjee and B.S. Gupta, editors.

CHAPTER I

POROUS STRUCTURE AND LIQUID FLOW MODELS

PRONOY K. CHATTERJEE
Nutech International Co., 331 McDowell Drive, East Brunswick, NJ 08816 (USA).

BHUPENDER S. GUPTA
College of Textiles, North Carolina State University, Raleigh, NC 27695-8301 (USA).

Contents

1. INTRODUCTION

The absorbency phenomena are characterized by the mode and the extent of transport of liquid into an absorbing material. For absorption to occur, the main driving force must come from the intrinsic liquid attraction capacity of the material itself; while an externally imposed force, such as gravity or pressure, may play a secondary role in affecting the absorption process. The absorbency phenomena are in general limited to systems where there is affinity between the liquid and the absorbent. However, there are many other factors which have a significant influence on the absorbency phenomena. Vapor sorption is an indication of the affinity between the molecules of the absorbent and of the absorbed. This is characterized by the equilibrium adsorption isotherms.

The driving force for absorption of the bulk of the liquid into an absorbing material, which is in most cases a porous medium, is the capillary pressure. The absorption mechanism is traditionally interpreted as a flow through a system of capillary tubes using standard capillary flow equations. The absorbing substrate is usually defined as a porous medium with interconnected pores of various sizes, where the flow is characterized by the existence of saturation gradient along the direction of flow. The unsteady flow of absorbency can be studied by applying Darcy's Law and making an analogy of diffusion type of transport which has been extensively used in soil physics.

An important aspect of absorbency of certain specific materials is the partial dissolution and swelling of the absorbing material itself. This phenomenon of liquid retention by swelling is becoming more important with the introduction of "superabsorbents" which can swell and retain absorbed liquid many times their own weights.

The techniques of measuring absorbency are designed to determine the amount of liquid absorbed. In some tests the special advances of the liquid are monitored. The measurement techniques can be broadly divided into spontaneous (demand) liquid absorption and liquid retention tests. They measure essentially the equilibrium absorbency reached by two different modes: absorption from dryness and exsorption from saturation.

The absorbent products are composites of materials, and like any other composites, the structure-property relationship is difficult to derive. The influence of hydrophobicity, hydrophilicity, repellency, water resistivity, absorption—exsorption hysteresis, porosity, swellability and other factors have to be taken into consideration in deriving the structure-property relationship of absorbent articles. Some investigation in this area is selectively discussed in the latter part of the chapter.

2. MOISTURE SORPTION

Even before an absorbent structure is in contact with the absorbed liquid, it is in contact with the vapor. Adsorption of water molecules takes place below a critical temperature, due to the Van der Waals' forces between the vapor molecules and the solid surfaces of the structure. Detailed discussion on adsorption of gases or vapors on solid surfaces is beyond the scope of this chapter. Fundamental discussions on the adsorption phenomena can be found in the literature on the subject [1-4].

The amount of moisture adsorbed by a given solid substrate depends on the vapor pressure and the temperature. The higher the vapor pressure, or the lower the temperature, the higher the amount adsorbed.

A plot of the amount of fluid adsorbed against vapor pressure at a constant temperature generates an "adsorption isotherm". At any given point in such an experiment, the system is at thermodynamic equilibrium, i.e., the chemical potential of the vapor is equal to that of the adsorbed film. An increase in vapor pressure will cause an imbalance in chemical potential, and more vapor has to be transferred to the adsorbed layer to restore the equilibrium. This is why the theories often treat adsorption as an equilibrium process and the isotherms obtained are often referred to as "equilibrium adsorption isotherms".

Langmuir [5] used the kinetic theory of absorption equilibrium to predict the isotherms. The so-called "Langmuir type" isotherm is characterized by the amount adsorbed approaching a limiting value, which corresponds to complete monolayer coverage.

Brunauer [6] identified five general types of isotherm as shown in Fig. 1. Type I is the Langmuir type. Type II is most common and reflects the occurrence of multilayer formation starting at some point (B) rather early in the adsorption process; "point B" often selected rather arbitrarily corresponds to the completion of monolayer coverage. Type III is somewhat rare. The leveling off in Types IV and V indicates saturation of fine capillary through condensation.

Adsorption by microporous solids has been found to involve hysteresis. If isotherms are obtained by decreasing the pressure from the saturation pressure, desorption will take place. The "equilibrium desorption isotherm" will be different from the adsorption isotherm. In general, the desorption curve shows a higher amount of moisture adsorbed than that of the adsorption curve at the same vapor pressure. The hysteresis effect has been explained by the capillary condensation theory, which is based on the hysteresis of interfacial tension [7-9].

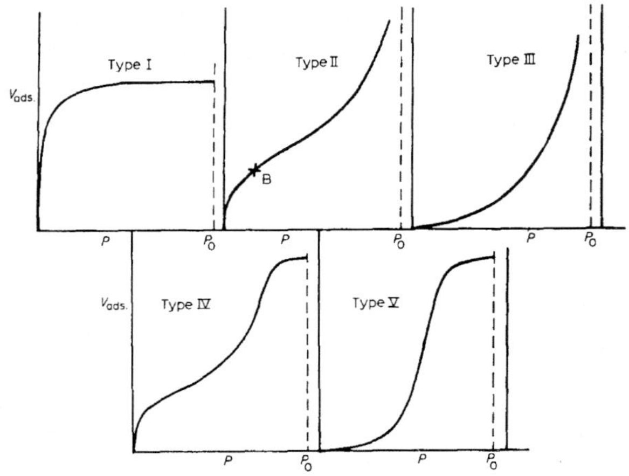

Fig. 1. Five types of adsorption isotherms according to Branauer [6].

Figure 2 shows the isotherms of kier-boiled cotton [10] where the abscissa represents the relative humidity in place of relative vapor pressure. The shape of these isotherms corresponds to Type II and is characteristic of cellulosic fibers. The amount adsorbed increases slowly at intermediate humidity and more rapidly at a relative humidity of about 70%. The area bounded by the two isotherms (adsorption and desorption) is the equilibrium area within which any combination and sequence of equilibrium adsorption and desorption will be confined.

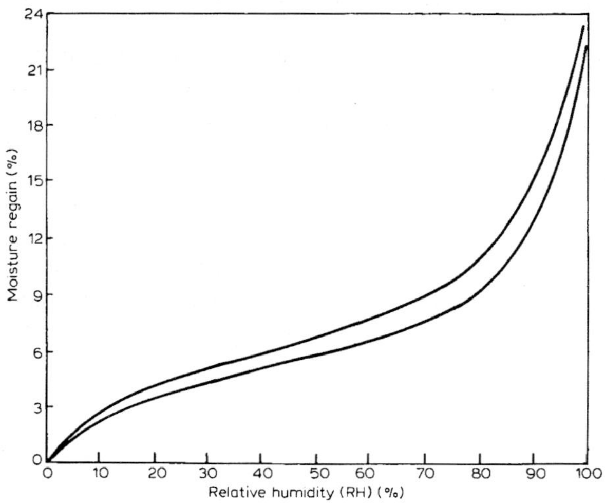

Fig. 2. Adsorption (lower) and desorption (upper) isotherms of kier-boiled cotton [10].

The adsorption phenomenon is useful in the measurement of surface areas of porous media since the amount of gas adsorbed on a solid surface is dependent on the total area of the surface. The Langmuir approach [5] and the more popular Brunauer, Emmett and Teller (BET) method [6] are useful for such a purpose.

In the case of absorbent fibers, e.g., cotton, pulp, rayon, etc., the moisture sorption is complex due to the interaction of moisture and the material. The moisture sorption phenomenon in those cases encompasses not only the relationship between regain and humidity, but also phenomena associated with hysteresis, heat effect, dimensional changes and elastic recovery effects owing to the limited swelling of fibers [11]. All these effects cannot be explained by a single theory.

All natural animal and vegetable fibers have chemical groups in their molecules, which attract water. Most synthetic fibers, on the other hand, contain few if any water attractive groups and this accounts for their low moisture sorption. In the case of hydrophilic fibers, the initial water molecules from the atmosphere will be adsorbed directly onto the hydrophilic groups. The subsequent ones will either be adsorbed on other available hydrophilic groups or form a new layer on top of the previous layer. The directly attached water molecules are probably firmly fixed on the fiber substrate, whereas the subsequent layers are more free to move. The crystallinity of fibers also plays an important role in moisture sorption. In the crystalline regions, the fiber molecules are closely packed together in a regular pattern and therefore water molecules do not penetrate into this region easily. However, if the basic molecule gets hydrated, the crystalline region will open up allowing water molecules to penetrate in. The moisture sorption then allows one way of estimating crystalline/non-crystalline ratio in hydrophilic fibers when the fiber molecule does not hydrate in the presence of water.

Urquhart [12] and Hermans [13] have put forward a theory of hysteresis based on molecular effects on cellulose structure by moisture sorption. The theory is based on the formation and interaction of cross-links in the amorphous region of cellulose with water molecules. It predicts that an initially dry specimen of cellulose will always retain a higher number of cross-links, resulting in less water adsorption than an initially wet specimen in the same atmosphere.

Limited swellability of hydrophilic fibers on moisture sorption is also an important aspect. The fiber molecules or microfibrils are somewhat pushed apart by the adsorbed water molecules. The resulting distortion of the fiber sets up internal stresses which influence the moisture sorption process. If there is a mechanical hysteresis of the fiber, it will accentuate the adsorption hysteresis.

Usually, at a low level of moisture sorption in cellulose fiber, the change in volume of the fiber is less than the volume of the adsorbed water, which is ascribed to the fitting of water molecules to the space available within the structure. As adsorption proceeds, the increase in volume becomes equal to the volume of water added, indicating that the water molecule is packed in the same way as liquid water and is spreading in the polymer structure.

At a very high humidity, the moisture is held as liquid water by the capillary forces in spaces between fibers or crevices on the fiber surface. The theory relating the equilibrium vapor pressure in a capillary to the relative humidity [11] predicts that, when given specimen is exposed to an atmosphere of a given relative humidity, there will be no tendency for water to evaporate from capillaries having radius below a critical value which can be defined by Kelvin's equation:

$$\ln\frac{p_a}{p_\infty} = -\frac{2\gamma M_w}{\rho_w R\tau a} \tag{1}$$

where p_a is the equilibrium vapor pressure over a concave surface (of water meniscus) of radius a, p_∞ is the equilibrium vapor pressure over a plane surface, γ is the surface tension of water, M_w is the molecular weight of water, ρ_w is the density of water, R is the gas constant and τ is the absolute temperature. Since p_∞ is defined as the saturation vapor pressure, the ratio p_a/p_∞ is known as relative humidity RH.

The theory also predicts that at higher relative humidities, water will be retained in larger capillaries increasing the moisture regain characteristic. The capillary theory was the first to provide a general explanation of moisture sorption over a wide range of humidities, particularly at high humidities.

Stamm [14] refuted the concept of capillary water, particularly in the case of wood or wood products, such as cellulose fiber. He remarked that there was plenty of evidence to indicate that only a small fraction of the water can be held in pre-existing capillaries, less than 2% of the cell wall volume of wood. He supported the idea of polymolecular adsorption involving the formation of a solid solution.

As described by Stamm [15], moisture is held by cellulose fibers by different ways and the sorption process goes through a gradual transition. The most firmly held water is the water of constitution. This water is released only when the temperature is raised to the degradation temperature of cellulose. Then there is monomolecularly held water, which is attached to the accessible surface of cellulose by hydrogen bonds. Excess water, if it is absorbed from an atmosphere of 20 to 90% RH, is polymolecularly held water in solid solution. A small fraction of the total amount of water is condensed in very fine pores which are less than 0.02 micron in diameter. There is no evidence that any condensed water in capillaries larger than 0.02 microns exists if the cellulose fiber is conditioned at an atmosphere of 90% RH or lower.

The structure of polymolecularly held water should be different from the structure of liquid water. The free liquid water exists in the form of associated molecules of water bonded with each other by hydrogen bonding. The thermodynamics of polymolecularly held water molecules are different from the liquid water because water molecules are restructured and preferentially oriented in the former case.

At 90% RH, if a relatively dry cellulose fiber is conditioned, the fiber will absorb 6 to 7 monomolecular layers of water, which can be considered as polymolecularly held water, plus a small amount of condensed liquid water in capillaries smaller than 0.02 micron in diameter [15]. Depending upon the supramolecular structure, as well as the fine structure of the cellulose substrate, the total amount of moisture could be as high as 15%, but still no liquid water would exist in capillaries larger than 0.02 micron in diameter.

The effect of temperature upon the adsorption of water vapor by cotton has been critically studied by Urquhart et al. [16,17]. A rise in temperature in the range of 10 to 50°C caused a decrease in equilibrium moisture content at all vapor pressure. The moisture content continues to drop if the temperature is raised above 60°C, provided the RH is maintained below 85%. Above 85% RH, however, the equilibrium moisture content increases. This sudden change in isotherms at higher temperature and relative vapor pressure has been attributed to the plasticization effect of cellulose by heat and moisture; consequently, the structure is opened up and more hydroxyl groups become available for water vapor adsorption [15].

Howsmon [18] deduced from available data that up to about 25% of the total water adsorbed by viscose rayon would be all bound water. The upper limit of the amount of bound water was measured by Carles and Scallan [19] using NMR spectroscopy; some of the reported values are: 0.23 g/g for ground sprucewood, 0.15 g/g for cotton and 0.3-0.33 g/g for sulfite pulp and kraft pulp.

Adsorption isotherms of natural fibers and proteins [20-22] are shown in Fig. 3. These are all Type II isotherms. Adsorption isotherms of certain hydrophilic polymers shown in Fig. 4 are similar to Type III adsorption [20,23-28].

In general, for natural polymers there is a one-to-one correlation between the number of polar groups and the number of water molecules adsorbed [18,20]. For synthetic polymers, except for those containing hydroxyl, carbonyl, or peptide groups, the ratio of water molecules adsorbed to polar groups is much less than unity [29]. Adsorption of non-swelling vapors was studied by a number of re searchers. Columbo and Immergut [30] studied the adsorption of benzene on cotton. Bhatia [31] studied the adsorption of butanol and carbon tetrachloride on cellophane. Tremaine and Gray [32] employed a gas chromatographic method to determine the isotherms for adsorption of n-decane, 1,4-dioxane and other vapors on cotton and ramie fibers. The results showed that the adsorption of non-swelling vapors takes place almost entirely on the external surface of the fibers and very little in the internal structure.

3. FLUID FLOW IN CAPILLARY TUBE

Even though absorbent fiber systems, e.g., paper towels, fluff pad and non-woven fabric have complex pore structures, the absorbency data on those materials are frequently treated using a simple capillary tube flow model. The information obtained from such treatment is useful for the qualitative characterization of the absorption process. The fundamental principle underlying such a treatment is outlined in this section.

The liquid moves into a porous medium by the capillary pressure, i.e., the differential pressure across the liquid-air interface due to the curvature of meniscus in the narrow confines of the pores. For wetting liquids, this pressure is effectively a pressure drop compared to the atmospheric pressure. The magnitude of the capillary pressure is commonly given by the Laplace equation as applied to idealized capillary tubes [1]:

$$p = \frac{2\gamma \cos\theta}{r_c} \tag{2}$$

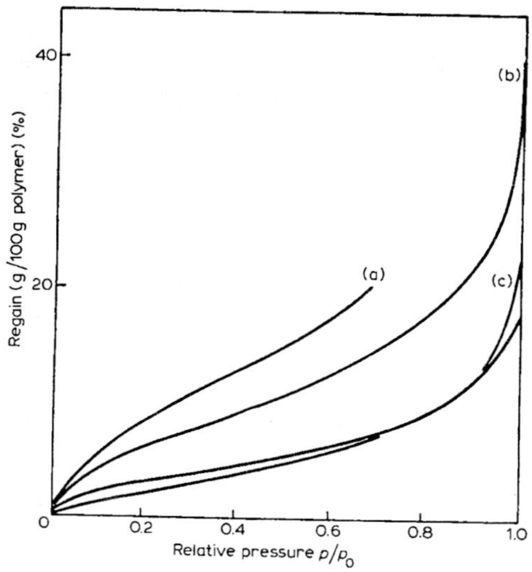

Fig. 3. Adsorption isotherms for the natural fibers and proteins [20]. (a) N.F. pectin at 29° C; (b) wool; (c) cotton; (d) secondary cellulose acetate at 30° C.

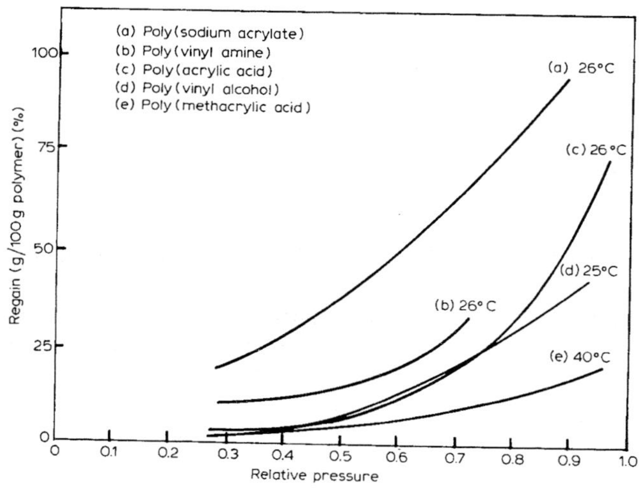

Fig. 4. Adsorption isotherms of more hydrophilic polymers [20, 23-28].

where r_c is the capillary radius, γ is the surface tension of the advancing liquid and θ is the contact angle at the liquid-solid-air interface. With an idealized tube structure, the Hagen-Poiseuille law for laminar flow through pipes can be employed. The law states that the volumetric flow rate is proportional to the pressure drop gradient along the tube [33,34]:

$$q = \left(\frac{r_c^{\,2}}{8\eta}\right)\frac{\Delta P}{L} \tag{3}$$

where q is the volume flux, r_c is the tube radius, η is the fluid viscosity, L is the wetted length of the tube, and ΔP is the net driving pressure (pressure drop across L). As applied to capillary rise, L would be the height of rise and ΔP would be $p - \rho_l g L$ for upward flow where p is the capillary pressure for a given capillary tube as illustrated by eq. 2, ρ_l is the specific gravity of fluid and g is gravitational acceleration. By replacing q by dL/dt (t is time), eq. 3 is transformed to the following differential equation [35]:

$$\frac{dL}{dt} = \left(\frac{r_c^2}{8\eta}\right)\left(\frac{p}{L} - \rho_l g\right) \tag{4}$$

When an equilibrium is reached, the upward capillary driving force p equals the weight of the column of liquid, the net force on the liquid is zero and the rising of the liquid stops. The equilibrium capillary rise can be expressed as:

$$L_{eq} = p / \rho_l g \tag{5}$$

The capillary rise between the time of initial contact and the final equilibrium was obtained by integrating eq. 4 as derived by Lucas and Washburn [35,36]:

$$\ln\left(1 - \frac{L}{L_{eq}}\right)^{-1} - \frac{L}{L_{eq}} = B_1 t \tag{6}$$

where B_1 is a constant equal to $r_c^{\,2}\rho_l g/8\eta L_{eq}$. According to eq. 6, the capillary rise approaches the equilibrium value asymptotically.

At low values of t, where L is very small compared to L_{eq}, eq. 6 can be represented by the following approximation, which is commonly known, as the Washburn equation [36]:

$$L = \sqrt{\frac{r_c \gamma \cos\theta}{2\eta}}\, t^{1/2} = k_0 t^{1/2} \tag{7}$$

where k_0 is a constant. This proportionality between L and $t^{1/2}$ has been confirmed experimentally in the liquid rise in capillary tubes [37], in the movement of liquid front during liquid imbibition studies [38,39], and in paper chromatography [40].

For horizontal flow, the gravity term in eq. 4 is zero and the solution is the same as the Washburn equation (eq. 7). The only difference in this case is that the capillary flow will go on indefinitely instead of approaching an equilibrium distance as in the case of vertical rise. In any case, the L vs. $t^{1/2}$ proportionality is supposed to hold true for both cases as long as L is small compared to L_{eq}.

In any fibrous structure, the capillaries are neither cylindrical nor all of them arranged in parallel. It is also doubtful whether the definition of the capillary radius, as implied in the Laplace equation, is the same as that defined in the Poiseuille equation. In the former case, r_c must be defined as the radius of wet and dry interface or simply the dry capillary radius (r_d), whereas in the latter case strictly represents the radius of wet capillary (r_w).

In cellulosic structures, it is well known that water influences the capillary dimension. If that influence is instantaneous, Chatterjee [41] suggested that the basic relationships in the Washburn equation should not be changed except that r_c in eq. 7 be changed to a function of r_w and r_d. By substituting r_w for r_c in the Poiseuille equation and r_d in the Laplace equation and rederiving the Washburn equation, it can be shown that r_c in eq. 7 transforms to (r_w^2/r_d) which may be defined as the effective radius r_e. Although the expression of k_o will be slightly changed by such a transformation, the relationship between L and t will remain the same.

If, however, the influence of water on the capillary dimension is not instantaneous but slow, the time and advancing distance relationship as expressed in the Washburn equation no longer remains the same as shown in the equation. Under such conditions, the Washburn equation is not valid. This situation may arise if the structure is composed of a highly swellable type of fiber. As the wicking proceeds in that structure, water continuously diffuses into the fiber structure resulting in a continuous change of the dimension on the wet capillaries. The radius of the wet capillary, r_w in the Poiseuille equation, and the radii function (r_e) in the modified Washburn equation become a function of time. Hence, the relationship between L and t as shown in eq. 7 does not hold true.

To some extent, all natural fibers swell in water and the swelling is always time dependent; therefore r_e is never a constant factor. However, as a first approximation, r_e may be considered a constant if the degree of swelling is not too high and the rate of swelling is fast. Perhaps cellulose fiber belongs to this category and, therefore, the mechanism of wicking in most cellulosic sheets is in close agreement with the Washburn relationship.

To test how closely the equation is applicable to the mechanism of wicking in a sheet consisting of wood pulp, let the Washburn equation be expressed in a general form as $L = k_0 t^m$ where k_0 and m are constants. A plot of $\log L$ versus $\log t$ should give a straight line with a slope, m, approximately equal to 0.5 where the mechanism fits with the Washburn equation or a curve where it does not fit with the equation.

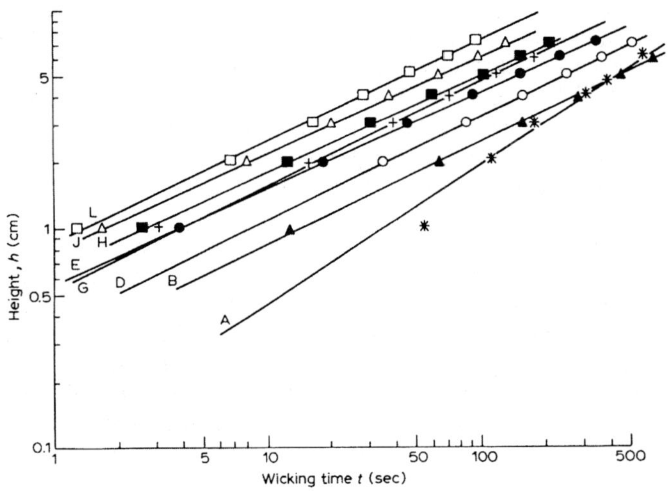

Fig. 5. Klemm test data for cellulose fiber sheet [41].

Figure 5 represents the typical plots of Klemm test [41] data for sheets containing cellulose fibers with hemicellulose content ranging from 4 to 8%. The constants, m and k_0 were calculated from the plots and it was evident that the wicking mechanism was in general agreement with the Washburn equation. In most cases, the slope, m, was between 0.46 and 0.48.

Since the slope is practically constant, the k_0 values could be used for the evaluation of the relative rate of wicking in different sheets. However, from the tests run with a large variety of sheets, it was found that the equation was not applicable in the case of a sheet containing a highly swellable type of fiber or for pulps having an excessively high amount of hemicellulose (over 15%).

At very small t (usually on the order of fractions of a second), there occurs a retardation in the capillary rise, i.e., the initial rise is slower than that expected from the Washburn equation. This has been observed by a number of investigators [37,42].

In eq. 4, as t approaches zero, the (initial) rate of advance of the meniscus dL/dt becomes infinitely large (due to $L = 0$). This anomaly is due to the fact that the Hagen-Poiseuille law, which deals with steady state flow, does not take into account the inertial effect. A more rigorous derivation by Szekely et al. [43] takes into account the kinetic energy change and the energy dissipation by the convergent flow at the entrance of the capillary tube. Thus, in place of eq. 4, the following differential equation is applicable:

$$\left(L+\frac{7}{6}r_c\right)\frac{d^2L}{dt^2}+1.225\left(\frac{dL}{dt}\right)^2+\frac{8\eta}{\rho_l r_c^2}L\frac{dL}{dt}=\frac{1}{\rho_l}(p-\rho_l gL) \qquad (8)$$

With the same initial condition, the solution of this equation can be obtained by numerical methods as was done by Szekely et al. [43]. The calculated results showed an initial "retardation" (which is in accord with the experimental values at very small *t*). For liquids having the same viscosity as water, this "retardation" lasts only a fraction of a second. After this initial period, the Washburn equation is in good accord with the experimental data.

These results reiterate the validity of the Washburn equation for the entire time period except at the very initial stage. For more viscous liquids, the retardation period becomes longer, only in such cases it may be worthwhile to use the more rigorous treatment in place of the Washburn equation.

It was also been observed that the curvature of the moving meniscus is deformed considerably from the static curvature [44,45]. Empirical corrections have been made to account for the changes in the observed contact angle with liquid flow velocity [45]. The contact angle tends to increase as the velocity of the meniscus increases, the tangent of the observed contact angle has been estimated to vary linearly with the meniscus velocity [46].

Recognizing the fact that an absorbent medium usually has pores of different sizes, a model of a bundle of capillary tubes of different radii has been considered [47]. In this model, each tube has a uniform radius throughout its length and the capillary rise in each tube occurs independently from all other tubes. Also according to this model, the liquid advance would be faster in the larger size tubes.

This model has the advantage of being simple. However, it does not describe realistically the fluid flow characteristics through porous media where the pores are interconnected and interdependent to form a three-dimensional network system. A fundamental flaw of this model is that for a given fluid, the tube radius predetermines both the driving force and the resistance of the flow; and since the resistance to the flow is related to the square of the tube radius (see eq. 3), the resistance in large-pore media would tend to be underestimated in comparison to the smaller pore media. The driving force and the resistance to flow, although related, are separate entities and have to be determined independently.

4. PORE STRUCTURE AND STEADY STATE FLOW THROUGH POROUS MEDIA

Absorbent materials such as paper, textiles, sponges, etc., are porous media with structures much more complex than that assumed in previous sections. The pores in such structures have different sizes and are interconnected in three-dimensional networks so that even for one-dimensional flow the fluid has to follow tortuous paths, rather than a straight line as in the case of a capillary tube. Also due to the different sizes of pores, the fluid located at partially saturated regions will experience a multitude of different "pulls" from different pores. The pore structure of the absorbing medium therefore has a significant influence on the fluid flow process.

Studies of fluid-flow through porous media are found in such diverse fields as soil physics, petroleum engineering, liquid filtration, paper chromatography, etc. In all such diverse systems, Darcy's law [48] has been applied for slow linear flows. Darcy's law is an empirical formula, which describes the kinetics of fluid flow through porous media in terms of the driving force gradient and the permeability of the medium.

There are, however, different ways one can look at the flow through porous media. It can be viewed as flow inside conduits. The Kozeny-Carman [49-51] approach used "hydraulic radius theory" to estimate the equivalent channel diameter of the conduit of any complicated shape. It can also be viewed as flow around submerged objects, which causes a drag resistance. Iberall's treatment [52] and others [53-56] estimated the flow resistance of the porous medium by adding up the drag resistance of the individual particles that make up the medium. The various concepts and theories on flow through porous media which have been developed over the years are summarized below.

4.1. Porosity

Pores are void spaces, which are distributed extensively throughout the volume of a porous medium. Porosity ϕ is the ratio of the void space in a porous medium over the total bulk volume of the medium. Porosity is thus a dimensionless quantity between 0 and 1. The fraction of the bulk volume that makes up the solid walls is thus 1 - ϕ.

If ρ_{bulk} is the bulk density of the porous medium and ρ is the density of the material making up the medium, then the porosity is given by

$$\phi = 1 - \frac{\rho_{bulk}}{\rho} \tag{9}$$

If the porous medium is made up of regularly packed particles of uniform size and shape, the porosity can be calculated from purely geometrical consideration. For example, a packed bed of spheres in "face-centered cubic" packing will have a calculated porosity of 0.259 [57].

Assuming there is no change in dimension, if the entire pore space is filled up with the absorbed liquid, the porous sample is said to be fully saturated and the maximum absorption capacity (C) of the sample can be defined as the mass of liquid absorbed per unit mass of dry solid medium (eq. 10 where ρ_l is the density of the flowing liquid).

$$C = \frac{\rho_l}{\rho} X \frac{\phi}{1-\phi} \tag{10}$$

If the void space is partially filled, the sample is only partially saturated, and the saturation level s is the fraction of the pore space filled by the liquid.

4.2. Pore Size Distribution

The total pore space consists of pores of different sizes. It is not possible to specify "pore size" geometrically, since the pores are irregularly shaped and are not well defined entities. The concept of pore size distribution is therefore equally nebulous. Scheidegger [58] offered a possible way out of the dilemma by defining "the pore diameter at any point within the pore space as the diameter of the largest sphere which contains this point and remains wholly within the pore space." This definition suggests that even in the more simple porous structures, short of a capillary tube, there exists a more or less continuous function describing the pore size distribution.

Following this definition, let $\alpha(\delta)$ be the pore size distribution (density function) describing the fraction of the total pore space with pore diameters between δ and $\delta + d\delta$. This is a dimensionless, volume-weighted distribution because it is a volume fraction.

Let $M(\delta)$ be the cumulative distribution describing the fraction of total pore space with pore diameter equal to or smaller than δ, then:

$$M(\delta) = \int_0^{\delta} \alpha(\delta) d\delta \tag{11}$$

Consider a porous medium uniformly and partially saturated by a wetting liquid. At equilibrium, only pores below a certain diameter δ_l will be filled because their capillary suction pressures are higher than the rest. The saturation s of the medium is therefore related to (equal to) $M(\delta_l)$.

Furthermore, each value of δ corresponds to a capillary pressure p given by eq. 2. The pore size distribution $M(\delta)$, therefore, corresponds to a "capillary pressure distribution function" $N(p)$. Thus, there is an "equilibrium-saturation-capillary pressure" relationship, $s = N(p)$ which is characteristic to a given porous medium and the fluid system.

In fact, there usually exists more than one such "equilibrium saturation-capillary pressure" relationship, depending on whether the fluid is being filled up or being withdrawn. Such hysteresis phenomena and their measurement are discussed in Chapter XI.

4.3. Darcy's Law

Darcy's law is an empirical formula based on an experiment performed by Darcy in 1856 [48]. Darcy's law is usually considered valid for a linear and slow steady state flow through porous media:

$$q = -K \frac{\Delta P}{L_0} \tag{12}$$

where q is the volume flux in the flow direction (i.e., volumetric flow rate per unit cross-sectional area of flow), ΔP is the net pressure head that causes the flow and L_0 is the length of the sample in the direction of flow. K, the proportionality constant, is the flow conductivity of the porous medium with respect to the fluid. The higher the value of K, the lower the flow resistance of the fluid, and vice versa.

K is often defined as k/η where k is the permeability of the medium, and η is the fluid viscosity. This way the effect of the medium alone, or the fluid alone, can be investigated separately.

In three dimensions, replacing q by v, the so-called "filter velocity" or sometimes "macroscopic velocity", the differential form of Darcy's law is:

$$v = -K \nabla P \tag{13}$$

where ∇ is the gradient operator ($\partial/\partial x + \partial/\partial y + \partial/\partial z$).

For one-dimensional flow, in the x direction, ∇P becomes $\partial P/\partial x$, and eq. 13 becomes:

$$v = -K\frac{\partial P}{\partial x} \tag{14}$$

In cases where gravity comes into play, e.g., for fluid flow in the vertical direction, the pressure term in eq. 14 should be replaced by the "piezometric pressure" defined by $\wp = P + \rho_l gh$ where h is the distance measured vertically upward from an arbitrarily chosen level [59,60].

Darcy's law is similar to Newton's law of viscosity, Fourier's law of heat conduction and Fick's law of diffusion, so the mathematical treatment of Darcy's law is readily available. The differential form of Darcy's law is more useful in unsteady state flow problems.

Equation 12 is often expressed in terms of the "friction factor" f defined as $-D_f\Delta P/\rho_l v^2 L_0$ and the "Reynolds Number" Re redefined as $D_f v\rho_l/\eta$. The relationship between friction factor and "Reynolds Number" is given by:

$$f = \left(\frac{D_f^{\,2}}{k}\right)\frac{1}{\mathrm{Re}} \tag{15}$$

where D_f is the effective average diameter of particles or fibers, and v is the flow velocity.

These dimensionless variables, f and Re, are common concepts of fluid mechanics [61]. Reynolds number is a measure of flow momentum and is often used as indicator for differentiating laminar and turbulent flow regimes.

4.4. Permeability

Darcy's law introduces the permeability k in order to characterize fully the porous medium as a flow resistor. The structural and geometrical factors such as porosity, tortuosity and specific surface area are all taken into account in the permeability factor.

Permeability is usually measured [60] by constructing a "plug" made of the porous medium of interest. A pressure difference is applied to cause a steady flow through the porous medium. From the flow rate measurement the permeability can be calculated by Darcy's law (eq. 12).

The device described above is an example of the steady state method. There are also permeameters that operate in unsteady state mode such as the so-called "falling head permeameter" [59,60].

A siphon type device [62,63] may be used if the porous sample is a flat sheet and if the liquid flow in the plane of the sheet has to be determined. In this set-up, the sample strip serves as a siphon, which steadily transports liquid from a reservoir of constant liquid level to a lower level. The driving force ΔP is the difference between the two liquid levels and L_0 is the total flow length of the sample.

However, it is not always possible to experimentally measure the permeability k. In many cases it is necessary to predict the permeability from theories, based on hypothetical models of porous structure, or from empirical correlations.

4.5. Kozeny - Carman Approach

The hydraulic radius theories typified by the work by Kozeny [49] and Carman [50,51] treat the flow through a porous medium as a conduit flow. The cross-sections of the conduits usually have non-circular shape; but using the "hydraulic radius" concept, the channel diameter, D_H, defined as four times the flow cross-sectional area divided by the wetted perimeter, can be derived as:

$$D_H = \frac{4\phi}{S_0(1-\phi)} \tag{16}$$

where S_O is the surface area of the channel per unit volume of the solid material making up the porous medium.

By applying a slightly modified version of the Hagen-Poiseuille equation to the flow through channels, the following Kozeny-Carman equation was obtained:

$$k = \frac{\phi^3}{k' S_O^2 (1-\phi)^2} \tag{17}$$

where k' is the "Kozeny constant" which includes a shape factor for the channel and a "tortuosity" factor, the latter being usually defined as the ratio of the length of the flow channel to the length of the porous medium. Carman suggested that empirically the value of the "Kozeny constant" be equal to 5 so that the equation was in better accord with experimental data. The Kozeny-Carman equation was found most applicable for beds of uniform spherical particles.

4.6. Network Models

Network models may be considered to belong to the conduit flow category. However, contrary to the wetted perimeter concept above, the treatment is more statistical in nature. Network models approximate the porous structure by networks of randomly distributed bands (not necessarily of cylindrical shape) and nodes through which the fluid flow occurs. These models usually require more detailed information on the pore structure of the absorbent material. Some successful correlations with experimental data on beds of sand and of uniform glass spheres were made by Lin and Slattery [64]. In general, these models have not been very successful in calculating permeabilities of beds of non-uniform particles.

A simplified network model was used by Dullien [65] to predict the permeabilities and other properties characterizing the flow of water through sandstones. The building block, or "component network", of this model is a cubic lattice network of capillaries each of which is composed of segments of randomly varied cross sections. For each component network i, the permeability k_i is given by:

$$k_i = \frac{\phi_i}{96} \frac{\left(\Sigma_j V_{ij} / D_j^2\right)^2}{\left(\Sigma_j V_{ij}\right)\left(\Sigma_j V_{ij} / D_j^6\right)} \tag{18}$$

where ϕ_i is the porosity of the component network; V_{ij} is the volume of the capillary segment of diameter D_j. The permeability k of the sample is the weighted sum of k_i shown below:

$$k = \Sigma_i b_i k_i \tag{19}$$

where b_i is the fraction of pore volume of the medium occupied by component network i. Reportedly, the permeabilities of 14 sandstone samples were predicted by the simple network model with an average error of ±23% [65].

The network models form the basis for the development of the percolation theory [66], which has important applications in petroleum engineering. A more detailed review of the network models can be found elsewhere [59].

4.7. Drag Theories

In drag theory the pore walls have been treated as obstacles to the fluid flow [58]. The drag of each element making up the wall was calculated by the Navier - Stokes equation, and the sum of all the drag is set equal to the flow resistance of the porous medium. The drag theory is believed to be more realistic for high porosity media.

Iberall [52] employed this approach for high porosity beds of randomly distributed cylindrical fibers of the same diameter D_f. The permeability is given by:

$$k = \frac{3}{16} \frac{\phi D_f^2}{1-\phi} \frac{2 - In\left[D_f v \rho_l / \eta\phi\right]}{4 - In\left[D_f v \rho_l / \eta\phi\right]} \tag{20}$$

This derivation indicates the dependence (however weak) of permeability on the flow velocity.

Brinkman [53,54] also used the same approach to derive the permeability of beds of spherical obstacles of the same diameter D_f as follows:

$$k = \frac{D_f^2}{72}\left(3 + \frac{4}{1-\phi} - 3\sqrt{\frac{8}{1-\phi} - 3}\right) \tag{21}$$

4.8. Other Correlations

The variation of k as a function of the porosity ϕ for random beds of various distributions of spherical particles was empirically found by Rumpf and Gupte [67] to be closer to $\phi^{5.5}$ than $\phi^3/(1-\phi)^2$ as in the Kozeny-Carman equation.

Of interest is the work by Davies [68] on fluid flow through fibrous webs, where the Reynolds number is less than 1 and the porosity is less than 0.98. The permeability is given by the following equation:

$$k = \frac{D_f^{\ 2}}{64(1-\phi)^{3/2}\left[1+56(1-\phi)^3\right]} \tag{22}$$

where D_f, in this equation, is defined as the effective fiber diameter.

Interested readers may also refer to the work of Kyan et al. [69] on the correlation of friction factor and fiber Reynolds number where the deflection of fibers is also taken into account.

All the methods and correlations above deal with steady state saturated flow. For unsteady state flow through partially saturated porous media, the localized permeability factor is likely to differ from the values obtained by the above theories. The permeability of partially saturated porous beds, is discussed in the next section.

4.9. Limitations of Darcy's law

Darcy's law is valid only for slow laminar flows. Deviation from the linearity of flow velocity vs. pressure gradient or, similarly, the linearity of friction factor vs. *1/Re*, occurs in the case of faster flow. The critical Reynolds number above which this deviation occurs is around 1. However, this critical boundary is not clear-cut, Scheidegger [58] found the value to be in the range of 0.1 to 75.

The classical correlations between porosity and flow in the non-laminar flow region with high Reynold number are available in the literature (see ref. 58, 59, 61,70).

5. UNSTEADY STATE FLOW THROUGH POROUS MEDIA

In most practical absorbency situations the liquid movement is an unsteady state flow where the porous medium is not uniformly and completely saturated, and where the liquid distribution throughout the medium changes with time.

At any given time, as liquid is absorbed into a porous medium, there is a saturation gradient in the medium along the direction of flow and this saturation gradient changes with time as the absorption process continues. Absorbency problems generally belong to this type of unsaturated flows. Quantitative observations of the saturation gradients were made in studies of capillary transport in filter papers [39,71-73]. Extensive studies on unsaturated flow can be found in the literature on soil physics [74-78]. All these studies recognized that unsteady flow in a porous medium is similar to a diffusion process (driven by concentration gradients) or a heat conduction process (driven by temperature gradients). Such an analogy is based on phenomenological observations as well as on the similarity between Darcy's law and the heat conduction and mass transfer equations.

5.1. General Equations for Unsteady State Flow

Consider a one-dimensional horizontal flow of an incompressible fluid in the x-direction through a porous medium of uniform cross-section. By combining the law of conservation of mass and the differential form of Darcy's equation, the following equation on saturation rate has been derived [72,73,78-81]:

$$\frac{\partial s}{\partial t} = \frac{\partial}{\partial x}\left[F(s)\frac{\partial s}{\partial x}\right] \tag{23}$$

F(s) in the above equation is the diffusivity factor and is often referred to in soil physics as "moisture diffusivity"; the diffusivity factor is related to the permeability k via [81,82] $F(s) = (k/\eta)\,(dp/ds)/\phi$.

The solution to this generalized equation (eq. 23) describes the unsteady state flow behavior in porous media in terms of saturation profiles along the length of the sample at various times. The saturation profiles can be obtained directly by solving eq. 23 [83]. When *F(s)* is not constant, which is usually the case, numerical methods are needed to solve the problem [84].

As liquid advances into a porous medium, the small pores with their greater capillary pressure will tend to fill first. They will be the last to empty when liquid is being withdrawn. Since at any location only pores up to a certain size are filled, the local flow is somewhat restricted to those filled pores. It follows that the local permeability would be dependent upon the local saturation. Equation 23 has been further extended to study two- or three-dimensional flow problems [85].

5.2. Idealized Examples with Constant Diffusivity Coefficient

Consider a one-dimensional case of eq. 23 where *F(s)* is a constant F_1, and the long strip of sample is originally dry and at time $t = 0$ one end of the sample contacts the fluid [79]. Equation 23 becomes:

$$\frac{\partial s}{\partial t} = F_1 \frac{\partial^2 s}{\partial x^2} \tag{24}$$

with the following boundary conditions:

at $t = 0$, $s\,(x, 0) = 0$; sample is originally dry;
at $x = 0$, $s\,(0, t) = 1$; one end of sample is contacting the fluid and is 100% saturated;
at large x, $s\,(x, t) = 0$; sample is dry far ahead of the advancing fluid front.

The solution to this problem is [61,79,83]:

$$s = 1 - erf\left(\frac{x}{2\sqrt{F_1 t}}\right) \tag{25}$$

where *erf* is the "error function". The main significance of eq. 25 is that s is a function of $x/t^{1/2}$; that means s would be a constant when $x/t^{1/2}$ is a constant. Rudd [79] pointed out that if s_v is the saturation at the visually observed advancing front and x_v is the visually observed advancing distance, then $x_v/t^{1/2}$ is a constant and is related to s_v. It follows that x_v is proportional to $t^{1/2}$ which is in accord with the common experimental observations of the wicking front advances of liquids in pulp webs or papers [38-40,71-73,79]. This also agrees qualitatively with the Washburn equation [36] on the proportionality between the wicking distance and $t^{1/2}$.

If the sample is long, the semi-infinite boundary condition of the previous example will apply over a longer period of time. As the fluid proceeds, it will eventually get to the other end of the long strip of sample (of finite length) and deviation from the above proportionality will occur.

To account for the finite length L_0 of the sample, the last boundary condition should be changed to:

at $x = L_0$, $\partial s/\partial x = 0$; no liquid entering or exiting the far end.

The solution to this problem is (cf. ref. 83, p. 101):

$$s = 1 - \frac{4}{\pi}\sum_{j=0}^{\infty}\frac{1}{2j+1}e^{-F_1(2j+1)^2\pi^2 t/4L_0^2}\sin\frac{(2j+1)\pi}{2L_0}x \qquad (26)$$

According to this relationship, if the amount of liquid is unlimited, then after a long enough time of flow the whole sample would be 100% saturated (as t approaches infinity, s is equal to 1) and the flow would practically come to a stop. Assuming F_1 = 0.20 cm2/sec (in the range of values for paper-like material) and L_0 = 20 cm, the saturation profiles for this example are illustrated in Fig. 6.

Both the examples cited above assume that the absorbing liquid is of unlimited quantity. Now if only a finite volume of liquid is absorbed into the sample, i.e., if the amount of liquid moving inside the porous medium is constant and no liquid is entering or leaving the system, the liquid in effect is redistributing itself into the system. The redistribution is done essentially by filling up the smallest unfilled pores in the drier regions and emptying the largest filled pores in the more saturated regions.

If the saturation profile at the onset of the redistribution process is $s_0(x)$, then the boundary conditions for the redistribution process are:

at $t = 0$, $s(x, t) = s_0(x)$; originally the saturation profile is $s_0(x)$;
at $x = 0$ and $x = L_0$, $\partial s/\partial x = 0$; no liquid entering or exiting at either end.

The solution to this problem, according to Carslaw and Jaeger (ref. 83, p. 101) is:

$$s = \frac{1}{L_0}\int_0^{L_0} s_0(x')dx' + \frac{2}{L_0}\sum_{x=1}^{\infty} e^{-F_1 j^2\pi^2 t/L_0^2}\cos\frac{j\pi x}{L_0}\int_0^{L_0} s_0(x')\cos\frac{j\pi x'}{L_0}dx' \qquad (27)$$

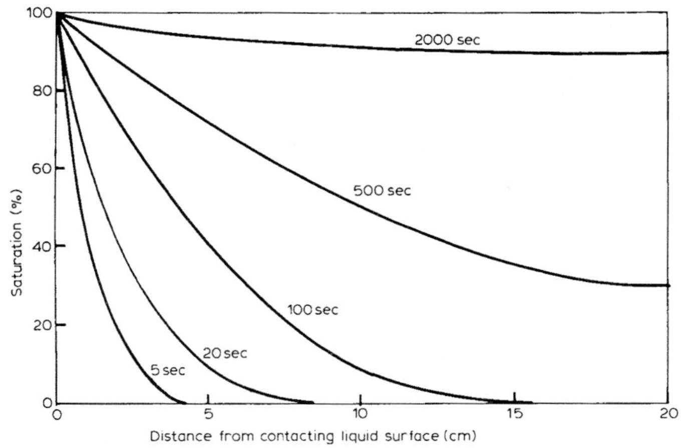

Fig. 6 Saturation profiles during absorption along the length of sample of finite length L_0 20 cm, according to eq. 26, F_1 0.20 cm^2/s.

where x' is a dummy variable for integration. Figure 7 shows the saturation profiles as the liquid redistributes itself from the initial state obtained from the 100-sec curve of Fig. 6.

This example also assumes there is no difference between advancing and receding behaviors, which is of course not true. Sorption hysteresis does occur and is discussed in chapter XI. If the hysteresis is ignored, after a long period of redistribution the saturation is uniform over the whole length of the sample (from eq. 27 as t goes to infinity, the 2nd term on the right hand side becomes zero and s is equal to a constant independent of distance). In reality, due to the hysteresis effect, such uniform saturation is never reached. At equilibrium, there would be more liquid at the "wet" end than at the "dry" end.

Many other types of boundary conditions are discussed by Carslaw and Jaeger [83], also by Crank [86] and Scheidegger [58].

5.3. General Properties of Semi-Infinite Radial Flow

In the first idealized example it was found that the distance moved by a liquid front of any given saturation is proportional to the square root of time. This behavior also holds true in a more general case where $F(s)$ is not a constant but a continuous function. As pointed out in the literature [72,73,78,80,86], if the boundary conditions are expressible in terms of the combined variable U where $U = x/t^{1/2}$ alone and not involving x and t separately (such as in the first idealized example cited above: for $U = \infty$, $s = 0$ and for $U = 0$, $s = 1$), then s is a function of U alone. For two- and three-dimensional radial transports, the radial advancement of the liquid front of saturation would vary in proportion to the square root of the wicking time.

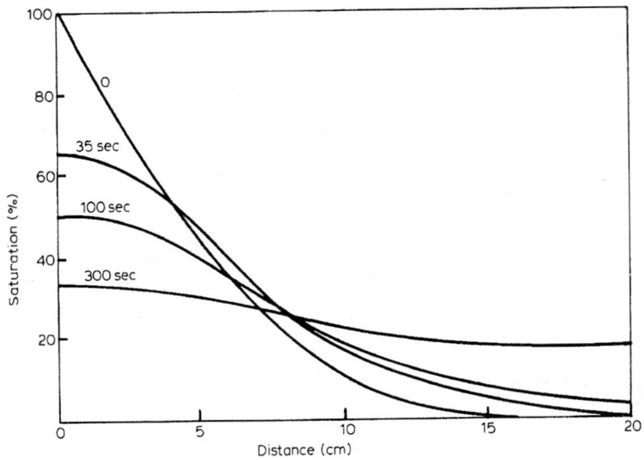

Fig. 7. Liquid redistribution in 20 cm-long sample according to eq. 27. F, 0.20 $cm^2/sec.$; initial saturation from Fig. 6 at 100 sec of absorption, the liquid source is removed and redistribution starts).

For one-dimensional flow, at any time the total volume V of fluid absorbed in the porous sample is equal to the area under a given saturation profile. If the flow cross-sectional area of the sample is A, then V is given by;

$$V(t)=A\,\phi\int_0^{\infty} s\,(x,\,t)\,dx \tag{28}$$

When the saturation fronts advance in proportion to $t^{1/2}$, the value of the integral is also proportional to $t^{1/2}$, and in such cases V is also proportional to $t^{1/2}$.

For two-dimensional flow, if the porous medium has a thickness T and if the fluid wicks radially outward from an area of radius a_0 at the center of the sample, then the total volume of fluid absorbed in the sample at any time t is:

$$V(t) = 2\pi T\phi\int_{a_0}^{\infty} L\,s(L,\,t)\,dL \tag{29}$$

If the radial advances are proportional to $t^{1/2}$ and if a_0 is infinitesimally small, the integral, and thus V, is proportional to t.

Similarly, for three-dimensional radial flow V can be shown to be proportional to $t^{3/2}$. To summarize, if m is the number of dimensions, V will be proportional to $t^{m/2}$. It follows that the volumetric absorption rate dV/dt would be proportional to $t^{m/2\,-1}$.

Figure 8 shows some of the experimental data on V versus t which Gardner and Mayhugh [78] have found to fit the theory well. The deviation from the theory is more likely

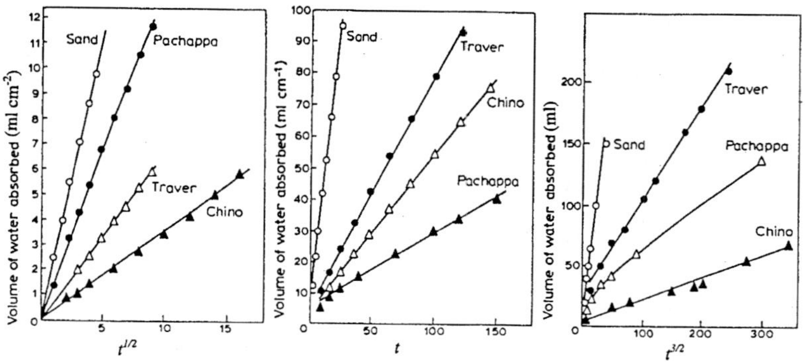

Fig. 8. Amount of water absorbed in beds of soil as function of time [78]; (a) one-dimensional flow, (b) two-dimensional radial flow, (c) three-dimensional radial flow.

to occur as one goes from one dimension to three dimensions which would be the case when the size of the liquid source is finite.

In the case of a small finite size liquid source, Phillip [85] showed that in both two- and three-dimension flows the absorption rates decreased with time to reach to a constant value

5.4. Unsteady State Flow With Non-Constant Diffusivity Coefficient

In the idealized examples, the diffusivity is assumed constant. In practice, however, the diffusivity, and the permeability, for unsaturated flows are dependent upon the saturation s. As discussed above, this is because at a given saturation only pores up to a certain size are filled and therefore only a fraction of the total pore space is used for transport.

A number of experimental methods were used to measure the permeability in partially saturated porous media. Experiments on two-phase (liquid-gas) flow through sand beds were carried out [87], where the electrical conductivity of the liquid phase at different liquid-gas ratios were measured and converted into permeabilities. Other experiments were carried out on the flow of water-oil through sandstones [88]. From these data it was suggested [89] that the permeability of partially saturated porous beds (so-called "effective permeability") varies approximately in proportion to the cube of the saturation s. This relationship was applied by Gillespie [90] to partially saturated filter papers during water imbibition. Nguyen and Durso [63] also used this cubic relationship in kinetic studies of water absorption in pulp fluff webs where good agreement between the calculated results and experimental wicking data was reported.

Gardner and Hsieh [91] measured the local velocities (via dye tracks) of water penetrating blotting paper in unsteady state flows. The relative permeabilities, calculated from the velocity data, are found to be extremely small throughout the range of saturation and to increase sharply near the 100% saturation level which agrees with the experimental results of Wyckoff and Botset [87].

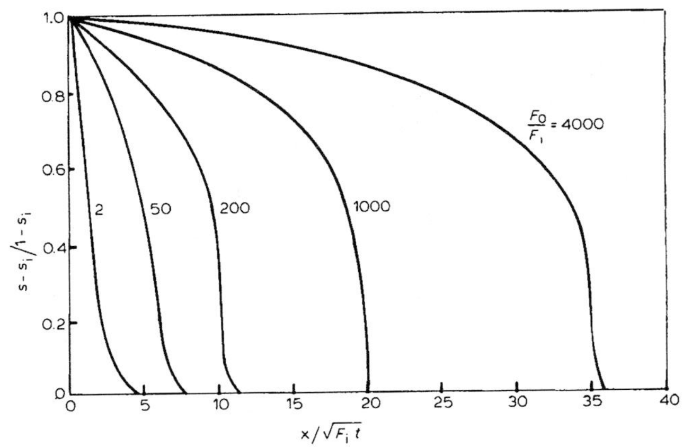

Fig. 9. Solution of the one-dimensional flow problem with diffusivity coefficient varying as in eq. 30, reference curves [78].

By devising a one-dimensional wicking test which simulates the semi-infinite boundary condition (long sample), if the saturation profiles are experimentally measured, the data can be used to back calculate the diffusivity as a function of saturation. This scheme was used to determine the diffusivities of fine pore materials such as filter papers [72,73] and beds of fine sands and beads [92]. The results from this method showed that the diffusivity reaches a maximum value at a saturation level lower than (but close to) 100%.

Based on the published data on the variation of diffusivity with saturation, Gardner and Mayhugh [78] proposed the following relationship for some types of soil:

$$F(s)=F_i \exp \beta(s-s_i) \tag{30}$$

where s_i and F_i are the initial saturation and the diffusivity corresponding to the initial saturation, respectively, and β is a positive constant, characteristic of the system. At 100% saturation $F_0 = F_i \exp \beta(1 - s_i)$, where F_0 is diffusivity coefficient for completely saturated porous medium.

In the case of a very long sample, the combination of variables described previously is applicable and the saturation becomes a function of only the combined variable U. Figure 9 presents a series of reference curves of dimensionless saturation $(s - s_i)/(1 - s_i)$ versus dimensionless combined variable $x/(F_i t)^{1/2}$ for porous media of different characteristic F_0/F_i. Each of these curves describes completely the function $s(x,t)$ for one-dimensional flow in semi-infinite porous sample.

Figure 10 shows the saturation profiles for horizontal wicking into Pachappa sandy loam using $F_i = 0.0005$ cm^2/sec and $F_0/F_i = 200$.

Ruoff et al. [72,73] studied unsteady flows in fine pore filter papers. The flows in these instances are slow enough to allow experimental measurement of the saturation profiles in the transient states.

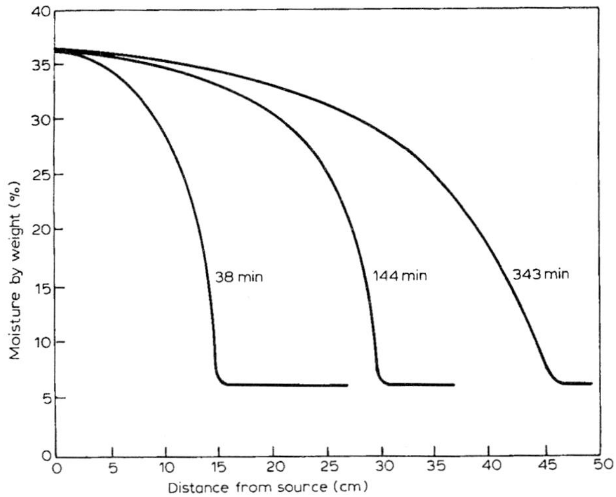

Fig. 10. Saturation profiles for water moving into Pachappa sandy loam in one dimension [78]. Profiles are calculated using Fig. 9 with $F_i = 0.0005$ cm^2/sec, $F_0/F_i = 200$.

The liquid concentration profiles in one-dimensional horizontal flow of water into filter paper is shown in Figure 11a. It is possible to show that the advancing distance at any saturation level varies in proportion to the square root of the time. It can be further observed that the profiles (shapes) for different times are in fact identical except for the difference in the distance scale. This was verified by the construction of the "reduced saturation profile"; the water content is plotted against the "reduced distance" which is the ratio of distance over the advancing front distance, and all the data points from the different profiles lay on a single curve, as shown in Figure 11b. Profiles of roughly the same shape were obtained by Everett et al. [39] and by Takahashi [71]. The recognition that all the concentration profiles at different times have identical shape was also made by Fujita [93].

In radial (two-dimensional) flow experiments, the same qualitative behavior was observed. The radial advancing distances for any saturation level are proportional to the square root of time just as in the case of one-dimensional flow. As a consequence, the saturation profiles at different times have identical shapes and the concept of "reduced saturation profile" is also applicable to the radial case.

5.5. Factors Affecting Unsteady State Flow

Gravity. Gravity imposes a ceiling on the height of the vertical capillary rise. In vertical flows, for a short period of time from the onset of imbibition, the absorbed liquid would still be relatively close to the liquid source and far from the final equilibrium state. In such cases gravity can be neglected. This fact has been used in imbibition studies where due to the ease of experiment, the wicking medium is hung vertically over the liquid source [38,39]. The Washburn equation (eq. 7) [36], which ignores gravity, was found to agree very well with these vertical wicking data.

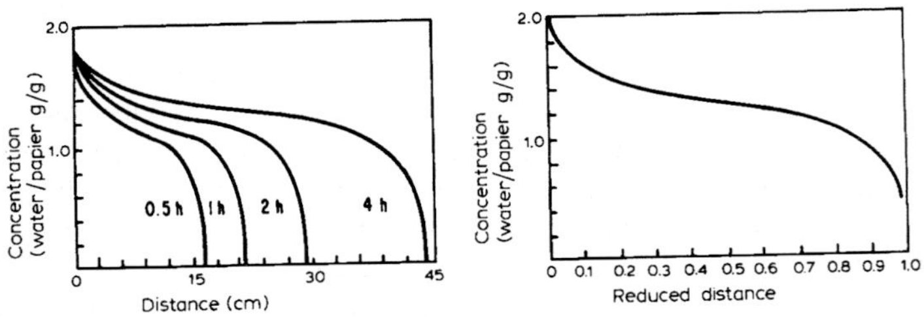

(a) (b)

Fig. 11. Experimental moisture concentration profiles for water penetration into filter paper in one dimension [72]; (a) actual profiles; (b) profile plot with reduced distance.

At large period of times from the onset of flow, gravity becomes an increasingly important factor. Gravity plays a major role in cases of large pore media such as uncompressed fluff pulp webs or foams.

The flow (eq. 23) with gravity included, becomes [81,82]:

$$\frac{\partial s}{\partial t} = \frac{\partial}{\partial x}\left[F(s)\frac{\partial s}{\partial x}\right] + E(s)\frac{\partial s}{\partial x} \tag{31}$$

where $E(s) = \rho_l g(dK/ds)/\phi$ and the x-axis is taken to be positive in the upward direction.

Compaction of a non-rigid porous medium would bring about an increase in capillary potential driving force (due to pore size reduction) and a decrease in permeability. Comparing the vertical rise of water in two beds of sand of different degrees of compaction [94], it was observed that the rise in the less compacted bed was faster initially but leveled off earlier. The rise in the more compacted bed started out slower but eventually ended up at a higher equilibrium level.

Initial moisture. It was found that liquid penetration into a porous medium which already has some initial moisture is faster than the case where the porous sample is completely dry.

By plotting the empirical rate factor, which is the slope of the weight of liquid absorbed versus the square root of time curve, as a function of the initial imbibing liquid content, Everett et al. [39] found that the rate factor went through a maximum at a low initial saturation. The improvement due to the presence of initial moisture can be attributed to an increase in the "spreading pressure" [95] in the case of a polar liquid advancing over a moderately polar solid substrate such as cellulose. At higher initial moisture, capillary condensation occurs in the smaller pores and the capillary pressure decreases.

Ijjas [94] recorded significant increase in penetration rate of water into sand beds when as little as 2% initial moisture is present.

Dimensional changes. A commonly encountered factor is the dimensional change of the bulk volume of the porous medium when it imbibes liquid. A well known example is fluff pulp webs which "collapse" when wet [96]; on the other hand, pre-compressed fluff webs or foams will swell when wet. These occurrences cause significant changes in pore sizes, pore size distribution and permeability of the porous medium.

Limited swelling of the pore wall, or of the fibers or particles making up the porous medium, commonly occurs. This is the case of most cellulosic materials. As Chatterjee reported [41], the wicking distance versus time relationship $L = k_0 t^m$ for wood pulp sheets was obeyed with m not far from 1/2. It appears that the limited dimensional change affects primarily the value of the proportionality k_0, which is a measure of the linear wicking rate. However, excessive swelling would drastically affect the L and t relationship.

The compressional effects on pulp fluff and water system were also studied by Painter and Nguyen [97] who reported m to be close to 1/2 for both linear advance L and adsorbed volume V. The wet-swelling webs were found to be at considerably higher average saturation levels than the wet-collapsing webs. However, the volumetric rate was found to be highest when the density of the pulp fluff web was in the vicinity of a "critical density" where the dimensional change of the web upon wetting is small. This is in agreement with Aberson's earlier work [38].

Transient behavior of sub-second time. If it is desired to study the transient flow behavior at sub-second times from the initial time ($t = 0$), the differential equation 23 as such may be inappropriate. A more rigorous derivation by Szekely et al. [43] for the transient fluid flow into a capillary tube should be used where the kinetic and other energy loss terms are included. The results showed that there is actually a response delay at sub-second time where the velocity starts from 0, increases to a maximum and then as normally expected, tapers off. For liquids with water-like viscosity, beyond such extremely short time, the kinetic term was found to become insignificant and can be ignored. For more viscous liquids, the delay time becomes longer and a more rigorous treatment may be needed.

Non-Newtonian fluid. When the liquid involved is a non-Newtonian fluid, the viscosity is no longer a constant but varies with the shear stress. The study of flow of viscoelastic fluid through porous media can be found elsewhere in the literature [98-101]. In the case where the fluid is blood at room conditions, such as the case of gauze or bandages in use, the viscosity of the fluid varies naturally with time as well.

Structural factors. There is also the effect of the "quality" of the pore structure in terms of pore connection, continuity of the pore size distribution and macroscopic homogeneity of the pore structure. Poor pore connection for advancing fluid is usually associated with larger pores, especially when they are in clusters. The advancing fluid may avoid the large pores leaving air bubbles in its wake. For receding fluid, isolated clusters of small pores tend to retain isolated pockets of undrained fluid.

Anisotropy of the pore structure is a common occurrence. In anisotropy media, the permeability value is different in different directions and this has to be taken into account in two and three-dimensional flows.

6. DIFFUSION AND SWELLING

Besides surface adsorption and pore filling via capillary flow, the absorbed liquid can be retained by partial solubilization of the absorbing material. This causes swelling which in

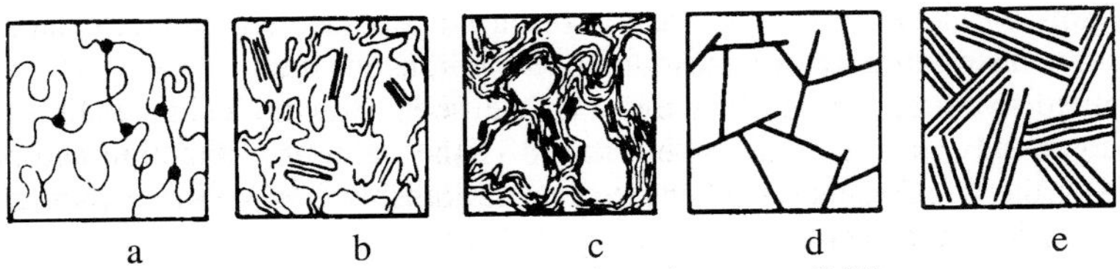

Fig. 12. Five models of swollen gel structures [104].

many cases can involve a ten- or hundred-fold increase in volume, such as in the case of the "superabsorbents".

Swelling is a diffusion phenomenon driven by the affinity of the molecules of the swelling material for the molecules of the contacting fluid. Five general models of swollen gel structure [102-104] are shown in Fig. 12. The first diagram (a), a classical cross-linked network, has a swelling limit controlled by a balance between the thermodynamic forces due to polymer-solvent interactions and the entropic force of coiled polymer chains. The second model (b) is similar since a "pseudo-cross-link" is present in the form of an insoluble crystalline phase; and may give rise to model (c) characterized as a polymer-diluent matrix. The "house of cards" structure (d) and the paracrystalline state (e) are colloidal in character rather than macromolecular. Charge distributions at the ends or edges of rigid anisotropically shaped particles are what give rise to model (d); which can also spontaneously order into a structure like model (e). The phase separation phenomenon in model (e) is concentration dependent and addition of diluent or changes in potential can bring about a reversible change in the interparticle spacing [105].

Current absorbent materials are more commonly depicted as model (a) or (b). Most commercial "superabsorbents" are polyelectrolytes which have been insolubilized by cross-linking or by grafting and their structures with respect to these models are further discussed in a later chapter.

6.1. Equilibrium Swelling

Swelling of cellulosic fibers. Cellulose fibers can undergo limited swelling by the absorption of interactive liquid. As more interactive liquid comes in contact with fibers, a fibrillar unzipping type of mechanism occurs [106]. As swelling begins, more hydroxyl groups become accessible to accommodate more liquid, which opens up the structure and causes more swelling. However, if the interactive liquid is just water, inaccessible hydroxyl groups present in the amorphous region become available, but the bulk of those in the crystalline regions remain inaccessible throughout the process.

It is generally accepted that qualitatively most of the areas available for water are in the non-crystalline regions and that the crystalline lattices serve as a kind of "pseudo-cross-links" which impose an elastic constraint (and thus a limit) on the swelling process. However, according to some authors [18,107], quantitatively there exists no correlation between the equilibrium water swelling and crystallinity.

At the onset of absorption, however, a slight volume contraction can be observed [18,108]; this was attributed to the free void volume in the dry cellulosic fibers.

Natural fibers, such as pulp, cotton, wool, etc., as well as the regenerated cellulose fibers, will swell to different degrees with water. The amount of water retained by the swollen fibers varies from about 6% to over 100% based on the dry weight of fibers [108].

Most of the swelling of the fibers occurs in the transverse direction and hardly any along the length of the fibers [109]. For fibers conditioned at 100% relative humidity, a very large additional increase in swelling occurs upon immersion in water.

Because of the anisotropy of fiber swelling, the amount of water retained by swelling can be determined by microscopic measurement of the change in fiber cross-section [110]. Laser beam diffraction has also been used to measure fiber diameter [111]. Another common method for determining the swollen water is the centrifugal method [112,113] which removes most of the inter-fiber liquid, and the amount of liquid retained in the swollen matrix is determined by weight.

Swelling of polymer network. A three-dimensional network of polymer in contact with a solvent, for example vulcanized rubber in benzene, will strive to dissolve in the liquid. In the process the polymer chains uncoil to more open conformations so that each chain can maximize its contact with the solvent molecules, and thus swelling of the network takes place. If there is no cross-linking, the swelling would continue until a viscous solution is obtained. But in a cross-linked network as the chains between cross-links become increasingly elongated, an elastic retractive force develops. The swelling will stop at an equilibrium point when the swelling force is balanced by the retractive force. Networks of water soluble polymers give rise to the so-called "hydrogels". Examples of these range from the highly swollen "superabsorbents" to the slightly swollen soft contact lenses.

The swelling equilibrium is dependent on the entropy of dilution, the heat of dilution, and the entropy of the polymer network. For a given polymer-solvent system the equilibrium swelling is a function of the cross-link density. The relationship involving the equilibrium swelling ratio ε, defined as the ratio of the final swollen volume to the original unswollen volume of the network, was derived by Flory [114]. If the cross-link density is not too high and the solvent is a good solvent, the following approximate relationship was obtained:

$$\varepsilon^{5/3} = BM_c/(1 - 2M_c/M) \tag{32}$$

where $B = (0.5 - \mu)/\rho_p v_s$ is a constant for a given polymer-solvent system, μ is a dimensionless parameter expressing first neighbor interaction for solvent with polymer, ρ_p, is the polymer density, v_s is the molar volume of the solvent, M_c is the average molecular weight per cross-link, and M is the average molecular weight of the analogous uncross-linked polymer.

According to eq. 32, if the cross-link density is high, the swelling ratio is approximately proportional to $M_c^{3/5}$.

Swelling of polyelectrolyte network. If the polymer network is that of a polyelectrolyte, the swelling forces are increased by the ionic repulsion of like charges on the polymer chains. This further expands the network and increases the equilibrium swelling. An example is a cross-linked network of partially or fully neutralized poly(acrylic acid) or poly(methacrylic acid).

In cases where the solvent is an electrolyte solution rather than pure water and where the concentration of the mobile ions throughout the gel-solvent system is high enough, the charge screening effect becomes important. The mobile ions which screen the fixed charges on the polymer chains reduce the electrostatic repulsion of the latter and thus reduce the equilibrium swelling. The equilibrium swelling ratio in this case is dependent on the electrolyte concentration as well as the degree of ionization.

The approximate relationship for dilute solution akin to that leading to eq. 32, assuming that the electrolyte concentration is not excessively high, is as follows [114]:

$$\varepsilon^{5/3} = \left(\frac{i^2}{4v_m^2 I_O} + B \right) M_C /(1 - 2M_C / M) \tag{33}$$

where v_m is the molar volume of the monomeric unit and I_0 is the ionic strength of the external solvent solution and is given by $I_0 = \zeta_+ \zeta_- (C_+ + C_-)/2$ with ζ_+ and ζ_- being the valence of the electrolyte cation and anion and C_+ and C_- being the external concentration of electrolyte cation and anion.

This equation will reduce to eq. 32 when $i = 0$ (i.e. the polymer chains are not ionized) or when $I_0 = \infty$ (i.e., when the electrolyte concentration in the solvent is very high, the charge screening effect becomes complete). In either case the system behaves essentially like a nonionic system.

A more rigorous treatment was made by Katchalsky et al. [115] who derived an "equation of state" of the gel where the degree of ionization is added to the three traditional variables: volume V_g, pressure P and temperature τ. A correction factor is included in this equation for highly swollen gels where the deviation from the Gaussian distribution of chain length becomes significant.

For the case of zero degree of ionization, the equation of state is as follows:

$$\frac{\lambda V_g^{2/3}}{Z} = \frac{1}{2Z} + V_g \ln \frac{V_g}{V_g - 1} - 1 - \mu / V_g \tag{34}$$

where Z is the number of monomers between neighboring cross-links and λ is the correction factor cited:

$$\lambda = 1 + \frac{0.6eV_g^{2/3}}{Z - eV_g^{2/3}} \tag{35}$$

where e = number of monomers per statistical element in Kuhn's theory [116]. For moderate degree of swelling, $\lambda = 1$ and eq. 34 reduces to eq. 32. For $i > 0$ the equation of state of the gel becomes:

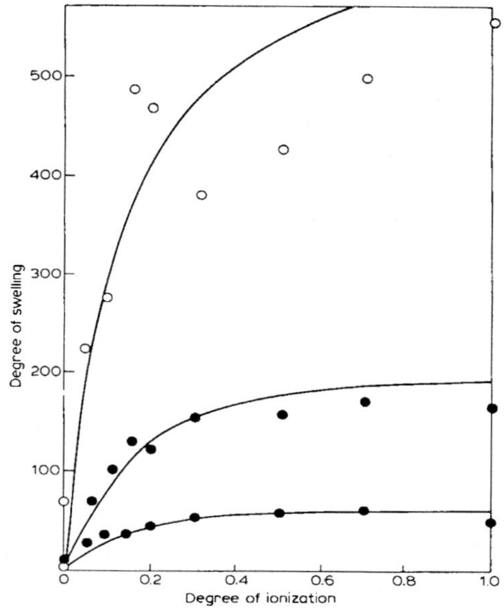

Fig. 13. Variation of the equilibrium swelling rate of poly(methacrylic acid) with degree of neutralization by sodium hydroxide [115]. The curves correspond to different degrees of cross-linking by divinylbenzene (1. 2. 4% respectively, for upper, middle and lower curves).

$$\frac{\lambda V_g^{2/3}}{Z} = i\left[1+\beta\left(\frac{iV_g}{l}\right)^{1/2}\right] \tag{36}$$

l = number of monomoles of the cross-linked polymer and β is a complex, slowly varying expression given in Katchalsky's paper [115].

Experimental results on the swelling in water of poly(methacrylic acid) gels of different cross-link densities and of varying degrees of ionization are shown in Fig. 13. The approximation curves were calculated by eq. 36 using e = 10. The decrease in equilibrium swelling of the same gel in approximately inverse relationship with the ionic concentration of the solvent (akin to eq. 33) was observed by Michaeli and Katchalsky [117].

Numerous studies on swelling of hydrogels can be found in the literature on ion exchange. Relevant theoretical treatment as well as experimental data can be found in Helfferich's book [118] and others [119,120].

6.2. Diffusion Kinetics

Swelling involves diffusion of the solvent into the polymer network. The classical treatment used Fick's diffusion equation [121]:

$$\frac{\partial c}{\partial t} = \nabla(D \nabla c) \tag{37}$$

where c is the solvent concentration, and D is the diffusion coefficient. When swelling occurs, D is not a constant but varies with c as the polymer network opens up.

For a one-dimensional case, eq. 37 has the same form as eq. 23. In such a case, it was pointed out that the amount of liquid absorbed $V(t)$ varies in proportion to the square root of time. This is true at least in the initial period and regardless of how D varies with c. A plot of V versus $t^{1/2}$ will give a straight line at the initial period until V reaches about half of the equilibrium value where the curve begins to level off [122]. This "normal" or Fickian behavior is observed experimentally in solvent diffusion into soft, rubbery materials. An example is the sorption of benzene into polyisobutene [122].

With glassy polymers, however, "anomalies" were observed: the sorption curves are sigmoid in shape. This "non-Fickian" behavior was reported in the sorption of water by cellulose [123,124] and by poly(vinyl alcohol) [125].

If the non-Fickian condition is extreme, a simple limiting case for such swelling occurs. This case was recognized by Alfrey, Jr. et al. [126] who termed it "Case II" diffusion. Case II diffusion is characterized by (1) a sharp advancing boundary between the swollen gel and the glassy unperturbed solid, and (2) a constant rate of advance of this boundary, which means a linear relationship between the initial weight gain by swelling and time.

The stress involved can be most clearly observed in Case II diffusion. The growing gel is in a state of compressive stress exerted by the unswollen part. Inversely, the swollen gel applies a tensile stress on the unswollen glassy part which in some cases can cause fracture of the latter [126]. Due to this interplay of forces, swelling occurs mainly along the direction of diffusion [127]. For fiber like geometry, more swelling occurs in the radial direction, and the dominant tensile stress is in the axial direction.

The presence of a sharp boundary between the swollen gel and the unperturbed solid was shown by Thomas and Windle [128] as resulting from the concentration dependence of the viscous flow rate of the glassy polymer.

Many authors have been able to correlate their swelling rate data by adding a linear term to the Fickian relation as follows [129-131]:

$$V = K_1 t^{1/2} + K_2 t \tag{38}$$

A plot of $V/t^{1/2}$ versus $t^{1/2}$ would give an indication of the degree of superposition of the two limiting cases. If the plot is a horizontal line, a purely Fickian behavior is obtained. A straight line passing through the origin would indicate a purely Case II diffusion.

7. STRUCTURAL MODELS FOR CHARACTERIZING PORE VOLUME AND PORE SIZE

7.1. Introduction

For a given fluid and fiber system, the two primary parameters that govern absorbency are the pore volume of the fabric, per unit mass or per unit volume of the constituent material, and the pore size. The former affects the total volume of fluid absorbed, or the absorbent capacity of a fabric, and the latter, as noted in several places in the chapter, affects the rate at which a fluid is imbibed into the structure. The values of these parameters must be known in order to understand and predict the absorbency behavior of a product. Accordingly, models are considered in this section that could be used to characterize the values of these parameters.

Fibrous materials used in absorbent products are usually in the form of bats or sheets prepared by one of the many methods available for converting fiber masses into nonwovens. The bats may be made by an air laying technique, a carding/cross-lapping procedure or a wet laying method, these leading to webs of different orientations and compactions. The primary web so produced may then be bonded for strength and handling, if required, by needling, hydroentangling, thermal, or chemical methods. Although the structure may have some discontinuities at local levels, the fabric can be treated at the macroscopic level as a continuum for most modeling purposes. An exception would be a layered material in which the layers may differ from each other in structure and composition. In such cases, each layer may be treated separately.

The fabrics, or individual layers of a laminated structure, may be composed of a mixture of two or more materials. The latter may be two or more different fibers, or one or more fibers and a resin. The fibers used may have different sizes (diameters or linear densities), cross-sectional shapes and specific gravities. Further, the structures may vary in terms of the fractions of different components as well as in terms of the over all areal densities (mass per unit area). Finally, these may be compressed and compacted to various degrees during manufacture and/or subjected to different external pressures during use.

Accordingly, the web composition, in terms of the type, size and the fraction of different materials used, the web areal density, and the packing fraction are some of the important factors that could affect pore structure and should be considered when developing models to characterize it.

It is understood that in absorbent products, fluid is absorbed in the interstitial space between fibers as well as into the internal structure of fibers. In systems containing regular materials, however, the amount absorbed into the internal structure is usually only a small fraction of the total. Most of the fluid acquired is in the spaces between fibers and held by capillary forces. An estimate of pore volume will, thus, provide a measure of the amount that can be expected to be absorbed. In fabrics containing highly swellable materials, on the other hand, such as superabsorbent, a significant fraction could also be absorbed into the internal structure. The total amount of fluid absorbed would, thus, equal the sum of the two. However, it is expected that such materials swell by local exchange between fiber and water molecules. Accordingly, pore volume assessed by subtracting volume of dry fibers from that of final (wet) web will give an estimate of the amount absorbed in these materials as well.

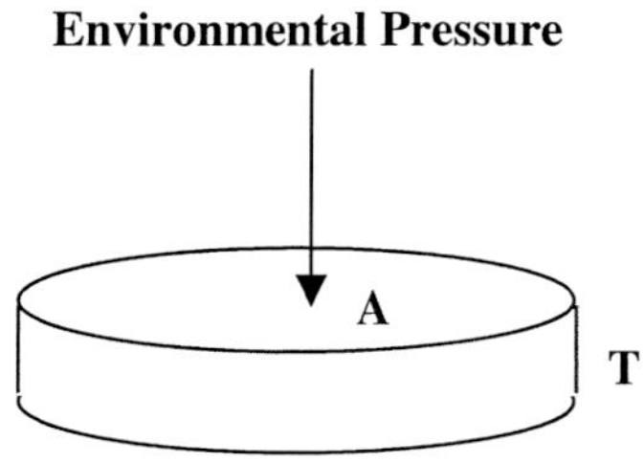

Figure 14. Die cut specimen under imposed environmental pressure.

The only amount that will need to be added to the calculated value will be the one that diffused into the internal structure and did not cause comparable swelling, i.e. the fluid simply filled holes inside the fibers. This amount could be expected to be very small and neglected. Thus, one of the parameters that need to be modeled for a web is the pore volume.

The model for the rate involves fluid properties, cosine of the contact angle and pore radius (see eq. 7). Other factors, such as the distribution of pore size, the shape of pores, the tortuosity and the orientation of flow channels, and the swelling characteristics of fibers, no doubt, also affect the rate [132] but they are not part of a known model and their effects are not well understood. Collectively, these factors are expected to lower the rate over the one predicted by Washburn's model. This is discussed in section 5.2, Chapter III. Of the factors mentioned, the one that is affected by the structure is the pore size. Accordingly, that is the second parameter that needs to be modeled and predicted for a web.

7.2. Pore Volume

One can model pore volume of a fabric that may be a single component material or a blend containing multiple components. The model presented is for the latter [133]. The analysis considers a fabric element cut to area A, whose dry mass is W and thickness, under imposed pressure P, is T (Figure 14).

Mass of fiber i in the element, $W_i = Ww_i$

Volume of fiber i in the element = W_i/ρ_i

Total fiber volume in element, $V_f = W\Sigma w_i/\rho_i$

Total pore (air) volume in element, $V_a = AT\text{-}V_f$

In these, i is the index representing individual components and w is the mass fraction and ρ is the density of the individual components. Total pore volume of the element divided by the mass of the element gives specific air volume, V_s:

$$V_s = \frac{V_a}{W} = \frac{AT}{W} - \Sigma\frac{w_i}{\rho_i}$$

The second part of the right hand side of the above equation represents the reciprocal of the weighted average density of the components. If ρ_{av} represents this quantity, then the above equation for V_s can be rewritten as follows:

$$V_s = \frac{AT}{W} - \frac{1}{\rho_{av}} \tag{39}$$

where,

$$\rho_{av} = \left[\Sigma \frac{w_i}{\rho_i} \right]^{-1} \tag{40}$$

The quantity calculated by eq. 39 gives the maximum interstitial space or absorbent capacity per unit dry mass of fabric under the imposed pressure (Figure 14). This quantity, in units of cm^3/g, when multiplied by fluid density, ρ_l, should represent the absorptive capacity in units of (g fluid/g fiber).

In terms of packing fraction, ψ, defined as the ratio of fiber volume to web volume, V_s is given by equation 42:

$$\psi = \frac{V_f}{AT} \tag{41}$$

$$V_s = \frac{AT - \psi AT}{W} \tag{42}$$

or, since $W = \rho_{av}\, \psi\, AT$,

$$V_s = \frac{1}{\rho_{av}} \left[\frac{1-\psi}{\psi} \right]$$

Also, because $\psi = (AT - V_a)/AT$,

$$V_s = \frac{1}{\rho_{av}} \left[\frac{V_a / AT}{1 - V_a / AT} \right] \tag{43}$$

For a constant value of ψ or V_a/AT, the above equations show that:

$$V_s \propto \rho_{av}^{-1} \tag{44}$$

Equations 42 and 43 are interesting expressions for specific air volume or absorbent capacity, which has the units of volume of fluid absorbed per unit mass of fabric (cc/g), the units in which absorbent capacity is usually expressed. These show that for a given packing fraction (eq. 42) or for a given volume of fluid absorbed per unit geometrical size or volume of fabric used (eq. 43), the lower the fiber density, the greater will turn out to be the value of absorbent capacity.

As found by Gupta (138), the packing fractions noted have ranged from 0.025 to 0.1, or 2.5 to 10%, depending upon the material, the fabric weight (areal density), and the bonding methodologies used. Figure 15 illustrates the impact the fiber density can have on capacity, when expressed in the units of cc fluid/g fiber, for various values of packing fraction. Figure 16 shows the relative values of the absorbent capacity of different materials expressed as a ratio with the value of a cellulosic material for any given value of packing fraction. Accordingly, the lower density materials provide an advantage in applications of fibers in absorbent products provided all other characteristics, in particular, surface wettability and mechanical properties, are equivalent. From the work presented in Chapter III, it should be clear that the surface wetting characteristics of cellulosic fibers make them highly suitable for use in absorbent products. The challenge, thus, lies in developing next generation polymeric materials that have the densities of polyolefins but the hydrophilicities of celluloses.

A model for absorbent capacity that accounts for the difference in fiber densities will be the one in which the volume of fluid absorbed varies only with fiber packing. Such a model will be given by the product of specific air volume and fiber density, as shown below:

$$V_{so} = V_s \rho_{av} . \tag{45}$$

V_{so} represents the volume of fluid absorbed per unit volume of fibers (cc fluid/cc fibers). This dimensionless parameter can be expected to more truly represent the absorbency behavior of a fabric and more effectively correlate with the structure. This parameter will be particularly valuable when behaviors of materials differing broadly in terms of density are compared. Substituting for V_s from eq. 39 one gets the following for the quantity, V_{so}.

$$V_{so} = \frac{AT\rho_{av}}{W} - 1 \tag{46}$$

In terms of porosity, ϕ ,defined in section 4.1, V_s and V_{so} are given as follows:

$$V_s = \frac{\phi}{\rho_{av}(1-\phi)} \tag{47}$$

$$V_{so} = \frac{\phi}{1-\phi} \tag{48}$$

7.3. Pore Size

Pore size is an important parameter for absorbent materials as it affects the rate at which a fluid flows into or thorough a capillary network. This is evident from the Hagen-Poiseuille (eq. 3) and the Washburn (eq. 7) models. The availability of a structural model for this parameter will serve two highly useful purposes: it will allow the value of pore size to be computed for a structure and fed into a flow model for predicting the rate, and it will

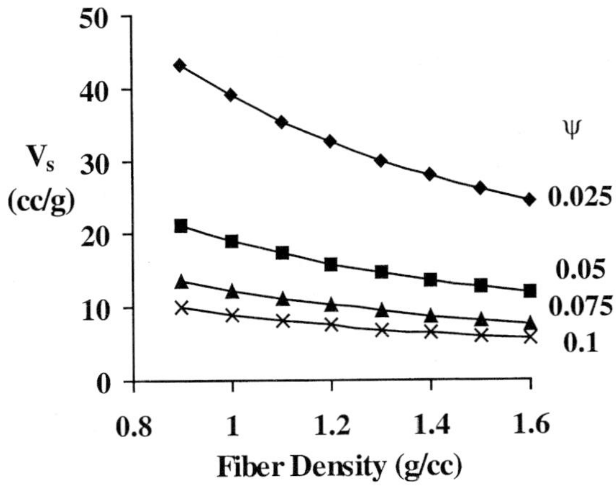

Figure 15. Effect of fiber density and packing fraction on specific air volume (or absorbent capacity).

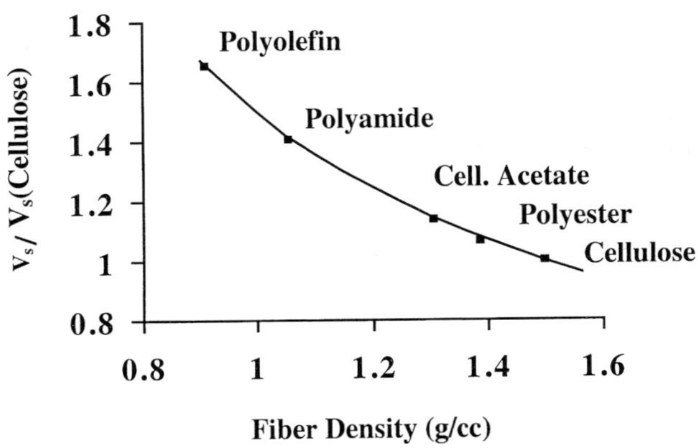

Figure 16. Effect of fiber density on the ratio of the specific air volume of a web containing a given fiber to that of a web containing cellulose.

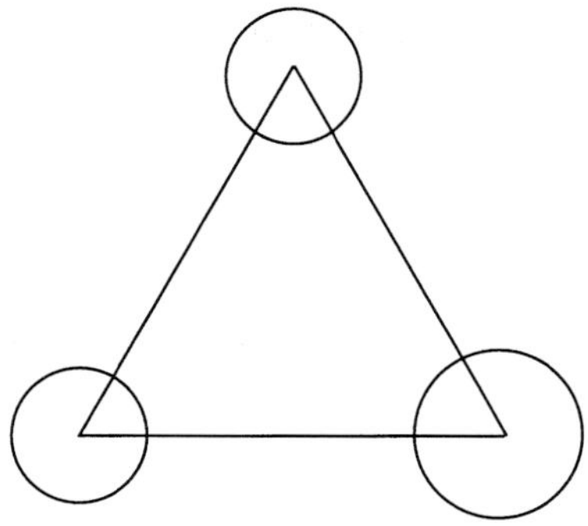

Figure 17. Capillary formed in space enclosed by three fibers lying at the apexes of an equilateral triangle.

illustrate the factors that play roles and provide specifications for engineering structures that have desired characteristics.

In developing the model, two cases are considered [133]: a simple hypothetical one that involves parallel arrangement of fibers, and a more complex but realistic one that involves random arrangement of fibers. The models considered are for a web containing a blend of multiple materials. Special cases are considered in sec. 7.4, in which the general models are applied to structures containing one to three different types of fibers and to structures that also contain an adhesive [134].

7.3.1. Parallel arrangement model

It is assumed that the fibers are distributed uniformly throughout the structure, they are arranged parallel to each other, three fibers enclose a space and define a capillary, and the latter has the shape of an equilateral triangle of width Y with the fibers lying at the apexes (Figure 17). If n_i is the number of fibers of type i out of 3, and d_i is the corresponding linear density, then

$$n_i \propto w_i/d_i \tag{49}$$

$$\Sigma n_i = 3. \tag{50}$$

The volumes per unit length of various quantities are calculated as follows:

Volume/unit length of the triangle = $\left(\sqrt{3}/4\right)Y^2 \cdot 1 = \left(\sqrt{3}/4\right)Y^2$ (51)

Volume/unit length of n_i fibers = $\left(n_i d_i / \rho_i B_o\right)$ (52)

In this, B_o is the constant whose value is determined by the base length associated with the linear density used. For example, if d is the denier, and all parameters are in the c.g.s. system of units, then B_o will have the value 9 X 10^5. Each fiber contributes 1/6th of its volume to the volume of the triangle. The volume per unit length of the triangle (the capillary unit cell) occupied by the fibers is S and is given by:

$$S = \frac{1}{6B_o}\left[\Sigma\frac{n_i d_i}{\rho_i}\right] \tag{53}$$

The packing factor for the triangular element is given by the following equation:

$$\psi = \frac{4S}{Y^2\sqrt{3}} \tag{54}$$

The value given by eq. 54 must equal the packing factor of the fabric element given by eq. 41. Equating the two yields the following value for the length, Y:

$$Y = \left[\frac{4}{\sqrt{3}}\cdot\frac{ATS}{V_f}\right]^{1/2} \tag{55}$$

Capillary radius, r, is defined as the radius of a circle whose area equals the unoccupied area of the triangle. Thus:

$$\pi r^2 = \frac{\sqrt{3}}{4}Y^2 - S \tag{56}$$

or,

$$r = \left[\frac{1}{\pi}\left(\frac{\sqrt{3}}{4}Y^2 - S\right)\right]^{1/2} \tag{57}$$

For hypothetical maximum packing, shown in Figure 18, the capillary radius will be as follows:

$$r = R_f\left[\frac{\sqrt{3}}{\pi} - \frac{1}{2}\right]^{1/2} = 0.227R_f \tag{58}$$

where R_f is fiber radius.

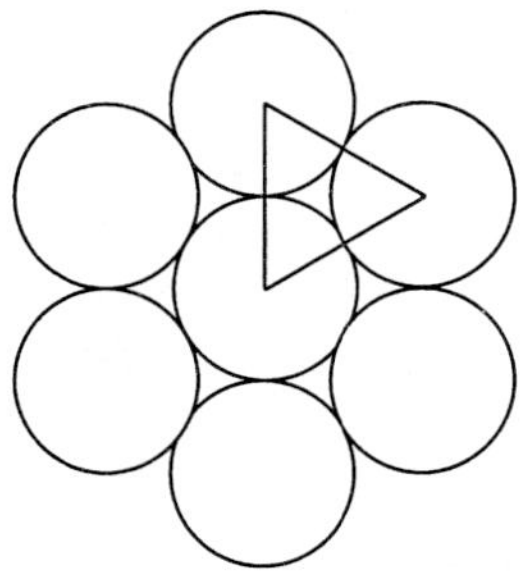

Figure 18. Capillary in hexagonal close packed structure.

By making appropriate substitutions for Y and S in eq. 57, one can show the detailed structure of the geometrical model for the pore radius.

$$r = \left[\frac{1}{6\pi B_0} \left\{ \frac{AT\rho_{av}}{W} - 1 \right\} \left\{ \sum \frac{n_i d_i}{\rho_i} \right\} \right]^{1/2} \tag{59}$$

By combining eqs. 46 and 59, one can also express r by the following simpler equation:

$$r = \left[\frac{V_{so}}{6\pi B_0} \sum \frac{n_i d_i}{\rho_i} \right]^{1/2} \tag{60}$$

7.3.2. Random arrangement model

The capillaries are not straight and parallel, as assumed in sec. 7.3.1., but tortuous and randomly oriented, and they are interconnected, i.e. a given capillary may not be bound by the same fibers from point to point along its length. It is assumed, however, that each fiber is surrounded by free space that is proportional to its own volume (Figure 19). An interstitial space is still bounded by three fibers, and each fiber contributes one-sixth of its free volume at each point along its length to the volume of a capillary.

$$\text{Free volume/fiber volume} = \frac{AT - V_f}{V_f} = \frac{AT}{V_f} - 1 \tag{61}$$

$$\text{Fiber volume per unit fiber length} = \frac{d \cdot 1}{\rho B_0} = \frac{d}{\rho B_0} \tag{62}$$

$$\text{Fiber free volume per unit fiber length} = \left[\frac{AT}{V_f} - 1 \right] \left[\frac{d}{\rho B_0} \right]$$

$$\text{Associated area perpendicular to fiber axis} = \left[\frac{AT}{V_f} - 1 \right] \left[\frac{d}{\rho B_0 \cdot 1} \right]$$

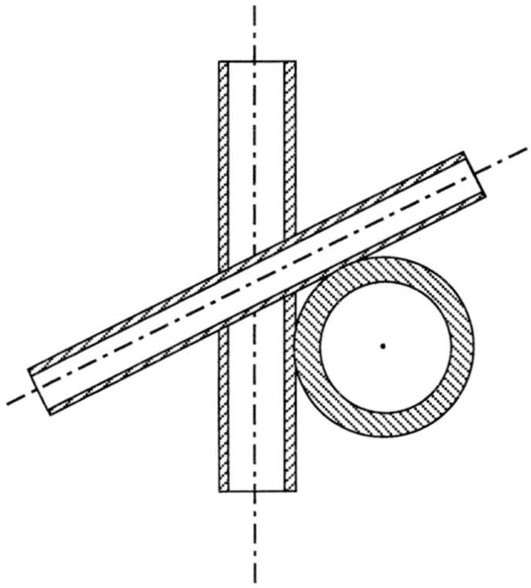

Figure 19. Interstitial space bounded by three fibers randomly arranged.

Area each fiber contributes to interstitial space = associated area divided by 6,

i.e., $\left[\dfrac{AT}{V_f}-1\right]\left[\dfrac{d}{6\rho B_0}\right]$

The fibers' contribution to interstitial space is from n_1 fibers of type 1, n_2 fibers of type 2, etc. Therefore, the total area of interstitial space bounded by three fibers but made up of different materials is given by eq. 63.

$$\text{Area of capillary} = \left[\frac{AT}{V_f}-1\right]\left[\frac{1}{6B_0}\sum\frac{n_i d_i}{\rho_i}\right] \tag{63}$$

or, by combining eq. 63 with 53,

$$=\left[\frac{AT}{V_f}-1\right]S \tag{64}$$

Equation 64 for the interstitial area bounded by three fibers randomly arranged, exactly equals eq. 56, developed for the unoccupied area of the triangle for the case of the parallel arrangement of fibers. Thus, for either the parallel or the random arrangement of fibers, the two-extreme structures considered, the capillary radius is given by eq. 59. For any other arrangement of fibers, therefore, the capillary radius can also be given by the same equation.

7.4. Special Cases

7.4.1. One component fabric

Many absorbent structures contain only one type of fiber. In such a case, two equations characterizing specific pore volume (39 and 46) and pore size (59) simplify to the following:

$$V_s = \frac{AT}{W} - \frac{1}{\rho} \tag{65}$$

$$V_{so} = \frac{AT\rho}{W} - 1 \tag{66}$$

$$r = \left[\frac{1}{6\pi B_0} \left\{ \frac{AT\rho}{W} - 1 \right\} \frac{3d}{\rho} \right]^{1/2} = \left[\frac{V_s d}{2\pi B_0} \right]^{1/2} \tag{67}$$

Most interesting among the above is the equation 67 which shows how simply pore size is related to pore specific volume. According to this equation, one can show, pore size is the radius of a capillary whose volume per unit length corresponds to the mass of one-half fiber of unit length (1cm).

7.4.2. Two component fabric

These may contain two different fibers, or a fiber and a low melt polymer. The purpose of the former will be to combine properties or attributes of two different materials, not adequately provided by a single material, e.g. wettability and resiliency, and that of the latter will be to have a structure that could be bonded by heat.

A. Two different fibers

The only unknowns for using equations 39, 46 and 59 are the magnitudes of n_1 and n_2, i.e. the number of fibers out of 3 belonging to type 1 and 2. These are determined as shown below [133].

From Equation 49,

$$n_1 / n_2 = w_1 d_2 / w_2 d_1 \tag{68}$$

$$n_1 = 3 w_1 d_2 / (w_1 d_2 + w_2 d_1) \tag{69}$$

$$n_2 = 3 - n_1 \quad (70)$$

B. One fiber and an adhesive

This will also cover a fabric containing a regular and a low melt fiber, the latter after melting coating the fiber and serving as an adhesive. The designations used for the adhesive are w_{ad} for mass fractions and ρ_{ad} for density. It is assumed that ρ_{ad} is the density of the adhesive in the final or bonded state, and the adhesive or the molten polymer uniformly coats the surface of the regular fiber. The pickup by the fiber will increase the linear density from d to d', and change the density from ρ to ρ'. The new values are plugged in the equations pertaining to one component fabric (65-67) and the quantities V_s, V_{so} and r determined. The values of these parameters are as follows [134]:

$$\rho' = \frac{\rho \rho_{ad}}{w \rho_{ad} + w_{ad} \rho} \quad (71)$$

$$d' = \frac{d}{w} \quad (72)$$

7.4.3. Three component fabric [134]

These may include three different fibers or two different fibers and an adhesive.

A. Three different fibers.

Examination of equations 39, 46 and 59 indicates that the only quantities needed to be determined to characterize such a structure are the values of n_1, n_2 and n_3, i.e. the number of fibers out of 3 belonging to each type. The values of these quantities can be shown to be given as follows:

$$n_1 = (3w_1d_2d_3)/(w_1d_2d_3 + w_2d_1d_3 + w_3 d_1d_2) \quad (73)$$

$$n_2 = (3w_2d_1d_3)/(w_1d_2d_3 + w_2d_1d_3 + w_3d_1d_2) \quad (74)$$

$$n_3 = (3w_3d_1d_2)/(w_1d_2d_3 + w_2d_1d_3 + w_3d_1d_2) \quad (75)$$

B. Two different fibers and an adhesive or a low melt material.

As before, it is assumed that the adhesive or the molten polymer will uniformly coat the surfaces of regular fibers. The amount picked up by different fibers will be proportional to their surface areas. The net result will be an increase in linear densities, from d_1 and d_2 to d_1' and d_2', and increase in mass fractions, from w_1 and w_2 to w_1' and w_2', and a change in densities, from ρ_1 and ρ_2 to ρ_1' *and* ρ_2', respectively, for components 1 and 2. The values of the new quantities are first plugged in equations 69 and 70 to estimate the values of n_1 and n_2 and then the values of all these quantities are used in equations 39, 40, 46, 59, to characterize the needed parameters. In the treatment given below, A_{si} represents the surface area of fiber i in the fabric specimen and A_{st} represents the total surface area of all fibers in the fabric. The quantities related to adhesive have the subscript "*ad*". The values of A_{si}, A_{st}, and of the changed quantities, ρ_i', d_i', w_i' can be shown to be as follows:

$$A_{si} = 2\pi \cdot \textit{radius fiber } i \cdot \textit{length fiber } i$$

$$= 2\pi\, Ww_iB_0/(\pi d_i\rho_iB_0)^{1/2} \tag{76}$$

$$A_{st} = \Sigma A_{si} \tag{77}$$

$$\rho_i' = \frac{w_i + w_{ad} \cdot A_{si} / A_{st}}{(w_i / \rho_i) + w_{ad} \cdot A_{si} / (\rho_{ad} \cdot A_{st})} \tag{78}$$

$$d_i' = d_i[1 + w_{ad} \cdot A_{si} / (w_i \cdot A_{st})] \tag{79}$$

$$w_i' = w_i + w_{ad} \cdot A_{si} / A_{st} \tag{80}$$

7.4.4. Four component fabric – three fibers and an adhesive [134]

This is the most complex structure considered in this treatment. Firstly, the values of the changed parameters, ρ_i', d_i' and w_i', are determined using equations 78-80 for the three fibers given the coating. In the second step, the new quantities d_i' and w_i' are used in equations 73 to 75 to estimate the magnitude of n_1, n_2 and n_3. Finally, the quantities n_1 to n_3, ρ_1' to ρ_3', w_1' to w_3' and d_1' to d_3' are used in equations 39, 40, 46, and 59, as before, to characterize the parameters of interest.

7.5. Estimation of Porosity Related Parameters in Absorbent Structures

As shown in sec. 7.3.1, the estimate of pore size for maximum hypothetical packing is $0.227R_f$. The corresponding value of the specific air volume for such a structures is $0.12/\rho$, i.e. its value varies from approximately 0.08 for cellulosic fibers to 0.13 for polyolefins. It is interesting to note that the hypothetical minimum value for cellulose corresponds to the value of moisture regain found under standard atmospheric conditions. The value of V_s noted in actual fibrous webs is substantially greater. For mechanically bonded structures - needled and hydroentangled – the specific air volume, or the maximum absorbent capacity (*C*), has been found to range between 10 to 20 cc/g [132, 135-137]. The value in some of the spun bonded structures has been noted to be around 5cc/g [138]. The actual value, as clear from the results in Chapter III, varies with the wet mechanical properties of the fiber, fiber size and shape, web areal density, bonding method and level, and the environmental pressure under which tests are conducted. The fibers most widely used are cotton, rayon, polypropylene and polyester. The denier for most materials used is about 3, except cotton, for which the value is

Table 1.
Calculated values of packing factor ψ, porosity ϕ, pore size r and ratio of pore size to fiber size r/R_f, for different materials and observed ranges of capacities, V_s.

Material	ρ (g/cc)	d	R_f (cm)x10³	V_s (cc/g)	Ψ (x100)	Φ (x100)	r (cm)x10³	r/R_f
Cotton	1.5	1.5	0.60	5	11.76	88.24	1.15	1.92
				10	6.25	93.75	1.63	2.72
				15	4.26	95.74	2.00	3.33
				20	3.23	96.77	2.30	3.83
Rayon	1.5	3.0	0.84	5	11.76	88.24	1.63	1.94
				10	6.25	93.75	2.30	2.74
				15	4.26	95.74	2.82	3.36
				20	3.23	96.77	3.26	3.88
Polyester	1.38	3.0	0.88	5	12.66	87.34	1.63	1.85
				10	6.76	93.24	2.30	2.61
				15	4.61	95.39	2.82	3.20
				20	3.50	96.50	3.26	3.70
Polypropylene	0.96	3.0	1.05	5	17.24	82.76	1.63	1.55
				10	9.43	90.57	2.30	2.19
				15	6.49	93.51	2.82	2.69
				20	4.95	95.05	3.26	3.10

around 1.5. For these materials and structures, the values of packing factor, ψ (eq. 42), porosity, ϕ (eq. 47), pore size, r (eq. 67), and the ratio of pore size to fiber size, r/R_f, assuming fibers are circular in cross-section, were calculated and are given in Table 1.

Accordingly, in typical mechanically bonded absorbent structures, used in studies, the porosity is 0.9 or greater (90% or more of the volume is air), and the pore size is from 2 to 4 times the fiber size. This is instructive, as such low values of pore size could not generally be speculated without modeling as done here. Thus, although the web is largely air, the pore size is still only about 2 to 4 times the fiber size. In spun bonded and resin bonded materials, in which the absorbent capacity found is of the order of 5 cc/g, the porosity ranges between 0.82 to 0.89 and, therefore, the pore size ranges between only 1.5 to 2 times the fiber size.

8. FLOW RATE MODELS FOR FIBROUS WEB

Considered now are the models that could be used to characterize the rate of flow when a fluid is imbibed by a specimen in one of several ways.

8.1. Linear horizontal wicking

The most basic model used is that due to Hagen-Poiseulle [33, 34], eq.3. For capillary assisted flow, capillary pressure, given by the Laplace eq. 2, is substituted for the pressure drop, ΔP, in eq. 3. Integration from $L = 0$ to $L = L$, leads to the Washburn model given by eq. 7, repeated below. According to this equation, the movement of a fluid front through a porous strip (fig. 20A) is proportional to the square root of time, i.e. the rate decreases with passage of time, but the flow continues, till presumably the rate of evaporation equals the rate of absorbency.

$$L = \left[\frac{r_c \gamma \cos\theta}{2\eta} \right]^{1/2} t^{1/2} \tag{7}$$

The factors affecting the rate of flow are the pore size, the fluid surface tension and the cosine of the contact angle, directly, and the fluid viscosity, indirectly.

8.2. Vertical Wicking

As the fluid rises (Fig. 20B), the pressure generated, $L\rho_l g$, opposes the capillary pressure, and, therefore, this term must be subtracted from the latter as done in eq. 4, sec. 3. One could expect the fluid to rise to an equilibrium height in this case as it could rise only to a level at which the capillary pressure, p, is balanced by the gravitational pressure, $L_{eq}.\rho_l.g$ (eq. 5). Integration leads to a more complex solution; however, as pointed out in sec.3 that if $t << t_{eq}$, or $L << L_{eq}$, expansion of eq. 6 and elimination of higher order terms, reduces the equation to the one given by Washburn model (eq. 7). One can show that for a typical pore of about 2 x 10^{-3} cm, and θ of 30°, and water as the fluid, the equilibrium height reached will be about 64 cm and the time taken to reach it, if no diffusion or evaporation occurs, will be more than 10 minutes. In most wicking tests, however, the heights targeted are of the order of a few centimeters and the times considered as the upper limit are of the order of 2 minutes. These values are small and justify the elimination of the effect of gravity and the use of eq. 7 for characterizing the rate in the vertical wicking as well. A recent article by Miller [139] showed that if the wicking height exceeds about 10% of the equilibrium height, the rate of flow starts to deviate noticeably from the one predicted by eq. 7. Accordingly, in situations of this type, the fuller model, represented by eq. 6, must be used.

8.3. Areal or Volumetric Flow from Limited Source

When a liquid drop of a given size is placed on a porous material, it will be imbibed and spread under the influence of capillary forces, provided the contact angle is less that 90°. The process occurs in two phases (fig. 21). In the first, the drop saturates the material directly underneath the area covered by the drop and the immediate surrounding till the drop disappears. In the second, the fluid is pulled by the smaller capillaries from the larger and the area stained or wetted increases. This increase occurs as a function of the time, the

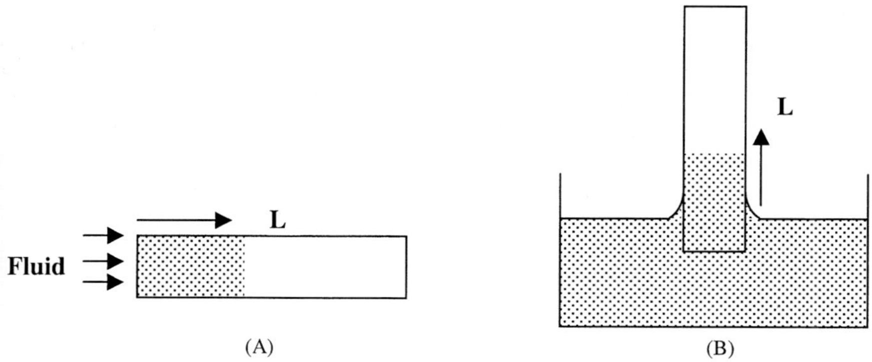

Figure 20. Linear wicking. (A) Horizontal, (B) Vertical

volume of the drop and the characteristics of the material. For volume of drop, V_D, large enough to fully saturate the area lying underneath through the thickness, the work of Gillespie (140) led to a model, for wetted area, A_W, expressed in a generalized form by Kissa (141), as follows:

$$A_W = K_0 \left[\frac{\gamma}{\eta} \right]^u V_D{}^m t^\ell \qquad (81)$$

In this K_0 is the capillary sorption coefficient given by eq. 82.

$$K_0 = \frac{27\pi b k_s \cos\theta}{8T^2 c_s{}^3} \qquad (82)$$

The theoretical values of the exponents are shown to be 0.33, 0.66, and 0.33, respectively, for u, m and ℓ. In eq. 82, b is a constant, k_s is the permeability of the substrate, θ is the contact angle, T is the thickness of the substrate and c_s is the saturation concentration of the liquid in the substrate. For the first phase absorption of the drop, the behavior was shown to be given by Washburn model, i.e. eq. 7, so that the values of the exponents u and ℓ were 0 5. and that of m was 0 [141,142]. For the second phase absorption, if the fluid did not diffuse and the material did not swell, the works of Kissa (141) and of Kawase et. al. (142), from investigations on a number of woven fabrics, gave the values for the exponents, u, m and ℓ, as approximately 0.30, 0.7 and 0.30, respectively. If a fluid did penetrate the fiber structure, the value of ℓ tended to be lower (i.e. the rate slower) and that of m tended to be higher.

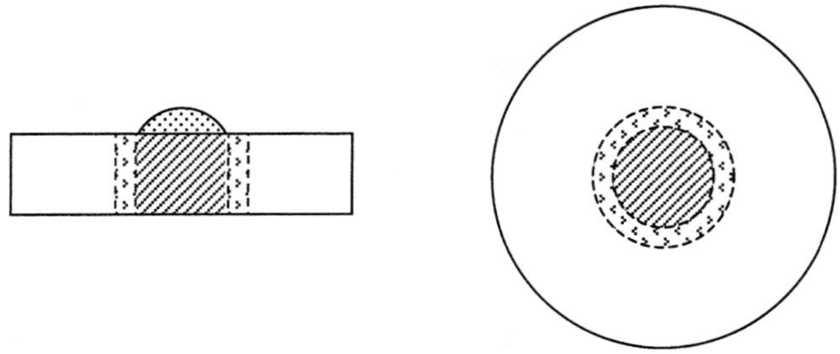

Figure 21. Areal flow from a source of limited supply, such as a drop.

8.4. Volumetric Spreading from Unlimited Source

Washburn's model describes linear flow along a channel, most accurately when the channel is oriented horizontally. In many of the controlled commercial tests, however, the fluid is presented at a point in the middle of a circular specimen, which then spreads radially outward as seen in Figure 22. Flow continues till the specimen is saturated and then terminates nearly abruptly. A device widely used to characterize the behavior in this mode is the Gravimetric Absorbency Testing System, or the GATS, discussed in detail in Chapter XI.

Washburn's equation, i.e. eq. 7, can be modified to apply to radial flow, as it occurs in this case. If the liquid front advances to distance L in time t, and the penetration only takes place through the pores, the volume of fluid imbibed in this time period is given by

$$V_l = \pi L^2 T(1-\psi) \tag{83}$$

where $(1-\psi)$ is the fraction of web volume available for fluid flow in, or the porosity of, the specimen. Substituting for ψ from eq. 41 and replacing V_f with W/ρ, one gets

$$V_l = \pi L^2 T\left[1-\frac{W}{AT\rho}\right] \tag{84}$$

By substituting for L^2 from eq.7, replacing Washburns ideal capillary radius, r_c, by the average pore radius, r, measured on a specimen, and dividing both sides with mass, W, and time t, we get the following equation for the specific flow rate, or the rate of absorbency, Q, for radial flow from unlimited source:

$$Q = \frac{V_l}{Wt} = \frac{\pi r\gamma\cos\theta}{2\eta}\left[\frac{T}{W}-\frac{1}{A\rho}\right] \tag{85}$$

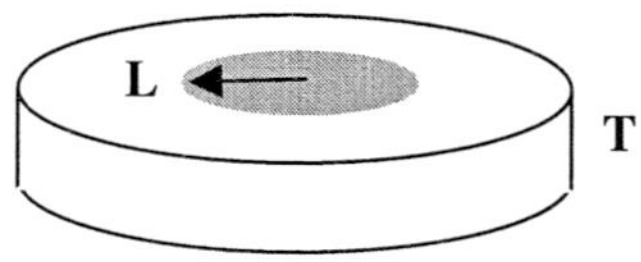

Figure 22. Radial flow from a source of unlimited supply in a circular die-cut specimen from a point in the middle.

Generalizing this equation for a structure that may have a blend of two or more materials, one gets the following equation for the rate:

$$Q = \frac{\pi r \gamma (\cos\theta)_{av}}{2\eta}\left[\frac{T}{W} - \frac{1}{A\rho_{av}}\right] \qquad \text{[cc(fluid)/g(fiber)} \bullet \text{sec]} \tag{86}$$

where, $(\cos\theta)_{av} = \Sigma w_i \cos\theta_i$ (87)

Equation 85 shows that the volumetric absorption of fluid per unit mass, in the radial flow set up, is linearly proportional to time, as against one-half power of time for linear wicking, and even a smaller power of time for spreading from a limited source. The equation also shows that, for a given fluid and fiber system, the rate is not only affected by pore size, but also by the fabric bulk, or the thickness per unit mass, *T/W*. Capacity increases by an increase in the bulk alone, but the rate increases by an increase in both the bulk and the pore size. Equation 59 shows that pore size is itself positively affected by the bulk. Therefore, any change made in the material properties or the process that leads to an increase in the bulk should lead to an increase in both the capacity and the rate, but the fractional increase in the latter could be expected to be greater than that in the former.

The c.g.s. units of the rate, as given by eq. 86, will be (cc fluid/g fiber-sec). Alternatively, the rate may also be expressed as a function of the volume of fibers, instead of the mass, as done for the capacity through the parameter V_{so} (eq. 46). The new quantity, which will have the units (cc fluid/cc fiber-sec), is represented by Q_0, and will be given by:

$$Q_0 = \frac{\pi r \gamma (\cos\theta)_{av}}{2\eta}\left[\frac{T}{W} - \frac{1}{A\rho_{av}}\right]\rho_{av} \qquad \text{[cc(fluid)/cc(fiber)} \bullet \text{sec]} \tag{88}$$

8.5. Application of Flow Models to Fibrous Web

Several models have been presented for characterizing the rate, depending upon the mode in which fluid is imbibed. They lead to different rates, as judged by the values of the exponents on time *t* in equations 7, 81 and 85; however, clearly, all models involve similar parameters as the factors affecting the behavior.

Accordingly, any of the methods discussed, i.e. the methods based on assessing the behavior in horizontal wicking, vertical wicking, radial spreading from a limited source, or radial spreading from unlimited supply, could be used to relatively rank the performance of different materials. However, for generating quantitative data on a product for a specific type of application, the test method used must simulate the actual use conditions.

9. ACKNOWLEDGEMENT

This chapter is a thoroughly revised version of Chapter II (by P. K. Chatterjee and H. V. Nguyen) in the previous edition of this monograph, entitled "Absorbency", edited by P. K. Chatterjee and published by Elsevier Science Pub. in 1985.

10. GLOSSARY

a Radius of water meniscus.

a_0 Radius of area in the center of specimen initially wetted by fluid (≈ radius of the fluid delivery hole).

A Area of the sample; also cross-sectional area perpendicular to the main flow direction in linear flow.

A_{si} Surface area of fibers of type i in a fabric specimen.

A_{st} Total surface area of fibers in fabric specimen ($=\sum A_{si}$).

A_w Wetted area of specimen.

B Defined as $(0.5 - \mu)/\rho_p v_s$, is a constant for a given polymer-solvent system.

B_o Constant, whose value is determined by the base length associated with the linear density.

B_1 A constant equal to $r_c^2 \rho_l\, g/8\eta\, L_{eq}$, in Lucas-Washburn equation.

c Solvent concentration in case of solvent diffusion into a polymer network.

c_s Concentration of liquid in substrate at saturation.

$(\cos\theta)_{av}$ Average value of cos θ in a fabric containing a blend of different materials.

C Absorbent capacity of a porous sample (capacity to fill up all the pore space, volume of fluid per unit mass of conditioned fiber (cc fluid/g fiber).

C_0 Absorbent capacity of a porous sample, volume of fluid absorbed per unit volume of fiber (cc fluid/cc fiber).

C_+, C_- External concentration of electrolyte cation and anion, respectively.

d Fiber linear density.

d' Linear density of coated fiber.

D Diffusion coefficient for diffusion of solvent into a polymer network, usually a function of c.

D_f Effective average diameter of particles or fibers making up the porous sample; diameter of cylindrical fibers (Iberall's equation); diameter of spherical beads (Brinkman).

D_H Average channel diameter defined by the hydraulic radius concept.

e Number of monomers per statistical element in Kuhn's theory.

f Friction factor, dimensionless.

F Diffusivity coefficient for liquid penetration into a porous medium, usually a function of s.
F_i F at initial state.
F_0 F for completely saturated porous medium.
F_1 A constant value for moisture diffusivity used in idealized examples.
g Gravitational acceleration.
h Distance measured vertically upward from an arbitrarily chosen datum level (for defining "piezometric pressure").
i Degree of ionization of the polyelectrolyte multiplied by the valency of the fixed ionic group on the polymer chains; i would be the degree of ionization when this valency is 1.
I_0 Ionic strength of the external electrolyte solution, defined as $\zeta_+ \zeta_-(e_+ + e_-)/2$.
k Permeability of the porous medium, unit = cm^2.
k_0 Proportionality constant in Washburn equation.
k' Kozeny constant.
K Flow conductivity of the porous medium with respect to fluid involved.
K_0 Capillary sorption coefficient in Kissa model.
K_s Permeability of substrate.
K_1, K_2 Proportionality constants involved in superposition of Fickian and Case II diffusion.
l Number of monomoles of the cross-linked polymer.
L Wetted length of capillary tube, or wetted radius of a sample in radial flow.
L_{eq} Equilibrium capillary rise.
L_0 Length of the sample in the main flow direction.
M Molecular weight of the polymer making up the network.
M_c Molecular weight per cross-linked unit.
M_w Molecular weight of water.
n_i Number of fibers of type i out of 3 making up a capillary ($\sum n_i = 3$).
p Capillary pressure.
p_a, p_∞ Equilibrium vapor pressures over concave surface of water meniscus of radius a, and over a plane water surface.
ΔP Macroscopic pressure head driving force.
q Volume flux (cm^3/sec area) of fluid in the main transport direction.
Q Rate of absorption (cc fluid/g fiber-sec).
Q_0 Rate of absorption (cc fluid/cc fiber-sec).
r Average capillary radius.
r_c Radius of a capillary tube.
r_d, r_w Dry and wet capillary radius.
r_e Effective capillary radius ($= r_w^2/r_d$).
R The gas constant.
Re Reynolds number, dimensionless.
R_f Fiber radius.
s Local degree of saturation of a porous sample, dimensionless and varies from 0 to 1.
s_i Saturation at initial stage.
s_0 Initial saturation profile before liquid redistribution.
s_v Visually observed saturation level.

S Volume per unit length of capillary unit cell occupied by fibers (parallel arrangement).
S_o Surface area of the flow channel per unit volume of the solid material making up the porous medium.
t Time variable.
T Sample thickness.
U Combined variable equal to $x/t^{1/2}$ for linear transport.
v Filter velocity of fluid in the main transport direction equal to volume flux (or volumetric rate per unit cross-sectional area of sample) in the same direction.
V Volume of liquid absorbed or present in the porous sample at any given time.
V_a Total air (pore) volume in a sample.
V_l Volume of liquid absorbed in a specimen.
V_D Volume of drop in spreadability study of Gillespie.
V_f Total fiber volume in a sample.
V_g Volume of swollen gel.
V_s Specific air volume in fabric (air volume per unit fiber mass).
V_{so} Specific air volume in fabric (air volume per unit fiber volume).
w_{ad} Mass fraction of adhesive in adhesive or polymer bonded structures.
w_i Mass fraction of component i in blend.
W Dry (conditioned) mass of fabric specimen.
x, y, z Distances in the three rectangular coordinates.
Z Number of monomers between neighboring crosslinks.
α Density function describing pore size distribution.
γ Surface tension of the liquid being absorbed.
δ Pore diameter in pore size distribution function.
ε Equilibrium swelling ratio of hydrogels = swollen volume/dry volume.
η Viscosity of liquid.
θ Contact angle of liquid-solid-air interface.
λ Correction factor to account for deviation from Gaussian distribution of highly swollen polymer network.
μ Dimensionless parameter expressing first neighbor interaction for solvent with polymer.
v_m Molar volume of the monomeric unit.
v_s Molar volume of the solvent.
ξ_+, ξ_- Valence of the electrolyte cation and anion, respectively.
ρ Density of the solid material of the porous medium; density of fiber.
ρ' Density of coated fiber.
ρ_{av} Average fiber density $(=\sum w_i\rho_i)$
ρ_{ad} Density of adhesive used in bonding.
ρ_{bulk} Density of the bulk of the porous medium.
ρ_l Density of the liquid.
ρ_p Density of the polymer making up the cross-linked network.
ρ_w Density of water.
τ Absolute temperature.
ϕ Porosity of the porous medium, given by the ratio of air volume to bulk volume.
ψ Packing factor (fiber volume per unit fabric volume).

11. REFERENCES

1. A. W. Adamson, Physical Chemistry of Surfaces, J. Wiley, New York 1967.
2. J. H. de Boer, The Structure and Properties of Porous Materials, D. H. Everett & F. S. Stone (Eds.), Vol. 10, Butterworths, London, 1958.
3. S. Brunauer, The Adsorption of Gases and Vapors, Vol. I, Princeton University Press, Princeton, NJ, 1971.
4. R. Defay and I. Prigogine. Surface Tension and Adsorption (translated by D. H. Everett), Longmans Green, London, 1966.
5. I. Langmuir, J. Am. Chem. Soc., 40 (1918) 1361.
6. S. Brunauer, P. H. Emmett and E. Teller, J. Am. Chem. Soc., 60 (1938) 309.
7. J. H. de Boer, The Dynamical Character of Adsorption, Clarendon Press, Oxford, 1953.
8. D. N. Broad and A. G. Foster, J. Chem. Soc., (1945) 366.
9. P. C. Carman and F. A. Raal, Proc. Roy. Soc., A, 209 (1951) 59.
10. A. R. Urquhart and A. M. Williams, J. Textile Inst., 17 (1926) T38.
11. W. E. Morton and J. W. S. Hearle, Physical Properties of Textile Fibers, the Textile Institute, Manchester, 1993, p. 223-264.
12. A. R. Urquhart, J. Textile Inst., 20 (1929) T125.
13. P. H. Hermans, Physics and Chemistry of Cellulose Fibers, Elsevier, Amsterdam, 1949, p. 197.
14. A. J. Stamm, TAPPI, 40 (1957) 761.
15. A. J. Stamm, Wood and Cellulose Science, Chapter 8, Sorption of Water Vapor by Cellulosic Materials, The Ronald Press Co., New York, NY 1964, 142-165.
16. A. R. Urquhart, W. Bostock and N. Eckerall, J. Text. Inst., 23 (1932) T135.
17. A. R. Urquhart and A. M. Williams, J. Text. Inst., 15 (1924) T132, T433, T559; 16 (1925) T155; 28 (1958) T159.
18. J. A. Howsmon, in Cellulose and Cellulose Derivatives, Part 1, 2nd edit., F. Ott, H. M. Spurlin and M. W. Grafflin (Eds.), Interscience, New York, 1954, p. 393.
19. J. E. Carles and A. M. Scallan, J. Appl. Pol. Sci., 17 (1973) 1855.
20. J. A. Barrie, in Diffusion in Polymers, J. Crank and G. S. Park (Eds.), Academic Press, New York, p.259.
21. F. A. Bettleheim, C. Sterling and D. H. Vollman, J. Pol. Sci., 22 (1956) 303.
22. R. Jeffries, J. Text. Inst.. 51(1960) T339, T399, T44I.
23. G. H. Peterson, J. Pol. Sci., 28 (1958) 458.
24. B. Katchman and A. D. McLaren, J. Am. Chem. Soc., 73 (1951) 2124.
25. L. J. T. Hughes and D. B. Fordyce, J. Polymer Sci., 22 (1956) 509.
26. I. Sakurada, A. Nakajima and H. Fujiwara, J. Polymer Sci., 35 (1959) 497.
27. P. M. Hauser and A. D. McLaren, Ind. Eng. Chem. Ind. Edn., 40 (1948) 112.
28. A. W. Myers, J. A. Myer, C. E. Rogers, V. T. Stannett and M. Szwarc, TAPPI, 44 (1961) 58.
29. A. D. McLaren and J. W. Rowen, J. Pol. Sci., 7 (1951) 289.
30. E. A. Columbo and E. H. Immergut, J. Polymer Sci. C, 31(1970)137.
31. R. K. S. Bhatia, Indian J. Chem., 9 (1971) 1279.
32. P. R. Tremaine and D. G. Gray, J. Chem. Soc., Faraday Trans. I, 71(1975) 2170.
33. G. Hagen, Ann. Phys. Chem., 46 (1839) 423—442.
34. J. L. Poiseuille, C. R. Acad. Sci. Paris, 11(1840) 961, 1041; 12 (1841) 112.
35. R. Lucas, Kolloid Z., 23 (1918) 15.
36. E. W. Washburn, Phys. Rev., 17 (1921) 273.
37. E. J. Legrand and W. E. Rense, J. Appl. Phys., 16 (1945) 843.
38. G. M. Aberson, The Physics and Chemistry of Wood Pulp Fibers, D. H. Page (Ed.), TAPPI, (1970) 282.
39. P. H. Everett, J. M. Haynes and R. J. L. Miller, in Fiber Water Interaction in Paper Making, Trans. Symp. 1977 Oxford, the British Paper and Board Federation, London, Vol. II (1978) 519.
40. B. J. Ackerman and H. G. Cassidy, Anal. Chem., 26 (1954) 1874.
41. P. K. Chatterjee, Svensk Papperstidning, 74 (1971) 503.
42. R. L. Peek, and D. A. McLean, Ind. Eng. Chem., Anal. Ed., 6 (1934) 85.
43. J. Szekely, A. W. Neumann and Y. K. Chuang, J. Colloid Interfac. Sci., 35 (1971) 273.

44. A. M. Schwartz, C. A. Rader and E. Huey, Adv. Chem. Ser., 43 (1964) 256.
45. W. Rose and R. W. Heins, J. Colloid Sci., 17 (1962) 39.
46. V. Ludviksson and E. N. Lightfoot, A. I. Ch. E. J., 14 (1968) 674.
47. H. L. Ritter and L. C. Drake, Ind. Eng. Chem., 17 (1945) 782.
48. H. Darcy, Les Fontaines Publiques de la Ville de Dijon, Dalmont, Paris, 1856.
49. J. Kozeny, Royal Academy of Science, Vienna, Proc. Class I, 136 (1927) 271.
50. P. C. Carman, J. Soc. Chem. Ind., 57 (1938) 225.
51. P. C. Carman, Flow of Glass Through Porous Media, Butterworths, London, 1956.
52. A. S. Iberall, J. Res. Nat. Bur. Stand., 45 (1950) 398.
53. H. C. Brinkman, Appl. Sci. Res. A, 1 (1947) 27.
54. H. C. Brinkman, Appl. Sci. Res. A, 1(1948) 81.
55. R. A. Mott, Some Aspects of Fluid Flow, Edward Arnold, London, 1951, p. 242.
56. J. Happel, A.I.Ch.E.J., 5 (1959) 174.
57. J. Hrubisek, Kolloid-Beihefte, 53 (1941) 385.
58. A. E. Scheidegger, The Physics of Flow Through Porous Media, University of Toronto Press, Toronto, 1947.
59. F. A. L. Dullien, Porous Media Fluid Transport and Pore Structure, Academic Press, New York, 1979.
60. J. Bear, Dynamics of Fluids in Porous Media, American Elsevier, New York, 1972.
61. R. B. Bird, W. E. Stewart and E. N. Lightfoot, Transport Phenomena, J. Wiley, New York, 1960.
62. S. Hedvat, Capillary Flow Studies on Fibrous Substrates, Ph. D. Thesis, Princeton Univ. 1980.
63. H. V. Nguyen and D. F. Durso, TAPPI Proc., 1983 Int. Dissolving and Specialty Pulps Conf., Boston, 1983, p. 137.
64. C. Y. Lin and J. C. Slattery, A. I. Ch. E. J., 28 (1982) 311.
65. F. A. L. Dullien, A. I. Ch. E. J., 21(1975) 299.
66. J. M. Golden, Water Resource Res., 16 (1980) 201.
67. H. Rumpf and A. R. Gupte, Chem. Eng. Tech., 43 (1971) 367.
68. C. N. Davies, Proc. Inst. Mech. Eng., B, 1 (1952) 185.
69. C. P. Kyan, D. T. Wasan and R. C. Kintner, Ind. Eng. Chem. Fundamentals, 9 (1970) 596.
70. I. F. MacDonald, M. S. El-Sayed, K. Mow and F. A. L. Dullien, Ind. Eng. Chem. Fundamentals, 18 (1979) 199.
71. A. Takahashi, Kagaku, 14 (1950) 41.
72. A. L. Ruoff, D. L. Prince, J. C. Giddings and G. H. Stewart, Kolloid Z., 166 (1959) 144.
73. A. L. Ruoff, G. H. Stewart, H. K. Shin and J. C. Giddings, Kolloid Z., 173 (1960) 14.
74. A. Klute, Proc. Soil. Sci. Soc. Am., 16 (1952) 144.
75. E. E. Miller and R. D. Miller, Proc. Soil. Sci. Soc. Am., 19 (1955) 267.
76. A. Klute and G. E. Wilkinson, Proc. Soil. Sci. Soc. Am., 22 (1958) 278.
77. E. C. Childs and N. Collis-George, The Control of Soil Water, in Advances in Agronomy, Vol. 2, A. G. Norman (Ed.), Academic Press, New York, 1950, p. 234.
78. W. R. Gardner and M. S. Mayhugh, Proc. Soil. Sci. Soc. Am., 22 (1958) 197.
79. D. F. Rudd, J. Phys. Chem., 64 (1960) 1254.
80. M. J. Denton, J. Chromatog., 18 (1965) 615.
81. H. Schubert, in Fiber Water Interaction in Paper Making, Trans. Symp. 1977 Oxford, The British Paper and Board Federation, London, Vol. II (1978), p. 537.
82. L. A. Richards, Physics, 1 (1931) 318—333.
83. H. S. Carslaw and J. C. Jaeger, Conduction of Heat in solids, 2nd edition. Clarendon Press, Oxford, 1959.
84. D. U. von Rosenberg, Methods for the Numerical Solution of Partial Differential Equation, American Elsevier, New York 1969.
85. J. R. Phillip, Trans. UNESCO Symp. on Water in the Unsaturated Zone, 1966, p. 503.
86. J. Crank, Mathematics of Diffusion, Clarendon Press, London, 1956, p. 148.
87. R. D. Wyckoff and H. G. Botset, J. Appl. Phys., 7 (1936) 325.
88. J. T. Morgan and D. T. Gordon, J. Pet. Tech., 22 (1970) 1199.
89. P. J. Jones, World Oil, 129 (1949) 170.
90. T. Gillespie, J. Colloid Sci., 14 (1959) 123.
91. W. H. Gardner and J. C. Hsieh, Proc. Soil. Sci. Soc. Am., 20 (1956) 157.
92. R. R. Bruce and A. Klute, Proc. Soil Sci. Am., 20 (1956) 458.

93. H. Fujita, J. Phys. Chem., 56 (1952) 625.
94. I. Ijjas, Trans. UNESCO Symp. on Water in Saturated Zone 1966, p. 547.
95. F. M. Fowkes, in Chemistry and Physics of Interfaces, Am. Chem. Soc., 1965.
96. A. A. Burgeni and C. Kapur, Textile Res. J., 37 (1967) 356.
97. E. V. Painter, and H. V. Nguyen, 185th ACS National Meeting. Cellulose, Paper & Textile Division, Seattle, Wash., March 22, 1983.
98. R. J. Marshall and A. B. Metzner, Ind. Eng. Chem. Fundamentals, 6 (1967) 393.
99. M. Barbosa, C. Rangel and B. Mena, J. Rheol., 23 (1979) 281.
100. J. A. Deiber and W. R. Schowalter, A. I. Ch. E. J., 27 (1981) 912.
101. P. V. Choksi, E. E. Spaeth and J. A. Shiff, INDA Technical Symposium, 1977, p. 29.
102. P. H. Hermans, in Colloidal Science II, H. R. Kruyt (Ed.), Elsevier, Amsterdam, 1949.
103. E. V. Beebe, R. L. Coalson and R. H. Marchessault, J. Polymer Sci. C-13 (1966) 103.
104. R. H. Marchessault, M. Dube, J. St. Pierre and B. Fisa, in Fiber Water Interaction in Paper Making, Trans. Symp. 1977, Oxford, The British Paper and Board Federation, London Vol. II (1978) 795.
105. P. J. Flory, Proc. Roy Soc. (London), Ser. A. 234 (1956) 73.
106. P. K. Chatterjee and C. M. Conrad, J. Polymer Sci., 11(1967)1387.
107. R. H. Marchessault and J. A. Howsmon, Textile Res. J., 27 (1957) 30.
108. J. M. Preston and M. V. Nimkar, J. Textile Inst., 40 (1949) 674.
109. G. E. Collins, J. Textile Inst., 21(1930) T311.
110. J. M. Preston and D. Gupta, J. Text. Inst., 38 (1947) T60.
111. J. V. Brancik and A. Datyner, Text. Res. J., 47 (1977) 662.
112. E. F. Thode, J. G. Bergoni and R. E. Unson, TAPPI, 43 (1960) 505.
113. G. Jacquelin, Svensk Papperstidning, 20 (1963) 801.
114. P. J. Flory, Principles of Polymer Chemistry, Cornell University Press, Ithaca, 1967.
115. A. Katchalsky. S. Lifson and H. Eisenberg, J. Polymer Sci., 8 (1952) 476.
116. A. Katchalsky, 0. Kunzle and W. Kuhn, J. Polymer Sci., 5 (1950) 283.
117. I. Michaeli and A. Katchalsky, J. Polymer Sci., 23 (1957) 683.
118. F. Helfferich, Ion Exchange, McGraw Hill, New York, 1962.
119. J. A. Kitchener, in Ion Exchangers in Organic and Biochemistry, C. Calmon and T. R. E. Kressman (Eds.), Interscience Publishers, Inc., New York, 1957, p. 36.
120. M. Rinaudo, in Charged Gels and Membranes Part I, E. Selegny (Ed.), Reidel Publishing Co., Boston, 1976, p. 91.
121. J. Crank and G. S. Park, Diffusion in Polymers, Academic Press, New York, 1968, p.1.
122. G. S. Park, Ref. 121, Chapter 5, p. 141.
123. A. C. Newns, Trans. Faraday Soc., 52 (1956) 1533.
124. L. Westman and T. Lindstrom, J. Appl. Polymer Sci., 26 (1981) 2561.
125. F. A. Long and L. J. Thompson, J. Polymer Sci., 15 (1955) 413.
126. T. Alfrey, Jr., E. F. Gurney and W. G. Lloyd, J. Polymer Sci., C, (1966) 249.
127. J. Crank, J. Polymer Sci., 11(1953)151.
128. N.L. Thomas and A. H. Windle, Polymer, 23 (1982) 529.
129. H. L. Frisch, T. T. Wang and T. K. Kwei, J. Polymer Sci., A-2, (1969) 879.
130. H. B. Hopfenberg, L. Nicholais and E. Drioli, Polymer, 17 (1976) 195.
131. T. Haga, J. Appl. Polymer Sci., 26 (1981) 2649.
132. B. S. Gupta and C. J. Hong, TAPPI Journal, 77 (1994) 181.
133. B. S. Gupta, TAPPI Journal, 71, No. 8 (1988) 147.
134. B. S. Gupta, to be published.
135. B. S. Gupta and T. H. Hammond, INDA Technical Conference, U.S.A., 1980, 88-103.
136. B. S. Gupta and C. J. Hong, INDA, INJ, 7, No. 1 (1995) 34.
137. B. S. Gupta, INDA Technical Conference, 1998, U.S.A., p. 21.1.
138. B. S. Gupta, unpublished results.
139. B. Miller, INDA, INJ., 9, No. 1 (2000) 35.
140. T. Gillespie, J. Colloid, Sci. 13 (1958) 32.
141. E. Kissa, J. Colloid Interface Sci. 83 (1981) 265.
142. T. Kawase, S. Sekoguchi, S. Furjii, and M. Minagawa, Textile Res. J. 56 (1986) 409.

Absorbent Technology.
P.K. Chatterjee and B.S. Gupta, editors.

CHAPTER II

SURFACE TENSION AND SURFACE ENERGY[1]

ANTHONY M. SCHWARTZ
Nutech International Co., 331 McDowell Drive, East Brunswick, NJ 08816 (USA).

Contents

1. INTRODUCTION

Absorption, defined in our restricted sense as the spontaneous physical imbibition of a liquid by a contacting solid, can occur by either of two processes. In the first process or mode the pores of the solid, the openings through which the liquid enters, are of molecular dimensions and the liquid enters by diffusion. The solid is macroscopically continuous but has the molecular structure of a cross-linked network and is typically but not necessarily polymeric. The molecules of the solid have sufficient affinity for the liquid to cause the network to expand progressively and allow the liquid to diffuse inward from the phase boundary interface. This effect is similar in many ways to a dissolving action in which the solid acts as solvent and the liquid as solute. The swelling of rubber by benzene is an example of this type of absorption, which is generally referred to as swelling and is discussed elsewhere in this treatise. In the second mode of absorption the solid is either continuous and macroscopically porous or discontinuous but composed of closely adjacent particles. The pores or interparticulate spaces are occupied by gas or vapor at the ambient pressure which in

[1] Reproduced from "Absorbency", ed. P. K. Chatterjee, Elseview Science Publ., Amsterdam, The Netherlands, 1985 (Chapter III).

most cases is atmospheric. The liquid enters the pores by bulk convection that cannot be accounted for by pressures external to the system. This effect is referred to as capillary absorption or wicking. In this discussion these two terms will be used interchangeably, although popular usage sometimes restricts the term wicking to imbibition that occurs in opposition to gravity or other external pressures. Imbibition of water by a sand bed or a polyurethane or rubber sponge or a glass fabric are examples of absorption by pure wicking. In many practical systems absorption occurs by a combination of swelling and wicking. In such systems the substance of the solid is significantly swellable and the arrangement is porous or close packed particulate. Examples of systems that absorb water by wicking and swelling are natural sponge, rayon fabric, humus soil, and ordinary bread. In most absorbent systems the wicking aspects are easily discernible and separable from the swelling aspects.

The driving forces in wicking derive from the free energies of the phase interfaces in the system, and it is with these energies that the present discussion is concerned. For wicking to become a significant practical effect it is necessary that the interfacial areas in the system be large in comparison with the bulk volumes. In porous systems the area of the pore walls must be large as compared with the volume of the solid. In particulate systems the particles must be small in at least one dimension and must be closely spaced. This latter condition ensures that the ratio of solid-liquid (SL) interfacial area to liquid volume will be large. These requirements follow from the fact that interfacial free energy like other forms of energy is an extensive quantity. It is, however, conveniently expressed and handled mathematically as specific interfacial free energy, i.e. energy per unit interfacial area, usually symbolized by a small greek gamma *(γ)* with subscript to indicate the interface being referred to. Wicking is one example of the more general set of phenomena termed 'capillarity". Capillarity encompasses all the dynamic and kinematic effects and behavior of phase interfaces. It should be noted at this point that the phase interface, i.e. the boundary between two bulk phases, appears to be geometrically sharp and it can be so regarded when solving practical wicking problems. At the molecular level, however, the boundary is a region of special properties, comprising molecules from both bulk phases and having a statistical thickness in the range of a few molecular diameters. Some theorists treat this region as a distinct phase and refer to it as" the interfacial phase" or "the interphase". In this chapter the expression "phase interface" simply means the macroscopic boundary between two bulk phases, and except where indicated says or implies nothing about its special nature.

There are three phase interfaces of interest in capillary absorption: liquid-vapor (LV), solid-vapor (SV), and solid-liquid (SL). A fourth interface, the liquid-liquid (L_1L_2) is of interest where the pores of the solid are initially filled with a liquid (L_1) rather than with a gas. If this mass is submerged in a second liquid (L_2) that is immiscible with L_1, the L_2 may start to displace L_1 from the pores. Such an action is fully anologous to ordinary absorption in which the displaced fluid is a gas. One set of laws governs the displacement of a fluid from a solid surface by a liquid regardless of whether the displaced fluid is liquid or gaseous. A common example of L_1 displacement by L_2 is the washing of an oil soaked rag by an aqueous detergent solution. In many situations the behavior of a liquid interface and a gas or vapor interface will be identical. The term " fluid", abbreviated Fl, will then be used. For example the designation SFl includes both SV and SL. The fifth type of phase interface, solid-solid (S_1S_2) is important in problems of adhesion, friction, and electron transfer but not in capillarity, and will therefore not be considered further in this discussion. The terms "surface" and "interface" will often be used interchangeably as they are in ordinary technical usage. At other times when an explicit distinction is required for clarity or rigor it will be made. The terms "vapor" and "gas" are used interchangeably except where the context requires one or the other. Before proceeding to consider the individual interfaces and their coordinated effects in absorption a brief review of the generally applicable physical principles and terminology can save considerable needless repetition.

2. FUNDAMENTALS OF FORCE AND ENERGY APPLICABLE TO PHASE INTERFACES

The tendency of an isolated mass of liquid to minimize its surface area by forming spherical drops, and the rise of a liquid in a narrow tube. both manifestations of surface tension, have been recognized since ancient times. The first modern mathematical treatments of this phenomenon were elaborated separately by Laplace. Young, and Gauss in the early nineteenth century. These investigators used the concepts and terminology of classical Newtonian mechanics. The term "tension" then as now indicated a tensile force per unit area, i.e. the opposite of a " pressure" or compressive force per unit area. Since the surface tension of a liquid was recognized to be a force exerted parallel to and in the surface of the liquid, i.e. in two rather than three dimensions, it would have to be expressed as a force per unit length normal to the force vector: dynes/cm or in SI units Nm^{-1}, denoted herein by a small greek sigma (σ). Since by definition energy is equal to $\int f ds$, where f is force and s is the distance through which it acts, the integral $\int \sigma ds$, where s is measured parallel to and in the surface, and in the direction of the force vector has the dimensions of energy/cm2. This quantity γ is the specific surface free energy. σ and γ are dimensionally identical and mathematically equivalent. There is of course a real difference between the two quantities with regard to physical meaning. The specific surface free energy γ is the more general and more useful quantity since it is applicable to solid as well as liquid surfaces and interfaces. There is no experimental evidence that any force σ acts parallel to and in the plane of the solid-vacuum (SVac) interface. As discussed later in this chapter, one of the conditions for the existence of a tension σ in an interface is that at least one of the phases sharing that interface be a liquid possessing the property of fluidity. Since solids are by definition not fluid there is no reason to believe that a tension σ exists in the SVac interface. The phrase "solid surface tension" is not infrequently encountered in the literature. In this writer's view, however, it has a confusing connotation; the mathematically synonymous term "specific surface free energy of solid" (γ_{SVac}) should be used. The reality of σ as a force per unit length on the LV interface is illustrated in Fig. 1 by a drop of water hanging from a pipette tip. The weight W of liquid below the neck of the hanging drop is supported by the force σp where p *is* the perimeter at the neck.

The mechanical approach to capillarity and surface tension antedated the thermodynamic approach by several decades, just as the discovery and application of the gas laws antedated the thermodynamics of heat engines. Gibbs is generally credited with first applying thermodynamics to interfacial phenomena, especially of multicomponent systems. Detailed consideration of the relevant equations will be presented separately when discussing the individual phase interfaces. The concept of interfacial free energy, however, is ubiquitous and fundamental in capillarity. Interfacial free energy is the excess free energy a system possesses by virtue of its phase boundaries. The term "free energy" is used in

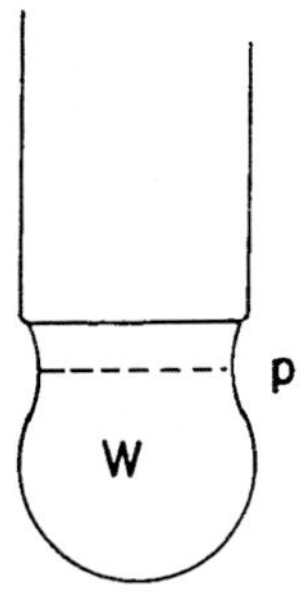

Fig. 1. Liquid drop hanging from a pipette tip. Weight *W* of liquid below perimeter *p* is supported by the force $\sigma_{LV} * p$. p = perimeter in cm. σ_{LV} = surface tension in dynes/cm.

thermodynamics to designate either of two closely related functions: the Helmholtz function *F* or the Gibbs function *G*. The Helmholtz function is defined as:

$$F=U\text{-}TS \tag{1}$$

where *U* is the internal energy of the system, *S* is the entropy content, and *T* the absolute temperature. The Gibbs function is defined as:

$$G = U\text{-}TS+PV = F+PV = H\text{-}TS \tag{2}$$

where *P* is the pressure, *V* the volume and *H* the enthalpy. Accordingly:

$$\mathrm{d}F = \mathrm{d}U - T\mathrm{d}S\ - S\mathrm{d}T \tag{3}$$

$$\mathrm{d}G = \mathrm{d}U - T\mathrm{d}S - S\mathrm{d}T + P\mathrm{d}V + V\mathrm{d}P\ = \mathrm{d}H - \mathrm{T}\mathrm{d}S -\ \mathrm{SdT} \tag{4}$$

Helmholtz called *F* the "free energy" of a system because in a reversible isothermal process $-\Delta F$ (the greater than infinitesimal decrease in *F)* is the energy that is released and can be converted to mechanical work. It is important to note that ΔF does not necessarily require a corresponding change in *U*. The source of energy may be the heat reservoir that keeps the process isothermal. A decrease in the Gibbs function, $-\Delta G$, is equal to the total energy released in a reversible isothermal and isobaric process minus the $P\Delta F$ energy. It is of particular value in describing the energy transformations that accompany phase changes. The question arises as to whether "interfacial free energy" refers to an *F* function or a *G* function. In practically all capillarity calculations it makes no significant difference which function is meant. Thermodynamic rigor, however, dictates that γ designate specific interfacial *F*. One physical reason for this is that the vapor pressure of a liquid does vary with the curvature of the LV interface, although the variation does not become significant until the curvature reaches a very high value. Similarly, high pressure of insoluble gas above a liquid surface lowers the surface tension. For planar liquid interfaces at vapor equilibrium the excess free energy γ contains no *PV* terms and the distinction between *F* and *G* vanishes [1].

To exemplify the concept of interfacial free energy as an excess quantity of F associated with a phase boundary, visualize two spheres of the same size and same liquid at the same temperature and pressure; one a spherical portion of the interior of a larger mass of liquid, the other suspended or in free fall in its own vapor, as shown in Fig. 2. Sphere x, an arbitrarily chose region not bounded by a second phase, possesses the free energy $F_x = U_x - TS_x$ as defined in eq. 1. Sphere y, bounded by a vapor phase, possesses the free energy $F_y = U_x - TS_x + \gamma_{LV} \bullet A_{LV}$ where A_{LV} is the area of the LV interface. For a one component, two phase system with negligibly curved interface such as a large drop of liquid in equilibrium with its vapor, the general equation for an infinitesimal change in internal energy is:

$$dU = TdS - PdV + \gamma dA + \mu dN \quad (5)$$

where n *is* the chemical potential and N the number of mols in the system. Equation 5 can be generalized to include phase boundaries other than the liquid-vapor and to include multicomponent systems. With regard to the interfacial free energy term, it serves as both a statement of existence and a formal definition. The driving force in any closed mechanical or thermodynamic system is the tendency to minimize the free energy. For a system to do mechanical work, i.e. to change its configuration spontaneously, dF must be negative. When $dF=0$ the system is at equilibrium and is static. It is important in capillarity problems to bear in mind that d F of the whole system, not just the phase interfaces, must be negative for energy to be released. If other potentials in the system oppose and outweigh the interfacial potentials the total interfacial free energy of the system, F_{if}, will generally not have reached its theoretical minimum value when dF has vanished. The ever present example of this situation is the effect of gravity on the shape of liquid menisci and on the movement of liquid in a capillary tube or a textile fabric wick. Since the three specific interfacial free energies of concern in capillarity are functions of state we can write:

$$dF_{if} = \gamma_{LV} * dA_{LV} + \gamma_{SL} * dA_{SL} + \gamma_{SV} * dA_{SV} \quad (6)$$

Equation 6 is the basic formula for calculating capillary motions and configurations in the absence of gravity or other interfering potentials, for example in orbiting space vehicles or in horizontal earthbound systems having a small vertical dimension. The addition of a term to

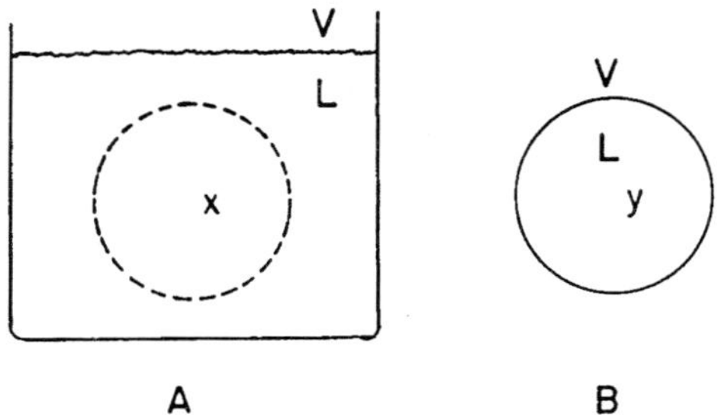

Fig, 2, Contribution of interfacial free energy to total free energy L = liquid. V = vapor. Free energy of interior mass X, F_x, $= F_y - \gamma_{LV}$* Area LV.

take care of the gravitational potential is conceptually simple but can involve some highly irksome calculation. The same is true for electrical effects which next to gravity are the most likely perturbations to be encountered in capillary systems. In the following discussions except where otherwise stated it will be assumed that outside potentials are absent. Chemical effects, i.e. a change in composition of the components of the system, will also be assumed absent.

The specific interfacial free energy functions γ all decrease in value as the temperature of the system rises. For the liquid-vapor system YLV vanishes at the critical temperature. For the liquid-liquid system γ_{LV} vanishes at the consolute temperature. The three specific interfacial free energies involving a solid phase, γ_{SVac}, γ_{SV} and γ_{SL} cannot and do not reach zero before the bulk solid phase ceases to exist as such. They do, however, decrease with increasing temperature. The rates of decrease of all γ functions with temperature can be evaluated precisely by a variety of experimental methods; some mechanical, some calorimetric, and some physicochemical.

The specific interfacial entropy, like γ, is an excess function. By definition it is the derivative of the specific interfacial free energy with respect to temperature and is equal to $-d\gamma/dT$. This function is seldom, invoked in surface thermodynamic calculations and does not have any generally accepted special symbol. Similar considerations apply to the chemical potential $\mu_i = \partial F/\partial N_i$, where N_i is the number of moles of component i in the system. The function μ_i, becomes important when the fluid phase(s) of the system contains more than one component.

2.1. Intermolecular Attraction

Interfacial free energy has its source in the universally present forces of intermolecular attraction. These are the forces represented by the constant "a" in the Van der Waals equation of state for real gases:

$$\left(P + \frac{a}{V^2}\right) * (V - b) = RT \tag{7}$$

They are the same configurational forces responsible for the condensation of gases of liquids and liquids to solids when the opposing kinetic forces (manifested macroscopically by the temperature) get weak enough to be overcome. These forces are generally considered to be of two different types referred to respectively as dispersion forces and polar forces. The dispersion force of attraction is present between any neighboring pair of molecules regardless of their chemical composition. It is ascribed to an interaction between the electromagnetic fields generated within each of the molecules by the orbiting electrons. Between two individual molecules this force of attraction varies as the inverse sixth power of the distance separating them. Between macroscopic planar solid surfaces the dispersion force varies as an inverse lower power of the separation distance, theoretically as the inverse third power if the molecules in the surface are regarded as contiguous.

The polar force of attraction results from one or another type of charge localization within each of the molecules. This force can accordingly be due to ion-dipole, dipole-dipole, or induced dipole interactions or to any combination of these. Some investigators emphasize the importance of acid-base or acceptor-donor interaction across an interface as a source of interfacial free energy, and make a distinction between these acid-base forces and other polar forces [2-4].

These intermolecular forces acting both across a phase interface and/or within each of the phases near a phase interface are the direct cause of capillary fluid motion. The phrase "capillary fluid motion" is here used in its most general sense. It includes not only wicking and spreading, i.e. motion of a fluid along a solid surface. but also the assuming of a uniformly curved contour by an LV interface and a spherical shape by a discrete unattached mass of liquid. The molecular mechanisms through which these interfacial forces bring about the fluid motion at all SFl and LFl interfaces are discussed hereinafter when each of these interfaces is considered separately. It is again emphasized that in this treatment perturbing forces of gravity, electricity et al. will be assumed absent except when they are specifically introduced. Thus the LV interface of a mass of liquid in free fall, where we can examine unmodified surface tension effects, has the form of a sphere. If the radius of this sphere is sufficiently large any section of the surface can be regarded as planar.

3. THE LIQUID-VAPOR INTERFACE

To visualize the mechanisms by which the intermolecular forces of attraction tend to make an LV interface minimize its area it is helpful to examine at least qualitatively the molecular structure of liquids [5]. The liquid phase is referred to as condensed, which means that it has the property of coherence. Liquids respond elastically to pure tensile stress. The molecules at all times remain within a certain limited distance of their nearest neighbors. In this respect liquids resemble solids. which also have the property of coherence, and contrast with gases or vapors. In gases the statistical distance between any molecule and its nearest neighbor is limited only by externally exerted pressure. In a liquid, however, the nearest neighbors of any molecule are continuously changing identity. This is another way of saying that molecules in the liquid state can diffuse to unlimited distances, just as they can in a gas but cannot in a solid. This freedom to diffuse also explains in molecular terms the fluidity of a liquid, the property of having no permanent resistance to shear or, in rhelological terminology, a zero yield point in shear. Several molecular models, all quite similar to one another, have been proposed to explain these part solid-like part gas-like properties. One such theory [6] postulates a structure in which most of the molecules are separated by spaces smaller than a molecular diameter but at about every tenth molecule a space or hole occurs large enough to accommodate a molecule. Any molecule bordering that hole can move into it, leaving an exactly similar hole in the molecule's previous location. In this manner the holes and the molecules move freely, continuously and, in the absence of localized shear stresses, randomly throughout the liquid. The ratio of about ten molecules per hole corresponds to the approximate 10% difference in molar volume between the liquid phase and solid phase of many substances at the melting point. This ratio is, of course, statistical. In any localized region at any point of time the ratio may be considerably greater or less than 10 to 1. Other models of the liquid state used in statistical mechanics provide similarly for both the coherence and the fluidity of liquids.

The average molecular spacing in a liquid, which necessarily includes the space across the holes, is at any given temperature and at mechanical equilibrium a constant characteristic of the material. Its value is determined by the balance between the intermolecular forces of attraction and the much stronger and shorter range forces of repulsion that come into play when the molecules get close enough for their electron clouds to overlap. If a tensile stress is applied to a liquid the resulting strain corresponds to an increase in the average molecular spacing. The restoring force is a measure of the tendency of the molecules to resume their characteristic spacing. The tensile and compressive (negative tensile) stress-strain behavior of a bulk liquid is for small strains quite similar to that of an elastic solid. For both materials the restoring force is directly proportional to the displacement, i.e. to the extent of the change in average molecular spacing. With even this very sketchy and abbreviated picture of liquid structure the molecular basis of surface tension can be examined more meaningfully.

Consider a molecule of a single component liquid at mechanical and thermal equilibrium somewhere in the interior remote from the surface. The time averaged force field acting on this molecule due to the attraction forces exerted vis-a-vis its neighbors is symmetrical and has the same magnitude in all directions. This follows from the fact that the density of neighboring molecules is the same in all directions. Because this interior molecule sits at the center of a spherically symmetrical force field the direction in which it moves will be determined solely by momentum exchanges with colliding molecules and will therefore be random.

The molecular picture near the LV interface is quite different. With regard to geometry the surface itself can be sharply defined. There is an outermost layer of molecules whose average distance from their underlying neighbors is little if any greater than it would be in the interior. We can refer to this as the geometric surface layer of the liquid. Outward from this layer at an average distance of several molecular diameters (far enough for the force field of the liquid phase to be of negligible magnitude) are molecules in the gaseous or vapor phase. The intervening space is occupied by molecules in transit, either evaporating from the liquid phase to the vapor or condensing in the opposite direction. The texture of the geometric surface layer is necessarily rough due to thermal jiggling but it is easily distinguished from the vapor phase and from the transit region by its closer intermolecular spacing. The geometric surface is the outer boundary of a thin region of liquid that possesses greater free energy per mole than is possessed by bulk interior liquid.

Consider a liquid molecule that in its random thermal motion is approaching the geometric surface. As it moves closer to this surface the number of interior neighbors remains constant but the number of exterior liquid neighbors diminishes rapidly. The force field acting on this molecule becomes increasingly unsymmetrical with the net force directed back toward the interior. This net inward force is strongest in the geometric surface but its magnitude is still significant at a considerable distance inward. We can accordingly describe a dynamic surface layer which averages considerably thicker than a single molecular diameter, and may be as thick as three or four molecular diameters. The molecules in this dynamic layer are continually being attracted inward and their average residence time in any volume element of the layer is less than it would be in a corresponding volume element of the interior. Accordingly the molecular density in this layer is less than in the interior, which is to say that the average distance between molecules is greater. The dynamic surface layer is therefore in a continual state of tensile strain. Surface tension is the restoring force corresponding to this strain. Surface tension can also be regarded as the intensity factor of a potential mechanical energy, the specific interfacial free energy of the LV interface.

Some aspects of this molecular picture of surface tension merit further comment. In the classical thermodynamic treatment of surface tension, first propounded by Gibbs, the entire region between interior liquid and interior vapor is called the "dividing surface". It essentially includes the region we have described as the dynamic surface layer, but is treated mathematically as a continuum [7,8].

Statistical mechanical treatments of surface tension necessarily start with models of the liquid state. Although rigorous in their development these treatments are limited by lack of assurance that the models are valid [9].

From the viewpoint of rheology and theoretical mechanics surface tension is described in terms of the stress or pressure tensor. The nine component tensor that describes stress in a three dimensional condensed phase can be contracted to include only those components tangential to the surface. This is possibly justifiable on the basis that the dynamic surface layer maintains constant thickness regardless of changes in the LV area.

It is evident that the phenomenon of surface tension results from the fact that the liquid state of matter possesses two salient properties: coherence which is characteristic of solids, and fluidity which is characteristic of gases. One description of surface tension often seen is that of a surface "skin" tending to contract after the manner of a stretched rubber membrane. This skin picture implies that the molecules in the surface layer remain there in a continually strained condition. The skin in contracting or expanding would presumably slide over the underlying liquid. This picture imputes a lack of fluidity to the surface molecules, denying them their gas-like property. It is an experimental fact that surface tension is a function of state, i.e. of chemical composition, pressure, and temperature. For a single component liquid at constant pressure and temperature the surface tension is constant regardless of whether or not the surface area is being changed. This constancy reflects the constant extent of strain in the dynamic surface layer. But in a stretched solid membrane the restoring force is proportional to the strain, increasing as the strain increases. In contrast, when an LV interface is increased or decreased in area the thickness of the dynamic surface layer poorly defined though it be on the interior side, remains constant. We can accordingly say with full assurance that there is no solid-like skin at the LV interface.

The question arises as to the molecular mechanisms involved when an LV interface contracts or expands. Consider first of all a static surface in mechanical equilibrium, for example the spherical surface of a liquid mass in free fall or, more familiarly, the surface of a pool flattened by gravity. Since the surface is neither contracting nor expanding the number of molecules arriving at it from the interior in any given time period must equal the number returning from it to the interior. The attenuated molecular density in the surface layer results from the difference in average velocity between the arriving and the departing molecules. Consider now an ellipsoidal droplet of free falling liquid. Fig. 3A. This droplet will spontaneously assume the spherical shape of Fig. 3B. The ultimate cause of this change in shape is that the surface in the X region contracts and the surface in the Y region expands. At X more molecules leave the surface per unit time than arrive at it. At Y in the same time period increases. A solid membrane has a well defined thickness. When the membrane is stretched this thickness decreases and when the membrane is allowed to contract the thickness more molecular arrive at the surface than leave it. An LV interface contracts by losing substance to the interior without changing structure; and can be expanded only by acquiring material from the interior whereupon the added material assumes the surface structure. Now these arriving and departing molecules that determine the expansion or contraction of the dynamic surface layer need only diffuse the distance of a few molecular

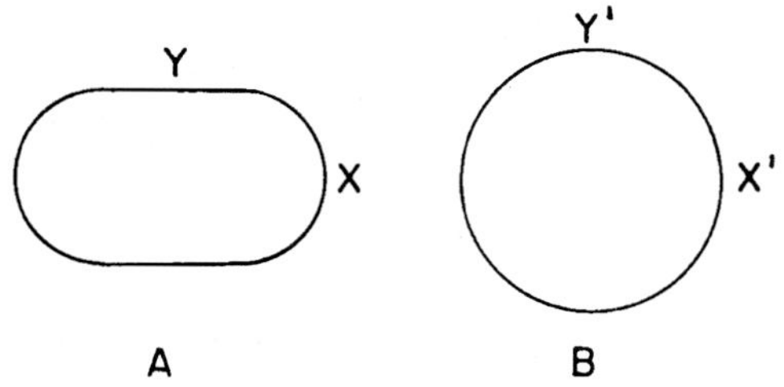

Fig. 3. Drop of liquid changes shape from ellipsoidal to spherical by surface expansion at *Y* to *Y'* and contraction at *X* to *X'*.

diameters to be either in the dynamic surface layer or in the interior. The time necessary to diffuse this distance in a liquid of ordinary viscosity is a matter of microseconds or less. Adjustment within the layer to equilibrium molecular density occurs within a similarly short period. The LV interface, always tending to contract, is capable of contracting almost instantaneously, and will in any event contract as rapidly as opportunity allows. The contracting dynamic surface layer pushes ahead of it bulk liquid to other parts of the system. In the example of Fig. 3 the bulk liquid under *X* is pushed into the region under *Y*. This effect, incidentally, could have been duplicated by a stretched membrane or skin encapsulating the ellipsoidal droplet, but ability to cause this effect is the only resemblance between such a skin and the dynamic surface layer of an LV interface.

From the viewpoint of hydrodynamics the ellipsoidal drop of Fig. 3 goes to the spherical shape because the hydrostatic pressure in the liquid under *X* is greater than that in the liquid under *Y;* and the sphere remains stable because the pressures under *X'* and *Y'* are equal. Since it is continuously tending to diminish in area, a curved LV interface will exert a pressure against its concave side. Conversely, in order to cause a planar LV interface to curve one must push against it, whereupon the side being pushed becomes concave. To illustrate this effect we can again invoke the similarly behaving stretched membrane. Figure 4A shows a drum consisting of a hollow pipe with membranes stretched across both ends. The pressure inside is P_1, equal to the pressure outside. If air is sucked out of the drum as at B, so that pressure P_2 is less than P_1 the drum heads are sucked inward. If, as at C, the drum is inflated so that P_3 is greater than P_1, the drum heads bulge outward. In B and C the pressure on the concave sides of the stretched drum heads is greater than the pressure on their convex sides, and in A, where the stretched drum heads are planar, the pressure on both sides is the same.

Referring again to Fig. 3A the pressure in the liquid under *X* is greater than atmospheric, but the pressure under *Y* is also greater than atmospheric. The pressure under *X*, however, is greater than the pressure under *Y* because the curvature of the surface is greater at *X* than at *Y*.

The quantitative relationship between the curvature of an LV interface and the pressure difference across that interface caused by the tension in it is:

$$P_1 - P_2 = \gamma\left(\frac{1}{r_1} + \frac{1}{r_2}\right) \tag{8}$$

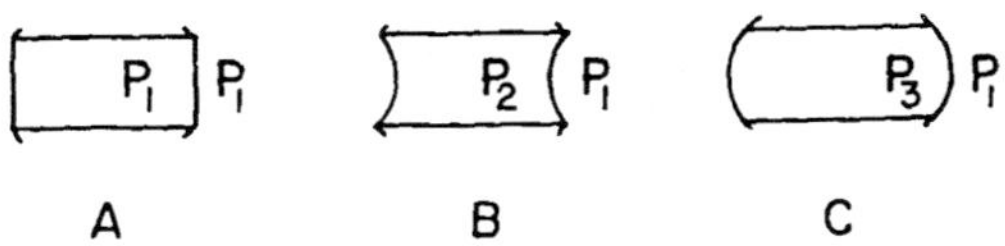

Fig. 4. Pressure differential across a drum head. At B, $P_1 > P_2$. At C, $P_3 > P_1$.

where P_1 is the pressure on the concave side and P_2 the pressure on the convex side, γ is the surface tension, and r_1 and r_2 are the principal radii of curvature of the surface. Equation 8 is known as the Laplace equation of surface tension and is one of the fundamental equations of surface physics. It is quite general and applies to any localized areas of arbitrary size on the interface. It can therefore be used to analyze pressure gradients under wavy or irregular surfaces. Provided that an LV interface is shaped only by its own tension, and is not influenced by outside forces, the condition for it to be in mechanical equilibrium is that $(P_1 - P_2)$ be constant over the whole interfacial area.

For spherical interfaces r_1 and r_2 are equal and equation 8 becomes:

$$P_1 - P_2 = \frac{2\gamma}{r} \tag{8a}$$

where r is the radius of the spherical surface. In practical wicking problems nonspherical liquid fronts are far more common than spherical ones. The liquid fronts in a textile wick, for example, are most often saddle shaped, in which case r_1 and r_2 are not only different but have opposite signs. In such situations eq. 8a is valueless and eq. 8 must be used if indeed the force approach rather than the energy approach is used to solve the problem. These two approaches and the distinction between them are discussed later in this chapter.

Equation 8 can be derived mathematically by the equations of classical mechanics. A simple and possibly more revealing derivation is directly applicable for eq. 8a, and applicable in principle for eq. 8. Going again into a zero gravity environment we start with a mass of liquid at equilibrium and therefore spherical. We are going to inject more liquid into this sphere by means of the piston in stem S as shown in Fig. 5. The outside pressure, exerted both on the sphere and on the back of the piston is P_2. The pressure inside the liquid is P_1. At any value of the sphere's radius r the area A is $4\pi r^2$ and the volume V is $4\pi r^3/3$*$dA/dr = 8\pi r$ and dV/dr $4\pi r^2$. Therefore $dA/dV = 2/r$. As shown previously, the incremental energy needed to expand this LV interface is equal to γdA. But at any stage of the operation this energy is furnished by the piston acting against the pressure P_1. Designating as P_x the pressure we must put on the piston to drive an incremental volume of liquid into the sphere, $P_x + P_2 = P_1$. $P_x = (P_1 - P_2)$ = the pressure difference across the curved interface. The incremental energy expended is $(P_1 - P_2)*dV$ and

$$(P_1 - P_2)dV = \gamma dA \tag{9}$$

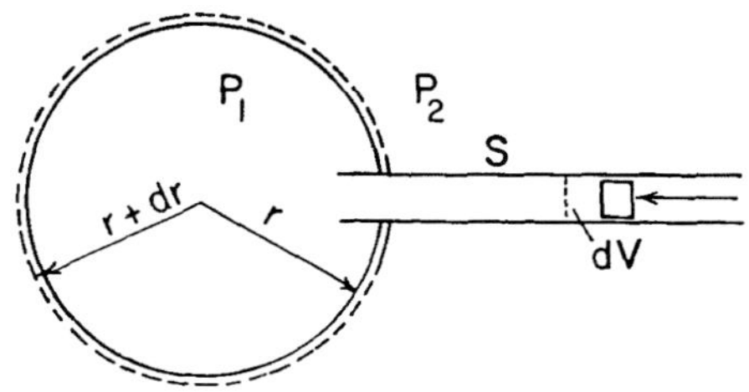

Fig. 5. Mechanical verification of Laplace equation for spherical mass of liquid. $(P_1 - P_2)dV = \gamma_{LV}\, dA$.

Substituting into eq. 9 the value for *dA/dV we* have:

$$(P_1 - P_2) = \frac{2\gamma}{r} \tag{8a}$$

Equation 8a and the more general eq. 8 simply reflect the geometric fact that the ratio of the area of a surface zone to the volume under that zone increases in direct proportion to the curvature of the zone. The above derivation also emphasizes the equivalence of surface tension and specific surface free energy. This equivalence can be used to greatly simplify some otherwise difficult practical problems in capillarity [10].

Equations 8 and 8a are valid for values of *r* down to the 100 Å range. Below this the mass of the dynamic surface layer becomes an appreciable fraction of the total mass of liquid under it. In practice this value of *r* corresponds to paths of migration in which diffusional flow is difficult to distinguish from ordinary convective flow.

3.1. Temperature Effects

With few exceptions, some of them dubious, the surface tension of pure liquids decreases with increasing temperature. This relationship is surprisingly linear over fairly long ranges. As the temperature nears the critical point, however, the linearity breaks down rapidly. Early work in liquid state theory was much concerned with the surface tension/temperature relationship in its quantitative aspects. Although of little practical applicability in absorbency problems this γ-T relationship does shed light from a different angle on the more fundamental relationship between surface tension and the internal structure of liquids.

If one gram-mole of a pure liquid is allowed to assume its equilibrium spherical shape the area of that sphere is called the molar surface area, A_0. By analogy with the molar volume of gases, the molar surface areas of all pure liquids under constant temperature and pressure conditions should contain the same number of molecules. This latter statement implicitly assumes that the molecular shapes and orientations are comparable. On this basis the product of the molar surface area by the specific surface free energy should be the same for all pure liquids. This quantity, the molar surface free energy. γA_0, can be written in terms of the molar volume V_0, which is equal to *M/D*, the molecular weight divided by the density. Calling *k* the proportionality constant that relates the volume of a sphere to its area:

$$\gamma A_0 = \gamma\left(\frac{M}{D}\right)^{2/3} \cdot k$$

Like γ itself the molar surface free energy has a remarkably constant temperature coefficient below the critical range, i.e.:

$$\frac{d}{dT}\left[\gamma\left(\frac{M}{D}\right)^{2/3}\right] = K$$

Integrating, and correcting for the experimentally noted breakdown of the relationship near the critical point:

$$\gamma\left(\frac{M}{D}\right)^{2/3} = K(T_C - T - 6) \tag{10}$$

where T_C, is the critical temperature and 6 is the experimentally determined number of degrees below T_C, at which the linear relationship ceases. Equation 10 is known at the Eotvos-Ramsay-Shields equation. *K*, the slope of the line, has the value of 2.12 for many so-called "normal" liquids, mainly hydrocarbons, chlorocarbons, liquified inert gases, etc. From a more modern point in view these are substances in which in intermolecular forces are of the dispersion type. Deviations from this value of *K* can be quite large, and were originally considered evidence of molecular association in the liquid state. More recent evidence has shown that this is not always the explanation. The value of *K* evidently depends on several factors, including the specific types of intermolecular force responsible for the liquid's cohesion. *K* values and variations from the numerical term 6 have been determined experimentally for many liquids and are available in the literature. Along similar lines relationships between *y* and molar heats of vaporization have been elaborated.

Regardless of its quantitative aspects the variation of surface tension with temperature is of great importance in practical capillarity problems. Referring to eq. 8 it is evident that a change in the value of γ can change the value of (P_1— P_2), the capillary pressure, even when the curvature remains constant, and the effect will be the same as if the curvature were changed. When the surface is planar and the value of γ changes locally, i.e. when a surface tension gradient is set up, the effect is somewhat different. It has been pointed out that molecular density in a high γ region of the dynamic surface layer is lower than in a corresponding low γ region of the same pure liquid. Molecules will now tend to move within the dynamic surface layer from the low γ region to the high γ region. The high γ region shrinks not by becoming flatter but by being transformed into a region of lower γ. The migrating molecules will also, by virtue of the viscous property of liquids, drag in their wake molecules from the adjacent interior. The macroscopic result is a ripple or ridge of liquid moving parallel to the surface, in and just under the surface, at the advancing front of the low γ region. This convection of liquid caused by a surface tension gradient is called the Marangoni effect. Even slight temperature differences can cause sufficient differences in surface tension to initiate Marangoni convection. It is a common cause of uneven drying in fabrics, and is especially pronounced and troublesome when the liquid is a solution having a non-volatile solute.

3.2. Stability of the LV Interface

If a falling stream of liquid becomes too thin or too slow it breaks up into drops. This and the intermittent drip of water from a leaky faucet are the commonest examples of the instability of a liquid surface. When the surface of a mass of liquid is extended sufficiently while the volume remains constant the mass will eventually separate into two or more masses. This instability of the surface is due to surface tension, and more particularly to the characteristic constancy of surface tension regardless of the extent to which the surface area is increased. The examples of the falling stream and drippy faucet involve not only surface tension but also gravity and an increasing volume of liquid. They both have been treated extensively and thoroughly in the classical literature of surface physics, including many of the appended general references. In most capillarity problems the effect of gravity on interfacial stability is negligible, even when its well recognized effect on flow rates and volumes is dominant. In capillary systems the ratio of LV interfacial area to the volume or mass tending to change that area is very high. Since gravity acts on mass the gravitational effect on the LV interfacial shape in these systems is small. This effect is illustrated by the meniscus shapes in small vs. large diameter horizontal tubes, as shown in Fig. 6. In tube A, having an internal diameter of I mm or less the meniscus is essentially spherical. In the larger tubes B and C the meniscus, while still stable, departs progressively from the spherical shape due to the effect of gravity. When the tube is sufficiently large, as in D, the meniscus collapses completely. Since cross sectional areas of the liquid channels in capillary systems are usually quite small the LV fronts are shaped by their *y* values and of course by the configuration of the channel walls.

To illustrate the limits of LV surface stability consider (again in zero gravity) the model system consisting of a fixed volume of liquid bridging two parallel plates, as in Fig. 7. This mass of liquid, having minimized its surface-to-volume ratio and having a contact angle of 90 ° against the plate material is cylindrical in shape with length 5 and radius /-. As explained later, the contact angle is a characteristic constant of the system under isothermal-isobaric conditions, and was set at 90° in this diagram simply for convenience. The following argument is valid for any contact angle although the calculations would be more complicated. In the shape shown at *A* the surface is quite stable. As we pull the plates apart, increasing the separation distance 5 and correspondingly decreasing the cylinder's radius *r,* the surface remains stable until a certain critical ratio of *s* to *r* is reached. If the plates are pulled farther apart the liquid surface yields and the mass divides into two portions, each clinging to one of the plates. These masses rapidly assume the hemispherical shape shown at C. This shape provides minimum surface-to-volume ratio while maintaining the 90 ° contact angle. The critical ratio at which the liquid bridge will pull itself apart was determined by Rayleigh in the last century. The derivation he worked out used the Laplace equation (eq. 8) and the theory of surface waves. At the critical ratio even the slightest surface disturbance will start a standing wave that increases in amplitude and causes the surface to yield [11].

An alternative method of determining the critical *s/r* ratio for stability is to determine when the surface-to-volume ratio of the system will be decreased by a yielding and separation of the surface. Referring to Fig. 7, the LV area of the liquid cylinder is $2\pi rs = A_1$.

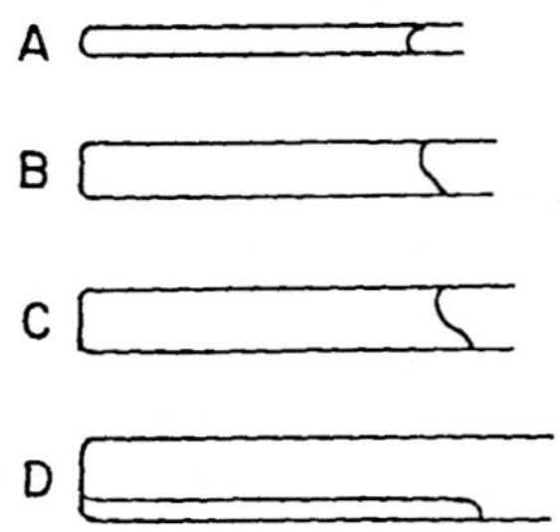

Fig. 6. Gravitational distortion of LV surface as area-to-volume ratio decreases.

The volume is $\pi r^2 s = V_1$. $A_1 = V_1$. For the spherical shapes $A_2 = 4\pi R^2$ and $V_2 = 4/3\pi R^3$, and $A_2/V_2 = 3/R$. Equating $A_1/V_1 = A_2/V_2$ at the critical ratio, and recalling that $V_1 = V_2$ at all times, we can solve for the critical s/r ratio. When $s/r \geq 4.5$ the cylinder of Fig. 7B will spontaneously transform to the two hemispheres of Fig. 7C. This thermodynamic treatment [12] is based on the concept of γ as a surface free energy tending toward minimization. It is generally applicable to all problems of LV surface stability, and is usually easier to apply than the classical Rayleigh-Laplace treatment. This classical treatment, which regards γ as a surface tension, is mathematically difficult for all but the simplest surface geometries. It does, however, elucidate the physical mechanisms (standing surface waves) involved in surface yielding, which the thermodynamic treatment does not.

Surface yielding occurs frequently and inevitably in wicking and in removing liquid from fibrous materials by wringing or squeezing. It is an important aspect of these processes that must be taken into account when predicting or interpreting their results. Surfaces can yield in some situations where the channel wall separation remains constant. As an example, if the contact angle of the liquid on the solid plates of Fig. 7A were zero the liquid would spread on each of the plates, and eventually would pull itself apart even though the separation distance s remained constant.

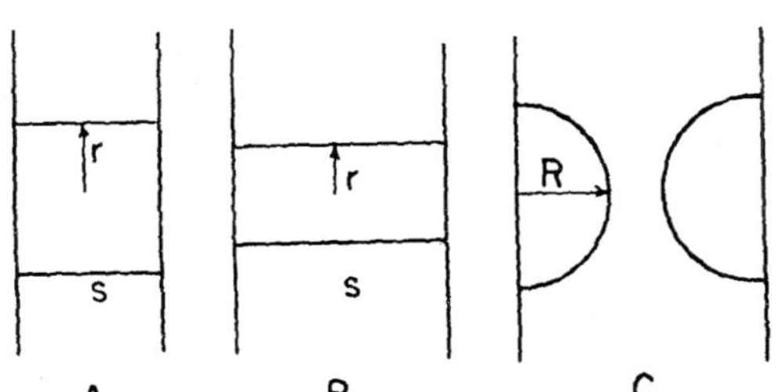

Fig. 7. Instability of cylindrical liquid surface when s/r ratio exceed critical value.

3.3. Measurements of Surface Tension

For purposes of the present discussion the various methods of measuring surface tension need be considered in outline form only. There are available several excellent and complete treatments of this subject that describe thoroughly both its theoretical and experimental aspects [13,14]. The commonly used methods for measuring the surface tension of pure liquids depend on either of two principles. The first is to measure, directly or indirectly, the pressure differential across the LV interface of a surface of known curvature and apply eq. 8 or 8a. The second involves actual extension of the surface and a measurement of the force necessary to bring about that extension.

Of the pressure differential methods the most widely used is that of capillary rise. A perfectly cylindrical glass tube of small, constant, and accurately known internal diameter is dipped vertically into a reservoir of the liquid to be measured. The reservoir should be wide enough so that its surface is flattened to a plane or near plane by gravity as in Fig. 8. It is highly preferable that the liquid have a zero contact angle $(\theta=0)$ against the walls of the capillary tube. If the contact angle against glass is not zero (a rare situation except for molten salts or metallic liquids) it must be known with great accuracy. The internal diameter of the tube should be small enough for the meniscus to have spherical form, and if $\theta = 0$ the meniscus will be a hemisphere with radius of curvature equal to the radius of the tube. For liquids of ordinary density the tube should have an internal diameter in the range of 0.5 mm or less. The liquid is allowed to rise until it has stopped completely, and height h is measured. Applying eq. 8a the pressure P_2 just under the liquid surface in the reservoir is equal to P_1 the atmospheric pressure. The pressure P_3 just under the meniscus is equal to $P_1 - 2\gamma/r$ where r *is* the radius of the tube. This pressure is balanced by the hydrostatic head of the liquid, which is equal to ρph; ρ being the difference in density between the liquid and the atmosphere and g the acceleration due to gravity. Accordingly:

$$\rho gh = \frac{2\gamma}{r} \tag{11}$$

from which γ can be calculated.

Another of the pressure differential methods for measuring surface tension is the drop weight or drop volume method. The weight and size of a drop of liquid that first forms and then drops from the end of a tube is a function of the diameter of the tube and the surface tension of the liquid. Instruments that provide for forming drops slowly under well controlled conditions, and either weighing them or measuring the drop volume, are called stalagmometers. The weight of a drop falling from a carefully ground tip is given by the empirically determined expression:

$$W = 2\pi r\gamma \cdot f\left(r / V^{1/3}\right)$$

where V *is* the drop volume and $f(r/V^{1/3})$ is a non-analytic function of r and $V^{1/3}$. Since $W= mg$, where m is the mass of the drop:

$$\frac{mg}{2\pi r} / f\left(r / V^{1/3}\right) = \frac{mg}{r} \cdot F \tag{12}$$

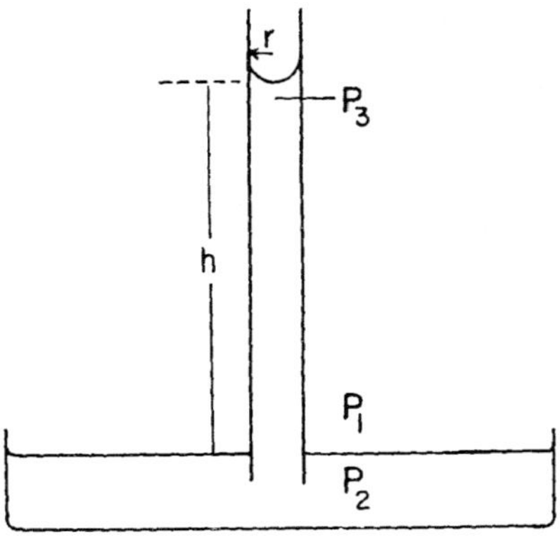

Fig. 8. Capillary rise against gravity, $\theta_{eq} = 0$. $P_1 = P_2; P_3 = \left(P_1 - \frac{2\gamma}{r}\right); \frac{2\gamma}{r} = \rho g h$.

where $F = 1/2\pi f(r/V^{1/3})$. Tables relating F to V/r^3 have been published by Harkins and Brown [15] and are accepted as standard. Using an accurately made and calibrated stalagmometer, and measuring *W* and *V* and *r*, γ can be calculated from eq. 12 and the tables of *F*. If *W* is measured *V* can be calculated if ρ, the density of the liquid, is known.

A third widely used pressure differential method, maximum bubble pressure, consists in dipping a tube to a measured depth under the surface of the liquid and forcing air down the tube to blow a bubble. The pressure of air required increases at first as the bubble grows. It reaches a maximum and then decreases as the bubble grows beyond a certain critical size. If the tube orifice is small enough the maximum pressure occurs when the bubble is hemispherical. In practice corrections have to be applied for orifices of practically usable size. Tables have been worked out relating γ to the maximum bubble pressure for various orifice sizes, depth of immersion and density of the liquid. The method is simple, rapid and sufficiently accurate for all but the most exacting requirements.

The sessile drop and pendent drop methods for measuring surface tension depend on the departure from spherical form that gravity causes in a weighed drop of a non-spreading liquid that either rests on a flat horizontal plate or hangs from the underside of such a plate. The density of the liquid must be known. The profile of the drop is obtained by photography or projection, or more frequently just the height and base diameter of the drop are measured. In 1883 Bashforth and Adams [16] published tables relating the geometric form of these drop surfaces, shaped by the opposing forces of surface tension and gravity, to the surface tension. These tables, the result of a truly heroic effort in the days before the computer, have been checked and modified and adapted for practical surface tension measurements in both the sessile and pendent configurations as well as in the bubble pressure method.

Direct measurement of the force necessary to extend a liquid surface is by far the most widely used method for measuring surface tension. In the most widely used instrument, the duNuoy tensiometer, a platinum ring of known diameter is pulled through the liquid surface and the maximum force needed is measured by means of a torsion wire. Platinum is

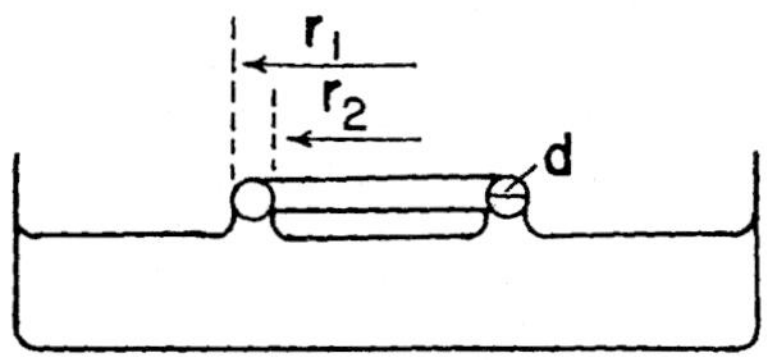

Fig. 9. Pull of duNuoy ring through liquid surface. Detaching force = $2\pi\gamma(r_1 + r_2)$.

used because it is stable, easily cleaned by flaming, and because it has a zero contact angle against most liquids. In theory a zero contact angle is not necessary but in practice it is. The exact theory of the ring pull method is complicated, and correction factors must be used to get absolutely accurate results. The elementary theory is simple and quite adequate for most purposes. A smooth solid immersed in a liquid against which it has a zero contact angle is subject to a pull of γ dynes along every centimeter of the solid-liquid-air contact line (also referred to as the three phase contact line or TPL). Figure 9 shows an enlarged cross section of a wire ring being pulled vertically out of a liquid surface shortly before it detaches. At this point the TPL coincides with the maximum diameter of the wire, d, so the pull of the liquid on the wire is vertically downward. This pull is equal to $2\pi r_1\gamma + 2\pi r_2\gamma$. If d is small relative to r_1, and r_2 the pull is approximated by $4\pi r\gamma$. Since the pull is measured directly and r is known γ is readily calculated.

In the Wilhelmy plate method, widely used in research laboratories, a thin rectangular plate is used instead of a ring and an automatic balance is used instead of a torsion wire. The plate is usually made of roughened platinum although any easily cleaned material of zero contact angle can be used. To minimize edge effects, which have a perturbation effect, the plate should be thin enough so that the TPL length can be taken as equal to twice the plate width. A more critical discussion on Wilhelmy technique is available in Chapter XI.

There are methods of measuring surface tension that depend neither on pressure differential nor on direct pull. One of these depends on the fact that the wavelength of a surface wave or ripple is a function of surface tension, liquid density, and wave frequency. A wave of known frequency is initiated on the surface of a suitably large pool of liquid, conveniently by means of a tuning fork, and the wavelength is measured [17]. The formula for calculating γ, developed originally by Lord Kelvin, is:

$$\gamma = \frac{\lambda^3 D}{2\pi T^2} - \frac{g\lambda^2 D}{4\pi^2} \tag{13}$$

where λ is the observed wavelength, D is the liquid density, T is the period, and g the gravitational constant.

Another interesting method, especially useful for viscous liquids, consists in allowing a jet of air to impinge vertically against the surface. The depth of the depression formed is a function of the surface tension [18].

3.4. Multicomponent Liquids (Solutions)

A given mass of a pure liquid at constant temperature can decrease its free surface energy only by decreasing its total surface area, i.e. by assuming the form which exposes the least surface per unit of mass. This is because all the molecules in the liquid are of the same species and their force fields are identical. No matter which individual molecules are in the surface at any given instant, the pull on these molecules from the interior is always the same statistically.

In the case of a solution conditions are quite different. Consider, for example, a simple binary solution of a solute A in a solvent B. Within this solution, both species of molecules are free to migrate and exert forces of attraction on their immediate neighbors. In general, the fields of attractive force exerted by the molecules A will be different from those exerted by B. These force fields, as we have seen, are responsible for the free surface energy, i.e. they act to pull surface molecules into the interior. If the A molecules have stronger force fields than the B molecules, they will be pulled away from the surface at a greater statistical rate. The net effect will be a diminution in concentration of A molecules in the surface. In other words, the concentration of A in the surface will be less than in the bulk of the solution. Conversely if the A molecules have weaker force fields than B, they will tend to concentrate in the surface. The statistical accumulation in the surface of those molecules with weaker force fields results in a lowering of the free surface energy. Thus, a given mass of solution may lower its free surface energy not only by diminishing its total surface, but also by concentrating in the surface that component whose molecules have the weaker force fields. The tendency for these molecules to come to the surface is opposed by thermal agitation and osmotic or diffusion forces. The latter forces act to prevent a complete usurpation of the surface layer by the weakly attracting molecules.

The concentration of one component of a solution at a phase boundary is called adsorption. Qualitatively, if a solution has a lower surface tension than the pure solvent, the solute is positively absorbed in the surface. Conversely, if the solution has a higher surface tension the solute is negatively absorbed, i.e. it is more concentrated in the interior than in the surface. Most inorganic salts in aqueous solution are negatively absorbed. The surface tension of these solutions is higher than that of pure water. Some substances, such as sodium hydroxide for example, raise the surface tension of water markedly. Most water-soluble organic compounds lower the surface tension of water, and the surface active agents show this effect to an extreme degree.

The quantitative relationship between the degree of adsorption and the lowering of surface tension was deduced by Gibbs, using the methods of thermodynamics. The complete form of the Gibbs absorption equation for the change in surface tension at constant temperature due to adsorption in a system of i components is:

$$d\gamma = -\Gamma_1 d\mu_1 - \Gamma_2 d\mu_2 ... \Gamma_i d\mu_i \tag{14}$$

In this equation Γ_i is the "surface excess' of the ith component and μ_i is the chemical potential of the ith component. The surface excess is the concentration of the ith component in the surface phase minus the concentration in the interior. The chemical potential, μ_i, of the ith component is:

$$\mu_i = RT \ln f_i N_i + \mu_i^0 \tag{14a}$$

where f_i is the activity coefficient, N_i is the mole fraction and μ_i^0 is the chemical potential of the pure ith component.

For dilute binary solutions, where the activity coefficient of the solute is unity, eq. 14 reduces to:

$$\Gamma_2 = -C_2 d\gamma / RTdC_2 \tag{15}$$

where the subscripts refer to the solute (component number 2). This states that the surface excess of solute is proportional to the concentration multiplied by the rate of change of surface tension with respect to concentration. Equation 15 is the so-called approximate form of the Gibbs adsorption equation. It is much more widely used than the exact form (eq. 14).

In a binary or multicomponent solution the rates of diffusion of the different components will generally differ. Whenever a fresh LV interface is formed by convection of the interior liquid to the surface it requires a finite time for molecules to diffuse into or out of the dynamic surface layer and establish the equilibrium concentration in that layer. Accordingly it takes time for the surface to come to its equilibrium tension dictated by eq. 14. For pure liquids of low viscosity this time is extremely short, of the order of microseconds. For aqueous solutions of small non-amphiphilic molecules this time required for molecular relocation is somewhat longer but still quite short. The non-equilibrium values of surface tension that obtain during this time period are called dynamic surface tensions. In essentially all situations where liquid is flowing or moving by convection the surface is constantly being renewed, and the surface tension observed during that flow is a dynamic surface tension. Values of dynamic surface tension are most often measured by observing the oscillations of a stream of the liquid issuing from an elliptical orifice. Tension in the LV front of a liquid that is moving by capillary action is a dynamic tension. Except for solutions of surfactants and polymers, however, the rate of flow in capillary systems is not high enough to make the dynamic tension differ significantly from the static equilibrium tension. For many of the common surfactants, especially those with hydrophobic chains of 18 or more carbon atoms, the dynamic tension is much higher than the static, and a surface may take as long as several minutes after flow has stopped to regain its equilibrium tension. Since LV interfacial tension is one of the factors affecting wicking rate, a difference between the static and dynamic tensions can become important. For slow diffusing solutes the rate at which a freshly formed surface regains its equilibrium tension can be followed by the pendent drop or sessile drop technique.

In wicking systems there is always an SL as well as an LV interface. Adsorption of solute at the SL interface, discussed later, can impoverish the bulk solution to such an extent that the equilibrium LV tension may be changed. In systems where the ratio of SL interfacial area to liquid volume is high this effect can become dominant. A very common example is the wicking of an ionic surfactant solution in a fibrous or porous ion exchanging medium.

If in a binary solution the solvent and solute differ in volatility, and at least one of them is appreciably volatile, a strong Marangoni effect can be generated. Any localized change in the rate of evaporation will change the composition of the solution and therefore the surface tension at that spot. This is a commonly encountered phenomenon in capillary systems. The rate of evaporation just behind the TPL, where the liquid layer is thin, is greater than the rate in regions remote from the TPL. The more volatile component of the solution distills preferentially causing the compositions and therefore the surface tensions to differ in the two regions and Marangoni convection ensues. The climb of a ridge of liquid on the wall of a brandy glass is a widely cited example of this effect. Other less entertaining examples include the streaking, blotching and uneven drying of impregnated fabrics, and the ring around a grease spot that has been sponged with drycleaning solvent.

In aqueous solutions containing an ionized solute an electrical double layer is generally formed at the LV interface. It results from preferential adsorption of either the anion or the cation, which causes an unbalanced charge distribution in the dynamic surface layer. When the adsorption is strongly preferential, as it is for example with surfactants, the double layer capacitance becomes an important contributor to the total surface free energy. The theory of electrical effects at the LV interface is outside the purview of this discussion but is ably treated in some of the general references.

Surface tension effects in multicomponent solutions are similar in principle to those in binary solutions. Equation 14 is of course applicable, but in practice the data necessary to use it are seldom at hand. As in most interfacial tension effects experimental data for the system of interest must be developed first, and the theory to explain and interpret them applied afterwards.

A special type of binary liquid system is encountered quite frequently in practical capillarity systems, and merits special mention. This system consists of water or a salt solution with a monomolecular layer of an insoluble oily material spread on its surface. Systems of this type have been studied very extensively with regard to properties and behavior. Not all substances are spreadable to monolayers, and some that do form monolayers have no great effect on surface tension. Many proteins and hydrophilic polymers, for example, lower the surface tension only slightly when spread on water. The substances that form monolayers and do lower surface tension appreciably are generally amphiphilic. They possess a long fatty chain attached to a hydrophilic group that is not sufficiently hydrophilic to make the whole molecule water soluble. Most free fatty acids, alcohols, and esters have this property. As airborne or accidentally conveyed contaminants they are frequently encountered on aqueous surfaces that appear clean. The surface tension of water covered with a monolayer of this type diminishes as the monolayer is compressed but is usually closer to the surface tension of the fatty material itself than to that of water even at relatively low compressions. This effect can be explained on a molecular basis as follows: The fatty molecules, being insoluble, lie in the geometric surface and can move inward only transiently. Their hydrophilic ends, however, have a high affinity for water and thereby delay inward diffusion of water molecules adjacent to them in the dynamic surface layer. Molecular density of water in that layer is accordingly increased and the surface tension correspondingly lowered. In monolayer studies it is conventional to monitor the surface tension by means of a Wilhelmy plate inserted through the surface into the underlying water. In more precise studies the pressure exerted on the barrier by the monolayer is measured directly. This pressure, designated as π in practically all modern literature on this subject, is numerically equal to the difference between the surface tension of pure water and the surface tension of water covered with the monolayer. The results of monolayer studies are generally presented as plots of π vs. surface area.

Water that bears a monolayer will frequently start to wick or spread in a capillary system after the manner of a surfactant solution. Since the monolayer has such a small mass, however, it will rapidly be disrupted or removed by deposition on the channel walls, and the effect of low surface tension will be lost.

4. THE LIQUID-LIQUID INTERFACE

Emulsion science and technology is grounded in knowledge of the liquid-liquid (L_1L_2) interface and its behavior. The L_1L_2 interface is also of primary importance where one liquid is used to displace another from a solid surface, as it is in the enhanced oil recovery processes that depend on aqueous displacement. Neither of these systems, however, is generally considered an example of absorbency. The L_1L_2 interface may play a significant role in absorbency systems where the liquid is an unstable emulsion or an unemulsified mixture of two liquid phases, and liquid is displacing gas from the solid. A close look at such systems will often reveal that either the L_1V surface or the L_2V surface or both of them separately are operative in the gas displacement process. Accordingly the L_1L_2 interface is seldom of major importance in wicking.

There are both similarities and differences between the L_1L_2 interface and the LV interface. The similarities derive from the properties of fluidity and molecular diffusion that liquids and gases both possess. The differences reflect the divergent properties of gases and liquids. The gas phase has no coherence. Its molecular density under standard conditions is three orders of magnitude less than that of the liquid phase. Because of this molecular sparseness the gas phase exerts a negligible force of attraction per unit area across the LV interface, and the value of γ_{LV} is determined by intermolecular attraction from only one side of the interface. In L_1L_2 systems the value of γ_{L1L2} is determined by intermolecular attraction forces exerted from both sides of the interface.

Another difference between gases and liquids is that all gases, regardless of chemical species, are miscible. The solubility of two pure liquids in each other may vary from zero (e.g. mercury and water) to infinite. The extent to which L_1 and L_2 dissolve each other has a great effect on the interfacial tension γ_{L1L2}. If they are miscible there is no interface and $\gamma_{L1L2} = 0$. The greater the mutual solubility the lower is the value of γ_{L1L2}. For liquid pairs that have a consolute temperature the value of γ_{L1L2} decreases as the consolute temperature is approached and vanishes when it is reached. This effect is analogous to the disappearance of an LV interface at the critical temperature. In the following discussion it is assumed that the two liquid phases have been in contact long enough to have established solubility equilibrium and therefore interfacial tension equilibrium.

In L_1L_2 systems it is necessary to define what is meant by "contact" between the two phases. In a stable oil-in-water emulsion the oil droplets may appear to be in contact and in fact may be pressed together forcefully enough to distort their shapes (as they are in a creamed but unbroken emulsion) but they do not coalesce as they would if true oil-oil contact were established. Each of the oil droplets in this situation is still in virtual contact with the outer aqueous phase. We can define two liquid phases as being in contact if they share a single dynamic interfacial layer. In all L_1L_2 systems at solution and vapor equilibrium the specific interfacial free energy relationships are such that $\gamma_{L1V} + \gamma_{L2V} > \gamma_{L1L2}$.

The dynamic interfacial layer of an L_1L_2 system consisting of two molecular species A and B coincides physically with the Gibbs dividing surface of thermodynamic theory mentioned previously. This layer contains molecules of both species. There is free traffic of both A and B molecules across the dynamic interfacial layer but since both phases are internally saturated with their neighbor species this traffic affects neither the composition of the layer nor the value of γ_{L1L2}. The composition of the layer is governed by eq. 14. The total molecular density in the layer, which is a determinant of γ_{L1L2}, is lower than in either of the bulk phases.

The actual values of γ_{L1L2} in some systems can be extremely low, in the range of mdynes/cm. Methods for measuring γ_{L1L2} are generally similar to those used for γ_{LV}, but for very low interfacial tensions these methods are of little value. Instead there is used a procedure old in concept but recently revived and perfected, the spinning drop method [19]. A tube is filled with the higher density liquid L_1 and a drop of the lower density liquid L_2 is introduced. The closed tube is then rotated in a horizontal plane at controlled speed. Centrifugal force causes the L_2 drop to elongate in the radial direction, and the drop shape is plotted as a function of rotation speed. From the density difference and rotation speed the centrifugal force is calculated, and by applying eq. 8 the value of γ_{L1L2} can be calculated with great precision.

A curved L_1L_2 interface, like the LV interface, exerts a pressure against its concave side, and the magnitude of that pressure is given by the Laplace relationship, eq. 8. Emulsion droplets are bounded by L_1L_2 interfaces of high curvature.

When L_1 and L_2 contain solutes these solutes will in general be adsorbed at the interface and influence the interfacial tension. The equilibrium adsorption condition is given, as in all adsorption systems, by eq. 14. As at the LV interface, it may take appreciable time for the absorption equilibrium and the equilibrium value of γ_{L1L2} to be established.

5. THE SOLID-VACUUM AND SOLID-VAPOR INTERFACES

The solid vapor (SV) interface is of enormous technical importance as the locale of contact catalysis, and is the subject of a very extensive literature. The following discussion is limited to the free energy of this interface and of the solid-vacuum interface which it becomes when the vapor density is zero. These free energies appear as factors and terms in the basic equations of capillarity.

As a model solid surface we can consider the cleaved crystal face of a material that has negligible volatility. An elevation view of this model is shown in Fig. 10, with the surface layer. A, bordering on a vacuum. The molecules cannot diffuse. Their motion is limited to vibration about their positions in the lattice, a situation that corresponds to macroscopic rigidity. As in a liquid the surface is the locale of an excess of free energy per unit area, γ_{Svac}, over and above the free energy possessed by a similar molecular grouping in the interior. As in a liquid the source of this energy is the force of attraction between the surface region and the underlying regions. A layer in the interior such as layer X is being pulled equally by its neighboring layers W and Y and is spaced equidistant from them. The outermost layer A is only attracted inward. The second layer B is attracted inward by all underlying layers close enough to exert significant force (usually considered not more than about 3 or 4 molecular diameters distant) and is attracted outward only by layer A. Accordingly there is a net inward force on layer B but it is considerably less than the inward force on layer A. The net effect of these forces is to decrease the A-to-B spacing (and to a much less extent by B-to-C spacing) to somewhat less than the spacing between interior layers such as W and X. This decreased A-to-B spacing, shown greatly exaggerated in Fig. 10, has been demonstrated experimentally in some crystals. Since the molecules cannot diffuse there can be no physical tension in and parallel to the surface as there is in a liquid. Molecules in the edges of a whole crystal experience a greater inward force than surface molecules remote from the edges. This excess force results theoretically in a greater inward displacement of the edge molecules, as shown at Z in Fig. 10. Adsorption, solubility, and chemical reactivity data confirm that crystal edges and apexes have greater free energy than surface interiors.

The total surface free energy of a mass of condensed matter, solid or liquid, can be lowered in either of two ways. Since it is equal to γ X Area, we can either decrease the area or modify the surface in such a way as to decrease γ. A pure liquid in equilibrium with its own vapor (at constant T and P) has no possibility for lowering γ_{LV}. The area, however, can be and is minimized to the full extent allowable by other constraints. A multicomponent LFI system of constant mass can and does minimize its area, but it also can and does lower γ_{LFI} via adsorption. An ideal whole crystal bordered by vacuum has no possibility for lowering γ_{SVac}. Its surface area cannot be lowered by curving, but can and is lowered to a very limited extent by decreasing the A-to-B spacing on all surface planes, and the Z location on all edges and apexes. In short, it compresses its surface elastically, and that compression is the physical manifestation of the potential energy γ_{SVac}.

If the model crystal of Fig. 10 is bordered by a gas phase it can reduce its surface free energy by adsorbing the gas and forming a solid-gas (SV) interface. It is evident that for the same solid surface γ_{SV} is always less than γ_{SVac}. The numerical value of γ_{SVac} at fixed T and P depends of course on the chemical nature of the solid. The value of γ_{SV} depends additionally on the chemical nature of the gas. Using G (eq. 2) instead of F (eq. 1) the difference (γ_{SVac} - γ_{SV}) is equal to ΔH_{if} - $T\Delta S_{if}$, the changes in surface enthalpy and surface entropy per unit area, accompanying the adsorption. Since the γ value of the surface is lowered by the inclusion of a gas layer the A-to-B spacing of the crystal is correspondingly increased. With regard to molecular vs. thermodynamic definitions, the Gibbs dividing surface in the SVac system necessarily has the geometric surface of the solid as an outer boundary and the last crystal layer significantly mis-spaced from its neighbor as an inner boundary. In the SV system the Gibbs dividing surface also includes adsorbed gas out to the point where the gas density becomes constant. γ_{SV}, like γ values of all other phase interfaces, is the specific free energy of the Gibbs dividing surface. The density and thickness of the adsorbed gas layer depends very greatly on the difference between the system's temperature and the critical temperature (T_C) of the gas. If T_C is much lower than the system's temperature, as it is for the major atmospheric gases at room temperature, the density of the absorbed layer will be low and γ_{SVac} - γ_{SV} will be small. When T_C and the system's temperature are close to each other the density of the adsorbed layer will be high and γ_{SVac} - γ_{SV} will be large. The physical manifestation of γ_{SV} appears as a pressure exerted in the plane of the surface by the adsorbed gas layer. To visualize this effect more clearly consider the system composed of a planar solid surface in a vacuum surmounted by a tight fitting bell jar containing a gas, as in Fig. 11. The bulk gas in the jar is at pressure P, which is exerted equally in all directions. Adjacent to the solid surface is the adsorbed gas layer which has a greater density than the bulk gas above it and therefore exerts a greater pressure. But since this greater density results solely from attraction of the gas molecules by the solid the excess pressure is exerted only in the plane of the surface and only outward against the wall of the jar, as indicated by the arrows. The magnitude of this surface pressure, abbreviated π_{SV}, in dynes/cm is numerically

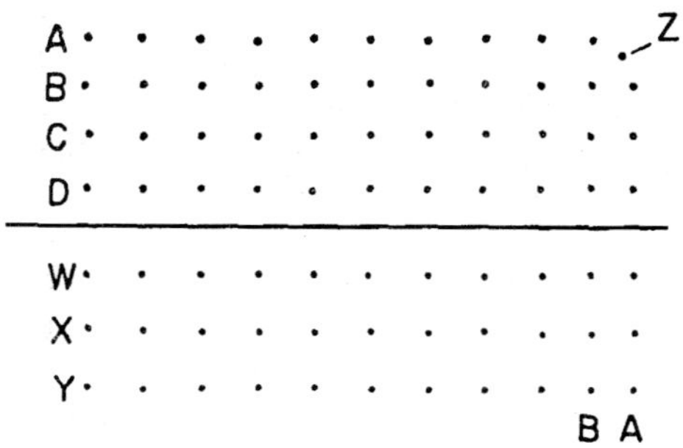

Fig. 10 Section through idealized solid. A-B spacing is less than interior (e.g. C-D or W-X) spacing. Edge Z displaced inward.

equal to γ_{SVac} - γ_{SV}, energy quantities expressed in ergs/cm^2. Since γ_{SVac} is a material constant of the solid it is evident that a low value γ_{SV} corresponds to a high value of π_{SV}, and vice versa.

It is appropriate at this point to emphasize the fundamental difference between the specific interfacial free energy γ and the specific interfacial force (dynes/cm) operative within the dynamic interfacial layer in the plane of the interface. In the LF1 systems the numerical values of γ_{LFl} and the specific interfacial force are identical, and therefore the force is always in the direction that tends to diminish the interfacial area, i.e. it is a force of tension. It is conventional to use the symbol γ as we have done for both the force and the energy. We introduced the other conventionally used symbol, σ, for the specific force that is surface tension, but have hardly used it. But force is a vector or, more precisely, a tensor, whereas energy is a scalar. In the SV system the numerical values of γ_{SV} and the specific force π_{SV} are not at all identical. The numerical identity of γ_{LFl} and π_{LFl} results from the fluidity of both phases. This fluidity (i.e. freedom of diffusion) allows all the potential energy to be released parallel to the interfacial plane. At the SV interface (and also the SL interface discussed below) only part of the energy is available for release parallel to the interfacial plane. When the SV or SL areas change part of the energy is released normal to the interface as a change in the A-to-B crystal layer spacing. From another point of view: unless we increase or decrease an SS area it is physically impossible to change the area of an SFl_1, interface without correspondingly changing in the opposite direction the area of a bordering SFl_2, or SVac interface. As noted previously, significant area changes in true SS interfaces, whether cohesive (SS) or adhesive (S_1S_2), occur with negligible frequency in practical absorbency systems.

6. THE SOLID-LIQUID INTERFACE

In analyzing energy and force relationships at the SL interface, and anticipating an analysis of the model SLV wicking system [20], we choose a model somewhat similar to the SV model. Figures 12A-C each show an insoluble solid S surmounted by a liquid contained in a flexible and stretchable bag such as might be made of thin rubber. The liquid fills the bag

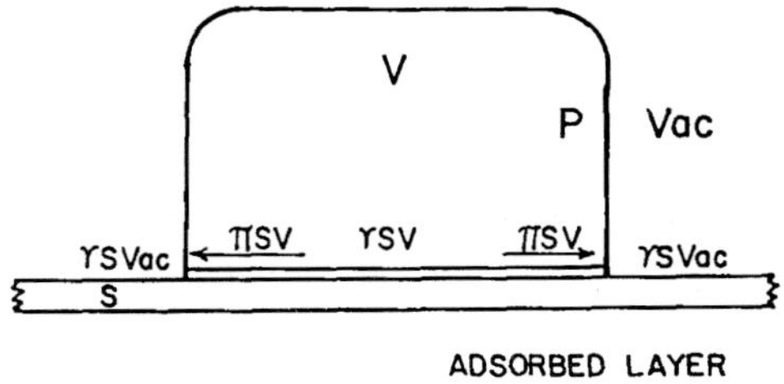

Fig. 11. Outward pressure of adsorbed gas layer on a solid surface. $\pi_{SV} = \pi_{SVac} - \gamma_{SV}$

completely. The bag's edge is constrained to remain in contact with the solid but slide freely over it forming a movable seal that separates the SL interface from the exterior SVac interface. The bag is also constrained to maintain the form of a spherical lens. Different liquids but the same solid are used in the three illustrations. The volumes of liquid are identical in all three illustrations. As at the SV interface the lateral molecular arrangement in the uppermost layer of the solid (layer A of Fig. 10) is unaltered. This layer is the laterally immobile physical surface on and against which the mobile liquid molecules diffuse. Liquid molecules on or near that surface are subject to attraction forces from both the solid and the interior liquid. They constitute a dynamic interfacial layer whose density in general will differ from that of the interior liquid. If these interfacial L molecules are attracted more strongly toward the solid than toward the interior liquid the interfacial layer will be denser than the interior liquid, and will exert an outward pressure.
This situation is illustrated in Figure 12A. The liquid L_1 forms a layer of excess density on solid S. This layer exerts an outwards pressure π_{SL1} against the confining barrier, which stretches outward until it is able to withstand the pressure. The magnitude of π_{SL1} is:

$$\pi_{SL_1} = \gamma_{SVac} - \gamma_{SL_1}$$

In Fig. 12B the liquid L_2 is the interfacial layer is attracted more strongly toward the liquid interior than toward the solid. It is lower in density than the interior liquid and tends to contract. It exerts a negative pressure, a tension, which stretches the bag by pulling its edges

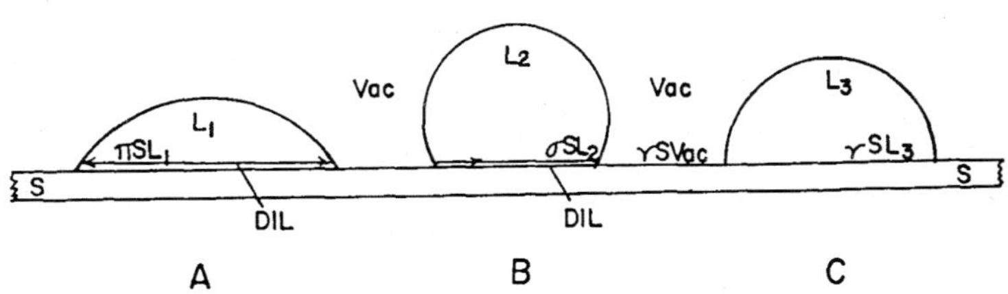

Fig. 12. Pressure of dynamic interfacial layer (DIL) of liquid on a solid. A. Outward pressure. π_{SL} positive. B. Inward tension. π_{SL} negative = σ_{SL}. C. $\pi_{SL} = \gamma_{SL} = 0$.

inward until the tension is balanced by the stretch of the bag. This tension σ_{SL2} has the magnitude:

$$\sigma_{SL_2} = -\pi_{SL_2} = \gamma_{SL_2} - \gamma_{SVac}$$

In Fig. 12C the interfacial layer of liquid L_3 is attracted equally by the solid and the interior liquid, and its density is the same as that of the interior liquid. In this situation:

$$\pi_{SL_3} = \sigma_{SL_3} = 0 \quad \text{and} \quad \gamma_{SVac} = \gamma_{SL_3}$$

The edge of bag C stands perpendicular to the solid surface and the bag surface is hemispherical, the geometrical shape of minimum area that will accommodate a fixed amount of liquid having one planar boundary. Because of this geometrical requirement, and because the liquid volumes in A, B, and C are equal, bag C has less surface area than either bag A or bag B.

The Gibbs dividing surface for both the SL_1 and SL_2 systems of Fig. 12 includes the uppermost crystal layer(s) of the solid and the dynamic interfacial layer of liquid. In the SL_3 system the Gibbs dividing surface is entirely within the solid. The surface free energy γ, corresponding to a physical stress in the Gibbs dividing surface, is positive, as it is in all phase interfaces at equilibrium. The sensible physical forces exerted in the plane of the interface, the forces that move the liquid in capillary system, reflect stress in the fluid portion of the Gibbs dividing surface. If the Gibbs dividing surface includes no fluid portion, as in the model SL_3 and the Svac systems, it can exert no interfacial pressure or tensions.

When the liquid portion of an SL system consists of more than one component (i.e. is a solution) there will in general be selective adsorption of the components. The concentration of each component in the interfacial liquid layer will differ from its concentration in the interior liquid. The quantitative relationship among these concentrations is given by eq. 14.

7. THE SOLID-LIQUID-VAPOR SYSTEM

Capillarity has many aspects. Capillary systems are encountered frequently and diversely in science and technology. The solid-liquid-vapor system is the system of wicking, the capillary mode of absorbency. In this system the forces deriving from the free energies of the LV, SV, and SL interfaces interact to produce the displacement of gas from a solid surface by a liquid. The purpose of this section is to consider these interactions and their results. The model SLV system is completely real and therefore differs in some aspects from the "thought experiment" system of Fig. 12. The vacuum is replaced by a gas, but if the gas or vapor is not strongly sorbed by the solid it will have little effect. As discussed earlier the LV interface, like the membrane, is always tensed but the tension is constant; it does not vary with the extent of stretch as it does in a stretched membrane. To develop the model SLV system we start with a plane solid surface and a fixed volume of a single component liquid, both in equilibrium with the vapor of the liquid, i.e. a three phase, two component system at constant temperature and pressure. As in the other model systems gravitational, electrical and other perturbing effects are absent. Initially the liquid and solid are not in contact, as shown

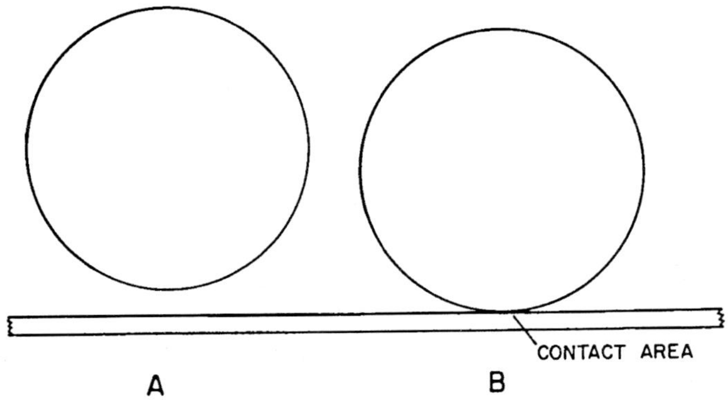

Fig. 13. Liquid drop contacting solid. A. Before contact. B. Initial contact, establishing Area SL which is stable and expands initially in all SLV systems.

in Fig. 13A. We now establish contact between them (Fig. 13B) creating an SL interface and a three phase boundary line (abbreviated hereinafter TPL). The SL interface, defined as having a Gibbs dividing surface that includes a dynamic interfacial liquid layer, has an initial area of a very few square molecular diameters, possibly 100 $Å^2$ or less. It is important to note that this interface will always form, if perturbing influences are absent, when the LV and SV surfaces come close enough for significant S-to-L intermolecular attraction to be generated. This follows from the fact that in all known SLV systems:

$$(\gamma_{SV} + \gamma_{LV}) > \gamma_{SL} \tag{16}$$

Among SL interfaces we distinguished among those tending to expand (SL_1 type), to contract (SL_2) type, or to do neither (SL_3 type). The newly formed miniscule SL interface of Fig. 13B will start to expand regardless of which type it is. This in an experimental fact. Its theoretical validity can be shown from eq. 16 and the geometry of the system, using thermodynamic principles [12]. The question is: how far will this SL interface spread before the system comes to equilibrium? Using the thermodynamic approach, the total interfacial free energy of the system, F_{if}, is:

$$F_{if} = \gamma_{LV} \bullet \text{Area LV} + \gamma_{SL} \bullet \text{Area SL} + \gamma_{SV} \bullet \text{Area SV}$$

and since the specific free energies of the interfaces are constants characteristic of the system:

$$dF_{if} = \gamma_{LV} \bullet \text{dArea LV} + \gamma_{SL} \bullet \text{Area SL} + \gamma_{SV} \bullet \text{dArea SV} \tag{17}$$

The liquid will move along the solid surface, with all three interfacial areas changing. as long as dF_{if} is negative. Because the solid surface is planar dArea SL = —dArea SV, and in the absence of gravity, etc., Area LV will maintain the form of a spherical zone. The angle

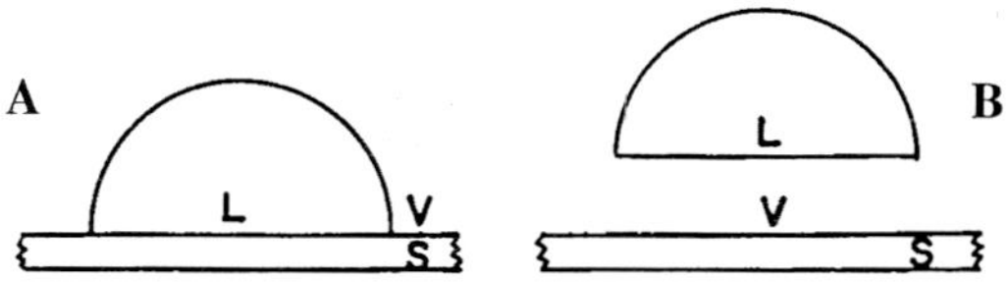

Fig. 14. Work of adhesion; schematic.

between the tangent of this zone and the solid surface at the TPL is called the contact angle. The contact angle is conventionally measured in the liquid. The system will reach equilibrium and become static when dF_{if} = 0. At this point the relationship among the γ values is such that:

$$\gamma_{SV} = \gamma_{SL} + \gamma_{LV} \cos\theta_{eq} \tag{18}$$

where θ_{eq} signifies the contact angle of the system at thermodynamic equilibrium. Equation 18 is called Young's equation or, by some authors, the Young-Dupre equation. Contrary to statements or implications in some of the literature Young's equation is not a force balance. It is a thermodynamic equation showing a valid and easily proved relationship among energies, i.e. scalar quantities. As such its validity is unquestionable. It is valid regardless of whether θ_{eq} is acute (cos θ_{eq} positive), obtuse (cos θ_{eq} negative), or 90 ° (cos θ_{eq} = 0). It is also valid when the force relationships are such that the liquid spreads indefinitely over the solid surface to the point where θ_{eq} = 0 and cos θ_{eq} = 1. Equation 18 is valid for all SLV systems at equilibrium regardless of geometry. It is a thermodynamic equation; and θ_{eq} is a thermodynamic parameter in the same sense that P, V, and T are the parameters of energy in a Carnot engine.

Equation 18 shows an energy equilibrium but tells nothing about the forces or molecular mechanisms that cause the system to approach this equilibrium. Before discussing these forces two more surface energy concepts should be examined, namely work of adhesion and work of cohesion. Consider the equilibrium SLV system of Fig. 14A and for convenience set the interfacial SL area at 1 cm^2. If we conceptually lift the liquid off the solid, as shown in Fig. 14B, we add to the system 1 cm^2 of new LV area and 1 cm^2 of new SV area. We also lose 1 cm^2 of SL area. The energy needed to effect this separation is called the work of adhesion, W_A and is:

$$W_A = \gamma_{LV} + \gamma_{SV} - \gamma_{SL} \tag{19}$$

By eq. 18, $\gamma_{SV} - \gamma_{SL} = \gamma_{LV} \cos\theta_{eq}$, and therefore:

$$W_A = \gamma_{LV}\left(1 + \cos\theta_{eq}\right) \tag{20}$$

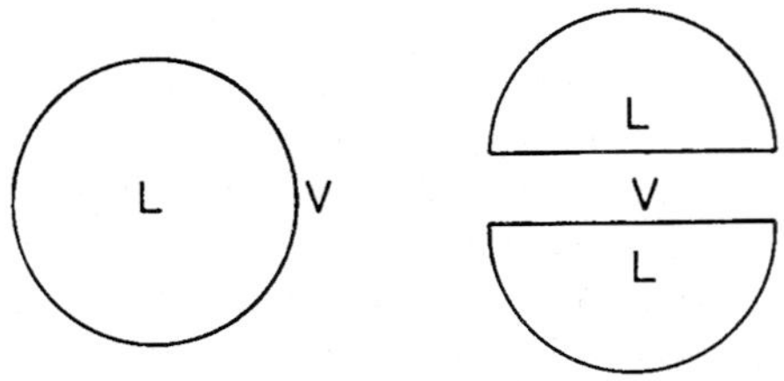

Fig. 15. Work of cohesion; schematic.

Comparing eq. 19 with eq. 16 it is evident that W_A is always positive. It is of some interest, although not pertinent to this discussion, that the analog of W_A in an SL_1L_2 system is not always positive. There are many practical SL_1L_2 systems in which $\gamma_{SL1} > \gamma_{SL2} + \gamma_{L1L2}$. These are the systems in which water will displace oil, or oil displace water, from a solid surface. In most such systems the liquids are multicomponent. Some three component SL_1L_2 systems of this character are known, however.

Considering now a liquid-vapor (LV) system, we can conceptually break the liquid into two parts over an area of 1 cm^2 as in Fig. 15. We thus add to the system 2 cm^2 of new LV area, each having a surface free energy of γ_{LV}. The energy necessary to effect this separation is called the work of cohesion, W_C of the liquid and is:

$$W_C = 2\gamma_{LV} \tag{21}$$

Comparing eq. 20 and 21 it is evident that the energy necessary to separate liquid from solid over 1 cm^2 in an SLV system for which $\theta_{eq} = 0$ is:

$$W_A = 2\gamma_{LV} = W_C \tag{22}$$

With the foregoing energy considerations as background we can proceed to analyze the forces operative in SLV systems, considering first those systems in which $180° > \theta_{eq} > 0°$, and later the special (and technically most important) systems in which $\theta_{eq} = 0°$.

7.1. Systems of Positive Contact Angle

Referring to Fig. 13B and eq. 16 we have seen that the energetics of the system dictates that the liquid must spread, i.e. that Area SL must increase to at least some extent after initial contact is made. The extent to which Area SL will increase is given ideally by Young's relationship, eq. 18. The forces that drive this increase, and all later liquid motion in an SLV system, interact with one another at the TPL, against which they operate. The TPL can be pictured physically as a rope of liquid, a very few molecular diameters thick, that can roll freely along the solid surface. It is an integral part of both the LV and SL dynamic interfacial layers although its molecular density and composition differ from that possessed by either layer in regions remote from the TPL. The SV dynamic interfacial layer also acts on the TPL. In systems where $\theta_{eq} > 0°$ this layer is best regarded as differentiated from the TPL by a fairly sharp boundary. In these systems π_{SV} is relatively small. It is, in fact, regarded by many investigators as negligible. A schematic picture of this situation is shown in Fig. 16, which shows equatorial sections of a sessile drop at equilibrium on a planar solid surface, and indicate the operative forces by vectors. Figures 16A shows a system for which $\theta_{eq} > 90°$. The pressure π_{SL} is positive, tending to push TPL outward. In Figure 16B $\theta_{eq} > 90°$ and π_{SL} is negative, i.e. it is a σ_{SL} tending to pull the TPL inward. The LV tension σ_{LV} acts on the TPL only via its component in the plane of the solid surface. It is evident that at equilibrium this component, $\sigma_{LV} \cos \theta_{eq}$, balances the resultant of the forces π_{SV} and π_{SL} (or σ_{SL}). The sessile drops of Fig. 16 are in static equilibrium because the forces acting on the TPL are balanced at all points around its perimeter. For these drops to move the TPL would have to increase or decrease in length, but the geometry of this system is such that the force balance determines the TPL length.

The sessile drop on a flat plate is not at all typical of the capillary systems of absorbency. The solid surfaces of absorbency are better modeled either by the exterior of a rod or the interior of a tube. Going again into zero gravity consider a mass of liquid L_1 into which we have inserted a solid rod S. The three interfacial forces have interacted to bring θ to its equilibrium value θ_{eq}. This situation is shown in Fig. 17A. The rod S in this diagram will be continuously sucked into the liquid by a force which we measure as equal to $\sigma_{LV} \cos \theta_{eq}$ multiplied by the circumference of the rod, C. The real force effecting this motion is $(\pi_{SL} - \pi_{SV}) \bullet C$, tending to push the TPL outward along the rod S, i.e. to increase the SL area. The rod of Fig. 17B, inserted into L_2, will be continuously pushed outward from the liquid because the force $(\sigma_{SL} + \pi_{SV}) \bullet C$ tends to pull the TPL inward, thereby decreasing the SL area. In both cases the resultant of the operative forces is equal in magnitude and direction to $\sigma_{LV} \cos \theta_{eq} \bullet C$, but σ_{LV} itself is no part of the force causing the relative solid-liquid motion in this system.

In a similar example we can insert a hollow cylindrical tube of S into the same large mass of liquid L_1, as shown in Fig. 18. The force $(\pi_{SL} - \pi_{SV})$ acts continuously and L_1 will move outward in the tube indefinitely.

It is interesting to compare this analysis of capillary motion, based on all three interfacial forces, with the classical picture of capillary rise (Fig. 8) that invokes only the Laplacian pressure. The unanswered question in the classical picture is: what causes the liquid to assume the characteristic contact angle that determines the surface curvature? Using 8a we calculate a driving pressure expressed (as force/unit area) in dynes/cm^2 cross section of

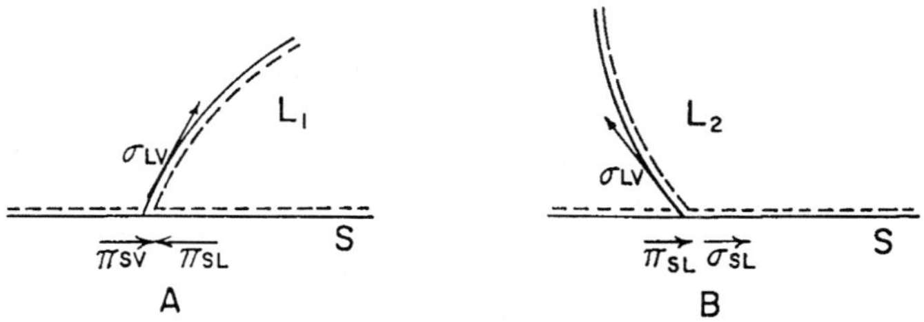

Fig. 16. Forces operative in the SL plane at the TPL. A. $\theta_{eq} < 90°$; π_{SL} positive. B. $\theta_{eq} > 90°$; σ_{SL} positive.

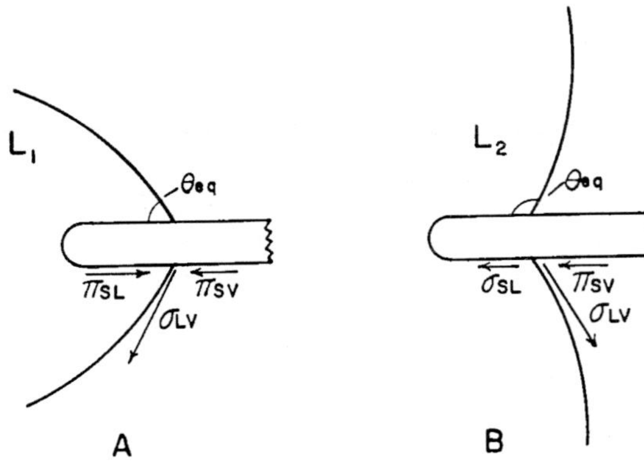

Fig. 17. Forces acting on a solid rod partly immersed in a larger mass of liquid. A. $\theta_{eq} < 90°$; rod is sucked in. B. $\theta_{eq} > 90°$; rod is pushed out.

the tube. In the analysis of Fig. 18 we calculate the total driving force as (π_{SL} - π_{SV}) multiplied by the length of the TPL; and the pressure is this force divided by the cross section of the tube. The results are, of course, identical. This again points up the fact that the surface tension σ_{LV} acts only on the liquid phase. It acts by itself to establish LV curvature. It acts in conjunction with the SL and SV interfacial forces to establish a contact angle.

The hydrodynamics of liquid motion at the TPL is an area of current study. Classical continuum mechanics are difficult to apply because small assemblages of molecules are not continua. But lack of a precisely valid model of the liquid phase makes molecular mechanics equally difficult to apply. Qualitative and quantitative attempts along this line have recently been made [21,22,23].

7.2. Systems of Zero Contact Angle

SLV systems for which $\theta_{eq} = 0$ are of special importance in absorbency for at least two reasons. First, these are the systems most commonly encountered in practice. Secondly,

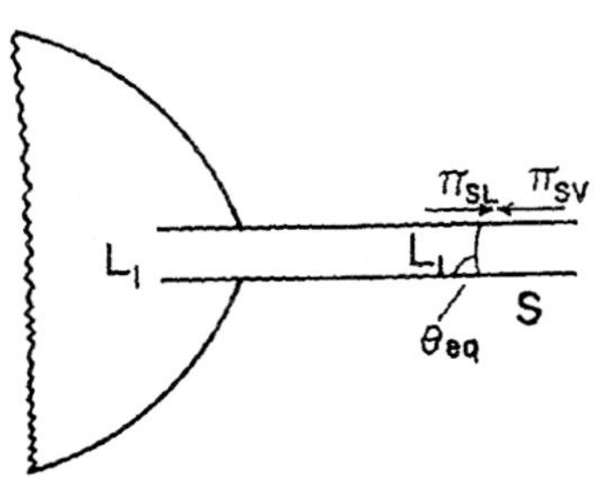

Fig. 18. Forces acting to pull liquid into a tube. $\theta_{eq} < 90°$.

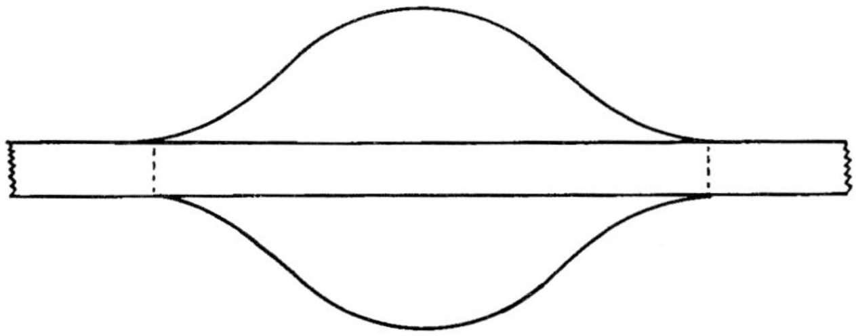

Fig. 19 Drop of liquid in static equilibrium on a solid cylindrical rod. LV interface has the form of an unduloid (elliptical cycloid) of revolution. θ_{eq} =0.

there are several frequently encountered geometries in which a zero contact angle can cause a liquid surface to yield completely. This is the situation, for example, when an isolated mass of liquid is trapped between two flat solid surface held apart at a fixed distance. The liquid will not yield if $\theta_{eq} > 0$.

A suitable static model for systems in which $\theta_{eq} = 0$ consists of a mass of liquid that has been brought into contact with a cylindrical rod. The liquid will spread around the rod and assume the form of an unduloid of revolution, as shown in Fig. 19. This is the shape for which the Laplacian pressure, eq. 8, is uniform over the whole LV area. With the added constraint that θ_{eq} =O it is the shape of maximum LV area-to-volume ratio. When $\theta_{eq} = 0$, $\pi_{SL} > \sigma_{LV}$ and may in some cases be much greater than σ_{LV}. However, when π_{SL} is large π_{SV} also becomes large by virtue of increased adsorption at the SV interface. Young's equation is valid for this system as well as for all SLFl systems. The force balance is:

$$\pi_{SL} = \sigma_{LV} + \pi_{SV} \tag{19}$$

Since cos $\theta_{eq} = 1$ the inward pull on the rod at each end of the unduloid is equal to σ_{LV} per cm of TPL. As in the systems of Fig. 17 and 18 the operative force per cm TPL exerting the pull is (π_{SL} - π_{SV}).

The molecular situation at the TPL in this system is somewhat different from that in systems of $\theta_{eq} > 0$. As stated previously, when $\theta_{eq} > 0°$ the SV dynamic interfacial layer is not regarded as an integral part of the physical TPL "rope": and this transition of the fluid component from the SV state to the LV or bulk liquid state is a first order transition [24]. When $\theta_{eq} = 0°$ the SV layer merges with the TPL just as the SL and LV layers do. The TPL rope is itself more diffuse than it is when $\theta_{eq} > 0°$. The TPL in this system can be regarded as a region in which the SV layer of adsorbed vapor merges with the liquid phase in a second order transition. In this system, as in systems where $\theta_{eq} > 0$, the LV layer pulls on the TPL rope. not on the solid surface.

If the liquid phase of a zero contact angle system consists of two or more components selective adsorption, adsorption rates, and Marangoni effects can seriously complicate the capillary behavior while equilibrium is being reached. At physicochemical equilibrium, however, the energy and force relationships are identical to those of a two component SLV system.

Fundamentals of surface tension and surface energy as they relate to the absorbency phenomena have been summarized in this Chapter. For further details, readers are suggested to review some selected books and articles which are cited here [25-32].

8. GLOSSARY

A	Area ($4\pi r^2$)
A_0	Molar Surface Area
D	Density of liquid
F	Helmoltz function for free energy
F_{if}	Total interfacial free energy of a system
f	Notation for function; also represents force
f_i	Activity coefficient
Fl	Fluid
G	Gibbs function for free energy
g	Acceleration due to gravity
γ	Specific interfacial free energy
Γ_i	"Surface excess" of the *i*th component
H	Enthalpy
h	Capillary liquid height
k	Proportionality constant relating volume to area
LV	Liquid-vapor interface
L_1L_2	Liquid (L_1)-Liquid interface
λ	Wave length
μ	Chemical potential
M	Molecular weight of liquid
N	Number of moles of a component in the system
Π_{SV}, Π_{SL}	Magnitude of surface pressure
P	Pressure
r	Radius of curvature; also, radius of a tube
ρ	difference in density between liquid and atmosphere
S	Entropy
SV	Solid-vapor interface
SL	Solid-liquid interface
S_1S_2	Solid (S_1)-solid (S_2) interface
s	distance
σ	Force per unit length
T	Absolute temperature
T_C	Critical temperature
θ	Contact agle
θ_{eq}	Contact angle at equilibrium
U	Internal energy of a system
V	Volume
V_0	Molar Volume

W	Weight
W_A	Work of adhesion
W_C	Work of cohesion
x, y	Arbitrarily chosen regions not bounded by a second phase

9. REFERENCES

1. F. C. Goodrich in E. Matijevic (Ed.). Surface and Colloid Science. Vol. 1. Wiley. New York. 1969.
2. F. M. Foukes and M. A. Mostafa. Ind. Eng. Chem. Prod. Res. Dev.. 17 (1978) 3.
3. W. B. Jensen. Chemtech December 1982, p. 755.
4. P. C. Stair. J. Am. Chem. Soc., 104 (1982) 4044.
5. H. N. V. Temperley and D. H. Trevena. Liquids and their Properties. Wiley. New York. 1978.
6. H. Eyring and M.S. Jhon, Significant Liquid Structures. Wiley. New York. 1969.
7. The Scientific Papers of J. Willard Gibbs. Vol. I (Dover Edition), p. 219 ff. Dover Publications. N.Y.1961. Original Edition. Longmans, Green. New York. 1906.
8. F. C. Goodrich and A. I. Rusanov. The Modern Theory of Capillarity. Akademie-Verlag. Berlin. 1981.
9. F. C. Goodrich in E. Matijevic (Ed.). Surface and Colloid Science. Vol. 3, Wiley. New York. 1971.
10. A. M. Schwartz and F. W. Minor. J. Colloid Sci.. 14 (1959) 572.
11. H. Lamb, Hydrodynamics. Chapter IX. esp. p. 471 ff. Cambridge University Press. 1932. Dover edition. Dover, New York. 1945.
12. A. M. Schwartz and F. W. Minor, J. Colloid Sci., 14 (1959) 584.
13. J. J. Jasper in I. M. Kolthoff and P. J. Elving (Eds.). Treatise on Analytical Chemistry Part 1. Vol. 7. 4611, Wiley. NY 1967.
14. J. F. Padday in E. Matijevic (Ed.). Surface and Colloid Science. Vol. 1. Wiley. New York. 1969.
15. W. D. Harkins and F. E. Brown. J. Am. Chem. Soc., 41 (1919) 499. International Critical Tables. Vol. IV, McGraw-Hill, New York. 1928.
16. F. Bashforth and J. C. Adams, An Attempt to Test the Theories of Capillary Action. Cambridge Univ. Press. 1883.
17. R.C. Brown. Proc. Phys. Soc. London, 48 (1936) 312.
18. A. H. Pfund and E. W. Greenfield. Ind. Eng. Chem. Anal. Ed.. 8 (1936) 81.
19. J. L. Cayias. R. S. Schechter and W. H. Wade in K. L. Mittal (Ed.) Adsorption at Interfaces. A. C. S. Symposium Series No. 8, p. 234. American Chemical Society. Washington. D.C., 1975.
20. A. M. Schwartz. The Dynamics of Contact Angle Phenomena. Advances in Colloid and Interface Science, 4 (1975) 349.
21. A. M. Schwartz in J. F. Padday (Ed.), Wetting, Spreading and Adhesion, Academic Press, N.Y.,, 1978, p. 9[illegible].
22. G. Navascues and M. V. Berry in J. F. Padday (Ed.), Wetting, Spreading and Adhesion p. 83, Academic Press, New York. 1978.
23. E. B. Dussan, V. Annual Review of Fluid Mechanics, II (1979) 371; Annual Reviews, Inc., Palo Alto, CA.
24. A. M. Schwartz, J. Colloid Interface Sci., 59 (1977) 477.
25. N. K. Adam, The Physics and Chemistry of Surfaces (Dover Edition), Dover Publications, N.Y. 1968. Original 3rd Edition Oxford University Press, 1941.
26. R. Defay, I. Prigogine, A. Bellemans and D. H. Everett, Surface Tension and Adsorption. Wiley, New York, 1966.
27. J. Frenkel, Kinetic Theory of Liquids, Oxford University Press, 1946.
28. J. R. Partington, An Advanced Treatise on Physical Chemistry, Vol. II, Longmans Green, New York, 1955.
29. A. M. Schwartz, Capillarity, Theory and Practice, Ind. Eng. Chem., 61 (1969) 10.
30. R. D. Void and M. J. Void, Colloid and Interface Chemistry Addison-Wesley, Reading, Mass. 1983.
31. A. W. Adamson, Physical Chemistry of Surfaces, 4th ed., Wiley, New York, 1982.
32. S. Wu, Polymer Interface and Adhesion, Marcel Dekker, New York, 1982.

Absorbent Technology.
P.K. Chatterjee and B.S. Gupta, editors.

CHAPTER III

FLUID ABSORPTION IN HIGH BULK NONWOVENS

BHUPENDER S. GUPTA
College of Textiles, North Carolina State University, Raleigh, NC 27695-8301 (USA)

Contents

1. INTRODUCTION

One of the major applications of disposable nonwovens is in absorbent materials, which constitute a broad range of products, including baby diapers, personal hygiene and adult incontinent pads, tampons, paper towels, tissues and sponges. Many of these articles, in particular diapers and sanitary pads, are highly engineered structures that contain several components, each performing an important but different function. The top layer, the cover sheet, which is in direct contact with the body, allows the fluid to pass through but ideally does not let it strike back, i.e. it acts like a one way valve. The next is a layer that serves to spread and distribute the fluid over a large area so that the capacity of the pad to absorb and hold fluid could be maximized. Following this, is the major component, the absorbent core, which exerts the force necessary to pull the liquid in, distribute it within the structure and hold it without releasing under normal external pressure. The outermost or back layer is the barrier sheet, which is a film or an impervious fabric that protects the user against leakage.

The component which is central to all absorbent products and which has been the subject of detailed studies is the absorbent core. This chapter is focused on the fluid imbibing and holding behavior of the absorbent core. Discussed are the methodology used in conducting tests, the models employed in predicting behavior, and the results obtained in a number of experimental studies. Also examined in a section at the end is the extent to which the models used are capable of accounting for the effects found.

Most of the scientific work concerning absorbency has been conducted using fibers of textile dimensions, i.e. fibers of length ranging from about 1 to 5cm. This is primarily due to the availability of textile fibers in a range of sizes and shapes suited for scientific studies, the ease of handling, and the ease of converting fibers into webs varying systematically in structure. It could be expected, however, that the effects found using these materials would generally be applicable to structures containing fibers of smaller sizes, such as fluff pulp, used in diapers and many other absorbent products.

The key requirement for absorbent core is the ability to imbibe rapidly and hold large amount of fluid under pressure. The total volume absorbed and held under pressure is largely determined by the interstitial space between the fibers, the absorbing and swelling characteristics of the material and the resiliency of the web in the wet state. The rate at which a fluid is absorbed is governed by the balance between the forces exerted by the capillaries and the frictional drag offered by the fiber surfaces. Additionally, gravity enters as an opposing force if the fluid rises against it. Accordingly, the net force imbibing fluid in a network is governed by the size and the orientation of flow channels, the surface properties of the fibers, and the properties of the fluid.

The size of the capillaries is affected by the thickness per unit mass and the resiliency of the web, and the size, shape and the mechanical properties of the fibers. The resiliency of the web is itself affected by the size, shape and the mechanical properties of the fibers, but the nature and the level of bonding between the fibers also significantly influence it. For absorbent core use, one of the common methods used for bonding is needling which has been shown to have a significant influence on absorbency behavior due to the positive impact it has on the orientation of flow channels and the resiliency of the structure.

In addition to the capillary characteristics, the chemical and physical properties of the absorbent and the absorbate also influence the rate. The chemical nature of the fiber and that of any topical treatment given to the surface account for the role played by the absorbent, whereas the surface tension, pH, electrolytic nature and the viscosity are some of the factors that account for the impact of the fluid.

Finally, the method employed in performing tests can be expected to be important. A fabric may be tested for horizontal spreading or vertical rise of fluid; the fluid may be delivered from a single hole, multiple holes, or from a porous plate; the hydrostatic head used may be positive, zero, or negative; and the environmental pressure imposed during testing could be large or small, depending on application.

All factors alluded to above can have a bearing on the absorbency performance of materials. Many of these have been included as variables in the past studies whose results are examined in this chapter.

2. METHODOLOGY

Two parameters of major interest in characterizing absorbency are the absorbent capacity and the rate of absorbency. These have been assessed using simple as well as more sophisticated methods. Among the former are the sink basket and the vertical wicking tests. In the sink basket test [1], a given mass of fiber material is packed in a wire gauze basket and dropped in fluid from a certain height. The time taken by the specimen to submerge completely is noted and used as a measure of the rate. The basket is removed, allowed to drain for a short period, and the weight of the wet specimen is determined. The amount of fluid absorbed is assessed and, when divided by the dry mass, is used as a measure of the absorbent capacity. The test is qualitative and the values measured, especially of the rate, are subject to significant errors. Also, the usefulness of the method is restricted mostly to determining the potential of a given fiber material, as compared to others, for applications in absorbent products. The method does not lend itself easily to studying the effects of structural factors and environmental conditions on absorbency.

In the vertical wicking test [2], the parameter assessed is the rate. A rectangular strip of fabric, usually 2.5 cm wide, is suspended from a cross bar over a reservoir containing the fluid. The bottom end is loaded slightly. The height is adjusted such that the bottom end is immersed in fluid to about 2.5 cm depth. The stopwatch is started and after a given interval the height to which the fluid is wicked is determined. In more involved tests, the length penetrated at lapse of different time periods is noted and plotted against time to characterize the behavior. Usually in such cases, videotaping or photographing and determining the length of strip wetted from the tape or the prints becomes necessary. The height reached increases with time but at diminishing rate and levels off to reflect the approach to equilibrium. Subjectivity enters in determining the level reached since the latter is not sharp but jagged. The test, although greatly subjective, nevertheless gives useful information about the overall capability of the fabric, influenced by both the fiber material and the capillary structure. It has been demonstrated that the rate when assessed near the beginning of the test, i.e. when the gravity effect is negligible, can be given by Washburn's [3] model and should correspond to the rate assessed on a horizontal strip under similar conditions [4,5].

In majority of studies, however, demand wettability type of device, in which a specimen of circular shape, with the fluid entering from below from a point in the middle, is used. The specimen is small enough so that absorbency starts (due to the presence of capillary force) as soon as the specimen is placed in position and terminates when the pores are filled up [6,7]. In this test, therefore, the end point is usually well defined, unlike found in vertical wicking or horizontal spreading from limited source (sections 8.2 and 8.3, Chapter I). Many versions have been used by workers in the field. An earlier device used by the author is shown in Figure 1. However, the one available commercially and is now widely utilized is known as the Gravimetric Absorbency Testing System, or the GATS [7]. A modified type used by the author in his studies is illustrated in Figure 2. A die cut sample of circular shape is positioned on specimen cell and centered over a hole from which fluid is delivered. A known weight is placed on the specimen to impose the required environmental

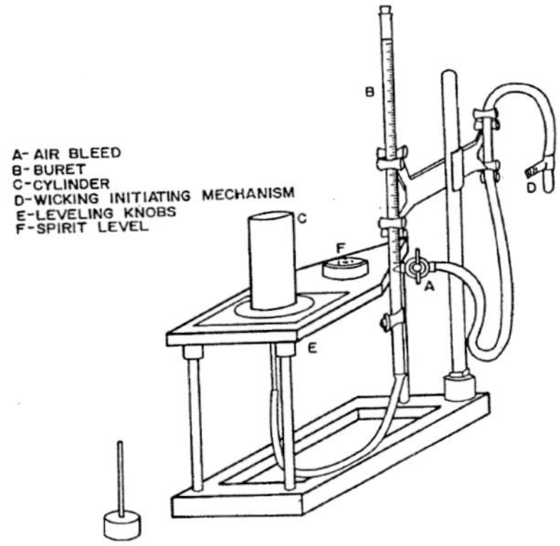

Fig. 1. Demand wettability device [10].

pressure. The fluid is transported from a reservoir resting on a sensitive balance, which records the amount of fluid flowing from the container. The level of the sample with respect to that of the fluid determines the hydrostatic head under which the test is conducted. In most absorbency tests, a zero or a slightly negative head is maintained. The device is equipped with two electromagnetic sensors, which measure the thickness of the specimen at two positions, diagonally across from each other, during the test. The signals from the balance and the thickness sensors are collected and displayed as a function of time (see Fig 7, given later). From the absorbency curve, the absorbent capacity, C (cc fluid/g fiber), given by the volume of fluid absorbed at equilibrium divided by the dry (conditioned) mass of the specimen, and the absorbency rate, Q (cc fluid/g fiber - sec), given by the slope of the absorbency curve divided by the dry (conditioned) mass of the specimen, are assessed. These parameters may also be expressed in terms of the volume, instead of the mass, of the dry (conditioned) fibers (sections 7.2 and 8.4, Chapter I). Symbols C_o (cc fluid/cc fiber) and

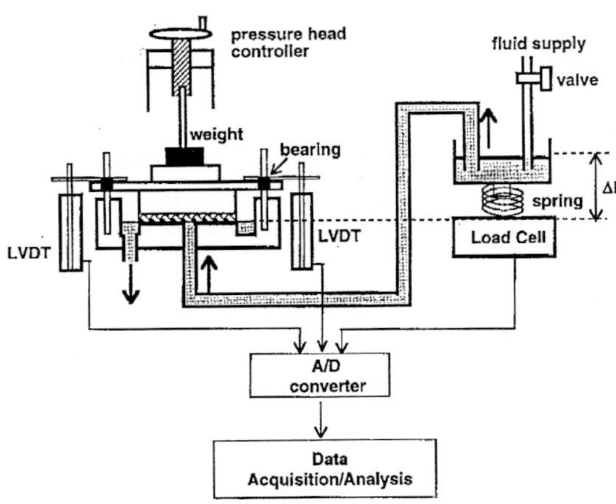

Fig. 2. The modified Gravimetric Absorbency Testing System (GATS) device [14].

Q_o (cc fluid/ cc fiber - sec) are used to represent the values if the denominator is not the mass but the volume of fibers in the test specimen.

3. THEORETICAL

Models have been presented in Chapter I that characterize the two parameters, *C* and *Q*, mentioned above. The one for the capacity is based on determining the total amount of interstitial space available for holding fluid per unit dry mass of fiber, V_s (eq. 39, section 7.2, Chapter I) or per unit dry volume of fibers, V_{so} (eq. 46, section 7.2, Chapter I). The equations for capacity are as follows:

$$C = V_s = A\frac{T}{W} - \frac{1}{\rho_{av}} \qquad \text{[cc(fluid) / g(fiber)]} \tag{1}$$

$$C_o = V_{so} = A\rho_{av}\frac{T}{W} - 1 \qquad \text{[cc(fluid) / cc(fiber)]} \tag{2}$$

where, $$\rho_{av} = \left[\Sigma\frac{w_i}{\rho_i}\right]^{-1} \tag{3}$$

In the above equations, *A* and *T* are, respectively, the area and the final thickness of the web (see Figure 14, Chapter I), *W* (*g*) is the mass of the dry web, w_i and ρ_i are, respectively, the mass fraction and the density of the different types of fibers in the web, and ρ_{av} is the weighted average density of the fibers in the web. For a one component material, $\rho_{av} = \rho$, where ρ is the density of the only fiber present in the fabric. In either of the equations 1 or 2, the only variable is the wet thickness per unit dry mass, *T/W*. Any factors of the study that affect this parameter should also affect absorbent capacity.

For absorbency rate, the equation used is the one given by Washburn-Lucas [3,9], but modified to apply to the webs in which fluid spread radially outward from a point in the middle (section 8.4, Chapter I). It is characterized by either of the following two equations depending upon the unit in which it is desired to be expressed:

$$Q = \frac{\pi r\gamma(\cos\theta)_{av}}{2\eta}\left[\frac{T}{W} - \frac{1}{A\rho_{av}}\right] \qquad \text{[cc(fluid) / g(fiber)-sec]} \tag{4}$$

$$Q_O = \frac{\pi r\gamma(\cos\theta)_{av}}{2\eta}\left[\frac{T}{W} - \frac{1}{A\rho_{av}}\right]\rho_{av} \qquad \text{[cc(fluid) / cc(fiber)-sec]} \tag{5}$$

where, $(\cos\theta)_{av} = \sum w_i \cos\theta_i$ (6)

In these, γ is the surface tension of the fluid, θ_i is the advancing contact angle of fiber *i* in the blend, η is the viscosity of the fluid, and *r* is the mean pore radius of the capillaries. For a one component fabric, $(\cos\theta)_{av} = \cos\theta$, where θ is the contact angle of the only fiber present in the fabric.

For a given fiber and fluid system, all parameters except mean pore radius and thickness per unit mass on the right hand side are constant. The value of *T/W* is expected to be determined by the structure of the web, the pressure under which measurements are carried out, and the wet resiliency of the fibers, and that of *r* is determined by the same factors, except that it is additionally affected by the size of the fiber.

The value of *T/W* was computed from the measurements of the conditioned mass *W* of the web prior to each test and of the final thickness *T* from the signals given by the thickness measuring sensors during the GATS tests. The value of *r* was predicted with a model due to Gupta (sec. 7.3, Chapter I) [8], given by Equation 7 as follows:

$$r = \left[\frac{1}{6\pi B_0} \left\{ A\rho_{av} \frac{T}{W} - 1 \right\} \left\{ \Sigma \frac{n_i d_i}{\rho_i} \right\} \right]^{1/2} \tag{7}$$

where, d_i is the linear density of fiber *i*, n_i is the number of fibers out of 3 belonging to type *i*, and B_0 is the constant whose value is determined by the base length associated with the linear density (*d*) used.

This model is based on the assumption that a capillary is bounded by three fibers, oriented parallel to each other or randomly, and the specific volume of the capillary unit cell equals that of the parent web. The three fibers that lie at the apexes of the triangle (Fig. 17, Chapter I) could belong to different fibers (maximum 3 considered), having different specific gravities and linear densities. The number of fibers of each type out of three is determined by the mass fraction of each in the blend and fiber linear densities.

For a single component fabric, the equation 7 reduces to equation 8, as follows:

$$r = \left[\frac{d}{2\pi B_0} \left(A \frac{T}{W} - \frac{1}{\rho} \right) \right]^{1/2} \tag{8}$$

For two component structures, used frequently in research projects involving absorbent materials, the values of n_1 and n_2 needed, are given by the following equations:

$$n_1 = \frac{3 w_1 d_2}{w_1 d_2 + w_2 d_1} \tag{9}$$

$$n_2 = 3 - n_1 \tag{10}$$

For more complex structures, i.e. fabrics containing 3 different fibers or fibers and an adhesive or a low melt material, the equations needed to calculate the required quantities are given in section 7.4, Chapter I.

According to equations 4 and 5, the rate of absorbency, in a web of given area, is affected by pore size, fabric thickness per unit mass, fiber density, fiber surface contact angle and fluid surface tension and viscosity. Any factors, fluid, fiber or fabric construction that influence the values of these parameters can also be expected to influence the rate.

4. RESULTS

4.1. Fiber Material

A number of fibers have been used in studies involving absorbent structures, these being a trilobal rayon, a regular crenulated rayon, cotton of several different sizes (micronaire values), and polyesters and polypropylenes of different cross-sectional shapes and linear densities. In most cases, fibers have been used as received; however, in limited studies the fibers had been stripped of the treatment and used in finish-free form. In one study, the fibers, which had been scoured, were given a known processing finish. Unless otherwise noted, the results given are for materials used in the as received form. Also, the results reported are generally in the conventional units of cc/g for capacity and cc/g-sec or cc/g-sec$^{1/2}$ for rate. However, as alluded to in sections 7.2 and 8.4, Chapter I, if the behaviors being compared were for materials differing substantially in density, then it was considered advisable to also express the results in the units of cc/cc for capacity and cc/cc-sec or cc/cc-sec$^{1/2}$ for rate to more effectively examine the effects.

An example of the impact the units can have on the results is shown in Table 1 in which the values given are for materials that have widely different values of density. In this table, ND refers to depth of needle penetration in mm, NI refers to needling intensity in needles/cm^2, HI refers to hydroentangling intensity in psi, and EP refers to environmental pressure in gram-force/cm^2. In going from the conventional (Part A) to the other (Part B) units for expressing capacity and rate, not only did the relative values among the three materials change but also in one case, the ranking changed.

The two main criterions that governed the relative performances of different materials were the resilience of the fiber, given by the cross-sectional size and shape and the mechanical properties of the fiber, and the chemical nature of the surface, which determined the degree of hydrophilicity or the value of the advancing contact angle.

Webs made of synthetic fibers whose surface lacked a hydrophilic character either did not absorb fluid at all, or absorbed it at low rates [10,11]. In the latter case, the capacity found was usually quite high, obviously due to high resiliency and, therefore, high pore volume supported by these materials. Blending a hydrophobic fiber with a hydrophilic produced similar results. In one of the studies, involving rayon and polyester, it was found that if the blend contained certain minimum amount of absorbing fiber, so that it attracted fluid, the capacity obtained was nearly the highest (Figure 3A). The effect on the rate was found to be mixed and could be traced to the change the blending produced on the values of the pore size, r, and the advancing contact angle, θ. An increase in the fraction of synthetic fiber could be expected to lead to an increase in r but also to an increase in θ, or a decrease in $\cos\theta$, the change in r and θ opposing each other in the effect they produced on the rate. In this study, the highest rate found was in the 100% rayon structures (Figure 3B).

Cross-sectional size and shape affected results as expected. Increase in size usually led to increases in both the capacity and the rate (Table 2), primarily due to the increase it produced on the bending rigidity of the fiber and, thus, on the resiliency of the fabric [12].

Table 1.
Absorbency results expressed in different sets of units. Materials: 3.3 denier trilobal cellulose acetate, 3 denier trilobal rayon and 3 denier polypropylene; web 40-120 g/m^2; ND 7 mm; NI 0-80 needles/cm^2; HI, 0-1000 psi; EP 12 gf*/cm^2; fluid 1% saline.

	Needled Fabrics		Hydroentangled Fabrics	
A.	Capacity (cc/g)	Rate (cc/g-sec)	Capacity (cc/g)	Rate (cc/g-sec)
Cellulose Acetate	18.6	3.13	15.1	1.36
Trilobal Rayon	15.9	3.80	10.3	2.02
Polypropylene	19.4	2.94	0.0	0.0
B.	Capacity (cc/cc)	Rate (cc/cc-sec)	Capacity (cc/cc)	Rate (cc/cc-sec)
Cellulose Acetate	24.2	4.07	19.6	1.77
Trilobal Rayon	23.9	5.70	15.5	3.03
Polypropylene	18.6	2.80	0.0	0.0

* gf is the force exerted by gravity on 1 gram mass. 1gf = 981 dynes or 9.81 x 10^{-3} N.

A change in cross-sectional shape from crenulated (roughly round) to trilobal in rayon led to significant improvements in absorbency performance (see results in Table 4). Two reasons offered for this were an increase in bending rigidity and an enhancement in surface wettability. Measurements of contact angle on the two fibers by the Wilhelmy technique (Figure 4) [13] showed that the advancing value in the trilobal material was much smaller than in the other and equaled the receding value which was nearly the same in all cellulosic fibers (Table 3). This showed that the fine capillaries formed by the longitudinal ridges of the trilobal shape (Figure 5) imbibed fluid, in the Wilhelmy test, further along the surface and hydrated the cross-section than expected in the fiber of smooth or round cross-section.

Table 2.
Effect of denier of polypropylene on absorbency in 50/50 blends containing polypropylene and 3 denier trilobal rayon. ND 10; NI 180; water [4].

Polypropylene	Absorbent Capacity (cc/g)		Absorbency (cc/g-sec)	
Denier	EP 12	EP 27	EP 12	EP 27
2.2	15.1	12.1	1.43	1.32
3.0	16.4	13.3	1.68	1.37
9.0	21.1	13.9	2.67	2.22

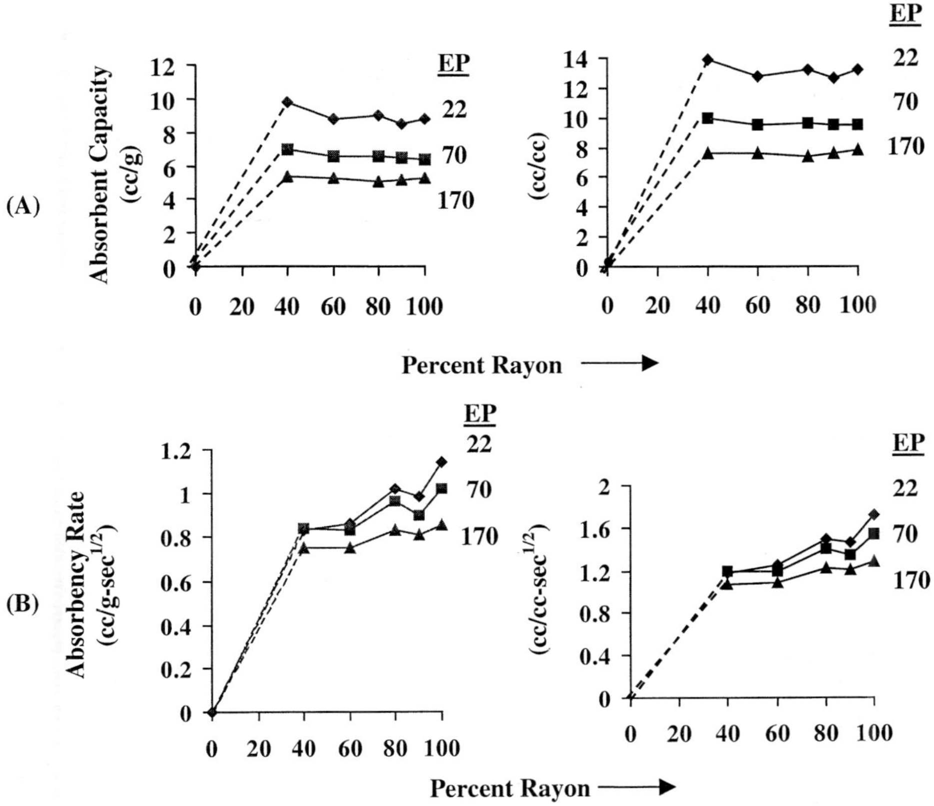

Fig. 3. Effects of the fraction of rayon and environmental pressure on absorbency in webs containing rayon and polyester. Absorbent capacity (A) and absorbency rate (B) are expressed in two different sets of units [10].

The results of a study in which the absorbency behaviors of cotton and rayon were evaluated and compared with each other are given in Table 4. Two types of cotton, a high micronaire fiber (5 micronaire or 1.8 denier nominal), CH, and a low micronaire fiber (2.8 micronaire or 1.0 denier nominal), CL, and two rayons, a trilobal fiber (3 denier), RT, and a regular crenulated, roughly round, fiber (3 denier), RR, were used. At any given pressure, the capacities of the two cotton samples were higher than those of the two rayon samples. The rates of the two cotton samples were also higher than that of the regular rayon but somewhat lower than that of the trilobal fiber. Among the two cotton samples, the higher micronaire fiber had relatively higher values of the parameters. Likewise, among the two rayons, the trilobal fiber had higher values of *C* and *Q*.

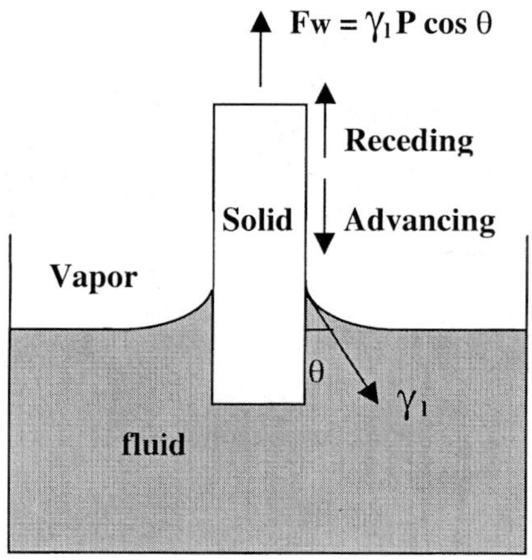

Fig. 4. Wilhelmy wetting force.

Support for most of these results is provided by the values of *T/W* and *r*, given in Table 5, and of tensile properties, given in Table 6. Increase in denier (cotton) or change in cross-sectional shape from round to trilobal (rayon) led to bulkier structures with higher values *T/W* and *r*. Interestingly, however, in spite of lower deniers, the two cottons had higher values of both parameters than that of RR, and this must be due to the former having significantly higher wet modulus than the latter. For the same reasons, the capacities of the two cottons were higher than that of RT. However, the rate of the latter was higher than those of the former. This was attributed to the fact that the trilobal fiber had a cross-sectional shape that enhanced capilarity and it also had a hydrophilic finish on the surface. In an experiment, discussed later (section 4.4), when finishes present on the surfaces, as received, were stripped off and a uniform soap finish (oleic acid based) was applied, the rate of the trilobal rayon dropped below that of the cotton.

Table 3.
Contact angles measured by the Wilhelmy method [13].

	Contact Angle (Degrees)	
Fiber	Advancing	Receding
Cotton (CH)	34.0	20.0
Trilobal Rayon (RT)	21.5	18.3
Regular Rayon (RR)	55.5	17.2

Fig. 5. Scanning electron micrograph of the cross-section of trilobal rayon fibers

In a more recent study, the absorbency behavior of webs containing a new polyester fiber 4 deep grooved, or 4DG, that has four grooves running along the length [16], of 6 denier and cellulosic fibers, CH and RT, were examined. The capacities increased by about 13% in cotton structures and 18% in rayon, when the blend contained 33% 4DG, and about 5% in either, when it contained 10% polyester. Blending polyester with cotton produced no effect on the rate, possibly due to polyester producing a positive effect on resiliency but a negative effect on surface wettability, the two effects canceling each other. Blending polyester with rayon, however, enhanced the rate; this must have been due to the former contributing significantly to fabric resiliency.

4.2. Environmental Pressure

Environmental pressure is determined by the force per unit area imposed on the material and varies from application to application and within an application from user to user. Under pressure, webs compress and undergo a decrease in thickness and, therefore, in pore volume and pore size. These cause a decrease in the absorption capacity and the rate. The degree to which a web compresses depends on web composition, bending rigidity of fibers (a function of fiber size, shape, density and tensile modulus [15]), arrangement of fibers in the web, the type and extent of bonding and the magnitude of pressure.

Table 4.
Values of absorbent capacity and absorbency rate for different cellulosic materials and environmental pressures [14].

	Capacity (cc/g)		Rate (cc/g-sec)	
Material	EP 12	EP 27	EP 12	EP 27
CH	13.91	10.75	0.87	0.61
CL	12.82	9.78	0.50	0.30
RT	12.41	9.36	1.04	0.71
RR	10.24	8.10	0.30	0.23

Table 5.
Equilibrium values of thickness per unit mass of web (of 31.68 cm^2 area) and mean pore size. NI 0; Fluid water [14].

Material	EP 12(gf/cm^2)		EP 27(gf/cm^2)	
	T/W (mm/g)	r(cm) x10^{-3}	T/W (mm/g)	r(cm) x10^{-3}
CH	4.26	2.01	3.45	1.79
CL	4.09	1.47	3.29	1.31
RT	3.95	2.51	3.10	2.20
RR	3.36	2.30	2.63	2.02

In a study involving needled fabrics containing blends of regular polyester and rayon, in which blend ratio, BR, ranged from 40/60 to 100/0 rayon/polyester, it was found [10] that all three major variables, namely, the blend ratio, the needling depth or intensity, and the environmental pressure, produced highly significant effects on absorbency (Tables 7 and 8). The factor having the greatest influence on absorbent capacity was environmental pressure; the effects of needling depth and blend ratio showed up at distant second and third positions, respectively. In the model of the rate, on the other hand, needling depth assumed the most important role, followed by the environmental pressure and the blend composition, in that order. The rate was also significantly affected by the two-way interactions. The most important among these was the product of environmental pressure and needling depth. Selected results from the study are illustrated in Figure 6.

In the study involving cellulosic fibers discussed earlier (Table 4), two levels of environmental pressure were used. The values of T/W and r, given in Table 5, clearly indicate that the values of these parameters, that directly affected the capacity and the rate, were appreciably lower at higher pressure.

Table 6.
Values of breaking stress σ (gf/denier), breaking strain ε, and secant modulus (gf/denier) [14] (*values in parenthesis represent standard deviations*).

Fiber	Dry			Wet (Water)		
	σ *(gf/den)*	ε *x10^{-2}*	σ/ε *(gf/den)*	σ *(gf/den)*	ε *x10^{-2}*	σ/ε *(gf/den)*
CH	3.60	8.38	42.9	4.17	10.52	39.6
	(1.07)	(2.54)		(1.41)	(2.54)	
CL	4.13	7.07	58.4	4.25	9.96	42.7
	(1.28)	(1.93)		(1.55)	(3.08)	
RR	1.66	29.58	5.6	0.76	12.56	6.1
	(0.21)	(3.56)		(0.08)	(1.33)	
RT	2.85	33.21	8.6	1.85	20.04	9.2
	(0.22)	(2.62)		(0.58)	(5.97)	

Table 7.
Analysis of variance results for capacity (cc/g) in 40/60 to 100/0 rayon/polyester blended materials [10].

Source	DF	Sum of Squares	Mean Square	F Value	PR>F
Model	44	611.5	13.89	110.7	0.0001
Error	180	22.6	0.12		
Correct total	224	634.1			
Source					
BR	4	15.7		31.2	0.0001
ND	2	18.0		71.8	0.0001
EP	2	570.1		2269.7	0.0001
BR*ND	8	2.8		2.8	0.0056
BR*EP	8	1.5		1.5	0.1587
ND*EP	4	0.4		0.8	0.5391
BR*ND*EP	16	2.9		1.5	0.1190

Capacity Mean: 7.14 (cc/g), CV (%): 4.96, $R^2 = 0.964$

4.3. Deformation of Webs during Absorption

Absorbent structures are usually composed of hygroscopic fibers, such as cellulose, which attract and imbibe fluid by capillary force into the interstitial spaces between the fibers. These materials also absorb fluid into their internal structure. This causes fibers to lose modulus and a web containing them to compress and give up a fraction of free volume when subjected to external pressure. Such loss in resiliency is undesirable for absorbent products and could be minimized by blending a non-absorbing fiber, such as polyester or

Table 8.
Analysis of variance results for the rate of absorbency (cc/g-sec$^{1/2}$) in 40/60 to 100/0 rayon/polyester blended materials [10].

Source	DF	Sum of Squares	Mean Square	F Value	PR>F
Model	44	10.28	0.233	68.5	0.0001
Error	180	0.61	0.003		
Correct total	224	10.89			
Source					
BR	4	1.22		89.1	0.0001
ND	2	6.82		999.3	0.0001
EP	2	1.19		174.2	0.0001
BR*ND	8	0.10		3.9	0.0003
BR*EP	8	0.23		8.3	0.0001
ND*EP	4	0.68		49.8	0.0001
BR*ND*EP	16	0.04		0.7	0.7762

Rate Mean: 0.94 (cc/g-sec$^{1/2}$); CV(%): 6.20; $R^2 = 0.944$

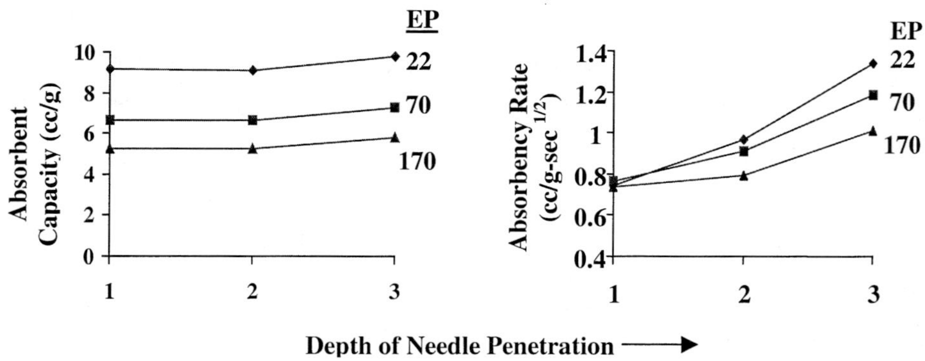

Fig. 6. Effects of needling depth and environmental pressure on absorbency in webs containing rayon and polyester blends [10].

polypropylene. Figures 7A to 7C show how the thickness of a web containing polypropylene (PP) and trilobal rayon (RT) changes during the absorbency tests conducted on GATS. Absorbency starts at the point marked by arrow when the weight is lowered and the web is pressed against the fluid delivery hole in the specimen cell. The dotted and broken lines show how the thickness, measured at two points diagonally across from each other, changes as a function of time. In webs containing 100% polypropylene fiber, there was no indication of any change; the web maintained its free volume, which was presumably filled with fluid at saturation. In the case of 100% rayon web, there was an extensive collapse in the structure that must have led to a high reduction in pore volume and in pore size. In structures containing blend of the two fibers, the collapse could be expected to be intermediate between the two.

In designing absorbent products such as diapers and sanitary pads, one of the aims is to reduce the size or weight without compromising the fluid holding capacity. This is usually accomplished in some structures by incorporating a percentage of superabsorbent polymer along with the main material in the core. In such instances, because of the enormous capacity of the superabsorbent to absorb fluid into its internal structure and swell while maintaining high gel strength, the thickness of the web could be expected to actually increase, as seen in Figure 7D.

The above results, thus, indicate that the final structure, the one in the wet state or at the end of the test, can be quite different from the initial, the one in the dry state or at the beginning of the test. These observations have an important bearing on modeling and predicting the behavior. In the model for the capacity, given by equations 1 and 2, the key factor is the thickness of the web per unit mass. Obviously, the value of the thickness used in the model must be the one assessed at the conclusion of the test. This requires a device that has the capability of recording the thickness of the specimen during the absorbency process.

Likewise, in the modeling of the rate, equations 4 and 5, both the thickness of the specimen per unit mass and the pore size appear in the numerator. The values of these two quantities change during the absorption process. In the webs containing regular absorbent fibers that swell only to a limited extent, the values of r and T/W are expected to decrease.

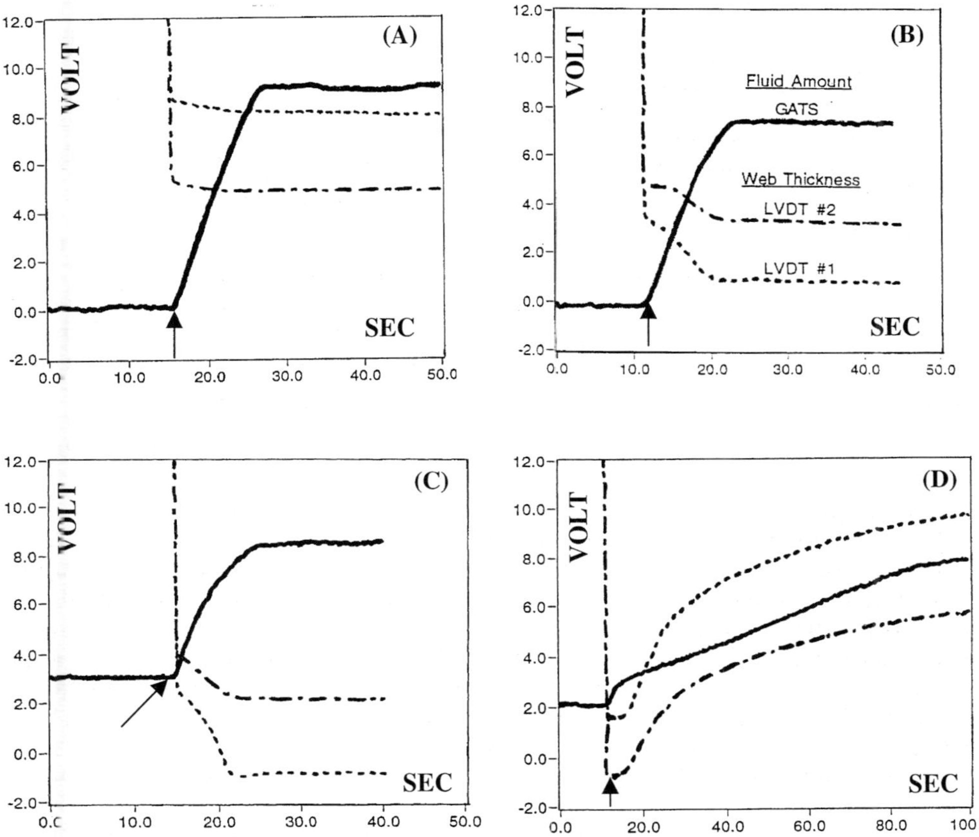

Fig. 7 Change in thickness of webs of different materials during absorbency tests on the GATS: (A) 100 % polypropylene, (B) 50 % polypropylene/ 50 % rayon, (C) 100 % rayon, (D) 30 % superabsorbent fiber/ 70 % polyester.

This change should be reflected in a decrease in the slope of the solid line in the GATS profile which is clearly seen in Figure 7C which represents the behavior of a 100% rayon web. In order to predict the rate at any point, therefore, the value of the thickness T must be obtained at that point. This value, along with the value of r calculated by using eq. 7, is substituted in equation 4 or 5. If the thickness does not change and the absorption profile is as seen in Figure 7A, the values of T and r assessed at any point in the absorption process could be used to estimate the rate.

4.4. Surface Finish

Fibers must usually have a finish before they can be converted efficiently into uniform products using the mechanical processing. The treatment applied is primarily to control friction and static electrification. Thus, synthetic fibers, as they are extruded, are given a topical application before they are collected as tow or wound as filaments. The finish given to the fibers that are particularly marketed for absorbent products usually contains a hydrophilic compound that enhances wettability. Natural fibers, such as cotton and wool, have waxes and other materials that cover the surface. These are undesirable impurities and are usually removed by a wet process. In the case of cotton, the process used also serves to bleach the material. A finish must then be applied before the fiber could be handled or converted into a product.

In most research studies of absorbency, fibers have been used as received, i.e. with the finish they came with. Since different manufactures use different formulations, the finish adds an uncontrolled variable which complicates the interpretation of results when different materials, or different sources supplying the same material, are involved. In a study involving cotton and rayon, the finish was removed by a scouring process and the fibers were given a uniform soap finish (oleic acid based). Webs were made on a model-carding machine, by hand feeding the opened stock at the back, and bonding by needling.

The results of that study are shown in Table 9. With the exception of regular rayon, the scouring and refinishing treatment produced little effect, if any, on the capacity of fibers. The effect on the rate, however, was negative and significant. Since all fibers had the same finish, the differences in the rates of different fibers could be assumed to be governed more or less by the differences that existed in their mechanical properties and cross-sectional sizes. A very significant change in the absorbency values of regular rayon with refinishing could be assumed to be due to the adverse effect the hot-wet treatment produced on the mechanical properties of the fiber.

In another study, cotton was also used in the finish free scoured form. Difficulties were encountered in processing the fiber into a uniform web, but priming the card by passing a regular fiber prior to each run of the scoured material allowed the formation of acceptable uniform structures. Presumably, the priming procedure lubricated the card wire and removed static build-up that allowed a finish free fiber to pass through. Needling also presented a problem but a similar procedure as used for carding alleviated the difficulty.

The results given in Table 10 compare the values obtained on the scoured and the ‘as received’ fibers. The results show, as expected, that the state of the surface did not produce a significant effect on the capacity but a highly significant effect on the rate. These results indicate that the surface of cotton free of impurities and finish is highly hydrophilic,

Table 9.
Comparison of results obtained on cellulosic materials when tested in the as received form and after scouring and refinishing with oleic acid. Web 100 g/m^2; unneedled; Fluid 1% saline. Results averaged over EP of 12 and 27 gf/cm^2.

	As Received		Refinished	
Material	C (cc/g)	Q (cc/g-sec)	C (cc/g)	Q (cc/g-sec)
CH	12.85	0.58	12.13	0.36
CL	12.08	0.32	12.06	0.29
RT	11.68	0.73	12.21	0.24
RR	10.09	0.17	6.46	0.02

Table 10.
Comparison of results obtained on cotton when tested in the as received (CH) and the scoured (CH1) forms. Web 100 g/m^2; NI 100 needles/cm^2; EP 12 gf/cm^2; Fluid 1% saline.

Fiber	State of Surface	Capacity (cc/g)	Rate (cc/g-sec)
CH	As Recieved	14.0	0.61
CH1	Scoured	14.2	2.71

presumably more hydrophilic than one containing a topically applied but unbonded hydrophilic finish. Thus, for absorbent applications a finish free natural cellulosic fiber presents a great advantage. However, the challenge could be expected to lie in overcoming the difficulty involved in economically fabricating products from such fiber, or vice-a-versa, removing finish economically and without affecting the structure from a fabricated product.

While the natural cellulosic, and presumably also the regenerated fibers, became more hydrophilic with the removal of topically applied finish, the synthetic fibers and cellulose acetate were found to become more hydrophobic with such treatment. This was the result noted when webs made from these materials were bonded by spun-lacing process in which the high-energy water jets could be expected to strip the surfaces of the finish. The process made cotton webs more absorbent [17] but cellulose acetate and polypropylene webs less absorbent (see results in sec. 4.5.2).

4.5. Bonding

4.5.1. Needled Structures

To produce structures for application as an absorbent core in many products, short fibers are either carded and cross-lapped into an oriented bat or air-laid into an unoriented one of desired weight. The webs so obtained are then bonded to provide mechanical integrity, necessary for meeting the handling and additional processing requirements, specific to the application.

In absorbent structures, a key concern is resiliency, i.e. the ability of the product to resist compression and maintain airspace for imbibing and holding fluid. Bonding, being one of the factors affecting this property, has been included as a major variable in the study. Among the methods available for the purpose, the most widely used one has been needling in which fibers are bonded by the passage of barbed needles through the thickness of the web. The process tends to consolidate the web while entangling the fibers. This means that while increase in needling could be expected to enhance properties, excessive needling could also break fibers and adversely affect resiliency and, therefore, absorbency. Accordingly, in absorbent studies, amount of needling has been used as a variable. It has been varied in terms of both the depth of needle penetration, i.e. the number of barbs or the length of needle penetrating through the thickness of fabric in each stroke, and the number of needles penetrating per unit area, i.e. the needling intensity.

In earlier studies, considered in this Chapter, needling has been conducted from both sides, with half of the total intensity given from each side [10,12,14]. In later studies, however, the process has been conducted from only one side. The results of one study have been given earlier in figure 6. They show that as the depth of needle penetration increased from 1 to 2 barbs, with needling intensity remaining constant, only small change took place in the values of the capacity and the rate. However, an increase in the depth from 2 to 3 barbs led to definitive increases in the values, with the increase in the rate being substantial.

In a study in which a given intensity, 100 needles/cm^2, was given either in one dose from one side, or in two half doses, with one-half from each side, the bonding from one side or both sides made little difference on the results. The results of another experiment, in which the intensity, with needling only from one side, was varied from 0 to 120 needles/cm^2, are presented in Table 11. The results show that both the capacity and the rate increased with needling. The increase in the rate, as also noted earlier, was generally greater than the increase in capacity. These observations could be attributed to the fact that needling improved resiliency that resulted in an increase in *T/W* and, therefore, also in *r*. While the increase in only the former affected the capacity, the increase in both affected the rate. In addition, the needling process was also expected to create channels, which could be expected to further enhance the value of the rate. In the results given in Table 11, the values seem to peak out by about 80 needles/cm^2. This was primarily because one of the materials used was scoured cotton, i.e. the fiber had no finish. During processing, this fiber resisted passage of needles and tended to break, especially if the amount of needling was excessive

4.5.2. Hydroentangled Structures

A second method used in bonding absorbent structures is hydroentangling (also known as spun-lacing), which is another mechanical process except that it is wet and the bonding is performed by high speed water jets impinging on a web carried over a perforated conveyor screen. A web in its passage through the system passes under several manifolds, each releasing high-speed water streams closely spaced across the width of the machine. The pressures at which the manifolds are operated can be controlled individually. Usually the pressure is either increased as one advances from the front to the back of the machine or is kept the same.

The equipment employed in the current study involved three manifolds. The pressures used ranged from 0 (control) to 1200 psi (~8.3 MPa). After entangling, the web was passed through a vacuum extractor (in the current work at about 5mm of Hg and 7.6 m/min linear speed) to extract excess water and then dried in a chamber wherein hot air was pulled through the web carried over a perforated cylinder. Air temperature and cylinder speed used were adjusted for the type and the weight of the material dried. In earlier studies, entangling was performed from both sides but in more recent investigations it was conducted from only one side.

The control referred to in the wet process is different from that related to the needled structures. In the latter, the control was simply the unneedled air laid or carded web, whereas in the former, an unbonded air laid or carded web was statically soaked in water and then taken through the spun-lacing process without the manifolds operating. In other words, the wetted web was passed though the hydroentangling unit with the water jets closed, it was vacuum extracted and through air-dried, as were the entangled webs.

Table 11.
Effects of needling intensity and web weight or areal density (g/m^2) on capacity. Materials: blends of scoured cotton (CH1) and 6 denier 4 DG polyester. N0, N1, N2, and N3 represent, respectively, 0, 40, 80, and 120 needles/cm^2 [18].

Fabric	Wt(g/m^2)	N0	N1	N2	N3	Avg.
		Absorbent Capacity (cc/g).				
CH1/4DG 100/0	40	17.3	18.0	18.5	21.6	18.9
	80	14.1	15.7	16.1	18.4	16.1
	120	13.7	13.6	15.0	14.0	14.1
	160	13.9	13.4	15.9	15.4	14.7
	Average	14.8	15.2	16.4	17.4	15.9
CH1/4DG 90/10	40	17.1	19.8	22.8	20.2	20.0
	80	15.6	16.7	17.1	14.1	15.9
	120	14.1	15.0	17.3	15.2	15.4
	160	14.4	13.6	14.8	14.7	14.4
	Average	15.3	16.2	18.0	16.0	16.4
		Absorbency Rate (cc/g-sec)				
CH1/4DG 100/0	40	2.60	2.04	2.82	2.03	2.37
	80	1.66	2.15	2.03	2.56	2.10
	120	1.42	1.87	1.84	2.06	1.80
	160	1.15	1.66	2.03	2.21	1.76
	Average	1.71	1.93	2.18	2.22	2.01
CH1/4DG 90/10	40	1.74	1.97	2.35	2.60	2.17
	80	1.80	1.94	2.34	1.24	1.83
	120	1.42	1.67	1.69	1.84	1.66
	160	1.24	1.34	1.59	1.44	1.40
	Average	1.55	1.73	1.99	1.78	1.76

Examples of results obtained are shown in Table 12. The materials used were 100/0 and 90/10 compositions of scoured cotton and 4DG polyester. In contrast to the results obtained with needling, the wet process involved in hydroentangling produced an adverse effect on absorbency. Generally, the greater the manifold pressure or the specific energy [17] used, the greater the decreases occurred in absorbent capacity and rate. Accordingly, one could conclude that the changes resulted from the web compacting during the process into a flattened sheet and bonding in that state during extraction and drying by hydrogen linkages. During rewetting, the bonds were likely to break but the fibers, largely set, were not expected to resilient back and cause the web to increase in thickness. Thus, with increase in hydroentangling energy, the values of both *T/W* and *r* decreased, which led to decreases noted in the values of the parameters.

A comparison of the results obtained on the needled and the hydroentangled structures (Table 11 and 12) show that the values of the absorbency parameters of the former were usually greater than those of the latter. This is more clearly seen from the results presented earlier in Table 1 in which the behaviors compared were of trilobal cellulose

Table 12.
Effect of hydroentangling intensity on absorbency. Materials: blends of scoured cotton (CH1) and 6 denier 4 DG polyester. Web 120 g/m^2, H0, H1, H2 and H3 represent, respectively, the structures entangled at zero (control), low (400 to 800 psi), medium (600 to 1000 psi) and high (800-1200 psi) entangling pressures [18].

CH1/4DG	H0	H1	H2	H3	Avg.
	Absorbent Capacity Values (cc/g)				
100/0	14.4	12.0	11.7	10.2	11.3
90/10	15.1	12.4	11.9	10.5	11.6
	Absorbency Rate Values (cc/g-sec)				
100/0	1.69	1.49	1.42	1.32	1.41
90/10	1.77	1.49	1.45	1.14	1.36

acetate, trilobal rayon and polypropylene [11]. The values corresponding to the needled structures were significantly greater than those corresponding to the hydroentangled materials. This indicates that the structures produced by the needling process were bulkier and more resilient than those produced by the spunlacing process. It will be instructive to examine the degree by which the values of the capacity and the rate, for the materials given in Table 1, changed (decreased) when one considered the hydroentangled structures over the needled. The results are presented in Table 13.

A lower decrease in capacity of cellulose acetate over rayon, in transition from the needled to the hydroentangled structures, was as expected, i.e., due to a relatively lower loss in resiliency. The decrease in the rate of the cellulose acetate fabric over that of the rayon was, however, greater. This was most likely due to a relatively greater change (increase) in the contact angle of the former that occurred due to the topically applied finish, expectedly hydrophilic, washing off during the spun-lacing process. An extreme example of this phenomena is seen in the case of polypropylene which was inherently hydrophobic and reverted to this state after the finish given was stripped off.

As compared to the absorbency values of the water jet entangled webs, those of the control (H0) were greater (Table 12). The relatively high value of the control indicates that the process of wetting, extraction and drying, through which the cellulosic (or the modified cellulosic) materials went, produced a structure, which had a balance of properties in terms of bulk and bonding. The bonds (hydrogen) were expectedly weaker and fewer; some broke

Table 13.
Percent change (decrease) in absorbency values when a given fabric was hydroentangled instead of needled.

Fabric	Capacity	Rate
Cellulose Acetate	19%	57%
Trilobal Rayon	35%	47%
Polypropylene	100%	100%

during rewetting and led to swelling. In contrast, in the water jet entangled webs (H1 - H3), the structures were in a collapsed and dense state and, therefore, the bonds were stronger and more closely spaced. Fewer broke during rewetting and the structures did not resilient back as much.

4.5.3. Thermally Bonded Structures

An alternative to mechanical bonding of absorbent structures is thermal in which webs containing hydrophilic fibers and low melt thermoplastic resins or fibers are bonded by heat. Since high bulk and resiliency are important in such structures, the most suitable way to achieve the desired results will be by bonding a carded or an air laid web, containing the mixture, with a through hot air system. The results of a study [11] in which bonding was carried out by this method are given in Table 14. In this investigation, the fraction of low melt polymer and the linear speed through the heating system, that determined the residence time, were varied. The materials used were 1.7 denier trilobal cellulose acetate and 4 denier low melt polyester, blended in ratios 85/15 and 70/30 cellulose acetate/low melt polyester. Bonding of the carded/cross-lapped webs was carried out in a hot air dryer (174°C) in which the web passed through the system over a rotating perforated drum. The residence time was varied by changing the drum linear speed.

It is seen that with an increase in residence time, the capacity increased in both structures. The change was about 14% in the 85/15 and 24% in the 70/30 blends. This was most likely due to a fabric becoming more effectively bonded and, therefore, more resilient, with increase in residence time. However, the difference between the average values of the capacity in the two blends was small (about 2.4% greater in 70/30), indicating that the presence of additional low melt fiber in the 70/30 structure did not significantly contribute to an increase in pore volume available for imbibing fluid.

The effects of the drum speed and the blend composition on the rate were most interesting. In contrast to the small effect the fraction of low melt polymer in blend produced on capacity, its effect on the rate was highly significant. On an average, the rate in the 85/15 material was more than twice of that in the 70/30 material. Two reasons given for the difference were that the 85/15 structure as compared to 70/30 had: (1) more hydrophilic polymer and (2) less blocked or interrupted channels for fluid flow. The second observation was substantiated by the effect the drum speed produced on the rate. While a decrease in the speed from 20 to 10 feet/minute, caused the rate to decrease in 70/30 material (~21%), due largely to molten polymer flowing into pores and partially blocking channels, it caused the rate to increase in 85/15 material (~ 19%) due mostly to increased bonding and, therefore, to increased resiliency.

On comparing the absorbency behavior of thermally bonded structures with those of the mechanically bonded ones, in particular the needled, the authors noted that the absorbency values found in the former were comparable to those found in the latter [11].

4.6. Areal Density

Absorbent products vary greatly in their weight per unit area, i.e. areal density, from as little as 2 g/m^2 found in lightweight tissues to as much as 200 g/m^2 or more found in absorbent cores of adult incontinent pads or large size diapers. A general goal of research is to develop light weight thin structures that are also highly absorbent. Accordingly, areal

Table 14.
Absorbency properties of through air thermally bonded structures containing 1.7 denier trilobal cellulose acetate and 4 denier low melt polyester fibers. Web 80 g/m^2, air laid; bonding temperature 174°C; fluid 1% saline [11].

Fabric	Drum Speed (ft/min)	Residence Time (sec)	Capacity (cc/g)	Rate (cc/g-sec)
Cellulose Acetate/ Low Melt Polyester Blend				
85/15	10	36	23.7	3.12
	15	24	22.0	2.91
	20	18	20.7	2.63
70/30	10	36	24.7	1.07
	15	24	23.3	1.33
	20	18	20.0	1.36

density has been included as one of the major variables in studies [11, 14, and 18]. Typical results found in an investigation have been given earlier in Table 11.

The results show that the effect of areal density was highly significant. The highest values of C and Q obtained were in the webs of the lowest weight used. As the weight increased, the capacity and the rate decreased but the greatest drop occurred with increase in weight from 40 to 80 g/m^2. The average changes occurring in transition from 40 g/m^2 to 80 g/m^2 were about 21% in C and 29% in Q, and those occurring in transition from the lightest (40 g/m^2) to the heaviest (160 g/m^2), used in the study, were about 34% in C and 54% in Q.

The results obtained could be accounted for by the effect areal density produced on the web thickness per unit mass, and the pore size (Table 15). According to the results, the webs of lower weight, which had higher values of T/W and r, were more resilient and compressed less when subjected to pressure than did the webs of higher weights.

Table 15.
Values of thickness per unit mass of web (of 31.7 cm^2) and pore size in materials of different areal densities. Materials: blends of scoured cotton (1.8 d) and 4 DG polyester of 6 denier. Web characteristics: carded/cross-lapped and needled; EP 12 gf/cm^2; Fluid 1% saline. (Results are averaged over needling intensities of 0 to 80 needles/cm^2).

Areal Density (g/m^2)	T/W (mm/g) 100/0	T/W (mm/g) 90/10	r (cm) x10^{-3} 100/0	r (cm) x10^{-3} 90/10
40	5.42	5.68	2.15	2.29
80	5.23	5.39	2.11	2.23
120	5.10	5.22	2.09	2.19
160	5.08	4.99	2.08	2.14

4.7. Fluid Properties

Absorbent products are expected to encounter a variety of fluids, ranging from one as simple as water to one as complex as menstrual. Furthermore, as the medical literature shows, the composition of body fluids is not constant but varies from person to person, and with the dietary habits and the age of the individuals [19]. The properties of fluid that influences the force of imbibition for a given capillary are the surface tension, the viscosity, and the contact angle, with the latter being an interaction parameter and determined by both the properties of the absorbent and the absorbate. Additionally, the chemical nature of the fluid vis-à-vis that of the fiber material determines the diffusional and the swelling characteristics of the fiber. A fluid that is a solvent for a fiber could lead to a low value of contact angle, and, therefore, to a high value of rate on this account; however, by diffusing into the fiber it could also disrupt molecular structure, which could lead to a loss in resiliency, decreases in pore volume and pore size, and, therefore, decrease in the rate. Therefore, with such an absorbate/absorbent system, the rate could increase or decrease or remain the same, depending upon the relative changes the interaction between the two produce on the surface properties and the bulk mechanical properties of the fibers.

For most purposes, a model used by the industry for representing body fluids has been 0.9 to 1 % saline solution. Typical results obtained on cellulosic materials are shown in Table 16. Addition of salt gave a small increase in the capacity but a somewhat greater decrease in the rate. The increase in capacity was due to a shielding effect the electrolyte molecules produce on the fixed charges of the fiber molecules [20]. This leads to a decrease in the penetration and, therefore, to a decrease in the tendency of the web to collapse under pressure. The decrease in the rate noted has usually been considered as being due to a decrease in the interaction (or an increase in contact angle) and an increase in the drag, i.e. due to an increase in the viscosity. In a study in which a series of fluids, including synthetic urine and menstrual fluid, were used, the adverse effect of viscosity on the rate was particularly evident [21].

4.8. Superabsorbent Fiber

For comfort as well as cosmetic reasons, many of the absorbent products in use must necessarily be limited in weight and bulk and yet continue to be effective in absorbing fluids over much of the working day, or the resting period, of the wearer. There has been a general tendency towards using the so-called superabsorbent material in such products as sanitary

Table 16.
Comparison between absorbency values obtained with 1% saline solution and water. Web characteristics: carded/cross-lapped 100 g/m^2; NI 0. (Results averaged over EP of 12 and 27 gf/cm^2) [14].

	Capacity (cc/g)		Rate (cc/g-sec)	
Material	water	1% saline	water	1% saline
CH	12.3	12.9	0.74	0.58
CL	11.3	12.1	0.40	0.32
RT	10.9	11.7	0.88	0.73
RR	9.2	10.1	0.27	0.17

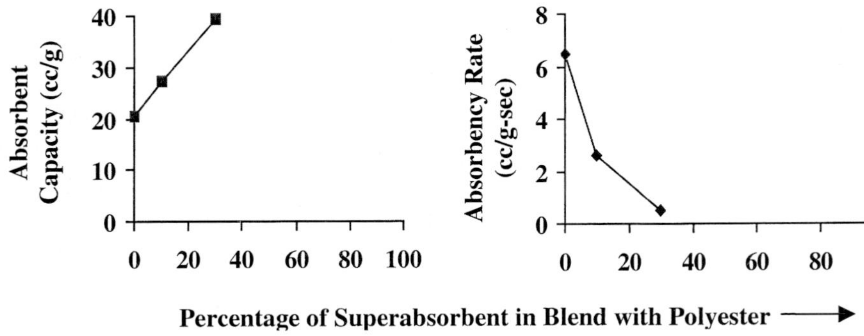

Fig. 8. Effect of the fraction of superabsorbent fiber in blend with polyester on absorbency.

napkins, baby diapers, and adult incontinent pads. Traditionally, superabsorbent used has been in the form of powder, or very short fibers, but in recent years the material has also become available in the form of staple fibers. The idea of using fibers is novel in the sense that the material could be controlled and handled better than possible with the powder. The superabsorbent fiber could be blended with the bulk and dispersed uniformly throughout the structure or positioned biasely at strategic points in the product. The superabsorbent materials have great capacity to swell and retain fluids many times their weight by chemical bonding. However, the absorbency phenomenon becomes complex as the molecular and supramolecular structures of materials, which control their swelling and gel strength and, therefore, the pore and the surface characteristics, change during the absorption process.

The results of a study [22] in which airlaid needled webs containing a blend of a polyester fiber (P) and a superabsorbent fiber (S), each of 3 nominal denier, in ratios ranging from 100/0 to 70/30, P/S, were used, are presented in Figures 8 and 9.

The results given in Figure 8 show that an increase in the fraction of the superabsorbent (S) caused a great increase in the capacity, as much as 100% with addition of 30% fiber, but a drastic decrease in the rate, from about 6 (cc/g-sec) for web containing 0 % superabsorbent to less than 1 (cc/g-sec) for web containing 30% of the material. The increase in the capacity was obviously due to an extra-ordinary ability of the fibers to absorb fluid into their internal structure and swell. The decrease in the rate could be considered as being due to (1) the transverse diffusion of fluid in the fibers that caused a loss in work and reduction in forward velocity, and (2) the swelling of fibers that caused a decrease in pore size.

Using saline instead of water as the fluid gave some interesting results (Figure 9). An increase in saline concentration from 0% to 2% caused the capacity to decrease but the rate to increase, the latter only by a small amount. This behavior was explained by the fact that the electrolyte solution produced a shielding effect on the fixed charge of the polyelectrolyte polymer and led to a reduction in the coulombic repulsion in the polymer network [20]. This restricted swelling and caused retardation in the continuing penetration of fluid into the fiber. As the concentration of salt in the solution increased, the gel strength decreased and so did

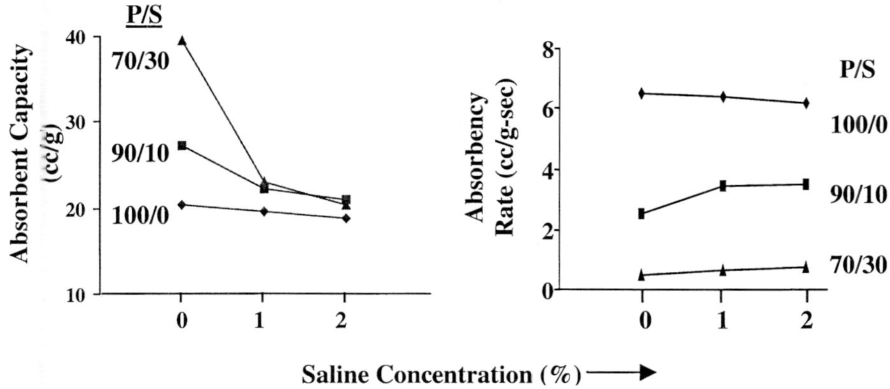

Fig. 9. Effect of saline concentration on absorbency in webs containing different percentages of polyester (P) and superabsorbent fibers (S) [22].

the ability of the web to expand against externally applied pressure. This gave a decrease in the capacity. The rate decreased to some extent with saline concentration in the 100% polyester web and the reasons for this have been given in section 4.7. In the blends, the rate increased but only by a small amount. As seen in the figure, the capacity at 2% saline in the webs containing the superabsorbent material was still greater than the value in web containing 0% material. This illustrates that the swelling was still present, although to a much lesser extent than at 0% saline. Accordingly, the change in the rate, with an increase in the salt concentration, was the resultant of the changes that occurred in absorbate/absorbent interaction, gel blocking, and diffusion.

4.9. Layering

Another important practical aspect of absorbency is the performance of layered structures. Absorbent products such as sanitary napkins, diapers and adult incontinent pads, contain a layer of hydrophobic material on top of hydrophilic core, primarily as a necessity for keeping the skin of the wearer dry. However, there also has been a general thinking that in such arrangement, i.e. with the hydrophobic material on front, in contact with fluid source, and the hydrophilic material immediately behind it, the tendency of the latter to attract fluid and that of the former to repel it may work together in a pull-push manner to efficiently draw the fluid into the structure.

The results from two studies both showing interesting effects are reviewed [10,18]. In one, two separately needled webs of polyester and rayon of approximately 129 g/m^2 were superimposed and needled together. This layered structure, and a second one, obtained by the same procedure but without needling the final composite, were tested in two different ways, in one case with the polyester side down, i.e. in contact with the fluid, and in the other with the rayon side down (Figure 10). The results obtained are given in Table 17 [10]. In the needled composite (Figure 10A), the sample that was tested with the polyester side down gave significantly higher rate of absorbency than did the sample tested with the rayon side down. Although the latter showed more absorption, the test proceeded relatively much

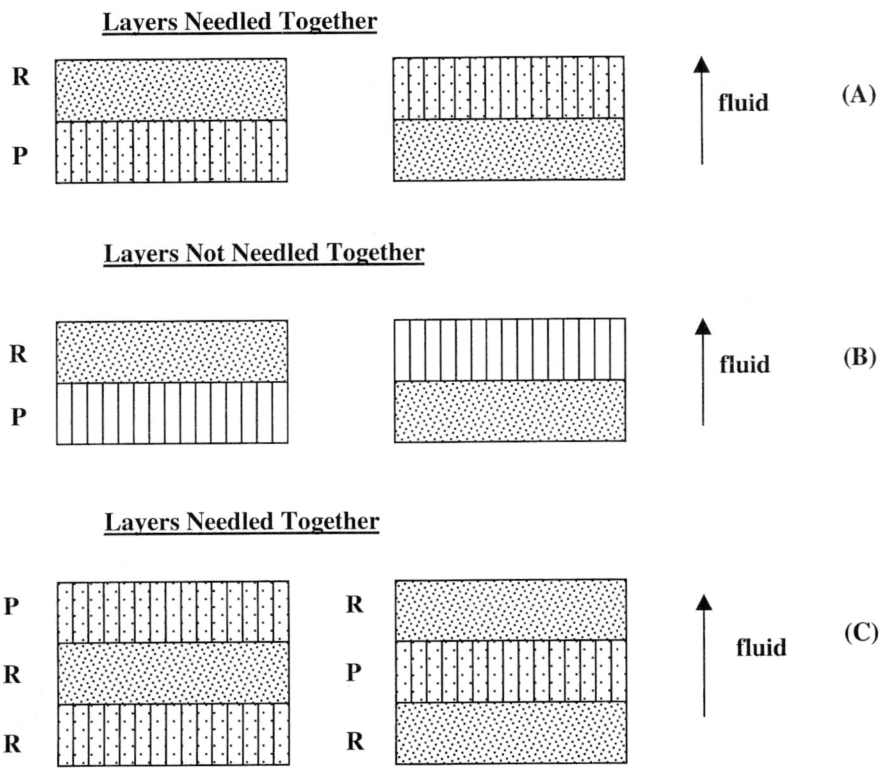

Fig. 10. Schematic showing arrangement of layers of polyester (P) and rayon (R) in laminated structures

slower – at about half the rate. In the sample in which the composite was not bonded (Figure 10B), the results showed that when the test was conducted with the polyester side down, no absorbency took place.

In a second part of the same study, a three layer needled composite was used (Figure 10C), polyester/rayon/polyester (P/R/P) and rayon/polyester/rayon (R/P/R). The results given in Table 18 show that the one with the polyester in the middle gave higher rate. Higher capacity in this structure was most likely due to greater fraction of rayon but higher rate

Table 17.
Absorbency values of two layered structures. Materials: regular polyester and rayon. Web: air-laid, 129 g/m^2; fluid water; EP 70 gf/cm^2; NI 20 needles/cm^2 individual layers and 80 needles/cm^2 composite structure [10].

Side, fluid	Needled together		Not needled together	
imbibed from	C (cc/g)	Q (cc/g-sec$^{1/2}$)	C (cc/g)	Q (cc/g-sec$^{1/2}$)
Polyester	3.76	1.09	0.0	0.0
Rayon	4.27	0.56	6.82	0.8

Table 18.
Absorbency values of three layered structures containing polyester (P) and rayon (R) layers. [10] (For specifications see legend in Table 17).

	Blend Patterns	
	P/R/P	R/P/R
Absorbent Capacity (cc/cc)	5.51	6.60
Absorbency Rate (cc/cc-sec$^{1/2}$)	0.93	1.09

could be assumed to be due to resilient polyester serving as an efficient passageway for transport of fluid. The results above show, however, that the channels in a hydrophobic material, with little ability of their own to attract fluid, were needed to be lined with a hydrophilic material for imbibing and transporting fluid.

The structures used in the above study involved layers of hydrophobic and hydrophilic materials, which were relatively thick in size, and of about the same weight (129 g/m^2). In most absorbent products in which layered structures are used, however, the layer of the hydrophobic material is very thin and serves primarily to keep the skin dry. In order to examine the nature of the results obtained in one such composite, layered structures were prepared by laminating a thin carded web (15 g/m^2) of 6 denier 100% 4DG polyester on top of a regular weight (120 g/m^2) web of 100 % cellulose or of 90/10 cellulose/polyester intimate blend. These were bonded by the needles or the water jets penetrating from the layered polyester side [18]. The tests of absorbency were also conducted from this side, i.e. the polyester side. For comparison, absorbency properties were also measured on webs that did not have the superimposed polyester layer, designated as "normal."

A summary of the results obtained is given in Table 19. The results show that while layering led to small and inconsistent effects on absorbency in the needled structures, it produced consistent and definitive effects on absorbency in the hydroentangled materials. Among the structures bonded by the latter process, the capacities and the rates obtained were lower in the layered than in the normal fabric, the average differences being 7% in the capacity and 33% in the rate.

These results point towards an important conclusion: the structures produced by the needling and the hydroentangling processes, used in this study, were quite different. In the hydroentangling process, the fibers did not move much through the thickness of the web.

Table 19.
Comparison of absorbency in normal and layered structures. Materials: primary web made up of scoured cotton CH1 (1.8 denier) and 4 Deep grooved 4DG polyester (6 denier), 120 g/m^2; superimposed layer made up of 100% 4DG polyester, 15 g/m^2; NI 120 needles/cm^2; HI 600 to 1000 psi; EP 12 gf/cm^2; fluid 1% saline [18].

		Needled		Hydroentangled	
		C (cc/g)	Q (cc/g-sec)	C (cc/g)	Q (cc/g-sec)
CH1/4DG 100/0	Normal	14.22	1.93	11.54	1.35
	Layered	14.55	1.83	10.84	0.75
CH1/4DG 90/10	Normal	15.84	1.73	11.55	1.30
	Layered	15.95	2.06	10.64	1.03

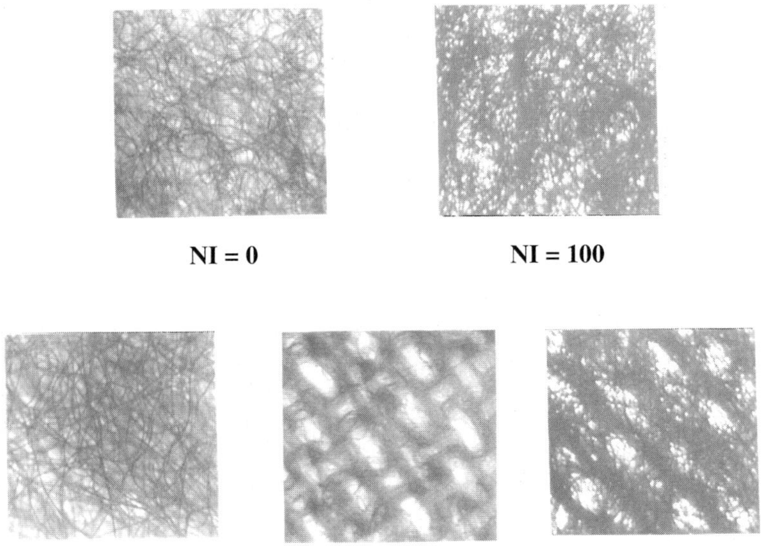

NI = 0 NI = 100

H 0 (Control) H 1 (Low) H 3 (High)

Fig. 11. Photomicrographs showing structures of needled (N) and hydroentangled (H) webs

They moved mostly laterally to allow water jets to penetrate – leading to large pores and dense packing of fibers around the peripheries (Figure 11). In laminated structures, the two layers remained largely separated and since the fluid entered from the polyester side, the rate, in particular, was adversely affected. In the needling process, on the other hand, the portions of the fibers caught by the barbs moved through the thickness in the Z-direction. It produced an integrated structure with the fibers from one layer passing through the other.

5. DISCUSSION AND COMPARISON WITH THEORY

The results presented in this chapter can be largely rationalized and understood by considering the effects the material, the fluid and the processing factors produced on the values of the parameters that make up the equations for the capacity (equations 1 and 2) and the rate (equations 4 and 5). In several instances, the theoretical values of C and Q were actually calculated and compared with those obtained experimentally. The accuracy of such predictions depended upon the accuracy with which the values of the parameters, T/W and θ, the latter being the advancing contact angle, could be measured. The demand wettability device used in some of the studies by the autnor was equipped with thickness measuring sensors, which recorded the thickness as a function of time during the absorbency process. The assessment of the contact angle was more difficult. The static methods rely on visually estimating the value and therefore involved an inherent judgement error. Moreover, the method could generally not be used effectively on fibers, or the fabrics made from them. The dynamic contact angle method, based on Wilhelmy principle, provided a more accurate means of estimating the value needed on single fibers. However, the method is tedious and required an extensive specimen preparation [13]. Therefore, measurements were made only on a few selected materials. Accordingly, in early works, the values of θ used have been those available in the literature and most likely measured by static procedures on polymeric films. In more recent works, especially involving natural cellulosic materials, the values used have been those actually assessed on fibers.

5.1. Absorption Capacity

For predicting the value of the capacity using equation 1 or 2, only the value of the parameter *T/W* was needed to be determined. The value of *W* was measured on each specimen prior to the GATS test and that of *T* was determined from the thickness profiles generated by the two thickness sensors during the test. The value of the capacity was assessed for many structures studied, including those (1) containing regular, synthetic and even superabsorbent fibers, (2) tested with different fluids, and (3) tested under different pressures. In almost every instant, the predicted value matched closely the measured value. Two examples are given in Figures 12 and 13.

5.2. Absorbency Rate

The rate of absorbency given by equation 4 or 5 is, however, a more complex parameter and affected by many factors. One is *T/W*, mentioned above, which is affected by fiber mechanical properties, fiber size, web areal density, and the type and level of bonding. The second is pore size, which is itself affected by *T/W* and, additionally, by the size and the density, in particular the former, of the fiber. If the fabric contains a blend, then the mass

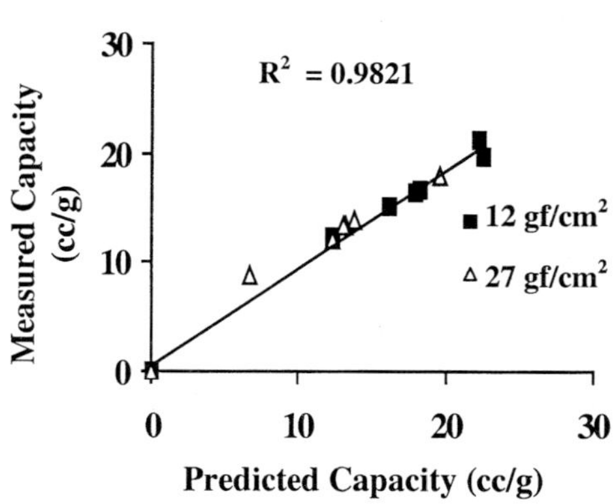

Fig. 12. Correlation between measured and predicted values of capacity in needled webs of polypropylene and trilobal rayon [4].

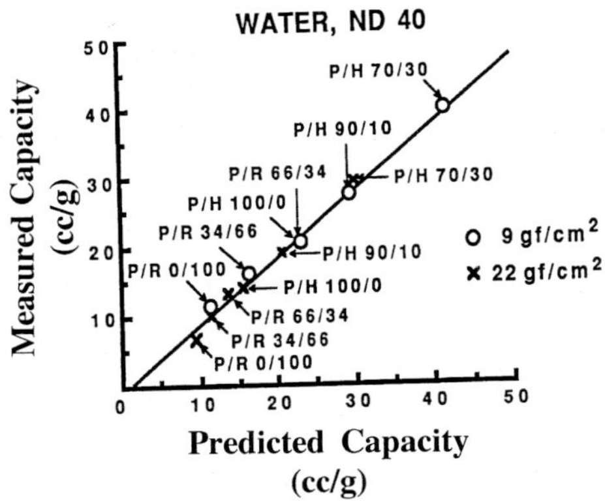

Fig. 13. Correlation between measured and predicted values of capacity. (Results representing different materials and test conditions are displyed together : P/R represents polyester / rayon blends and P/H represents polyester and hydrogel or superabsorbent fiber blends)

fractions and the sizes and densities of each component play the roles. A third is the orientation of flow channels, influenced by the process used in constructing webs. A fourth factor is the wettabillity of fiber surface, which is governed by the chemical constitution of the material, the nature of the surface finish and the cross-sectional morphology of the fiber. These collectively influence the value of $\cos\,\theta$ and, thus, the rate. The contact angle θ, however, is not wholly a fiber surface property. It is also affected by the fluid used. A fifth factor, therefore, is the properties of the fluid, among which the two most obvious ones are the fluid surface tension and the viscosity.

The properties of fluid play additional roles in absorbency. If the fluid penetrates the fiber it can cause swelling as well as a loss in resiliency. This can lead to a decrease in *T/W* and in pore size and, therefore, to a decrease in the rate. Additionally, in hydrophilic materials, transverse diffusion of fluid causes loss in energy and, therefore, a decrease in forward velocity. This can also result in a decrease in the rate. Thus, as compared to the synthetic materials, the cellulosic materials, and among the cellulosics, as compared to cotton, the rayon, can have lower rate due to these reasons since, in each pair, the latter absorbs more water and swells to a greater extent than the former, assuming all other factors remain the same.

To calculate the rate given by equation 4 or 5, the values of surface tension γ, the viscosity η, the advancing contact angle θ, and the pore size r, were needed. The values of γ and η were obtained from the literature. For contact angle, as pointed out, in most instances estimates from the literature were used. In the case of more recent studies, however, especially those involving cotton, the advancing values were measured on actual materials. The values of r were estimated using equation 7. One set of results is presented in Table 20. The materials used were blends of polypropylene and trilobal rayon. A value of θ was available for RT, but not for PP. Accordingly, knowing that the 100% PP webs absorbed fluid in the GATS tests, three values, less than 90°, were assumed. Rate was calculated using equation 5. One value given in the table, i.e. for PP (3)/RT (3) 0/100, 27 g/m^2, was omitted from consideration as it was anomalous, most likely caused by misrecorded value of *T/W* (the predicted value of capacity was also affected and omitted from the plot in Fig. 12). The results show that the predicted values were from half an order to one order of magnitude greater than the measured. In all other predictions of the rate as well, in which accurately assessed values of θ were used, the measured values tended to be half an order of magnitude or more lower than the predicted.

5.3. Structural Constant

The difference between the two values, predicted and measured, can be attributed to the difference that exists between the structure of the actual web and the one on which Washburn's model is applicable. It can be speculated that the reasons for the difference are that 1) the webs had pores of a range of sizes and shapes, which were also not bounded by solid material, where as the model assumed a single pore of circular shape, 2) the capillaries in the web followed tortuous paths, whereas the capillary in the model was straight, and 3) the fluid diffused in and swelled the fibers in the web, whereas the model assumed no such occurrence.

Table 20.
Comparison of measured and predicted values of absorbency rate in webs containing blends of polypropylene (PP) of 9 and 3 deniers and trilobal rayon (RT) of 3 denier. Web areal density 120 g/m^2; ND 10 mm; NI 180 needles/cm^2. Also given are the values of the structural constant, K. Values assumed: contact angles for RT 30°, and PP 70° (Q′), 60° (Q″) and 50° (Q‴); fiber densities for PP 0.96 g/cc and RT 1.5 g/cc.

Web Composition	Blend Ratio	Rate of Absorbency (cc/cc-sec) Meas. Q	Pred. (70°) Q′	Pred. (60°) Q″	Pred. (50°) Q‴	Structural Constant Pred. Rate/Meas. Rate K′	K″	K‴
EP = 12 gf/cm^2								
PP(9)/RT(3)	50/50	3.13	23.8	26.9	29.7	7.6	8.6	9.5
PP(3)/RT(3)	100/0	2.16	8.9	13.1	16.8	4.1	6.0	7.8
PP(3)/RT(3)	66/34	1.88	11.4	13.7	15.8	6.1	7.3	8.4
PP(3)/RT(3)	50/50	1.97	14.0	15.8	17.5	7.1	8.1	8.8
PP(3)/RT(3)	34/66	1.93	14.9	16.1	17.2	7.8	8.4	8.9
PP(3)/RT(3)	0/100	2.01	15.8	15.8	15.8	7.9	7.9	7.9
EP = 27 gf/cm^2								
PP(9)/RT(3)	50/50	2.60	11.7	13.2	14.6	4.5	5.1	5.6
PP(3)/RT(3)	100/0	1.66	7.3	10.6	13.7	4.4	6.4	8.2
PP(3)/RT(3)	66/34	1.62	7.1	8.5	9.8	4.4	5.2	6.0
PP(3)/RT(3)	50/50	1.60	8.8	9.9	10.9	5.5	6.2	6.8
PP(3)/RT(3)	34/66	1.60	10.1	10.9	11.6	6.3	6.8	7.3
PP(3)/RT(3)	0/100	1.70	6.7	6.7	6.7	4.0	4.0	4.0

The departure between the structures of the actual and the ideal capillary networks can be accounted for by including an empirical constant, K, termed the "*structural constant*," in the equation of the rate:

$$Q = \frac{Q'}{K} = \frac{\pi r \gamma \cos\theta}{K 2\eta}\left(\frac{T}{W} - \frac{1}{A\rho_f}\right) \tag{11}$$

The value of K, given by the ratio of the rate predicted by the classical model to the one measured with a device, will be 1 if the structure of the actual capillary network matches that of the ideal. A value of K greater than 1 will indicate that the actual structure departs from the ideal, on which the model applies, and the greater the value the greater the departure.

One can expect that the value of this parameter will vary with the swelling characteristics of the fiber, the porosity of the web, and the structure of the capillary network in terms of the orientation and the distribution of flow channels. An increase in the ability of fibers to allow fluid to diffuse into the internal structure and swell, that can result in a change in porosity, tortuosity of channels, and pore size distribution, will be expected to lead to an increase in the value of the structural constant. A generally consistent result noted in Table 20 is that an increase in environmental pressure led to lower value of K. Likewise, an increase in bonding or decrease in areal density gave lower values of the parameter (Table 21). A significantly higher value of the constant was found for webs containing rayon (~9) than those containing cotton (~5) [18].

Based on the concepts presented, one can conclude that the lower value of K found in (1) cotton compared with that in rayon is due to relatively less diffusion of fluid into the internal structure and swelling in cotton than in rayon, (2) thinner and more highly bonded webs is due to relatively more prominent and less tortuous channels, and (3) more highly compressed structures, is due to relatively narrower distribution of pore sizes and more prominent and better bounded pores.

Table 21.
Effect of needling and areal density on the values of the structural constant, K, in webs of 1.8 denier scoured cotton, CH1. EP 12 gf/cm^2, fluid 1% saline, ND 10 mm, contact angle 34°. For values of N0, N1, N2 and N3, see Table 11.

Areal Density	Structural Constant, K				
(g/m^2)	N0	N1	N2	N3	Avg.
40	3.7	5.2	3.7	5.1	4.4
80	5.1	4.8	5.0	4.0	4.7
120	5.8	4.9	5.0	5.3	5.2
160	6.7	5.1	5.0	5.0	5.4
Average	5.3	5.0	4.7	4.8	5.0

5.4. Final Comment

The effects of fiber material, fabric construction, fluid and testing related factors found on the absorbency behavior of nonwovens can be rationalized by the models developed based on classical theories. The concept of a structural constant whose value reflects the degree by which the structure of an actual capillary network departs from that on which the classical model for the rate applies has been presented. The factors affecting its value have been discussed. From the results discussed in sections 5.1 to 5.3, it should be clear that the value of capacity can be monitored and predicted effectively by simply measuring thickness per unit mass and using eq. 1. In order to predict the rate of absorption using eq. 11, however, not only are the values of the parameters θ, T/W and r, needed to be measured or estimated, but, in addition, the value of the structural constant K is needed to be determined. At present, a model that can characterize and predict the value of the structural constant does not exist, but is clearly required given how important the role the rate of absorption plays in determining the success of an absorbent product.

6. ACKNOWLEDGEMENT

The work reported in this chapter was supported by funds from a number of sources, including Cotton Incorporated, Dow Chemical Company, and the organized research budget of the College of Textiles of the North Carolina State University. I gratefully acknowledge these supports. The graduate students who participated in the work were Ms. Ann Crews, Ms. Terry Hall Hammond, Dr. Cheol-Jae Hong and Dr. Hyun Suk Whang. To these former students, now my associates, I extend my thanks and best wishes. I take this opportunity to thank my friend and professional colleague, Dr. Pronoy K. Chatterjee, my co-editor, for the pleasure of working with him on this book and for his technical, intellectual and enthusiastic association throughout the undertaking. And finally, I express my love to my companion and wife, Dr. Vasudha Gupta, for her understanding and support, both literary and moral, during the writing of the various chapters of the book and the completion of this project, and to my children, Sumi, Apu and Anoopum, who were always there to give a hand when needed!

7. GLOSSARY

4DG Four deep groove polyester, 6 denier [16].

A Area of the sample; also cross-sectional area perpendicular to the main flow direction in linear flow.

B_0 Constant, whose value is determined by the base length associated with the linear density.

C Absorbent capacity of a porous sample (capacity to fill up all the pore space, volume of fluid per unit mass of conditioned fiber (cc fluid/g fiber).

C_0 Absorbent capacity of a porous sample, volume of fluid absorbed per unit volume of fiber (cc fluid/cc fiber).

CA Cellulose Acetate.

CH Cotton, high micronaire (5 micronaire, 1.8 denier); as received.

CH1 Same as CH, but scoured to remove surface finish and impurities.
CL Cotton, low micronaire (2.8 micronaire, 0.99 denier); as received.
CV Coefficient of variation, term used in statistical analysis of variance of data.
d Fiber linear density.
DF Degree of Freedom, term associated with the statistical analysis of variance of data.
EP Environmental pressure, the pressure under which absorbency tests are conducted, gf/cm^2.
F F-value, term associated with the statistical analysis of variance of data.
gf Gram force, the force exerted by gravity on 1 g mass. 1gf = 981 dyne, or $9.81x10^{-3}$ N.
H, HI Hydroentangling intensity, psi.
i Index used to represent a specific item in a series.
K Structural constant used in the rate of absorbency model, eq. 8. The value of K represents the degree by which the actual capillary network departs from the ideal on which Washburn's eq. 7, Ch. I, departs.
n_i Number of fibers of type *i* out of 3 making up a capillary ($\sum n_i = 3$).
N, NI Needling intensity, $needles/cm^2$.
ND Needling depth, mm.
P Polyester fiber.
PP Polypropylene fiber.
PR Probability, term used in statistical analysis of variance of data.
Q, Q' Rate of absorption (cc fluid/g fiber-sec).
Q_0 Rate of absorption (cc fluid/cc fiber-sec).
r Average capillary radius.
R^2 Correlation coefficient square, term used in statistical analysis of variance of data.
RR Rayon, crenulated or roughly round cross-section.
RT Rayon, trilobal cross-section.
S Superabsorbent fiber, abbreviation used for.
T Sample thickness.
V_s Specific air volume in fabric (air volume per unit fiber mass).
V_{so} Specific air volume in fabric (air volume per unit fiber volume).
w_i Mass fraction of component *i* in a blend.
W Dry (conditioned) mass of fabric specimen.
ε Tensile strain, or breaking tensile strain.
γ Surface tension of the liquid being absorbed.
η Viscosity of liquid.
θ Contact angle of liquid-solid-air interface.
$(\cos\theta)_{av}$
Average value of $\cos\theta$ in a fabric containing a blend of different materials.
ρ, ρ_i Density of fiber, density of fiber *i* in a blend.
ρ_{av} Average fiber density ($=\sum w_i \rho_i$)
σ Specific stress, gf/den.

8. REFERENCES

1. ASTM Method D 117-79, ASTM Standard Methods, ASTM, Philadelphia, PA, 1979.
2. INDA Test Method IST 10.3, Association of the Nonwoven Fabric Industry, Cary, NC.
3. E. W. Washburn, The Dynamics of Capillary Flow, Physical Review, 17(3), 273, (1921), 273.
4. C. J. Hong, Ph.D. Thesis, North Carolina State University, Raleigh, NC, 1993.
5. B. Miller, INDA, INJ., 9, No. 1 (2000) 35.
6. B. M. Lichstein, Proc. INDA Technical Symposium, U. S. A., 1974, p. 129.
7. E. V. Painter, INDA Technical Symposium, U. S. A., 1984.
8. B. S. Gupta, TAPPI Journal, 71 (1988) 147.
9. R. Lucas, Kolloid, Z., "Ueber das Zeitgesetz des Kapillaren Aufstiegs von Flussigkeiten," vol. 23, 15 (1918).
10. B. S. Gupta and T. H. Hammond, INDA Technical Conference, U. S. A., 1980, p. 88.
11. B. S. Gupta and E. W. Powers, Proc. Beltwide Cotton Conferences, National Cotton Council, 1 (2000) 764.
12. B. S. Gupta and C. J. Hong, TAPPI Journal, 77 (1994) 181.
13. H. S. Whang and B. S. Gupta, Textile Res. J., 70, No. 4 (2000) 351.
14. B. S. Gupta and C. J. Hong, INDA, INJ, 7, No. 1 (1995) 34.
15. W E. Morton and J. W. S. Hearle, Physical Properties of Textile Fibers, Textile Institute, Manchester, 3rd edtion, 1993, p.401.
16. W A. Haile and B. M. Phillips, TAPPI Journal, 78 (1995) 139.
17. T. F. Gilmore, N. B. Timble, and W. E. Morton, TAPPI Journal, 80 (1997) 179.
18. B. S. Gupta, Proc. INDA Technical Conference, U.S.A., 1998, p. 21.1.
19. D. S. Dittman (ed.), Blood and Other Body Fluids, Biological Handbook, Fed. Of Am. Societies for Exp. Biology, Washington, DC, 1961.
20. P. J. Flory, "Principle of Polymer Chemistry," Cornell University Press, Ithaca (1967), p. 565.
21. B. S. Gupta and A. L. Crews, "Nonwovens: An Advanced Tutorial," A. F. Turbak and T. L. Vigo (eds.), TAPPI Press, Atlanta, GA., 1989, p. 197.
22. B. S. Gupta and C. J. Hong, Proc. TAPPI 1993 Nonwovens Conference, TAPPI Press, Atlanta, GA, 1993, p. 59.

Absorbent Technology.
P.K. Chatterjee and B.S. Gupta, editors.

CHAPTER IV

INTRODUCTION TO COMPUTATIONAL MODELING AND ITS APPLICATIONS IN ABSORBENT TECHNOLOGY*

SACH KANGOVI
Simutel Associates, P.O. Box 252, Princeton Junction, NJ 08550, USA

Contents

1. INTRODUCTION

Absorption is a phenomenon where a liquid is transported into porous medium by immiscible displacement of air from pores. There are many important practical applications of this phenomenon and whole industries are built on the products based on this phenomenon. These industries and government are spending considerable amount of money on research and development in order to improve these products and also the machinery that build these products. However, there is an increasing emphasis on utilizing better tools of investigation to reduce the time to market. Computational modeling, when applicable, is a

---

*Based on author's experience in advanced computational field at Boeing, Johnson & Johnson, AT&T, Lucent Technologies and Drexel University.

powerful and cost-effective tool to study a problem with considerable accuracy in a short time. This chapter describes the application of computational modeling in investigating some typical applications in absorbent technology including penetration absorbency and pneumatic transport of fibers in a web-forming machine. This description includes the governing physical laws, techniques of modeling, validation of modeling results with that of experiments and the advantages of modeling over traditional empirical methods. Since the governing laws and modeling techniques differ considerably from one application to other, it was decided to divide the chapter based on applications instead of combining differing governing equations in one section and modeling techniques in another section.

Modeling, based on sound governing laws, is accurate, cost effective and efficient. For example, the modeling of penetration absorbency provided a tool to optimize absorbency by adjusting interfacial properties and selecting proper pore radius and depth. To achieve this same goal, it would take enormous number of experiments and associated data analysis. Similarly, modeling provided an effective, economical and convenient means of conducting parametric studies to optimize the performance of a complex web-forming machine and improve the quality of the end product. It would have taken enormous efforts and expenses to empirically optimize all these parameters because experiments would require building many such expensive machines to test variations in each parameter. Modeling made it possible to visualize particle trajectories, which were otherwise impossible to visualize by experiments despite the use of state of the art Fiberscope and Videoimagescope.

2. COMPUTATIONAL MODELING OF PENETRATION ABSORBENCY

2.1. Introduction to Penetration Absorbency

Absorption is a phenomenon where a liquid is transported into a porous medium by immicisible displacement of air from pores. Central to this phenomenon of absorbency is capillary action, which could be assisted or resisted by forces such as gravity and/or external pressure. Capillary action is a consequence of intermolecular attraction which, at a free surface, manifests in the form of interfacial free energy. This energy per unit interfacial area is called surface tension [1]. Surface tension is a unique property of a liquid and can be measured by any one of the numerous available methods. A common method used in many laboratories is called de Nuoy tensiometer.

Another parameter that influences the capillary action is called contact angle [2]. It is an interfacial property. Therefore, it depends on the absorbed liquid and absorbing porous medium. The magnitude of the contact angle depends on the interplay of solid-vapor, solid-liquid and liquid-vapor interfaces and is given by Young-Dupre equation [3]. For a given liquid and material of porous medium, contact angle can be determined directly by optical measurements. However, direct measurements are prone to errors because surface irregularities and contamination easily influence the contact angle. It is, therefore, usually determined by indirect measurements [3].

The capillary action is usually expressed in terms of capillary pressure as a function of surface tension, contact angle and pore radius. This relation is known as Laplace equation [4] and represents an equilibrium condition, in other words the capillary pressure given by Laplace equation is a hydrostatic pressure. In order to study the transition from the start to the final equilibrium state, one can idealize the pore as a tube and apply Hagen-Poiseuille's equation for laminar flow through tubes [5]. Combination of Laplace and Hagen-Poiseuille's equations results in an equation, which gives temporal variation of capillary rise. This relation is known as Washburn equation [6]. Around this fundamental framework additional complexities have been added and investigated [4].

A porous medium can be considered as a network of pores of myriad shapes and sizes in three-dimension and therefore the one-dimensional capillary analysis represented by Washburn's equation is not adequate. To study such cases, a diffusion analogy has been applied by combining conservation of mass and Fick's law of diffusion [4]. Accuracy of this method depends on empirical inputs. It is still a good tool to compare different absorbent materials.

The concept of capillary action, however, can be applied without any questionable assumption to a class of problems characterized as ***penetration absorbency***. In this class of problems, the porous medium does not have a three-dimensional network of pores and therefore wicking phenomenon is not as dominant as the absorbency due to penetration.

2.2. Governing Equations

In penetration absorbency, the porous medium can be modeled as a bundle of uniform circular capillary pores. Figure 1 illustrates the situation where liquid is entering a pore. The figure shows gravitational force, external pressure and surface tension acting on a drop of liquid. The total pressure acting downward is given by;

$$\Delta P = P_b + P_w + P_s \tag{1}$$

Where ΔP is the net driving pressure, P_b is external pressure, P_w is the pressure generated by weight of liquid drop and P_s is the capillary pressure. These are given by;

$$P_w = \rho g V / \pi (r_c)^2$$

and

$$P_s = 2\gamma \cos\theta / r_c$$

The Hagen-Poiseulle's equation [5] governing flow in the capillary is given below;

$$q = (r^2_c / 8\mu)(\Delta P / L) \tag{2}$$

Where q is the volume flux rate, r_c is capillary radius, L is the wetted length of the capillary, μ is co-efficient of viscosity, γ is surface tension and θ is the contact angle. Substituting ΔP

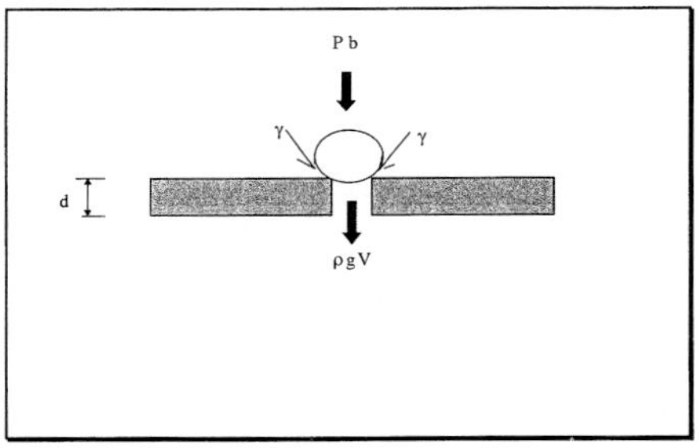

Figure 1. Liquid Penetration through a Pore

from equation 1, rearranging and integrating equation 2 over the thickness (d) of the absorbent media shown in Figure 1, we get the following;

$$\int_0^d L\,dL = \int_0^{t_p} \frac{r_c^2}{8\mu} \left(P_b + \rho g V/\pi(r_c)^2 + 2\gamma\cos\theta/r_c\right) dt \qquad (3)$$

Equation 3 represents a relation between penetration time, and pore dimensions including depth and radius for a given set of fluid and interfacial properties.

2.3. Modeling

A computer program [7] was developed to acquire fluid and interfacial properties from a user and then to calculate variation of penetration time with pore radius. The program was developed based on the process requirement specifications in structured language and other standard techniques like Data Flow Diagrams and Structured Design Charts [8]. The code is modular in construction with each module having many sub-modules. This ensures flexibility and is amenable to future expansion and ease of maintenance. The code is also easily portable to various platforms.

The solution to equation 3 is in a closed form and therefore it is not very compute intensive. However, the graphics does require certain minimum hardware capabilities. It is therefore essential that a proper system be selected to solve the problem. Hardware also must have high integer and floating-point processing capabilities combined with high overall throughput.

Numerical solutions of equation 3 usually generate large quantities of data and therefore a good scientific data visualization capability is essential in order to derive meaningful conclusions from voluminous data. The visualization capabilities of the system depend on the speed to draw two and three dimensional vectors and shaded polygons of a given pixel density. Other highly desirable characteristics of a good visualization system include lighting, rotation, pan and zoom capabilities. The visualization system must also

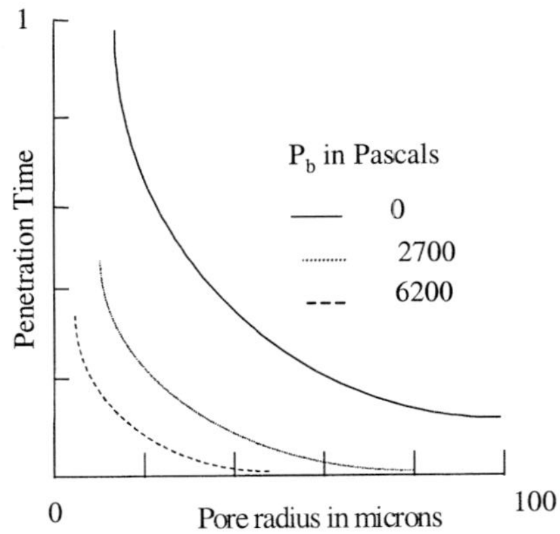

Figure 2. Influence of Pore Size

include the capabilities to quickly draw color hue and contour maps, vector plots and trajectories.

2.4. Results of Modeling

Variation of the penetration time, normalized by a reference time to make maximum value on Y-axis to be 1.0, with pore radius is shown in Figure 2. The results are for a typical liquid and porous material with known liquid and interfacial properties. Calculations were done at three representative external pressures. Thickness of the medium was kept constant. It can be seen that the penetration time decreases with increasing pore radius. Also, at a given pore radius the penetration time decreases with increasing external pressure. The results indicate that when external pressure is greater than zero, the penetration time asymptotically reaches very small values with increasing pore radius. Although these results are specific to a liquid and porous medium, general conclusions can still be derived. They also demonstrate the general capabilities of the computer-based modeling.

The variation of penetration time, again normalized by a reference time to make maximum value on Y-axis equal to 1.0, with contact angle is shown in Figure 3 for three different external pressures. When the contact angle is 90 degrees or greater and external pressure is zero, then there is no penetration of liquid. However, increasing external pressure causes penetration even at contact angles greater than 90 degrees. Penetration time becomes virtually independent of contact angle below 60 degrees. Such parametric studies help in arriving at an optimum contact angle.

Surface tension is another important factor governing penetration absorbency as explained in section 1.1. Its influence on the penetration time, normalized by a reference time, is shown in Figure 4. According to Laplace equation [4], other parameters remaining constant, capillarity is higher for liquids having higher surface tension. Higher capillarity

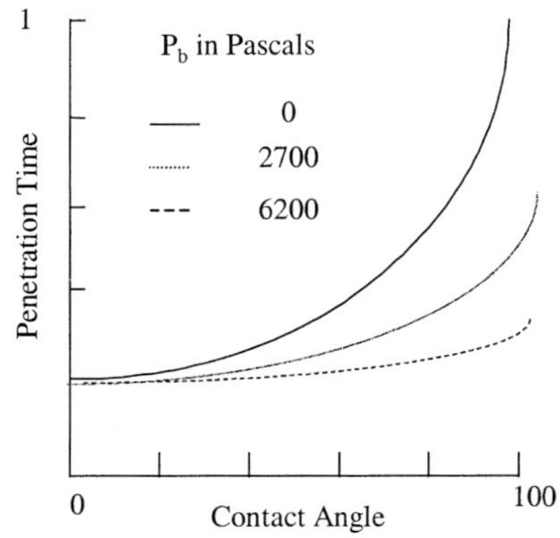

Figure 3. Influence of Contact Angle

means smaller penetration time and this fact is depicted in Figure 4. The figure also shows that for a liquid with surface tension equal to 0.05 newton/m, the penetration time is already small enough and increasing external pressure does not substantially reduce it further. For liquids with lower surface tension, however, increasing external pressure does have noticeable reduction in penetration time.

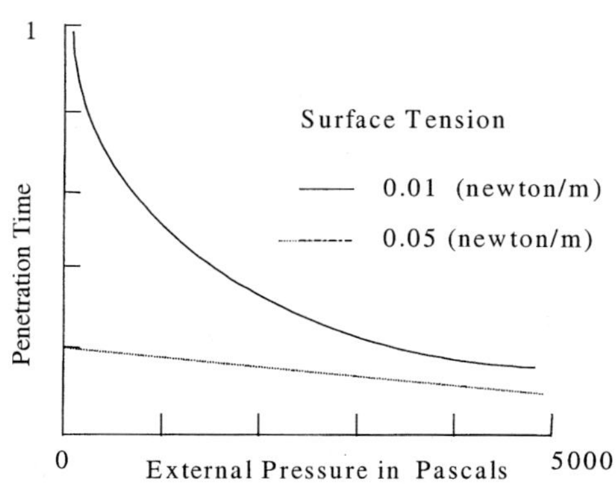

Figure 4. Influence of Surface Tension

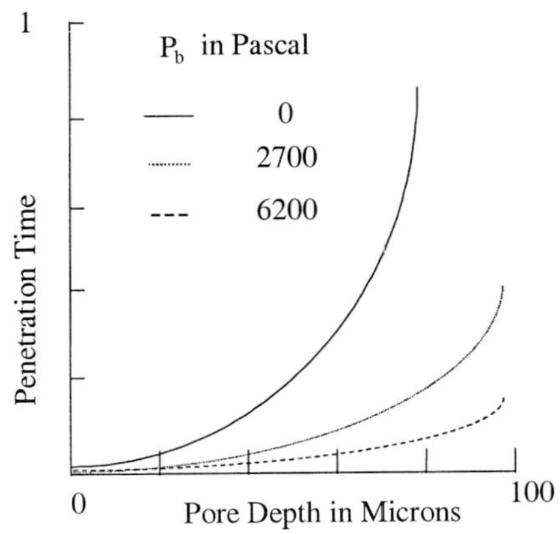

Figure 5. Influence of Pore Depth

Variation of penetration time with pore depth (d) is shown in Figure 5. Results show that the penetration time increases with depth. For a given pore depth, the penetration time reduces with external pressure.

Viscosity of liquid is another important parameter influencing the absorbency. Figure 6 shows that penetration time increases with co-efficient of viscosity but as in other cases the presence of external pressure makes the influence of viscosity less pronounced.

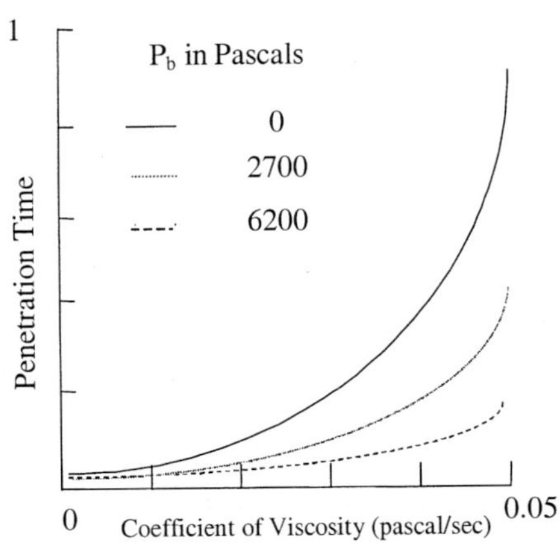

Figure 6. Influence of Co-efficient of Viscosity

Figures 2, 3 and 4 show that for a given liquid and interfacial properties, the penetration time reduces with increasing external pressure. This decrease is monotonic but asymptotic.

2.5. Validation of Modeling of Penetration Absorbency

Careful experiments [9] were conducted to study the effects of various parameters investigated in the modeling. In one set of experiments, the contact angle was changed by treating the absorbent material with surfactants while keeping the liquid and pore size constant. Rate of absorption and contact angles were measured directly by high-speed video camera coupled to a computer. In another set of experiments different liquids were tested to vary viscosity. Similarly, testing absorbent materials manufactured with different pore radius generated yet another set of experimental data. Experimental results compared extremely well with the trends shown by modeling and completed the validation of modeling.

2.6. Advantages of Modeling

Modeling proved to be far more convenient way of investigation than experiments. For example, in order to conduct experiments to study the influence of viscosity, many different liquids are required. The direct measurements of contact angle with high speed video camera and the analysis of data is tedious and prone to errors. Modeling, based on sound governing laws, is accurate, cost effective and efficient. It provides a tool to optimize the penetration absorbency by adjusting interfacial properties and selecting proper pore radius and depth. To achieve this same goal, it would take enormous number of experiments and associated data analysis.

3. COMPUTATIONAL MODELING OF PNEUMOPHORESIS IN A WEB FORMING MACHINE

3.1. Introduction to Fiber Transport

An absorbent material can be considered as a network of pores of myriad shapes and sizes in three-dimension to facilitate fluid absorption by diffusion. Such absorbent materials are commonly called web and the machines to manufacture these web materials are known as webbers or web-forming machines (Figure 7).

Most webbers utilize pneumatic transport and deposition of short fibers to form webs. Determination of the paths that fibers or particles follow in a given air flow field is of paramount importance in determining the quality of web. There are many other applications where pneumatic transport of particles is important. These applications include mechanical separators such as cyclones, spread of pollutants in atmosphere, dispersion of droplets in internal combustion engine cylinders, steam and gas turbines, industrial fluidized bed reactors and flow through respiratory tracts. General techniques of such studies also have impact on other areas of practical importance such as geometric optics, electron microscopy, mass spectroscopy, visible ultraviolet, x-ray or electron beam lithography where photons, ions or electrons are brought into focus [10].

Albrecht [11], based on ideal fluid assumption, first calculated trajectories of particles in fluid flowing past a transverse cylinder. He included inertial effects of particles in his calculations. Later Sells [12] included the effect of stagnation streamline. Kaufman [13]

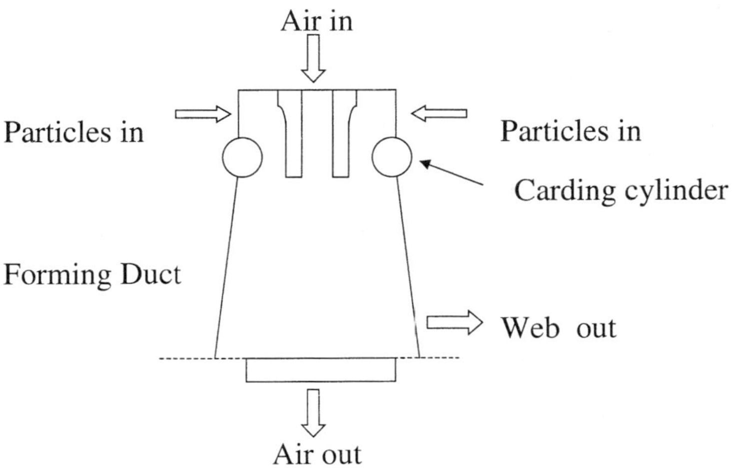

Figure 7. Schematic of a typical Webber Machine

added interception and diffusion effects of particles to ideal fluid calculations. Langmuir [14] calculated the trajectories by including the viscosity of fluid but neglected the inertia of particles in his viscous flow calculations. Later, Davies [15] included the inertia of particles in his viscous flow calculations. All these studies were based on first determining the flow field and then calculating trajectories by solving particle force balance equation. Particles were assumed to be spherical and Reynolds number [5] was assumed very low for the Stokes linear approximation [5] to drag coefficient to be valid. Morsi and Alexander [16] studied the effect of Reynolds number on the spherical drag coefficient and proposed several empirical relations between them, each valid for a different Reynolds number range.

Particles, in practice, are rarely spherical and investigations by Michaelides [17] showed that proper representation of particle drag coefficient is important in order to accurately determine the particle trajectories from the solution of particle force balance equation.

Stober et al [18] represented irregular particles by an aggregate of uniform spheres and proposed a relation between the equivalent aerodynamic diameter of the aggregate and constituent uniform spherical diameter. Sem [19] has described experimental methods to measure aerodynamic diameter of irregular particles. Knowing the aerodynamic diameter and particle density one can obtain the irregular particle's equivalent Stokes diameter [5], which is then substituted in the particle force balance equation. If, due to some reason, it is not possible to determine the aerodynamic diameter then the size of the irregular particle can be measured by microscopy. From this its volume and equivalent Stokes diameter can be calculated for a given terminal velocity. This is substituted in the force balance equation alongwith the spherical drag coefficient. On the other hand, if the actual coefficient of drag of irregular particle is available then it can be substituted in the force balance equation alongwith the particle's mass obtained from its volume and density. Clearly, it is important to accurately determine the particle density. In general, particle density, is different from the material or bulk density. For particles with porous surface, helium displacement method gives higher density than the mercury vapor displacement method and the latter is probably more accurate for pneumatic transport.

Owen [20] investigated effects of walls, gravity and shear on the particle transport. He also studied the effect of particles on turbulence of flow field. Arastoopour and Cutchin [21] proposed an experimental method to determine the particle-particle interaction term in a co-current gas flow, which can then be substituted in particle force balance equation. Hotchkiss and Hirt [22] studied particle trajectories in a transient, incompressible, viscous, 3-dimensional flow field by a numerical technique based on finite difference approximation of full Navier-Stokes equation [5].

In this section pneumatic transport of irregular particles is described. This corresponds to transport of fibers by air stream in a web-forming machine. There is no chemical reaction between particles and their carrier. The particle force balance equation includes gravity and a source term, which represents particle-particle interaction. However, diffusion terms have been neglected due to large particles traveling at higher velocities. The coupling between particle motion and flow field ensures influence of one on the other. Also, the interaction of particles with the boundaries is included in the calculations by assuming one of the four possibilities, which include reflection, escape, saltation and stagnation. The change in normal momentum due to particle reflection or saltation from a wall is accounted by the coefficient of restitution. Broom [23] has described a method to determine this experimentally for various particulate and wall materials.

3.2. Governing Equations

In order to apply the laws of conservation of mass, momentum and energy in their most basic form it is essential to first identify a system, which is a collection of matter of fixed identity and the surroundings of the system. Without this step, entities like mass, force, heat and work do not have basis. In terms of convenience, a system consisting of a fixed volume of space is better than a system of particular mass of fluid of fixed identity. This fixed volume is called control volume and the surface, which bounds the control volume is called control surface.

Figure 8 shows the control volume representation of the system under investigation. The system consists of a station where fibers are introduced, a suction pressure is applied to a vacuum box to draw air through openings at the top. The web formed by the deposited fibers comes out transversely at the bottom.

The conservation of mass in the absence of nuclear effects states that the mass of the control volume remains constant. This law, in its most basic integral form is given in equation 4. The conservation of momentum, equating the rate of change of momentum to surface forces (neglecting body forces, as there are no free surfaces), is given in equation 5. The conservation of energy, equating energy addition to the system with the work done by the system and changes in the internal and external energy level of matter in the system, is given by equation 6. In addition, some times other auxiliary laws come into picture such as perfect gas law given in equation 7.

$$\frac{\partial}{\partial t}\iiint \rho \, dV + \oiint \rho \, \bar{U} . \, d\bar{A} = 0 \qquad (4)$$

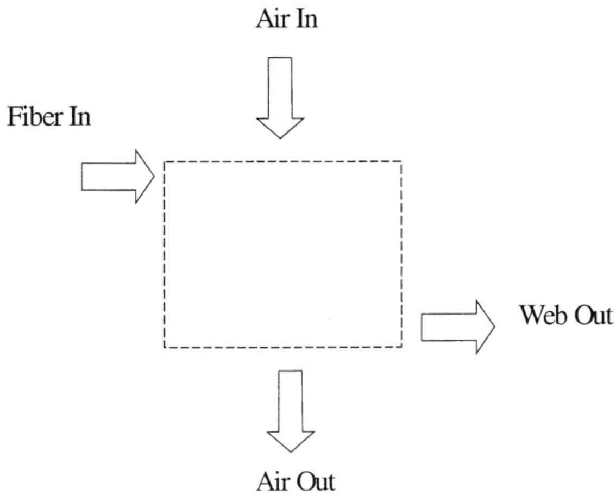

Figure. 8 Schematic of the System

$$\frac{\partial}{\partial t}\iiint \rho \bar{U} dV + \oiint \rho \bar{U} (\bar{U} . d\bar{A}) = - \oiint p d\bar{A} + \oiint \bar{\tau} dA \tag{5}$$

$$\frac{\partial}{\partial t}\iiint \rho h dV + \oiint \rho h (\bar{U} . d\bar{A}) = - \oiint dW_{total} + \oiint \bar{Q} . d\bar{A} \tag{6}$$

$$P = \rho RT \tag{7}$$

Where $\iiint$ is the volume integral, $\oiint$ is the surface integral, ρ is density, U is Velocity vector, h is enthalpy, R is universal gas constant, T is temperature, A is area, V is volume, Q is heat energy, p is pressure, τ is shear stress and W is work.

Applying these four equations [24] to the control volume we get a set of algebraic equations. Since the number of equations is equal to the number of unknowns, they can be solved uniquely. It should be noted that in many cases these equations are non-linear and the number of variables are greater than three and therefore the solution is in hyperspace which cannot be represented in 3-dimensional space. In such circumstances experimental data provides a valuable input to arrive at a starting point and also to check the analytical results.

It is, however, important to note that the solution of these equations provides a relationship between the boundary variables and does not give any information about the pneumatic transport of particles inside the web-forming machine and how that transport depends on the boundary variables. In order to get that information one has to resort to computational methods described in section 2.3 given below.

3.3. Modeling

The governing equations for the computations are called the Navier-Stokes equations [5]. These are obtained by integrating the integral form of the conservation equation for momentum around a fluid element and expressing shear stresses in terms of strain rates based on kinematics. Also, the fluid viscosity is assumed to obey Newton's law of viscosity for laminar flow cases and by eddy viscosity in turbulent flows. Eddy viscosity [5], in turn, is calculated by one of the many available phenomenological turbulence models. One of the popularly used turbulence models is called k-ε model. The Navier-Stokes equation and the differential form of mass and energy conservation equations are given below;

$$\frac{\delta}{\delta t}\rho + \nabla \cdot (\rho \bar{U}) = 0 \tag{8}$$

$$\frac{\delta}{\delta t}\rho \bar{U} + \bar{U}.\nabla (\rho \bar{U}) = -\nabla p + \mu \nabla^2 \bar{U} \tag{9}$$

$$\frac{\delta}{\delta t}\rho h + \nabla \cdot h(\rho \bar{U}) = \delta W_{shaft} + \delta W_{shear} - k \nabla T \tag{10}$$

Where, ρ is density, U is velocity vector, h is enthalpy, T is temperature, p is pressure, τ is shear stress and W is work.

There are many methods of converting these equations in to set of algebraic equations called discretizing. These methods are known as finite difference, finite element, finite volume and spectral methods. The discretized equations are simultaneous algebraic equations and can be represented in a tridiagonal matrix form. This is now amenable for solution with the help of digital computers with the help of many commercially available flow analysis software packages. The solution technique involves first generating the geometry of the domain of interest then converting the domain in to a grid and specifying proper boundary and initial conditions [24] and then solving the equations in an iterative way to get flow field.

Particle trajectories are obtained by solving the equation of particle motion in Lagrangian frame [5]. The particle force balance equation is given by;

$$\frac{d}{dt}(\bar{U}_P) = F(\bar{U} - \bar{U}_P + \bar{u}') + \bar{g} + \bar{s} \tag{11}$$

Where U_P is particle velocity, U is fluid velocity, u is fluid velocity fluctuation, g is gravitational acceleration, s is any other body force that may be present and F is a constant which is given by;

$$F = 18\mu \; C_D \frac{R_e}{24 \; \rho_P D_P^2} = \frac{1}{S_t} \tag{12}$$

Particle trajectories strongly depend on a non-dimensional number called Stokes number [15], which represents the ratio of particle kinetic energy and work done against viscous drag. As mentioned before, it is very important to accurately determine the particle density, size and coefficient of drag before solving the particle dynamics equation either coupled or uncoupled with the main flow equations.

Now we have a set of equations, which can be applied to model the fluid flow and fiber transport in a webber machine. This computational modeling allows us to study the effects of changes in the geometry of the machine's web forming chamber, the vacuum pressure in the chamber which in turn controls the fluid velocity and also to investigate the effects of fiber characteristics such as fiber length, density and concentration.

Numerical programs to solve fluid flow problems have matrices, each of them containing many arrays. Their sizes depend on the problem. They are particularly large for three-dimensional cases. It is therefore essential that a proper computer system be selected to solve such problems because a substantial disk space and memory are required. Hardware also must have a high integer and floating-point processing capabilities combined with high overall throughput.

As in previous example, these calculations usually generate large quantities of data and therefore a good scientific data visualization capability is essential in order to derive meaningful conclusions from data. The visualization capabilities of the system depend on the speed to draw two and three dimensional vectors and shaded polygons of a given pixel density. Other highly desirable characteristics of a good visualization system include lighting, rotation, pan and zoom capabilities. The visualization system must also include the capabilities to quickly draw color hue and contour maps, vector plots and particle trajectories.

3.4. Results of Modeling

Contours of a typical pressure distribution for a single-phase flow are shown in Figure 9. This was obtained by taking a slice in the middle. The pressure at the inlet drops from atmospheric to 1.25 kilo Pascals below atmosphere. As the flow accelerates, the static pressure drops rapidly. It starts to increase back again reaching a value of 2.49 kilo Pascals below atmosphere, which is equal to the suction pressure applied at the bottom of the system.

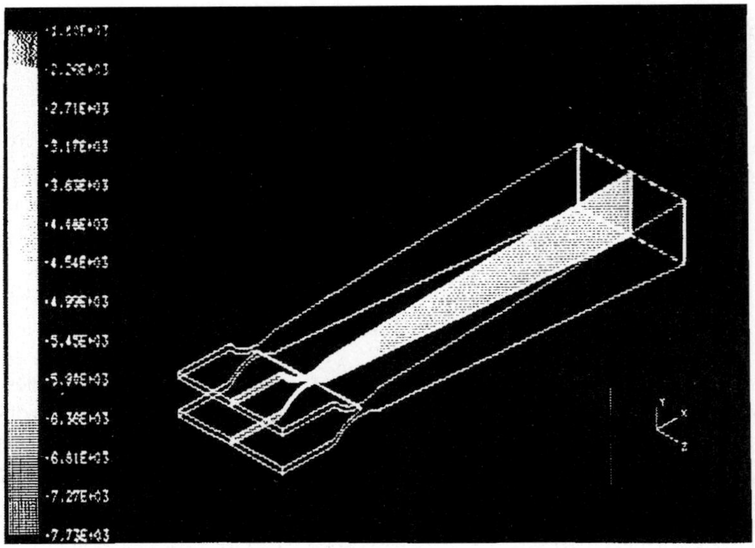

Figure. 9 Computed Pressure Contours

Pressure contours along any plane can be visualized by taking an appropriate slice. Results showed that the pressure distribution was fairly symmetric except near the inlet and outlet due to three-dimensional boundary layer effects.

The pressure contours for a two-phase flow with dispersed second phase of fibers show that they are generally higher than the single-phase case. Also, there is an asymmetry in the distribution near outlet in the two-phase flow case due to the traversing outlet.

Figure 10 shows the velocity distribution in x-direction. This is the U-velocity component. It is about 30 m/sec near inlet then accelerates to about 90 m/sec near throat and then decelerates to about 5 m/sec as it approaches the outlet. Image enhancement clearly shows the growth of boundary layer along wall. Velocity components in y and z directions namely, V and W components respectively can be similarly visualized along any selected plane.

The U-velocity component for the coupled case of two-phase flow when compared to single- phase case shows the influence of fibers as in case of pressure distribution. The velocity distribution is asymmetric near outlet due to traversing outlet. Velocity contours for both coupled and uncoupled cases indicate two high-speed jets and a mixing layer separating them. This mixing layer gradually fills the entire space as outlet is approached.

Contours of kinetic energy of turbulence for single-phase flow were calculated based on an inlet turbulence intensity of 5% and a characteristic length equal to inlet length. The kinetic energy reaches a maximum at the beginning of the mixing layer. As the mixing layer grows, the kinetic energy of turbulence decreases and also becomes uniform across the width of the system under consideration. At the outlet it is at a level which is close to that specified

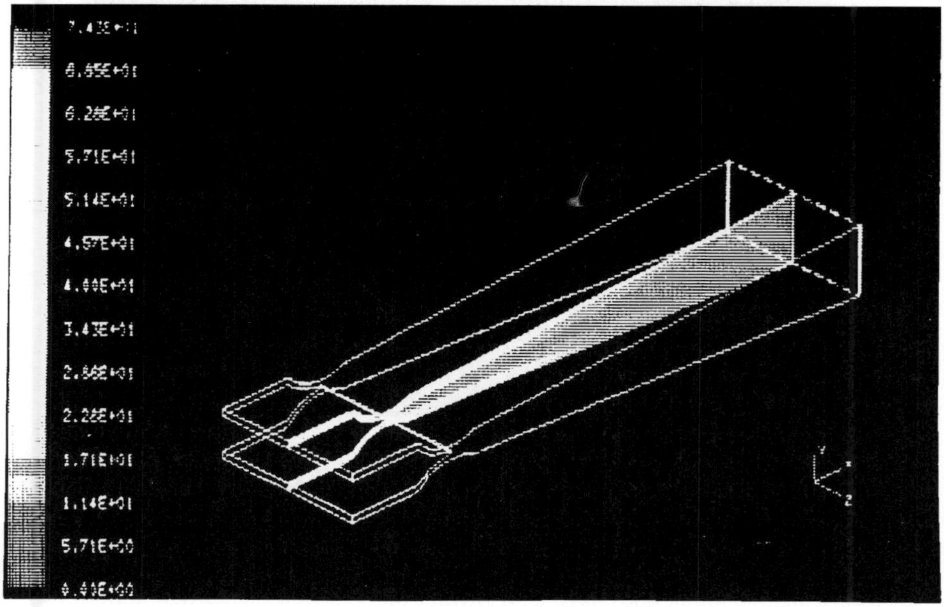

Figure. 10 Velocity Profiles from Modeling

at the inlet. Assuming isotropic turbulence, the kinetic energy information can be converted into turbulence intensity information. The maximum intensity is shown to be about 9% near the start of mixing layer.

Contours of kinetic energy of turbulence for two-phase flow increases from inlet to the beginning of the mixing layer and then decreases as the outlet is approached. Although the kinetic energy decreases near the outlet, the turbulence intensity increases due to lower mean velocity near the outlet. Also, the kinetic energy of turbulence contours exhibits asymmetry near outlet for two-phase flow just as the pressure and velocity distributions did due to the traversing outlet.

Particle trajectories were calculated for various cases, which include different flow conditions and particle properties. Typical trajectories for uniform particles of a given density and diameter, are shown in Figure 11. This figure also shows the effects of turbulence, which was included by stochastic approach. Particles were injected at discrete locations and these particles, due to turbulence, could take any one of the five trajectories shown in the figure. The mean paths were obtained by averaging the five tries. All these results are based on uncoupled calculations, which means that the continuous phase was first completely solved and then particle trajectories were calculated assuming no influence of the dispersed phase on the continuous phase. Similar uncoupled calculations were repeated with different coefficient of restitution. Comparison of these results indicates a significant influence of coefficient of restitution on particle trajectories.

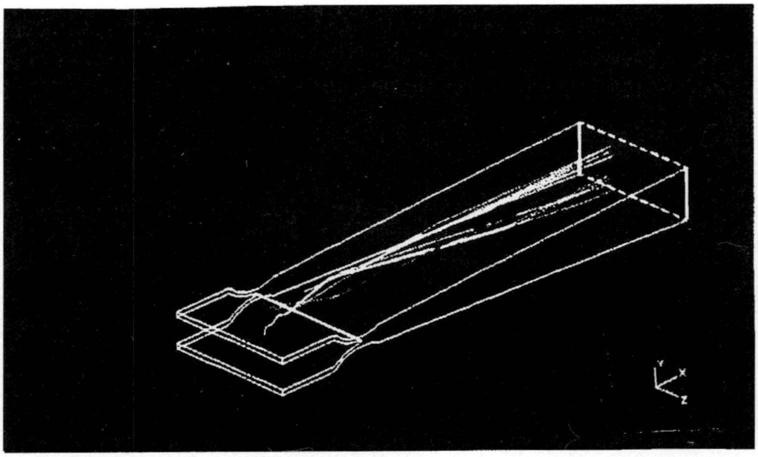

Figure. 11 Particle Trajectories from Computational Modeling

Calculations were repeated by coupling the continuous and dispersed phases. These results show that coupling does have a significant influence on the particle trajectories. It is therefore important to include coupling despite the fact that it increases computation time considerably.

3.5. Validation of Modeling

Pressure measurements were obtained [25] in the web-forming machine by several strategically located static pressure ports. These pressure ports were of diameter, which minimized aerodynamic effects and time lag. There were ports to measure temperature and also to mount flow visualization devices like Fiberscope or Videoimagescope. Pressure distribution inside the forming chamber was measured by traversing a static pressure probe. All the pressure ports and the traverse probe were connected to a multi-tube manometer with a least count of 0.01 inches of water. Observation of water columns of the multi-tube manometer did not show any low frequency, high amplitude fluctuations. This indicated a steady flow field inside the system.

A typical pressure distribution is shown in Figure 12. The static pressure is about 5 inches of water below atmosphere in the inlet section and then drops precipitously to about 30 inches of water near the start of fiberizing stations before increasing almost as rapidly to about 20 inches of water. The pressure stays almost constant for some distance downstream before dipping slightly due to change in the geometry and then rises again due to the stagnation region.

The pressure distribution obtained by computation is also shown on the same figure. It can be seen that despite complexities of the flow field, trends predicted by computation agrees well with experiments. Accuracy of the computation could be further improved by finer grid and complex turbulence models, however, each of them adds enormously to computation time especially in view of this being a three-dimensional problem.

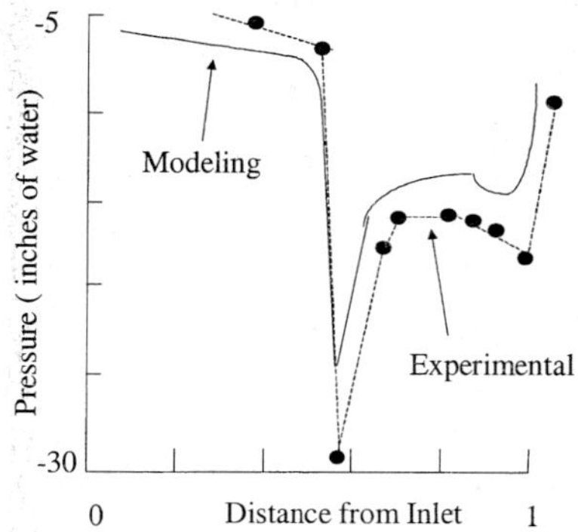

Figure 12. Pressure Distribution within Web-Forming Machine

Fiberscope and Videoimagescope were utilized to visualize flow of particles in the air stream. Two different fiberscope probes, both 11 mm in diameter, one with 120^0 and the other with 40^0 field of view were tried. Also, two different light sources, one a steady light source with 300 watt intensity and another a stroboscopic light source with a maximum of 3000 flashes per minute each of 1 watt intensity, were utilized. Fiberscope was connected to a Kodak Ektapro 1000 imager to record the pictures from 125 frames per second to 6000 frames per second. A special VHS tape was used for recording and Kodak Ektapro 1000 Motion Analyzer was used for playing back. Frozen frames could be printed to get hard copies.

Videoimagescope is a newer probe with CCD built at the tip to minimize image distortion and increase magnification. It is 12.6 mm in diameter and could be remotely articulated to move the tip around. It was used in conjunction with a camera control unit and the stroboscopic light source. The images were seen on a color video monitor and recorded on a regular VHS tape for replay.

Images obtained from Videoimagescope were clearer than those from Fiberscope. Rotating parts of the web-forming machine could be seen with excellent clarity. However, motion of individual fiber particles could not be seen due to low particle concentrations, small particle sizes and high air velocities.

3.6. Advantages of Modeling

It is evident that the performance of a complex machine such as this web-forming machine depends on many parameters which, in turn control the quality of the end product. It would take enormous efforts and expense to empirically optimize all these parameters because experiments would require building many such expensive machines to test variations in each parameter. Modeling, on the other hand, provides an effective, economical and convenient means of conducting parametric studies to optimize the performance and improve the quality of the end product.

Modeling also helped in identifying those parameters having strong influence on the fiber particle trajectory. Particle trajectory calculations indicated that turbulence and coefficient of restitution play a significant role in addition to particle density, coefficient of drag and flow field. Calculations also showed that coupling between the dispersed phase and continuous phase is important to both the flow field and particle trajectory.

Computations proved to peerless in visualizing particle motion, which could not be visualized by experiments despite the use of the state of the art Fiberscope and Videoimagescope.

4. CONCLUSIONS

Two important areas in Absorbent Technology namely, penetration absorbency and pneumophoresis of fibers in a web-forming machine were analyzed by computational modeling and the results were validated with experiments. This study showed that modeling, based on sound governing laws, is accurate, cost effective and efficient and far more convenient way of investigation than experiments. For example, in order to conduct experiments to study the influence of viscosity, many different liquids are required. The direct measurements of contact angle with high speed video camera and the analysis of data is tedious and prone to errors. Modeling also provides a tool to optimize the penetration absorbency by adjusting interfacial properties and selecting proper pore radius and depth. To achieve this same goal, it would take enormous number of experiments and associated data analysis.

Similarly, modeling provided an effective, economical and convenient means of conducting parametric studies to optimize the performance of a complex web-forming machine and improve the quality of the end product. It would have taken enormous efforts and expenses to empirically optimize all these parameters because experiments would require building many such expensive machines to test variations in each parameter.

Modeling also helped in identifying those parameters having strong influence on the fiber particle trajectory. These included turbulence, coefficient of restitution, particle density, coefficient of drag and flow field. Calculations also showed that coupling between the dispersed phase and continuous phase played an important role in determining flow field and particle trajectory.

Modeling made it possible to visualize particle trajectories, which were otherwise impossible to visualize by experiments despite the use of state of the art Fiberscope and Videoimagescope.

5. ACKNOWLEDGMENTS

The author would like to thank Dr. John Comerford, Mr. Ony Chinwala and Late Mr. Van Painter for valuable advice and encouragement. His gratitude is due to Dr. Pronoy Chatterjee for many useful discussions and also for the invitation to write this chapter. Thanks are also due to Mr. Eric Grald of Fluent, Inc., Prof. Arastoopour of Illinois Institute of Technology and Mr. Gerry Marshall for their ready help on many occasions. Thoughtful criticisms and helpful suggestions by many friends and colleagues, over the years, have greatly added to the material covered in this chapter. As always, the author is indebted to his wife Sita and daughter Shreya for all the support and encouragement and to his parents for always being there.

6. GLOSSARY

A	Area Vector
C_D	Coefficient of Drag
d	Depth of pore
D_P	Particle Diameter
g	Acceleration due to Gravity
h	Enthalpy
k	Thermal Conductivity
P_b	External Pressure
P_w	Pressure due to Weight
P_s	Capillary Pressure
Q	Heat energy
r_c	Capillary Radius
R	Gas Constant
Re	Reynolds Number
S_t	Stokes Number
t	Time
T	Temperature
u'	Turbulent Velocity Fluctuation
U	Velocity Vector
U_P	Particle Velocity Vector
V	Volume
W	Work
θ	Contact Angle
γ	Surface Tension
μ	Viscosity
ρ	Density
ρ_P	Particle Density
τ	Shear Stress

7. REFERENCES

1. Goodrich, F. C. and Rusanov, A. L., "The Modern Theory of Capillarity," Akademic-Verlag, Berlin, 1981.
2. Gould, R. F., "Contact Angle, Wettability and Adhesion," American Chemical Society, 443, Washington, D.C. 1964.
3. Miller, B., "Experimental Aspects of Fiber Wetting and Liquid Movement between Fibers," Chapter IV in Absorbency, edited by Chatterjee, P. K., Elsvier, 1985.
4. Chatterjee, P. K., and Nguyen, H. V., "Mechanism of Liquid Flow and Structure Property Relationship," Chapter II in Absorbency, edited by Chatterjee, P. K., Elsvier, 1985.
5. Schlichting, H., "Boundary Layer Theory," McGraw Hill, 6th Edition, 1968.
6. Washburn, E. W., Phys. Rev. 17-273, 1921.
7. Kangovi, S., "Apertured Cover Optimization," Unpublished Work, 1994.
8. DeMarco, T., "Structured Analysis and System Specifications," Prentice Hall, 1979.
9. Kangovi, S. and Abraham, G., " Dynamic Surface Wettability and Absorbency of Covers," Unpublished Work, 1995.
10. De La Mora, J. F., and Riesco-Chueca, P., "Aerodynamic Focussing of Particles in a Carrier Gas," J. of Fluid Mech., Vol. 195, pp. 1-21, 1988.

11. Albrecht, F., "Theoretische Untersuchungen uber die ablegerung von Staub und Luft und ihre Anwendung auf die Theorie der Staubfilter," Physik. Zeits. 32,48, 1931.
12. Sell, W., "Staubausscheidung an einfachern Korpern und in Luftfiltern," V. D. I Forscung. Heft, 347, 1, 1931.
13. Kaufmann, A., "Die Faserstoffe fur Atemschulzfilter," Z. Verein Deutsches Ing., 80, 593, 1936.
14. Langmuir, I., "Report on Smokes and Filters," Section I, U. S. Office of Scientific Research and Development, No. 865, Part IV, 1942.
15. Davis, C. N., "Separation of Airborne Dust and Particles," Proc. Inst. Mech. Eng., 1B(5), pp. 185-213, 1952.
16. Morsi, S. A. and Alexander, A. J., "An Investigation of Particle Trajectories in Two-Phase Flow Systems," J. Fluid Mech., Vol. 55, Part 2, pp. 193-208, 1972.
17. Michaelides, E. E., "On the Drag Coefficient and the Correct Integration of the Equation of Motion of Particles in Gases," J. of Fluids Eng., Transactions of ASME, Vol. 110, pp. 339-341, 1988.
18. Stober, W., Berner, A. and Blaschke, R., "The Aerodynamic Diameter of Aggregates of Uniform Spheres," J. of Colloid and Interface Science, Vol. 29, No. 4, 1969.
19. Sem, G. J., "Aerodynamic Particle Size: Why Is It Important?," TSI Quaterly, Vol. X, Issue 3, 1984.
20. Owen, P. R., "Pneumatic Transport," J. Fluid Mech., Vol. 39, Part 2, pp. 407-432, 1969.
21. Arastoopour, H. and Cutchin, J. H., "Measurement and Analysis of Particle-Particle Interaction in a Cocurrent Flow of Particles in a Dilute Gas-Solid System," Chem. Eng. Sci., Vol. 40, No. 77, pp. 1135-1143, 1985.
22. Hotchkiss, R. S. and Hirt, C. W., "Particulate Transport in Highly Distorted Three-Dimensional Flow Fields," Proc. Of 1972 Summer Computer Simulation Conference, San Diego, Calif., Vol. 2, Group 5, pp. 1037-1046, 1972.
23. Broom, G. P., "Adhesion of Particles to Fibrous Air Filters," Proc. Of the Filtration Soc., pp. 661-669, 1979.
24. Patankar, S. V., "Numerical Heat Transfer and Fluid Flow," Hemisphere Publishing Corp., 1980.
25. Kangovi, S., " Modeling of a Fiber Webber and Comparison with Experiments," Unpublished Work, 1993.

Absorbent Technology.
P.K. Chatterjee and B.S. Gupta, editors.

CHAPTER V

THE ROLE OF SURFACTANTS

JOHN C. BERG
Department of Chemical Engineering, Box 351750, University of Washington, Seattle, WA 98195-1750 (U.S.A.)

Contents

1. INTRODUCTION

Absorbency refers to the ability of a porous solid material to take up and retain under various conditions significant amounts of water or other liquids. As discussed more fully elsewhere in this monograph, absorbency depends upon the bulk properties of the solid (i.e., resistance to compression-expansion, shear and torsional distortion), the structure of the solid matrix (i.e., pore size, size distribution, shape, texture and roughness of the solid surfaces) and the bulk properties of the liquid (e.g., viscosity). It also depends critically upon the specific chemical makeup of the three interfaces involved in the absorption process: liquid-gas, solid-gas and solid-liquid. While surface active agents (surfactants) may play a minor indirect role in affecting the bulk material properties of the porous structure of the solid substrate, they play a major role in determining the properties of the interfaces involved. This chapter considers the manner in which interfacial properties influence absorbency and the manner in which surfactants influence the interfacial properties. These two aspects of the role of surfactants in absorbency are tightly interrelated, and even the simple question of whether absorbency is enhanced or retarded by a surfactant in a given situation often cannot be determined without consideration of details. Nonetheless, for a given solid matrix and given bulk liquid, the use of surfactants provides one of the most powerful means available to the practitioner for altering absorbency, either to promote it or to suppress it.

This chapter considers first the definition and general properties of surfactants, next the nature of their adsorption to both fluid and solid interfaces followed by a discussion of how such adsorption influences absorbency. It concludes by considering specifically the use of surfactants to promote absorbency and the future prospects for their further use and development.

A number of products are available whose end use is based upon their high absorbency. A brief list is given in Table 1. Surfactants have found frequent use in the preparation of such products, both to assist in the attainment of high absorbency and in its preservation over extended periods of shelf time. A second category of products might be

Table 1.
Examples of Products designed for High Absorbency

Baby diapers	Kitchen towels
Surgical dressings	Blotting paper
Absorbent cotton	Industrial wipes
Sanitary napkins	Oil cleanup absorbents
Incontinent pads	Fuel wicks

those for which absorbency per se does not constitute the end use, but for which high absorbency is essential. These would include all types of liquid-dispersible powders. Surfactants are widely used in the preparation of such materials, particularly agricultural chemicals such as wettable-powder pesticides and herbicides. A third type of product includes those in which a moderate level of absorbency is desirable. The most important of such materials are textiles, where surfactants are often used to impart soil release, antistatic and wearing comfort properties, all intimately related to moisture absorbency. Finally, surfactants are often used to reduce or eliminate a porous solid's absorbency to water or other liquids. Examples of the latter include the waterproofing of textiles, the sizing of cellulosic materials for paper and textiles and the waterproofing of masonry for weather protection. The present work will concern itself mainly with those situations where an enhancement of absorbency is desired, although the same principles apply regardless of the practical objectives.

2. SURFACTANTS

A surfactant may be defined as a substance which, when present at a low overall concentration in a system, exhibits a relatively high concentration at one or more of the system interfaces (i.e., high adsorption) and thereby significantly alters the free energy of such interfaces. Surfactant molecules are invariably constituted of one or more functional groups which are strongly attracted to the bulk solvent medium containing them (lyophilic groups) and one or more groups with little attraction for the medium (lyophobic groups). Furthermore, these groups must be adequately segregated from one another in the molecule. Thus, as shown in Fig. 1, l-hexanol is a surfactant at the aqueous interface against air because its hydrophilic hydroxyl group is clearly segregated from the hydrophobic hydrocarbon portion of the molecule, whereas glucose is not a surfactant because its hydroxyl groups and hydrocarbon portions are enmeshed together in the molecule. In the broadest sense, surfactants may be macromolecular or polymeric in that they may contain, for example, lyophilic groups only at adequately spaced locations on the molecule (block copolymers) or, in the case of a linear polymer, may have all their lyophilic functionality on the same side of the chain. While such molecules may fit our definition of surfactant, their use will not be considered in the present chapter. The use of such materials, particularly the polyelectrolytes, is considered elsewhere in the monograph. We shall confine our attention to oligomeric materials, generally in the molecular weight range between 50 and 1000. Surfactants are sometimes used as starting materials in the chemical modification of solid surfaces wherein

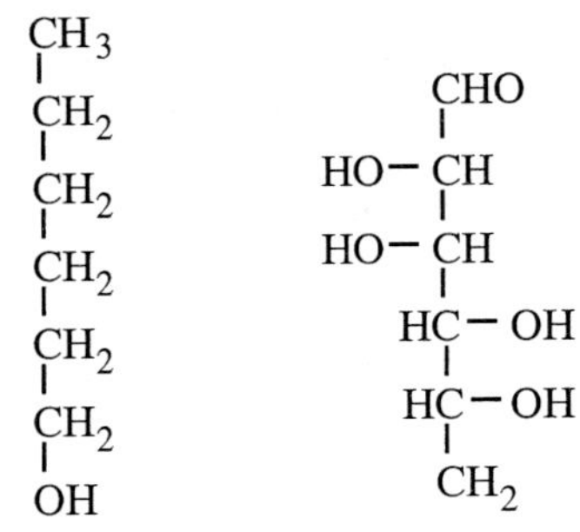

(a) (b)

Fig. 1. Molecular structures of (a) 1-hexanol, which is surface active, and (b) glucose, which is not surface active.

desired-functional groups are grafted to these surfaces. Surfactants used in such a context are also treated elsewhere in the monograph and will be largely excluded from the present chapter.

2.1. Classification of Surfactants

The chemical variety of surfactants available is enormous and growing rapidly. It is therefore useful to attempt to develop schemes of classification. It is common to classify surfactants in aqueous media in terms of the nature of their hydrophilic functional groups, although this is far from adequate for predicting the performance of the material in a given situation. The broadest subdivision is as follows:

2.1.1. Anionic surfactants

These are materials in which the dissolved surfactant bears a negative charge. The anions of alkali metal salts of fatty acids (soaps) are typical, as are the anions of long chain sulfonates, sulfates, and phosphates. Anionics constitute about three-quarters of current U.S. consumption of surfactants [1], their dominant position deriving in part from their generally lower cost than other surfactants.

2.1.2. Cationic surfactants

These materials bear a positive charge in solution and most often consist of amine salts or quaternary ammonium or pyridinium compounds. They are generally expensive to produce and account for only approximately six percent of current U.S. consumption [1]. There are nonetheless a number of applications to which they are uniquely appropriate, deriving principally from the fact that the preponderance of natural solid surfaces in contact with water at pH $\geq$ 5 are to some extent negatively charged.

2.1.3. Nonionic surfactants

This term usually refers to polyoxyethylene compounds, but includes also other materials such as sugar esters, fatty alkanolamides and amine oxides. The most common types are prepared by addition of ethylene oxide to compounds containing one or more active hydrogens, such as alkyl phenols, fatty alcohols, acids, mercaptans, amines, amides and polyols. These latter compounds, without ethylene oxide addition, are also sometimes considered nonionic surfactants. Nonionics make up most of the balance of surfactant consumption. Their almost limitless versatility is leading to a growth in their use, although their cost is often twice that of the corresponding anionic.

2.1.4. Amphoteric surfactants

These compounds contain functional groups (usually carboxyl, -COOH, and amino, -NH_2) which ionize in accord with the solution pH. At low pH values they are cationic, while at high pH-values they are anionic.

2.1.5. Zwitterionic surfactants

These relatively uncommon surfactants are those with both positive and negative charges in the hydrophilic portions of the molecule. Examples include sulfobetaine and long-chain phosphonyl cholines (lecithins).

Among the three major types of surfactants listed above, certain specific compounds have emerged as prototypes for their respective groups and have been investigated extensively in a large number of laboratories. Some are listed in Table 2. Many surfactants used in practice are much more complex in their structure than these "prototypes," but many general aspects of their behavior are represented well by the behavior of these specific compounds.

2.2. Hydrophobe Structure and Other Complexities

The examples of surfactants mentioned are simple in that their hydrophobes have all been straight-chain alkyl or alkylbenzene radicals, and the classification as to type has been based upon the hydrophilic groups alone. While many of the hydrophobes encountered in practice are of the above types, many other possibilities are also used, including branched-chain alkyl groups, alkyl naphthalene residues, rosin derivatives, perfluoroalkyl groups and polysiloxane groups. The latter two types are especially important as they often impart significant surface activity to surfactants in nonaqueous media. While the differences in properties of surfactants are generally more sensitive to hydrophile than to hydrophobe type, the latter play a highly significant role in the properties of surfactants used to promote absorbency.

Other complexities in surfactant structure also commonly arise. Many surfactants, both ionic and nonionic, contain multiple hydrophilic groups, the multiplicity itself imparting special properties. As important and growing class of surfactants consists of ionics (either anionic or cationic) which are polyoxyethylated to various extents. These compounds have significantly different properties from the simple ionics.

2.3. Surface Activity

The numerous property changes imparted by surfactants to systems which contain them are known collectively as their "surface activity." This generally includes their ability to lower surface tension, to promote wetting and spreading, to stabilize foams, sols and emulsions, to render detergency, etc. Not all of these aspects of surface activity are directly relevant to absorbency, so a general review of the various aspects of surface activity and their relationships to specific surfactant type and structure would not be appropriate here. The

Table 2.
Examples of Principal Surfactant Types

1. *Anionics*
   (a) Sodium stearate (lye soap)
   $C_{17}H_{35}COO^{-}Na^{+}$
   (b) Sodium dodecyl sulfate (SDS)
   $C_{12}H_{25}OSO_3^{-}Na^{+}$
   (c) Sodium dodecylbenzene sulfonate (SDBS)

   (d) Sodium dioctyl sulfosuccinate
   $C_8H_{17}OOCCHSO_3^{-}Na^{+}$
   $C_8H_{17}OOCCH_2$

2. *Cationics*
   (a) Hexadecytlrimenthylammonium bromide (CTAB)
   $C_{16}H_{33}N(CH_3)_3^{+}Br^{-}$
   (b) Dodecylpyridinium chloride

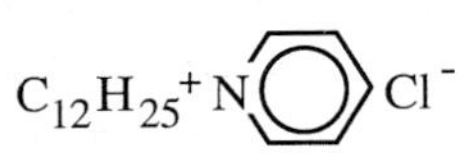

3. *Nonionics*
   (a) Alkylpolyoxyethylene alcohols
   $C_nH_{2n+1}(OCH_2CH_2)_xOH$
   (b) Polyoxyethylene alkylphenols

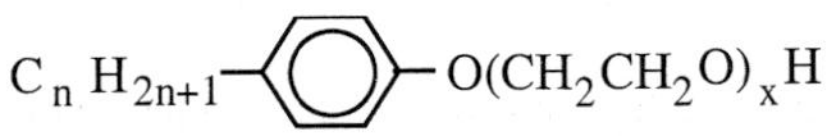

particular relationships that exist between surfactant type and absorbency are developed below. Excellent general treatments of surface activity can be found elsewhere [1-3].

3. GENERAL ASPECTS OF SURFACTANT ADSORPTION EQUILIBRIA

Surfactants influence absorbency in a number of different ways, but all are traceable to the fact that these molecules adsorb strongly at one or more of the interfaces in the system. Since the phenomenon of absorption refers to the displacement of air from a porous solid by a liquid (usually water or an aqueous solution), the interfaces involved are the liquid-air

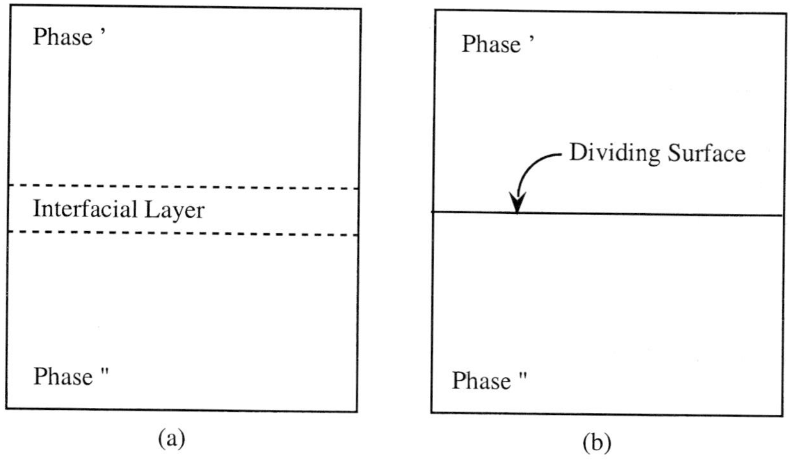

Fig. 2. Gibbs dividing surface model of a capillary systems. (a) Actual system; (b) Gibbs model.

surface at the advancing liquid front, the solid-air surface (which is being destroyed) and the solid-liquid interface (which is being extended).

3.1. Thermodynamic Definition of Adsorption

Adsorption refers both to the equilibrium difference which generally exists between the composition of an "interface," or "interfacial layer," and the bulk phases it divides and to the process by which such equilibrium distribution is achieved. In a multicomponent system, those components which are present in the interface in greater density than they are in the bulk phases (or at least in one of the bulk phases) are said to be "positively adsorbed," while components shunning the interface are "negatively adsorbed." Surfactants, by definition, are compounds capable of strong positive adsorption. At the outset in discussing adsorption equilibrium, one should make no appeal to any specific structure for the "adsorbed layer," but should instead define it in unambiguous thermodynamic terms. Following this, one may profitably look in some detail at the molecular structure of adsorption, its mechanisms and the means by which it influences absorbency.

In order to make the concept of adsorption quantitative, it is necessary to construct a model for a system consisting of a patch of interface and at least small portions of the bulk phases adjoining it. The most commonly-used of such models is that of the "Gibbs dividing surface," as pictured in Fig. 2, in which the interfacial layer is replaced by a mathematical surface (zero thickness) to which, nonetheless, masses of the various components in the system may be ascribed. The bulk phase compositions are taken as uniform right up to the dividing surface, and in order to assure mass equivalence between the actual system and the model, the needed amounts of each component to assure such equivalence are assigned to the dividing surface. Taken on a per unit area basis, these amounts are termed the "surface excesses". For example, Γ_2 = the surface excess of component 2 and has units of moles/area. It is evident that the various surface excesses are extremely sensitive to the (at least somewhat) arbitrary location of the dividing surface and as such are unreliable measures of any true distributions of the particular components between the interfacial layer and the bulk phases. Gibbs thus defined the "relative adsorption" of one component with respect to another, a quantity which proves to be invariant with respect to dividing surface location. $\Gamma_{2,1}$ is the relative adsorption of component 2 with respect to component 1 and is the surface excess of component 2 when the dividing surface is located such that the surface excess of component 1 is zero. It is common to designate component 1 as the "solvent" and to reckon relative adsorptions of the various solutes to it. For the interfaces between a liquid solution and a gaseous phase or a solid phase there is usually little ambiguity in designating the solvent. At any rate, when one refers to the "adsorption of a component at an interface," it is the relative adsorption, as defined above, that is (or ought to be) referred to.

Using material balances for the various components, one may show that, with complete generality [4]:

$$\Gamma_{i,1} = \Gamma_i - \Gamma_1 \left[\frac{C_i' - C_i''}{C_1' - C_1''} \right] \tag{1}$$

$$= \frac{1}{A} \left[(n_i - VC_i') - (n_1 - VC_1') \left(\frac{C_i' - C_i''}{C_1' - C_1''} \right) \right] \tag{2}$$

where C_1' is the molarity of component i in phase ', etc., n_i is the total number of moles of component i in the system, V is the total volume of the system, and A is the interfacial area. Note that the "system" refers to the capillary system, as shown in Fig. 2, consisting of interface of area A and some arbitrary but defined portions of the adjacent bulk phases. Examination of eq. 2 shows that the relative adsorption is indeed invariant with dividing surface location and can be evaluated, at least in principle, in terms of quantities accessible in the laboratory. Further discussion of the concept of relative adsorption may be found in ref. 4.

For the types of interfacial systems of interest in the study of surfactant effects on absorbency, eq. 1 and 2 may be simplified considerably. For both the solid-liquid and liquid-gas systems of interest, the dividing surface may be located without arbitrariness at the boundary between the phases, and the "system" may be defined to exclude either the solid phase (but not its influence) or the gas phase, respectively. This is true because both the liquid solvent and the surfactant solute *i* may be regarded as essentially absent from either the solid phase or the gas phase. In the latter case we assume both the solvent and the surfactant have negligible volatility. Adsorption thus occurs exclusively from the liquid phase to either type of interface. Then, taking phase ' to be the liquid solution and dropping the superscript, we have:

$$\Gamma_{i,1} = \Gamma_i - \Gamma \frac{C_i}{C_1} \tag{1a}$$

$$= \frac{1}{A} \left[(V^\circ C_i^\circ - VC_i) - (V^\circ C_1^\circ - VC_1) \frac{C_i}{C_1} \right] \tag{2a}$$

where superscript ° refers to conditions existing before any adsorption has taken place, and other quantities refer to conditions after adsorption equilibrium has been established. Equations 1a and 2a show clearly the competitive nature of adsorption from liquid solutions. It cannot in general be regarded as the deposition on the interface of a solute from a indifferent environment, but is instead the net outcome of competition between solute and solvent for the adsorbent area available. If the competition results in a tie, so that the concentration ratio of solute to solvent as the interface Γ_i/Γ_1, is the same as the concentration ratio in the bulk liquid phase, C_i/C_1, we see from eq. 1a that the adsorption is zero.

If the solution is highly dilute, as is often the case for surfactant solutes under conditions of practical interest, the above equations reduce further, viz.:

$$\Gamma_{i,1} = \Gamma_i \tag{1b}$$

$$= \frac{V}{A}\left[C_i^{\circ} - C_i\right] \tag{2b}$$

In this case, adsorption can be directly identified with the surface concentration of solute (adsorbate), but not otherwise.

In the context of our present assumptions, the gas-solid interface appears to be a trivial case. Surfactants certainly may be present at such interfaces, but since their presence within either the solid or the gaseous phase has been assumed negligible, adsorption equilibria in such systems appear not to be relevant. Equation 1b becomes rigorously true since any surfactant in such a system exists only at the interface. Films of surfactant at such interfaces may be formed by adsorption from a liquid solution and retained when the liquid recedes, as shown in Fig. 3a. They may also be formed by direct "plating" from the liquid gas interface as it recedes by a process described in detail by Langmuir and Blodgett and shown in Fig. 3b. It is sometimes postulated that a monolayer of surfactant may spread or diffuse on the surface spontaneously out onto the gas-solid interface from the interline with the liquid solutioin as shown in Fig. 3c, although unambiguous confirmation of such a process is still lacking. While such a process is thermodynamically feasible, it is likely that its rate will be very low. Finally, there are some cases of interest in which the surfactant has sufficient volatility to be deposited onto the solid directly from the vapor phase, as shown in Fig. 3d, but such vapor deposition is not usually an equilibrium adsorption process in cases of present interest.

3.2. Adsorption Isotherms

Adsorption equilibria are usually described in terms of "adsorption isotherms," which are relationships holding between the relative adsorption and the concentration of the corresponding component in the bulk phase at constant temperature, e.g., for a single-solute system:

$$\left[\Gamma_{2,1} = f(C_2)\right]_{\text{const T}} \tag{3}$$

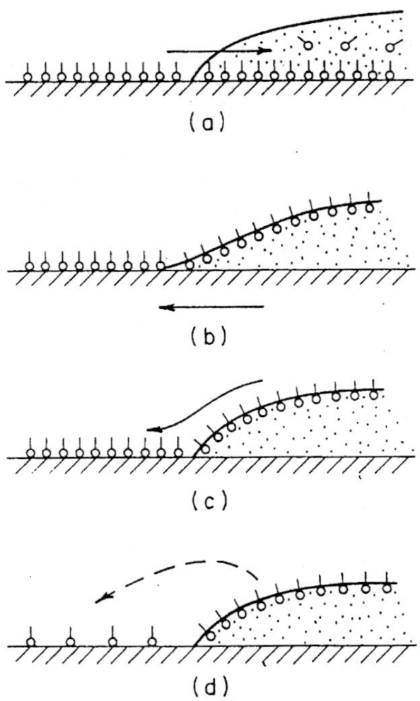

Fig. 3. Schematic representation of modes of formation of a surfactant film at an air solid interface. (a) Retained from liquid-solid interface as interline recedes, (b) plated from gas-liquid surface as interline recedes, (c) surface diffusion from gas-liquid surface (shown) or liquid-solid interface, (d) Evaporation from liquid followed by deposition.

For multi-solute systems, the adsorption isotherm will also depend on the other solutes present, i.e.,

$$\left[\Gamma_{2,1} = f(C_2, C_3, \ldots)\right]_{\text{const } T} \tag{4}$$

Many systems (perhaps most) of interest may be treated as effectively binary. A complete thermodynamic description of the adsorption equilibria consists of a family of isotherms covering the temperature range of interest. The "heat of adsorption" may be obtained from such a family of isotherms. The most commonly-used quantity is the "isosteric heat of adsorption," defined as the heat absorbed per mole of adsorbate at constant $\Gamma_{2,1}$. For dilute systems, this is given by:

$$Q_{\text{ads}} = -RT^2 \left(\frac{\partial \ln C_2}{\partial T} \right)_{\Gamma_{2,1}}, \tag{5}$$

which is a surface analog to the Clausius-Clapeyron equation. Heats of physical (as opposed to chemical) adsorption at the liquid-solid interface are usually negative and, for the case of adsorption from liquid solutions, usually quite small (less than a few kcal/mole adsorbed).

A fundamental difference between the study of solid-liquid systems and liquid-gas systems is that in the former, the isotherms may generally be obtained directly from laboratory measurements, but in the latter they may not. For solid-liquid systems, Equation 2a or 2b indicates that $\Gamma_{2,1}$ values may be obtained "stoichiometrically" by measuring the liquid phase solute content before and after adsorption. This can be done because it is usually possible to obtain the solid (the absorbent) in finely-divided and/or porous form so that *A* is very large. The total adsorption will then be correspondingly large enough so that a quantity such as $C_i^\circ - C_i$ will be measurable. If $\Gamma_{2,1}$ is to be obtained, one must have an independent measure of the specific area, $\sum$, of the adsorbent (usual units: m^2/g). This can usually be obtained by separate experiments involving nitrogen (or other gas) adsorption from the gas phase (BET method) or other techniques, but sometimes we must content ourselves with reporting isotherms in terms of, e.g., moles of solute adsorbed per gram of adsorbent. Typical solid adsorbents may have $\sum$-values ranging from a few to a few thousand m^2/g. When the adsorption is low, in particular when the total adsorbent area is low, adsorption measurements can be made by radioactively tagging the adsorbate and measuring the radioactivity in a given sample. This is most applicable to "irreversible adsorption" (as frequently is the case for the adsorption of polymers) so that all the free solute can be rinsed from the solution, and measured radioactivity may be identified with adsorbed material.

3.3. The Gibbs Adsorption Equation

For liquid-gas systems we usually do not have the option of obtaining large enough stable interfacial areas for a quantity such as $C_i^\circ - C_i$ to be measurable. For these systems, one can, however, easily measure the surface tension as a function of the solute (adsorbate) concentration in one of the bulk phases, i.e., we may obtain:

$$[\sigma = \mathcal{J}(C_2)]_{\text{const T}} \tag{6}$$

where σ is the surface tension. Techniques for measuring surface tension are discussed in many sources, e.g. ref. 7. We may rigorously derive the isotherm from the surface tension equation by making use of the Gibbs adsorption equation [8], which for a dilute nonelectrolyte binary system takes the form:

$$\left[\Gamma_{2,} = \frac{C_2}{RT}\frac{d\sigma}{dC_2}\right]_{\text{constT}} \tag{7}$$

In an electrolyte system, the right hand side of eq. 7 is multiplied by a factor $1/\alpha$ with $1 \leq \alpha \leq 2$ depending on the degree of ionization of the solute and the concentration of additional electrolyte.

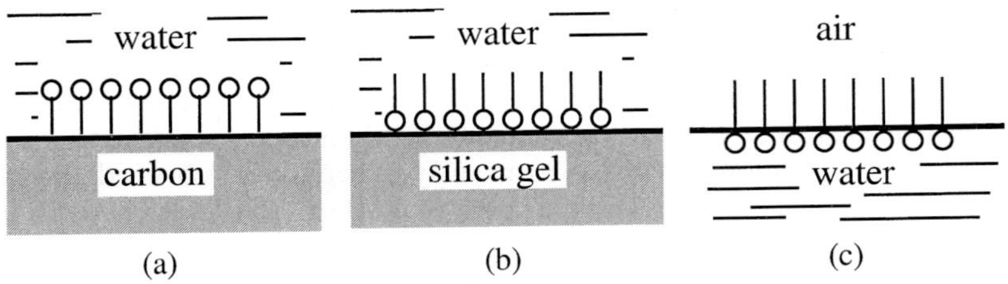

Fig. 4. Schematic examples of orientations of surfactant molecules at various interfaces. Vertical lines represent the apolar parts of the absorbing molecules, and circles represent the polar parts.

Thus adsorption isotherms may be obtained for either type of system, and the results of adsorption measurements permit us to make some useful preliminary generalizations about the phenomenon of adsorption.

The ability of surfactants to reduce the liquid surface tension is of vital importance in its own right to the role of surfactants in absorbency and will be discussed in more detail below.

3.4. Rules of Adsorption and General Isotherms for Dilute Systems

The most general statement concerning whether a solute will be positively absorbed from solution or not is Rehbinder's "rule of polarity equalization," according to which a solute will absorb at the interface between A and $\log t_w = A' - B' \log C_2$ if its presence tends to equalize the polarity between A and B [9]. This will be the case if the solute has a dielectric constant intermediate to that of A and B. Rehbinder's rule thus also explains the fact that amphipathic molecules, in particular surfactants, are strongly adsorbed: such molecules may orient themselves so that the more polar part of the molecule faces the more polar phase while the apolar portion faces the less polar phase. Some examples of surfactant adsorption are shown schematically in Fig. 4. We may thus conclude as our first generalization that all polar (hydrophilic) surfaces adsorb well surfactants from nonpolar or weakly polar liquids and conversely, the nonpolar (hydrophobic) surfaces adsorb well the surfactants from polar liquids (e.g., aqueous solutions). This is why polar adsorbents (silica gel, clays, etc.) are used for the adsorptive removal of surfactants from nonpolar media, and nonpolar adsorbents (carbon) for their removal from polar media.

A schematic diagram of the various types of adsorption behavior in accord with Rehbinder's rule is shown in Fig. 5. Our present interest lies in systems of Type (a) of Fig. 5 in the dilute range, i.e., dilute solutions of surfactants. For systems of this type, one often observes one of the following isotherms:

The Freundlich isotherm:

$$\Gamma_{2,1} = kC_2^{1/n}, \tag{8}$$

where k and n are constants. The constant n may be either greater or less than unity, leading to isotherms which are either concave or convex to the bulk concentration axis.

The Langmuir isotherm:

$$\Gamma_{2,1} = \frac{aC_2}{1 + bC_2}, \tag{9}$$

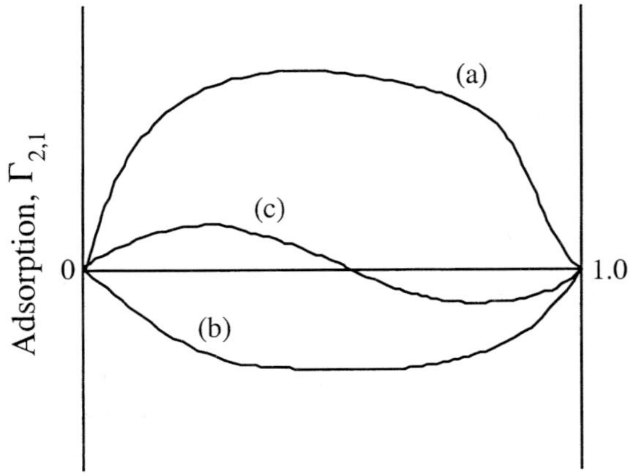

Fig. 5. Types of dependence of the relative adsorption of the solute (2) from a solution in a solvent (1) on bulk phase solute mole fraction. (a) Strong adsorption of the solute and weak (negative) adsorption of the solvent, (b) Weak (negative) adsorption of the solute and strong adsorption of the solvent, (c) Moderate-to-weak adsorption of both components.

where a and b are constants. All three types of isotherm are pictured schematically in Fig. 6. The Freundlich isotherm types are most often observed for adsorption from solution onto solids while the Langmuir isotherm is common to both liquid-solid and liquid-gas systems. Physically, the shape of the Freundlich isotherm with $n > 1$ for solid surfaces may be interpreted in terms of the latter's energetic heterogeneity, i.e., the highly energetic sites are covered first, etc., as bulk concentration of solute is increased.

The case of $n < 1$, is characteristic of rather weak adsorption of the individual surfactant molecules onto the solid, but strong lateral interactions between adsorbed molecules themselves once they become sufficiently concentrated on the surface. The special case of $n = 1$, wherein there appears to be constant partition of surfactant between the solid surface and the solution up to substantial concentration levels (most adsorption shows linear partition at sufficiently low concentration) is often cited [10] as a distinct mode of adsorption. It is quite commonly observed for the sorption of certain substances on textile fibers and on

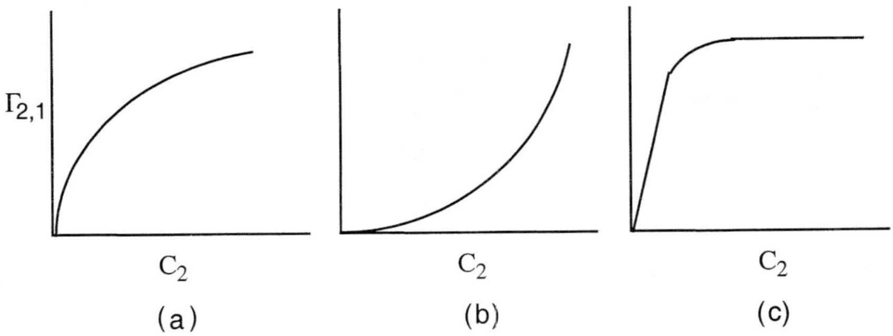

Fig. 6. Common isotherms for adsorption of surfactants from dilute solution. (a) Freundlich isotherms (eq. 8) with $n > 1$. (b) Freundlich isotherm with $n < 1$, (c) Langmuir isotherm (eq. 9).

other solids made up of crystalline regions separated by amorphous regions of much higher penetrability. The initial adsorption is thought to take place in the large pores of the amorphous parts and produces a swelling of the solid exposing new adsorption sites until the impenetrable crystalline regions are reached. The swelling then abruptly halts, and the isotherm becomes horizontal. Still another apparent type of isotherm for liquid-solid adsorption in which $\Gamma_{2,1}$ has a finite value at $C_2 = 0$, is sometimes cited. This "highly affinity" behavior is only apparent rather than real, and occurs for cases of extremely strong adsorption at low concentrations, as in the case of polymer adsorption.

The Langmuir isotherm, Fig. 6c, is often associated physically with adsorption onto an energetically uniform surface without lateral adsorbate interactions. The plateau is thought to correspond to a densely-packed adsorbate monolayer after whose formation there is no further adsorption. While physically reasonable, this interpretation should not be taken too literally. The mutual cooperation and cancellation of a variety of effects can lead to Langmuir-type adsorption.

3.5. The Adsorption Plateau

It is characteristic of adsorption of surfactants onto solids that regardless of the isotherm shape at very low concentrations, as bulk concentration is increased, the isotherm exhibits a plateau of quasi-plateau (where the slope of $\Gamma_{2,1}$ vs C_2 is quite low and much reduced from its value at lower concentrations). The bulk concentration of surfactant corresponding to the attainment of the plateau or quasi-plateau is often vital practical importance because it frequently corresponds to a condition of maximal surface activity in a variety of respects. Further addition of surfactant may not only fail to enhance surface activity but in fact may decrease it. Although numerous exceptions and complications exist, the attainment of the plateau usually coincides with the attainment of a close-packed adsorbate monolayer and with a threshold condition in the bulk surfactant solution frequently leading to the formation of aggregates (micelles) of surfactant molecules or ions containing usually between 50 and 300 monomer units. The concentration corresponding to the formation of such aggregates is termed the critical micelle concentration (CMC), and it can be determined experimentally in a number of ways [11].

Adsorption behavior subsequent to the first plateau or near-plateau in the isotherm can be quite varied and is usually interpreted in terms of the formation of (or failure to form) multilayers or aggregate structures ("hemimicelles") at the interface. Quite often one observes a maximum in surfactant adsorption which is not explainable in terms of solvent competition. While many interpretations have been given for such behavior [12], the most frequent cause is probably the presence of further components in the system. These may be very strongly adsorbed at low concentrations contributing to the apparent adsorption of the principal surfactant component, 2. After the attainment of the CMC, the micelles begin to compete for these components so that the total adsorption at the interface begins to decrease as the micelles become more numerous (with increasing C_2). This explanation parallels that given for the spurious surface tension minima observed for many surfactant solutions [13]. From a practical point of view, neither the details nor the explanation for the system behavior at bulk surfactant concentrations above those corresponding to the plateau (or near-plateau or maximum) are as important as knowing at what bulk concentration this occurs, because as stated earlier, this is generally where optimal surface activity is observed. For systems in which micelles form, this generally occurs at or near the CMC.

3.6. Surface Tension Behavior

Surfactant adsorption at the liquid-gas interface is much simpler than at the liquid-solid interface because the former is energetically homogeneous. The isotherm is almost always Langmuirian in type, and the attainment of the plateau is almost always somewhat ahead of the CMC. Recall that the adsorption isotherm for liquid-gas systems is derived from surface tension data via the Gibbs adsorption equation. The surface tension equation leading to the Langmuir isotherm is the Szyszkowski equation [14]:

$$\sigma = \sigma_o[1 - A\ln(1 + BC_2)], \tag{10}$$

where σ_o is the surface tension of the pure solvent, and A and B are empirical constants. For aliphatic surfactants, the constant A was found to be characteristic of a given homologous series (alcohols, amines, etc.), and B was dependent upon the member of the series. The Gibbs adsorption equation, of course, provides the relationship between the constants in the Szyszkowski equation and those of the Langmuir isotherm, viz.,

$$a = \frac{\alpha_o AB}{RT} \quad \text{and} \quad b = B. \tag{11}$$

At very low concentrations, the Langmuir isotherm is linear and the slope is $a = \frac{\alpha_o AB}{RT}$. Strong adsorption thus corresponds to large values of both A and B. At saturation adsorption (usually referring to a monolayer),

$$\Gamma = \Gamma_m = \frac{a}{b} = \frac{\sigma_o A}{RT} \tag{12}$$

so that the density of the monolayer packing is directly proportional to A.

When surface tension data for a surfactant solution are plotted on semilog coordinates, the result shown in Fig. 7 is obtained and may be interpreted as follows. The initial, nearly horizontal branch of the curve corresponds to the situation where is C_2 is so low that $BC_2 << 1$, and $\sigma \approx \sigma_o$. The subsequent downward-curving portion of the curve corresponds to the situation when neither BC_2 is negligible with respect to unity nor vice versa. The next straight-line portion of graph, sometimes crossing more than a decade in concentration, corresponds to $BC_2 >> 1$ so that:

$$\sigma = \sigma_o[1 - A\ln(BC_2)]. \tag{13}$$

Application of the Gibbs adsorption equation to eq. 13 shows that:

$$\Gamma_{2,1} = \frac{\sigma_o A}{RT} = \text{constant} \tag{14}$$

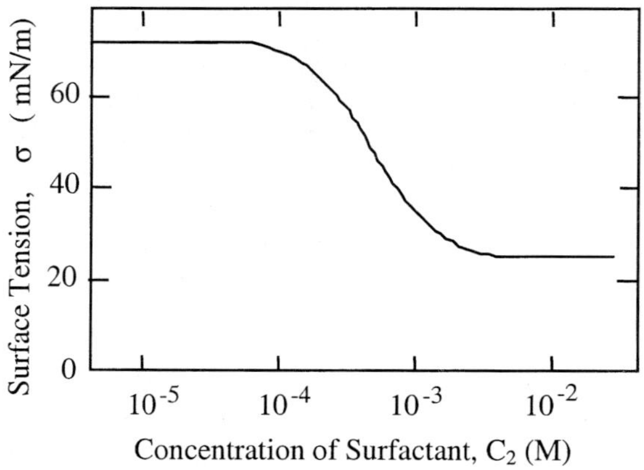

Fig. 7. Surface tension variation of a typical aqueous surfactant solution.

corresponding to saturation adsorption. It is interesting that most of the surface tension reduction occurs as the structure of the monolayer changes only minimally. The linear portion of the surface tension curve of Fig. 7 ends quite abruptly at the CMC, beyond which only slight changes in surface tension are observed (provided a surfactant contaminant is not present). The Gibbs adsorption equation in the form of eq. 7 is not applicable beyond this point since the micelles are a new species in solution. Typically the condition of saturation adsorption at the liquid-gas interface occurs at a concentration between one-fifth and one-third of the CMC.

3.7. The Critical Micelle Concentration

It is evident that the CMC is an important benchmark in describing the adsorption of a surfactant at either the liquid-solid or liquid-gas interface, and that in the latter case, the maximum surface tension reduction achievable (usually occurring at the CMC) is also an important parameter in describing the compound's surface activity in the particular system. These properties are among the most important ones to be examined in screening a particular surfactant for various applications, including its effect on absorbency.

Some brief generalizations concerning these properties for the various types of surfactants are possible. Fully ionized surfactants, either anionic or cationic, typically exhibit CMC's of between 0.1 and 0.001 moles/liter. The exact value depends for a given hydrophilic group on the size of the hydrophobe. The larger the hydrophobe, the lower the CMC. For straight-chain aliphatics, the CMC is approximately halved for each additional CH_2 group up to a total of about 18 carbons. Beyond this there is little change. The CMC also depends strongly on the counterion concentration in the solution due to their screening effect, which tends to counteract the lateral electrostatic repulsion encountered in the formation of the micelle. The CMC of a given ionic surfactant is divided roughly by three for each unit increase in the molality of univalent counterion. The dependence on polyvalent counterion concentration is much steeper. The effect of temperature on the CMC of ionic surfactants is ambivalent and usually fairly slight over modest temperature ranges. The CMC for a nonionic is usually two to three orders of magnitude lower than that of an ionic of corresponding molecular weight since there is no lateral electrostatic repulsion encountered in the formation of micelles. The dependence of the CMC on hydrophobe size is also generally greater than for ionics, being roughly a factor of three decrease for each additional methylene group. A phenyl group in the hydrophobe of either type of surfactant is equivalent to about 3.5 methylene groups. For polyoxyethylenated (PEO) compounds of a given hydrophobe, the CMC increases with increasing numbers of C_2H_4O (EO) units, since this increases hydrophilicity. The increase is much smaller than the decrease caused by adding a CH_2 group, however, and depends on the original size of the PEO chain. Typically the increase is of the order of 5-10% for each EO unit. The effect of added electrolyte is small for nonionics compared with its effect on ionics, but it does tend to decrease the CMC somewhat. The CMC of nonionics decreases sharply with an increase in temperature until at a temperature well below 100°C for most of them, they precipitate out as giant micelles. This temperature is termed the cloud point, and it is higher the higher the EO/hydrophobe ratio in the molecule. Organic additives may have a strong depressing effect on the CMC for either type of surfactant. Extensive compilations of CMC data may be found elsewhere [15,16].

Surface tension reductions achievable (at the CMC) for all types of surfactants range typically between 20 and 50 dynes/cm, the larger reductions being associated generally (but not always) with the more hydrophobic molecules. This represents a wide range in behavior, and while procedures have been developed for estimating the appropriate value for simple surfactants [17], it must generally be obtained experimentally.

It must be pointed out finally that not all surfactants are capable of forming micelles (although most do) or that the expected CMC lies above the monomer solubility limit at the temperature of interest. For these, the abrupt changes in the adsorption isotherm or surface tension equation associated with CMC are absent [18].

3.8. Amphipathic vs. Amphiphilic Adsorption

Although there are many specific differences from case to case, it is possible to make some further useful generalizations concerning surfactant adsorption. We may first of all divide it into two broad categories which might be termed amphipathic adsorption and amphiphilic (or specific) adsorption. Amphipathic adsorption is completely non-specific in nature and is primarily the result of the rejection of the lyophobic moiety of the surfactant molecule from the solution. In such a case, the solid adsorbent surface or the gaseous surface has no specific attraction for the lyophobic functional groups but merely provides a region for escape from the solution phase. Such adsorption is characteristic of surfactants adsorbed from an aqueous solution onto a hydrophobic solid surface (e.g., carbon, polyethylene, paraffin wax, Teflon) or at the interface against air. Examples are shown schematically in Fig. 4a and 4c, respectively. Amphipathic adsorption is the mode through which surfactants are generally employed to improve the absorbency of a material and is thus of special importance in the present context. It is not strictly correct to consider the driving force for amphipathic adsorption as simply the rejection of the hydrophobe (assuming we have an aqueous solution medium) from the aqueous phase. Indeed the attractive Van der Waals forces operating between a hydrocarbon group and water molecules are not much different from those operating between two hydrocarbon groups. The key fact is that the presence of the apolar hydrocarbon in the solution disrupts some of the strong hydrogen bonding that would exist between the water molecules in its absence. This is essentially the same driving force which leads to the formation of micelles, one which is self-evidently directly proportional to the size of the hydrophobe. For linear aliphatic hydrophobes, the relationship between amphipathic adsorption and related properties such as surface tension reduction is expressed by a simple law termed Traube's rule. It states, for example, that the bulk concentration required to attain a particular surface tension reduction by a surfactant of a particular head group type decreases by a constant factor for each methylene group added to the hydrophobic chain. This factor is about three for nonionic surfactants and closer to two for ionics. Similar rules apply for the extent of amphipathic adsorption at the solid-liquid interface [19]. For more complex hydrophobe structures, Traube's rule is less useful but still provides a qualitative guide to the strength of amphipathic adsorption.

In amphipathic adsorption, the hydrophobic portions of the adsorbate molecules are oriented toward the hydrophobic solid surface (or towards the air phase) where they are attracted to the wall and/or to each other by dispersion forces. The hydrophilic moieties are oriented towards the aqueous medium in which they may be ionized and/or hydrated. Adsorption of surfactants at the aqueous-air interface is always amphipathic. In some situations, the tendency to adsorb is best regarded as a combination of mutual attraction between hydrophobic groups by dispersion forces and their tendency to escape the aqueous medium. Such a combination of effects is sometimes termed "hydrophobic bonding". It is promoted by large hydrophobe size and ease of close hydrophobe packing.

An extreme and important form of amphipathic adsorption is that associated with the "precipitation" of the hydrophobe onto the solid surface which it may match so closely in structure that over a period of time (particularly if heated) it "co-crystallizes." The "bond" thus formed has a high degree of permanence and is the basis for a procedure used for the durable enhancement of the hydrophilicity of a variety of synthetic fibers using surfactants [20].

Amphiphilic adsorption of surfactants occurs as a result of specific attraction between functional groups of the surfactant molecule and the solid adsorbent. This type of adsorption does not occur at the liquid-gas interface. Unlike amphipathic adsorption it always depends on the specific chemistry of the solid, the surfactant and the solvent medium under the conditions of interest and may occur by a number of distinct mechanisms. Consider the case of adsorption from aqueous media. Since the hydrophobic portion of the surfactant molecule is not capable of specific chemical interaction with the adsorbent, amphiphilic adsorption depends on attraction between the hydrophilic groups of the surfactant and the solid surface. It is these groups, however, which are also attracted to the aqueous medium. Positive adsorption must thus be the result of preferential attraction of the adsorbent relative to the solvent water for the hydrophilic groups of the solute (sometimes with a significant assist from the amphipathic tendency to adsorb). The test of whether adsorption is amphiphilic, amphipathic or a combination of both (to be described later) is the orientation of the adsorbed molecule. If the hydrophilic groups are adjacent to and interacting with the solid, the adsorption must be regarded as amphiphilic. All the methods using surfactants to suppress absorbency and many of the newer and innovative methods for enhancing absorbency are based upon amphiphilic adsorption.

3.9. Mechanisms of Amphiphilic Adsorption

There are a number of different mechanisms for amphiphilic adsorption often operating together or to different extents at different stages in the particular isotherm. They also may act in concert with amphipathic adsorption or hydrophobic bonding leading to so-called "cooperative adsorption." The first three mechanisms discussed are electrostatic in origin, the fourth is a combination of electrostatic mechanisms with amphipathic adsorption,

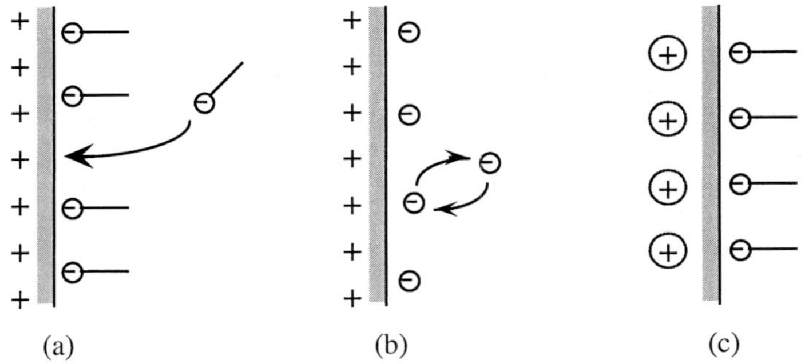

Fig. 8. Schematic representation of electrostatic modes of amphiphilic adsorption. (a) Ion pairing, (b) Ion exchange, (c) Image charging.

while the remainder involve the formation of generally weak and to some extent reversible chemical bonds between functional groups of the surfactant and the solid surface. The formation of strong covalent bonds between the adsorbate and the adsorbent (chemisorption) leads to permanent chemical modification of the solid-liquid interface, and is not considered in any detail in the present chapter. Finally, the purely physical interaction between permanent dipoles in the adsorbent and the adsorbate is usually too weak to cause amphiphilic adsorption [21].

3.9.1 Ion-pair

This type of adsorption, pictured schematically in Fig. 8a, occurs when surfactant ions enter the Stern layer, i.e., the layer of counter-ions adjacent to a charged surface in an electrolyte solution. The surfactant ions are charged opposite in sign to the solid surface. Such adsorption generally occurs only at very low bulk surfactant concentrations and in the absence of additional electrolyte. Prior to such adsorption, the Stern layer consists essentially only of hydrosyl ions or hydronium ions (depending on whether the solid surface is charged positively or negatively, respectively) which are incapable of neutralizing the solid surface charge by themselves. Such neutralization is effected in the diffuse or Guoy portion of the electrical double layer. Net adsorption of surfactant ions into the Stern layer tends to neutralize the surface charge and thereby to change the nature of subsequent adsorption. The reason that adsorption by ion pairing does not occur under conditions when additional electrolyte is present is that such electrolyte compresses the diffuse double layer so that the surface change is already effectively neutralized prior to surfactant adsorption. Adsorption by ion pairing is often extremely sensitive to pH since changes in pH can alter the state of ionization of both the surfactant and active sites on the adsorbent and may even change the sign of the charge on the solid surface. Many of the solid materials which are important in textile manufacture are protein fibers. These include both natural (e.g., wool and silk) and synthetic materials (e.g., nylon). Their surfaces contain both amino and carboxyl groups, as shown in Fig. 9, and are, therefore, amphoteric. At low pH, the carboxyl groups are in unionized form (COOH) and the amino group (and hence the surface) is positively charged (NH_3^+). Additional of a base causes a progressive ionization of the carboxyl groups (to

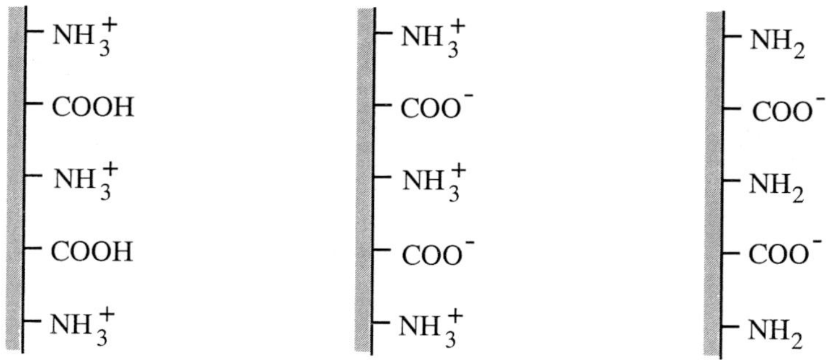

(a) (b) (c)

Fig. 9. Schematic of protein-type surface in contact with an aqueous solution: (a) low pH, (b) isoelectric point, (c) high pH.

COO^-) until at a certain pH, known as the isoelectric point, the number of positive and negative groups becomes equal, and the surface is zwitterionic. At higher pH's, the amino groups are neutral, and the surface becomes negatively charged. Protein surfaces at pH conditions well above or below the isoelectric point are examples of charged surfaces and are subject to ion pair adsorption. Most other textile, paper-making or absorbent materials, either natural or synthetic, bear a negative surface charge to some extent at all but extremely low pH's. These include cellulose (e.g., wood and cotton) or modified cellulosic materials (e.g., rayon), polythenes, polyesters, polyurethanes, etc. Thus ion pair adsorption is most often associated with cationics.

3.9.2. Ion exchange

This type of adsorption, pictured in Fig. 8b, often accompanies ion pairing and consists of an exchange of surfactant ions with smaller counter-ions in the Stern layer. It, by itself, does not effect a change in surface charge at the Stern plane and may occur even when there is a high concentration of electrolyte in the solution. It usually is an important mechanism of adsorption, however, only at low surfactant concentrations.

3.9.3. Image Charge Adsorption

Both anionic and cationic surfactants may be adsorbed to metal surfaces due to the mobility of the electron cloud in the metal [22]. Image charges are set up by interaction of the metal electrons with the charged end groups of the surfactant ions as shown in Fig. 8c. Nonionic surfactants are not strongly adsorbed on metals and are, therefore, often used for cleaning such surfaces without leaving a surfactant film behind.

3.9.4. Cooperative adsorption

After adsorption by ion pairing (usually accompanied by ion exchange) and an ionic surfactant onto an oppositely-charged surface has approximately neutralized the surface charge, the electrostatic driving force for adsorption has diminished to near zero and the slope of the isotherm decreases to near zero. As the bulk surfactant is raised further, however, one often observes a sharp increase in the slope of the isotherm, as shown in Fig. 10. Such adsorption represents the formation of a second layer or platelets of second layer

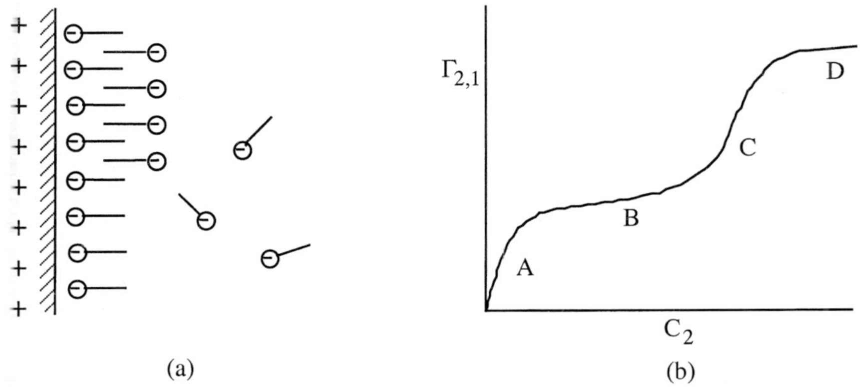

Fig. 10. (a) Schematic representation of cooperative adsorption, and (b) the resulting form of the isotherm: A (adsorption by ion pairing and ion exchange), B (slope decreases as monolayer fills), C (formation of hemimicelles, surface charge neutralized and reversed), D (slope again decrease due to electrostatic repulsion).

("hemimicelles") though amphipathic adsorption and hydrophobic bonding of additional surfactant to the layer already adsorbed. The surface charge is completely neutralized and then reversed in sign during this step, known as cooperative adsorption. As adsorption increases further, electrostatic repulsion now reduces the net driving force, and the isotherm approaches the horizontal. The surface has become effectively a dense bilayer of surfactant, and its charge is the same sign as that of the surfactant ion. Such adsorption has been widely observed for both cationics [23] and anionics [24] absorbed on mineral surfaces of opposite charge and on the surfaces of various synthetic protein-type fibers at pH's well above and below, respectively, their isoelectric points. Cooperative adsorption of cationics on negatively-charged cellulosic fibers has also been reported [25].

3.9.5 Brønsted acid-base interactions

An important type of specific interaction which can occur in adsorption is that between an acidic group on the surfactant ion or molecule and a basic site on the adsorbent, or vice versa. An acid, in the Brønsted sense, is any substance which can lose a proton to another substance (a Brønsted base) which has an unshared pair of electrons. The more weakly the proton is held, the stronger the acid, and the more weakly the unshared electron pair is held, the stronger the base. When the acid and base are ions of opposite charge, ion pairing adsorption leading to the acid-base interaction will produce two neutral substances and the ion pairing mechanism of adsorption is nullified. When either the acid or base is neutral and the other is an ion, the interaction merely transfers the charge from the absorbent to the adsorbate (or vice versa), and adsorption is not necessarily promoted. When the acid and base are both electrically neutral, however, the proton transfer will result in a pair of ions of opposite charge which will interact through a salt linkage resembling ion pair adsorption. An example would be the interaction of a carboxyl group of the adsorbate molecule (unionized fatty acid) with an amine group on a solid surface:

$$\text{Solid - NH}_2 + \text{HOOC - R} \rightleftarrows \text{Solid - NH}_3^+ \text{ - - - - - - - - - - } \overline{\text{O}}\text{OC - R}$$

Salt linkage

There would usually be only a narrow pH range, if any, in which both NH_2 groups and COOH groups could coexist.

3.9.6. Hydrogen bonding

Hydrogen bonding is one of the most important types of amphiphilic adsorption. It may be thought of as a particular case of Brønsted acid-base interaction when the relative electrical forces are such that the tendency of the proton to form covalent bonds with the acid and the base are roughly equal. Under such conditions, the proton may not transfer from one molecule to the other (if transfer does take place, it will be reversible), but instead be shared by both molecules leading to an effective bond between them. This may lead to strong adsorption by hydrogen bond surface complexation. Examples include the adsorption of free fatty acids onto polyester (A) [26] or onto Nylon 66 (B) [26]:

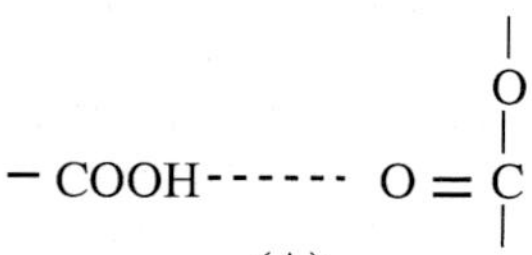

(A)

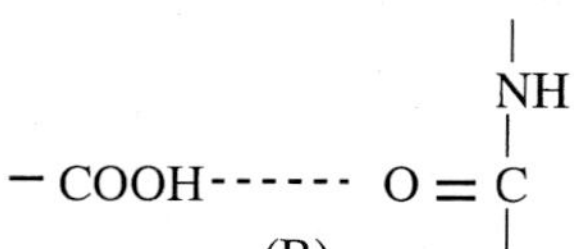

(B)

On the other hand, when the substrate has groups such OH, COOH, NH_2, etc., capable of furnishing a proton, surfactants containing a polyoxyethylene chain may be adsorbed by hydrogen bonding [26], e.g.:

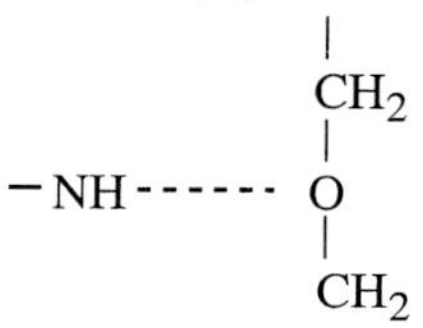

3.9.7. Lewis acid-base interactions; π-bonding

The Lewis concept of acid-base interactions focuses attention upon the unshared electron pair and does not require the presence of a proton to be transferred or shared. An acid is defined as any substance which can fill the valence shell of one of its atoms with an unshared pair of electrons from another (the Lewis base). The result of such an interaction may be a stable complex. A closely-related example is the case when the adsorbate contains electron-rich aromatic nuclei (π-electrons) which may interact with positively charged sites on the adsorbent [27] or vice versa.

The competitive nature of adsorption from liquid solutions is especially important to consider when acid-base interactions are involved. An acidic surfactant will adsorb into basic sites of an adsorbent only when (1) the solvent itself or other solutes in the solution are not equivalently strong bases so that the surfactant remains "tied up" by acid-base interactions in the bulk phase, and (2) the solvent or other solutes are not equivalently or more acidic than the surfactant so that they "tie up" the basic sites on the adsorbent. Another way of stating the above requirements is that acid-base adsorption is effective only when the solvent medium is neutral. This has been demonstrated by Fowkes [28]. Central to assessing the importance of acid-base adsorption in a given case is the relative strength of the various acid-base interactions possible. Fowkes has shown how these may be quantified using the correlations of Drago et al. [29,30]. The reader is referred to the referenced papers for the details of this important advance in the understanding of amphiphilic adsorption.

3.9.8. Cation salt bridging

Anionic surfactants may be bound to negatively-charged surfaces via polyvalent cations which form simultaneous salt linkages with both the solid and the surfactant ion. This is believed to be the role played by aluminum ions in fixing rosin sizes to papermaking materials under moderately acid conditions [31]. Other examples are the binding of anionic surfactants through calcium bridging to minerals such as bentonite [32] or of anionic dyes to proteinic or cellulosic materials using a variety of polyvalent cations [33].

3.9.9. Formation of covalent bonds

Chemisorption occurs when covalent bonds are formed between adsorbate and adsorbent, and while this is generally outside the scope of the present chapter, it is useful to give some examples and to admit that the distinction between chemisorption and surface complexation is not always clear cut. For example, the reaction between an amine group and a carboxyl group may lead under certain conditions to the formation of an amide linkage:

$$\text{Solid} --- NH_2 + HOOC - R \leftrightarrows \underset{\text{amide linkage}}{\text{Solid} - NH - OC - R} + H_2O$$

Similarly, the interaction between a hydroxyl group and a carboxyl group can lead to an ester linkage:

$$\text{Solid} - OH + HOOC - R \leftrightarrows \underset{\text{ester linkage}}{\text{Solid} - O - OC - R} + H_2O$$

Both reactions can be reversed (hydrolysis) under highly acidic or basic conditions. An ether linkage can be formed by the reaction between hydroxyl groups under the right conditions:

$$\text{Solid} - OH + HO - R \rightarrow \underset{\text{ether linkage}}{\text{Solid} - O - R} + H_2O,$$

but such a reaction is not easily reversed. Many other types of chemical reactions between adsorbate and adsorbent are, of course also possible. The literature on dyeing [33] contains a vast quantity of information on the mechanisms of adsorption at the solid-liquid interface.

It must be emphasized that Traube's rule, as started earlier, applies to no type of amphiphilic adsorption, and in some cases appears to apply in reverse, i.e., the larger the lyophobe, the less the adsorption [34].

4. THE LINK BETWEEN INTERFACIAL PROPERTIES AND ABSORBENCY

As stated earlier, surfactants often have a strong effect on absorbency because their adsorption alters the properties of the interfaces in the system. In order to quantify such

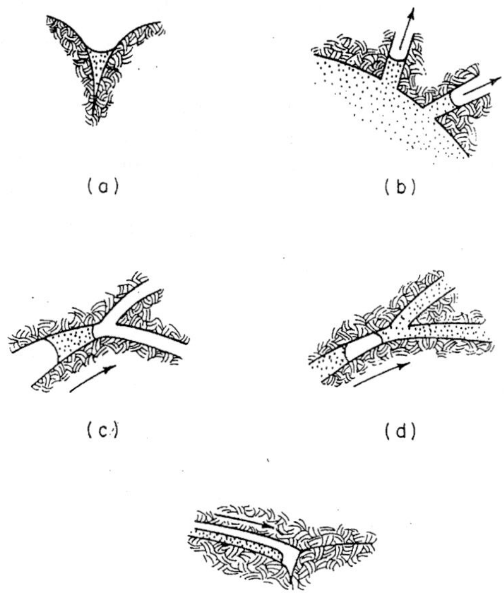

Fig. 11. Processes involved in absorbency. (a) Capillary condensation, (b) Penetration absorption, (c) Motion of liquid thread, (d) Expulsion of air bubble, (e) Surface wicking.

effects, we must examine the relationship that exists between interfacial properties and the processes involved in absorbency.

Absorbency refers to the uptake of liquid by a porous solid by several distinct spontaneous processes: (1) the condensation of liquid into the pores or crevices of a solid matrix from the vapor phase, (2) the penetration of liquid into a solid matrix when one side of the porous solid is in contact with a large liquid reservoir (penetration absorption), (3) the movement of small liquid masses (liquid "threads" or "indices") into the interior of a porous matrix, (4) the uptake of liquid into a porous solid which is totally immersed (immersion absorption) in a large liquid reservoir (involving the expulsion of trapped air pockets from within the solid), and (5) the motion of either large or small liquid masses along the macroscopically rough surface of a porous matrix, i.e., "surface wicking." These various processes are sketched schematically in Figure 11.

4.1. Capillary Condensation

Condensation of vapor into small pores or crevices on or within a porous solid at partial pressures of the condensing vapor above the vapor pressure of the condensate is known as a capillary condensation and occurs in accord with the Kelvin equation [35]:

$$P = P^{s} \exp \frac{2 v_{L} \sigma \cos \theta}{r RT} \tag{15}$$

where P is the partial pressure of the vapor condensing into a circular pore of radius r, P^s is the vapor pressure, v_L is the molar volume of the condensate liquid, σ is its surface tension, θ

is the contact angle of the condensate against the solid, R is the gas constant and T is absolute temperature. It is seen that capillary condensation occurs only when $\cos\theta > 0$ and is maximum when $\cos\theta = 1$ ($\theta = 0^{\circ}$), i.e., when the solid is wet out by the condensate. Moisture uptake by capillary condensation is an important aspect in the wearing comfort of textile fabrics. Cotton (particularly when not heavily sized) and other cellulosic fabrics are hydrophilic and produce a low contact angle with water, whereas silk, wool and many of the synthetic fabrics are hydrophobic and produce a large contact angle. Moisture uptake, in part by capillary condensation, in cotton leads to the greater comfort of these materials [36].

4.2. Penetration Absorption

The penetration of liquid into a solid matrix when contacted from one side with a large liquid reservoir is probably the most important mechanism of absorption in products whose end use is specifically that of an absorbent (see Table 1) and involves the displacement of air from the solid through open spaces in the matrix through the side not in contact with the liquid. This type of absorbency is measured by various wicking tests [37]. Considering first a single uniform circular pore of radius *r*, the linear rate of travel of the liquid front along the axis of the pore is given by the Washburn equation in differential from [38]:

$$\frac{dx}{dt} = \frac{r\sigma\cos\theta}{4\mu x} + \frac{r^2\rho g\cos\beta}{8\mu}, \tag{16}$$

Where x is the distance of travel from the liquid reservoir, μ is the liquid viscosity, ρ is the liquid density, g is the constant of gravitational acceleration and β is the angle between the direction of liquid movement and the downward vertical. The relevant contact angle θ to be used is the "advancing contact angle," to be discussed later [39,40]. The first term on the right hand side describes the spontaneous wicking effect while the second describes the resistance (if $90^{\circ} < \beta \leq 180^{\circ}$) or assistance (if $0^{\circ} \leq \beta < 90^{\circ}$) of gravity. The effect of gravity may be neglected if the ratio of the second to the first term on the right side of eq. 16 is negligible, i.e.,

$$\frac{r\,x\,\rho g\cos\beta}{2\sigma\cos\theta} \ll 1, \tag{17}$$

For fair-to-good wetting conditions ($\theta<60^{\circ}$) the above condition is generally satisfied for pore radii of a few microns or less, but any particular situation may be evaluated by examining eq. 17. Gravity may be important in some open or coarse-structured absorbent products designed to absorb and retain liquids against gravity [41].

If there is an externally-imposed pressure, Δp, relative to that which exists in the gas phase being displaced, eq. 16 should have an additional term on the right, viz.:

$$\frac{(\Delta p)r^2}{8\mu x} \tag{18}$$

Liquid can thus be forced to enter a capillary pore even when the wicking force may oppose it, i.e., when $\theta > 90^{\circ}$.

When gravity and external pressure effects are both negligible, eq. 16 equation becomes:

$$\frac{dx}{dt} = \frac{r\,\sigma \cos\ \theta}{4\ \mu\ x}\ , \tag{19}$$

or in integrated form:

$$x = \left[\frac{r\sigma \cos\ \theta}{2\,\mu}\right]^{1/2} t^{1/2} = kt^{1/2}\ , \tag{20}$$

the expression usually referred to as the Washburn equation.

After a brief initial period, termed a "wetting delay," (usually for $t < 1$ s and $x <$ a few mm), the format of the Washburn equation describes very well the rate of capillary penetration of pure liquids into complicated porous media, especially when the advancing contact angle is zero. The actual mean pore opening should not exceed about 250 μm [41], so that a capillary driving force can be developed. When the contact angle is greater than zero, the motion often occurs in a succession of short steps so that an average linear velocity is implied. The pore radius, r, should be replaced by r_e, the "wicking equivalent" pore radius, whose value takes into account the geometry of the pore spaces in the medium. Because of the tortuosity of the flow path, r_e is usually an order of magnitude smaller than the actual average pore opening in the medium [42]. The functional dependence of the wicking velocity on surface tension and viscosity given in eq. 20 has been verified [43], and wicking measurements in fiber networks combined with contact angle measurements on single fibers from these networks [44] shows that the predicted dependence on $\cos\theta$ is also retained.

Extensive investigation of the wicking of aqueous surfactant solutions of various types into fiber networks shows that the format of eq. 20 is retained for these cases as well so that it provides the basis for examining the role of surfactants in penetration adsorption [45,46]. Equation 20 suggests that absorbency of a given porous medium, as quantified by the Washburn slope k, depends directly on the $\sigma\cos\theta$. (It may be assumed that the viscosity μ of an ordinary dilute surfactant solution is the same as that of pure water.) It is useful to distinguish two situations: Case 1 ("high energy" solids), in which water and the surfactant solutions wet out the solid which makes up the porous medium, such that $\theta = 0°$ ($\cos\theta = 1$), and the absorbency depends solely upon the surface tension of the imbibing liquid, and Case 2 ("low energy solids"), in which the contact angle is finite so that both factors must be taken into account. It is evident that the addition of surfactants in Case 1 can have only a deleterious effect, since the presence of surfactants (sometimes drastically) reduces the surface tension. The effect of surfactants in Case 2 would appear to be ambivalent, as is revealed upon considering the relation between the contact angle and the interfacial free energies given by Young's equation, viz. [47]:

$$\cos\theta = \frac{\sigma_{SG} - \sigma_{SL}}{\sigma}\,. \tag{21}$$

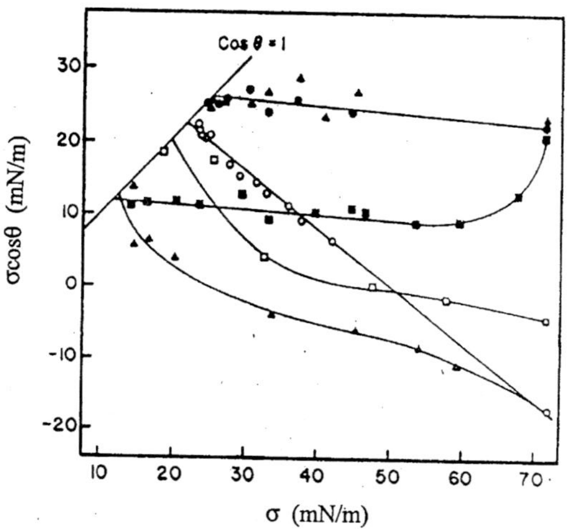

Fig. 12. σcosθ versus surface tension for various surfactant-solid systems. Key: AOT - Nylon 11 (▲); AOT - PMMA (●); AOT - Paraffin (O); PFO - Polyethylene (□); PFO - Paraffin (△); PFO - PMMA (■). AOT = Aerosol OT; PFO = Perfluoro-octanoic acid. From Ref. [48], by permission.

σ_{SG} and σ_{SL} are the interfacial "tensions" of the solid-gas and the solid-liquid interfaces, respectively, whereas σ is the surface tension of the liquid. Use eq. 21 yields:

$$\sigma \cos \theta = (\sigma_{SG} - \sigma_{SL}) . \tag{22}$$

The quantity (σ_{SG} - σ_{SL}) is known generally as the "adhesion tension," and eq. 22 shows how (for cases yielding finite contact angle) it may be evaluated from surface tension and contact and contact angle measurements. Substitution of eq. 22 into eq. 20 suggestions that in the case of less than complete wetting, wicking efficiency is *independent* of surface tension. Since σ_{SG}, the surface free energy of the dry surface, is unaffected by the presence of surfactants in the imbibing solution, the only effect the surfactant can have is associated with its ability to alter the solid-liquid surface energy, σ_{SL}.

Pyter *et al.* [48] investigated the quasistatic wetting of solids by surfactant solutions by measuring the equilibrium values of σcosθ vs. σ for several surfactant solutions against several smooth solid surfaces. Some of their results are shown in Fig. 12. They reveal thatquasistatic wetting (as evidenced by the value σcosθ) varied directly with surfactant concentration (i.e., inversely with σ) in all cases but one. This suggests that the adsorption decreased the value of σ_{SL}, which suggests that presence of the surfactant at the interface enhanced the compatibility of the water with the substrate. This is the expected result, since both surfactants used were anionic, and the solid surfaces were likely to be negatively charged in contact with water. Under these circumstances, one would expect the adsorption

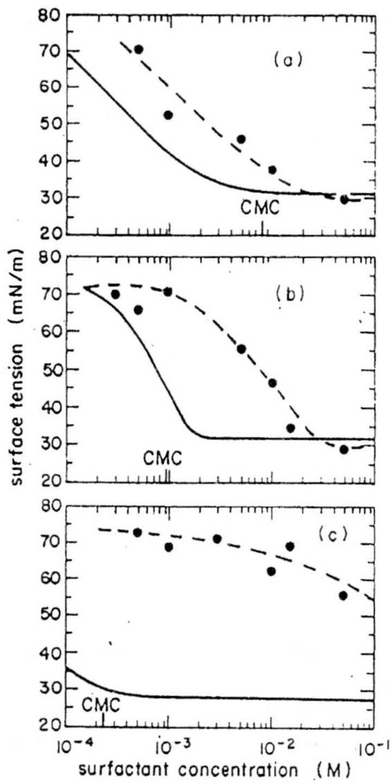

Fig. 13. Comparison of wicking-equivalent surface tension (---) with equilibrium surface tension (—) for aqueous solutions of (a) SDS, (b) CTAB, and (c) TX-100, imbibing into strips of Whatman No. 40 filter paper. From Ref. [46], by permission.

at the solid-liquid interface to be amphipathic, exposing hydrophilic ionic groups toward the water, decreasing σ_{SL}. The once exception was perfluoro-octanoic acid absorbed onto PMMA, which showed a small drop in wettability at very low surfactant concentrations, due probably to hydrophobic perfluoro tails lying flat on the surface under those conditions.

Hodgson and Berg [46] investigated the dynamics of wicking of several surfactant solutions into two types of fibrous media: strips of pure cellulose filter paper (Whatman No. 40) which was wet out ($\theta = 0°$) by water and all solutions (Case 1 above), and strips of thermomechanical pulp (TMP) sheets, which were only partially wet (Case 2 above). Some of the results for Case 1 are shown in Fig. 13. The Washburn slopes, k, in all cases were consistent with values of the surface tension higher than the equilibrium surface tension of the solution, until the surfactant concentration reached a value (dependent upon the surfactant) well above the corresponding CMC for the surfactant. Figure 13 shows the wicking-equivalent surface tension, σ^*, defined as:

$$\sigma^* = 72.0\left(\frac{k}{k_{\text{water}}}\right)^2 \ \text{mN/m}\,, \tag{23}$$

plotted as a function of the surfactant concentration together with the equilibrium surface tension values for (a) an anionic surfactant, sodium dodecyl sulfate (SDS), (b) a cationic surfactant, cetyl trimethyl ammonium bromide (CTAB) and (c) a nonionic surfactant, Triton-X 100 (TX-100). Despite the significant disequilibrium with respect to surfactant distribution, the Washburn rate law was always observed over the entire course of the imbibition, suggesting that a dynamic balance was established very quickly (and maintained) between the depletion of the surfactant from the meniscus by adsorption at the fiber surfaces and its replenishment by diffusion from the bulk solution. This is expected behavior because as the meniscus advances, both the required diffusion distance (proportional to x, which varies as $t^{1/2}$) and the diffusion path length (proportional to $\sqrt{Dt}$) are proportional to the square root of time. The magnitude of σ^* in each case is governed by the concentration of surfactant in the solution, the extent of adsorption at the fiber surfaces, the dependence of surface tension on concentration, and the diffusivity of the surfactant in solution. The meniscus depletion effect (σ^* - σ_{eq}) was least for the SDS solutions and greatest for the TX-100 solutions (as well as other large-molecule nonionics investigated). Furthermore, the concentration required for the convergence of σ^* and σ_{eq} was approximately equal to the CMC for SDS, but about 10 times the CMC for CTAB and over 100 times the CMC for Triton-X 100. The two principal reasons for the differences were differences in the extent of adsorption at the fiber surfaces and the surfactant diffusivity. High adsorption and low diffusivity favor large and persistent disequilibria. The difference between SDS and CTAB was due mainly to differences in adsorbed amount at the solid surface, their diffusivities in water being essentially the same both below and above their respective CMC's [49]. The closeness of their surface tension behavior with surfactant concentration is evident in Fig. 13. The much greater disequilibrium shown by TX-100 is attributable primarily to its low diffusivity, an order of magnitude lower than that for SDS or CTAB [50], who examined systematically the effect of chair branching in the surfactants (hence diffusivity) on the immersional absorbency of their solutions into porous materials.

In wicking experiments with strips of fibers only partially wet by the surfactant solutions neither the surface tension nor the contact angle was expected to be at is equilibrium value at the advancing meniscus. In these cases, the dynamic adhesion tension, (σ_{SG} - σ_{SL})*, was determined from the Washburn slopes according to:

$$(\sigma_{SG} - \sigma_{SL})^* = 72.0\left(\frac{k}{k_{water}}\right)^2 \text{ mN/m} \tag{24}$$

and compared with the equilibrium value obtain from the independent determination of the equilibrium value of $\sigma\cos\theta$. For both anionic and nonionic surfactants (cationics were not investigate) the dynamic adhesion tension was identical to the equilibrium value at vanishingly low surfactant concentration (equivalent to pure water), then with increasing surfactant concentration remained above the dropping equilibrium values until a sufficiently high concentration was reached, where the values again converged. The drop in adhesion tension, both the equilibrium and the dynamic values, is attributable to increases in σ_{SL}, suggesting (somewhat surprisingly, and in contrast to results of Pyter *et al.* cited above) an increase in surface hydrophobicity. A critically important factor determining whether a

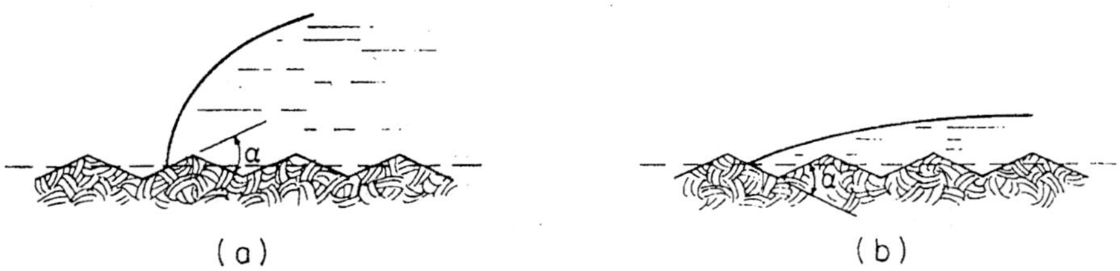

Fig. 14. Schematic representation of contact angle hysteresis due to surface roughness. (a) Advancing interline, (b) Receding interline. The intrinsic contact angle against the solid is the same in each case.

surfactant adsorbed at a solid-liquid interface increases or decreases σ_{SL} is its orientation in the adsorbed state, and often electrostatics play a key role [52,53]. When cationic surfactants adsorb to anionic surfaces (the more common situation for solid surfaces in contact with aqueous media) or for anionic surfactants adsorbing to positively-charged surfaces, the adsorption is in the “head-down” configuration, exposing the hydrophobic moieties to the aqueous medium and increasing σ_{SL}. The adsorption of ionic surfactants to like-charged solid surfaces yields “tail-down” adsorption and often a decrease in σ_{SL}. The divergence between the dynamic and equilibrium values of the adhesion tension is due to diffusion delay. The investigation of Hodgson and Berg [46] suggests that the use of surfactants to enhance the rate of penetration adsorption be made with caution, because for both the cases involving complete and incomplete wetting investigated, these rates were reduced.

A further caution which must be reiterated is that in dynamic wetting, Young’s equation (eq. 21) may be an oversimplification in that it is taken to imply a unique value for the contact angle dependent only upon the equilibrium "tensions" of the three interfaces meeting at the interline. What is actually observed in nearly all cases is a considerable difference (hysteresis) between the value observed for θ when the liquid is advancing over the solid surface (advancing contact angle) and that observed as the liquid retreats (receding contact angle). The advancing angle is always greater than (or equal to) the receding angle. The true origins of contact angle hysteresis are varied and complex involving at a minimum considerations of solid surface roughness and chemical heterogeneity [54,55]. Furthermore, it is often observed that the contact angle depends on the rate at which the interline is advancing or receding [56]. The subject of dynamic contact angle behavior is one of considerable current research [57-59], and there is little doubt that the presence of surfactants plays a significant role in determining the nature and magnitude of the hysteresis. The particular contact angle required to describe the wicking of liquid into a porous solid is the "dynamic advancing" angle, and it is perhaps naive to assume that this is the angle provided by the Young equation. In assessing the influence of surfactants on the contact angle, it is the effect on the dynamic advancing angle which must be sought.

Considering the simplest of the explanations which have been advanced for contact angle hysteresis, however, it is reasonable to use at least the functional form of the Young equation if account is taken of the surface geometry and correct values are assigned for the interfacial tensions. For roughened surfaces, the advancing angle would refer to the intrinsic contact angle, θ_o, augmented by the appropriate average slope of the surface asperities, as shown schematically in Fig. 14. (Receding angles would be reckoned similarly.) the advancing angle would be given by:

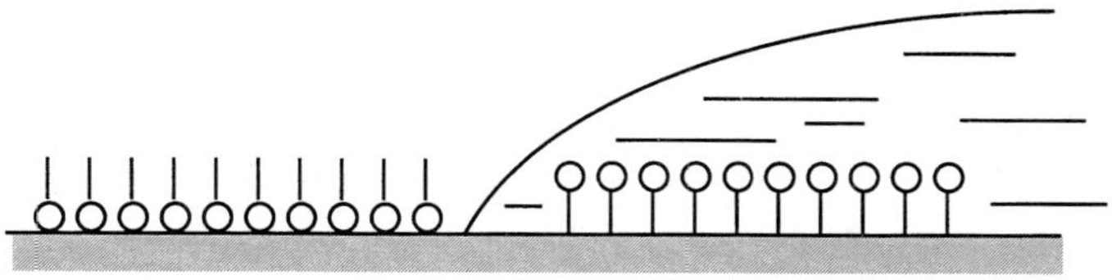

Fig. 15. Contact angle hysteresis due to adsorbed surfactant molecule reorientation upon interline movement.

$$\cos \theta_A = \cos(\theta_0 + \alpha) = \cos \theta_0 \cos \alpha - \sin \theta_0 \sin \alpha - (\cos \alpha) \cos \theta_0 \tag{25}$$

since α is generally quite small. Thus the form of eq. 18 is retained to within a multiplicative constant, $\cos \alpha$.

For surface consisting of a mixture of surface patches of different chemical composition, the advancing angle would refer to the patches of lower surface energy, i.e., lower values of $(\sigma_{SG} - \sigma_{SL})$, which would yield the higher value of the contact angle. Once again, the form of the Young equation is retained, with the appropriate significance being given to $(\sigma_{SG} - \sigma_{SL})$.

When a surfactant is present which can absorb from the liquid phase, hysteresis may be caused by its presence on the solid surface as the liquid retreats, as pictured in Fig. 3a., compared with its absence in front of an advancing liquid interline. Even when present at both the solid-gas and solid-liquid interface, the molecular orientation of the surfactant layer may be different for the two types of interface, as shown in Fig. 15. Thus, for example, a weakly absorbed nonionic surfactant at the interface between solid polyethylene and water will be oriented with the hydrophilic polyoxyethylene groups toward the aqueous phase, but when in contact with air, the molecules may overturn so that their lower energy hydrocarbon portions are outwards. In this case too, the form of the Young equation holds for both advancing and the receding contact angle, with the appropriate assignment of values to σ_{SG} and σ_{SL}.

If the structure of the porosity of the given solid is not such that it can be properly modelled as a bundle of uniform bore capillaries, the Washburn equation may not hold. For example, the appropriate form of eq. 17 for inward tapering circular capillaries of average radius r is [60]:

$$\frac{d\ell}{dt} = \frac{r\sigma\cos(\theta - \psi)}{4\mu}, \tag{26}$$

where φ is the angle of inward taper from the vertical. Optimum wicking would appear to correspond not to $\theta = 0^o$ but to $\theta = \psi$. The corresponding result for outward-tapering capillaries is

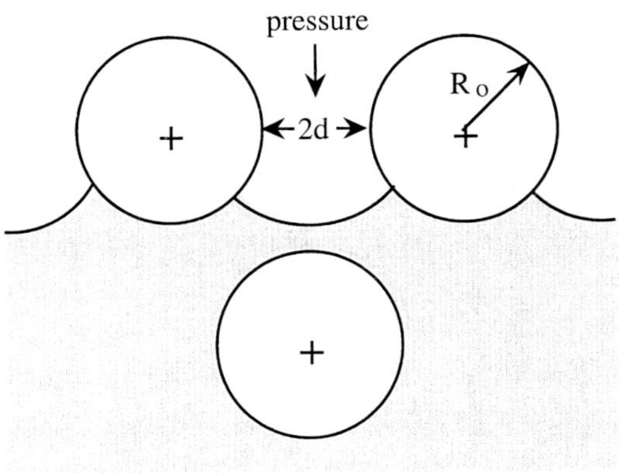

Fig. 16. Penetration of a liquid between horizontal

$$\frac{d\ell}{dt} = \frac{r\sigma\cos(\theta + \psi)}{4\mu}, \tag{27}$$

where ψ is the angle of outward taper from the vertical. Real porous matrices contain many inward and outward tapering regions, but if these are randomly and uniformly distributed, their effect appear to roughly cancel so that the simple form of the Washburn equation is applicable. One situation of great practical importance where they do not cancel is that presented by most textile yarns and woven fabrics. These may be modeled more closely as arrays of equi-distantly arranged parallel cylinders of radius R_0, spaced a distance $2d$ apart, with the cylinders in each layer being in echelon to those in adjacent layers, as shown in cross-section in Fig. 16. The array of cylinders is parallel to the undisturbed liquid surface. For this case, Baxter and Cassie [61] derived the result

$$P_{opp} = \sigma\left[R_o \cos\theta + \sqrt{(R_o + d)^2 - R_o^2 \sin^2\theta}\right], \tag{28}$$

where P_{opp} is the capillary pressure opposing entry of the liquid into the fiber mat of cylinders. Equation 28 is an important result because it indicates that penetration of liquid into the material will be favored by either a lower surface tension or a lower contact angle (by cylinders themselves) or both. Surfactants which have simply the effect of lowering the surface tension of the liquid without affect (σ_{SG} - σ_{SL}) may thus encourage penetration absorption into materials of this structure.

4.3. Motion of Liquid Threads

A liquid thread in a capillary with gas on both sides of it, as shown in Fig. 17, will move from left to right whenever the pressure difference $p_1 - p_2 \equiv \vec{\Delta} p$ is positive and remains positive when the liquid moves. If the pressure on each meniscus from the gas phase is the same, this pressure difference is given by:

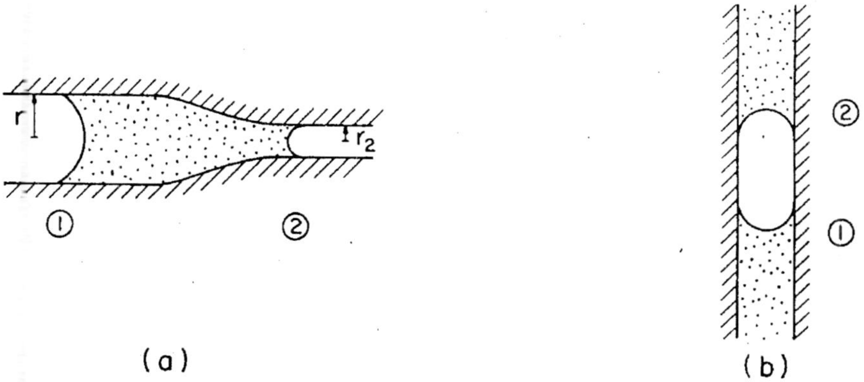

Fig. 17. Finite fluid elements in capillaries in porous solids: (a) liquid thread, (b) gas bubble.

$$\vec{\Delta} p = 2\sigma \left[\frac{\cos\,\theta_2}{r_2} - \frac{\cos\,\theta_1}{r_1} \right], \tag{29}$$

where subscripts 1 and 2 refer to conditions at the left and right of the thread, respectively. Motion of the thread is generally not affected by gravity. If the surface tension and contact angle are the same at both ends, and $\theta < 90^\circ$ (i.e., the solid is wet by the liquid), the liquid will flow spontaneously toward the smaller end of the capillary or into the smaller pores. On the other hand, as the liquid starts to move, the upstream meniscus will assume a configuration consistent with the receding contact angle, i.e., $\theta_1 = \theta_R$, while the downstream or leading meniscus will have the advancing angle, i.e., $\theta_2 = \theta_A$. Since $\theta_A > \theta_R$, contact angle hysteresis will always oppose such absorption of finite liquid masses into the interior of a porous solid. A surfactant will tend to reduce both σ and the contact angles, again with canceling effects, but in assessing the net effect of surfactants on absorption of finite liquid masses, one must consider not only the influence on contact angle and on surface tension, but also on the contact angle hysteresis. If the receding contact angle is already zero in the absence of surfactant, as is often the case even for only slightly hydrophilic solids, the pressure tending to move the liquid in toward the smaller capillaries becomes

$$\vec{\Delta} p = 2\sigma \left[\frac{\cos\,\theta_A}{r_2} - \frac{1}{r_1} \right] \tag{30}$$

If the advancing contact angle is at least proportional to the expression given by the Young equation, i.e.,

$$\cos\theta_A = k\frac{\sigma_{SG} - \sigma_{SL}}{\sigma}, \tag{31}$$

substitution into eq. 25 gives:

$$\vec{\Delta} p = 2k(\sigma_{SG} - \sigma_{SL}) - \frac{2\sigma}{r_1}, \tag{32}$$

from which it is seen that a decrease in surface tension alone without a counterbalancing increase in (σ_{SG} - σ_{SL}) will actually decrease the tendency of the liquid to move into the interior of the porous solid.

4.4. Immersional Absorption

The situation occurring when a porous mass is totally immersed is one in which liquid wicks into the solid from all directions temporarily trapping pockets of air. This is the type of absorption most important in the cleaning, dyeing, finishing or impregnation of textile fabrics. Absorbency in this mode is measured by the well-known Draves [62] or "sinking" test, * as well as others, discussed in more detail elsewhere in the monograph. The process by which such a pocket of air is removed is similar to that of the motion of a liquid thread and is pictured in Fig. 17 for the case of a bubble in a vertical, uniform-bore capillary. The bubble seeks to move upward by buoyancy but is held back by the forces due to contact angle hysteresis. The net upward pressure becomes:

$$\vec{\Delta} p \uparrow = \frac{\rho g V}{\pi r^2} - \frac{2\sigma}{r}\left[\cos\theta_R - \cos\theta_A\right], \tag{33}$$

where ρ is the density of the liquid (the air density is neglected), g is the gravitational constant and V is the volume of the air pocket. Again if $\cos\theta_R = 1$, as would often apply, and in any event would pertain to the final detachment of an air bubble, and if $\cos\theta_A$ is proportional to the Young expression:

$$\vec{\Delta} p \uparrow = \frac{\rho g V}{\pi r^2} + 2k\frac{(\sigma_{SG} - \sigma_{SL})}{r} - \frac{2\sigma}{r} \tag{34}$$

Here it is clearly evident that a decrease in surface tension will assist in immersion absorption, even in the absence of increases in (σ_{SG} - σ_{SL}).

4.5. Spreading Wetting vs. Wicking

In the penetration mode of absorption, the process may be impeded or even interrupted if the advancing liquid encounters large air pockets. These may be overcome if the liquid spreads as a thin film over the surface of the solid, replacing the solid-gas interface

* The Draves test is commonly, but somewhat misleadingly, referred to as a "wetting" test. Wetting (i.e., low contact angle) is a necessary but not sufficient condition for absorption. The Draves test actually measures the rate of immersional absorption.

with equal areas of solid-liquid and liquid-gas interface. The thermodynamic driving force this process, termed spreading wetting, is the spreading coefficient, S [63]:

$$S = \sigma_{SG} - \sigma_{SL} - \sigma \tag{35}$$

A positive value for S also causes the spreading out of liquid masses over the top of a porous solid prior to its inhibition. The condition for spreading wetting is more demanding than that for wicking. A value of $\theta_A = 0^o$ gives $S = 0$, and any finite value of the contact angle yields $S < 0$. Surfactants may produce spreading wetting by reducing the surface tension, reducing σ_{SL}, or a combination of these effects.

4.6. Surface Wicking; General driving Force for Interline Movement

A liquid may spread along grooves or rugosities on a surface even if it does not spread on a smooth surface of the same solid. The driving force for such surface wicking depends on the geometry of the grooves as well as the surface tension of the liquid and free energies of the solid-gas and solid-liquid interfaces.

Schwartz [64] has proposed a general thermodynamic approach to all wicking and spreading phenomena for cases when $S \leq 0$. The driving force for the movement of a liquid interline normal to itself a distance d*s* along a solid surface is given by:

$$\frac{dF}{ds} = \sigma \left[\frac{d}{ds}(A_{LG}) - \cos\theta \frac{d}{ds}(A_{SL}) \right], \tag{36}$$

where F is the total free energy of the system, and A_{LG} and A_{SL} are the liquid-gas and solid-liquid interfaces, respectively. It is important to note that the advancing contact angle is required. The interline will advance over unwetted solid surface if dF/dA is negative. Equation 36 provides the basis for a given wicking geometry, of sorting out the relative importance of the surface tension and the contact angle to the wicking process. For penetration absorption, $dA_{LG}/ds = 0$, and it is evident that the driving force for this type of wicking is proportional to $\sigma\cos\theta_A$ as indicated earlier using the Washburn equation. For a liquid thread advancing from a circular pore where the radius is large into a region where it is small, eq. 36 may be written to account for the difference in contact angle between that at the advancing interline and that at the receding interline:

$$\frac{dF}{ds} = \sigma \frac{dA_{LG}}{ds} - \sigma\cos\theta_A \frac{dA_{SL}^A}{ds} - \sigma\cos\theta_R \frac{dA_{SL}^R}{ds} \tag{37}$$

At the advancing meniscus, $dA_{SL}^A/ds = 2\pi r_A$ and at the receding meniscus, $dA_{SL}^R/ds = -2\pi r_R$, where r_A and r_R are the capillary radii. Substitution into eq. 37, noting that $dA_{LG}/ds = 0$, leads to

$$\frac{dF}{ds} = 2\pi\sigma\left(r_R\cos\theta_R - r_A\cos\theta_A\right) \tag{38}$$

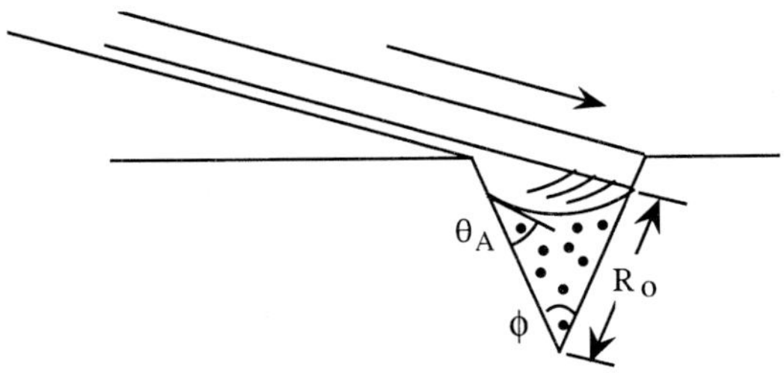

Figure 18. Surface wicking in an angular groove.

The net capillary pressure propelling the thread forward is obtained by dividing each term on the right hand side of eq. 38 by the appropriate cross-sectional area of the capillary, i.e. πr_A^2, respectively. This recovers eq. 28.

We may use eq. 36 to compute the conditions for a simple type of surface wicking, as shown in Fig. 18. Consider liquid from a large reservoir moving in the straight-edged horizontal trough at a depth R_0 measured along each side. The angle of the trough bottom is ϕ, measured in radians. Assume R_o and ϕ sufficiently small that the liquid surface may be regarded as a portion of a right cylinder. Under such conditions, advance of the liquid along the channel gives $dA_{LG}/ds = R_o\phi$ and $dA_{SL}/ds = 2R_o$. Equation 36 then gives:

$$\frac{dF}{ds} = \sigma R_o(\phi - 2\cos\theta_A), \text{ or more simply : } \theta_A < 90 - \frac{\phi}{2} \qquad (39)$$

The critical condition for surface wicking is thus that the liquid surface in the groove be concave upward, and once again the process is favored by low contact angle and high surface tension. Similar computations based on eq. 36 may be used to obtain surface wicking conditions for other geometries, as those pertaining to the grooves between parallel cylindrical textile fibers, etc.

4.7. Structural Changes in the Porous Solid

The foregoing discussion has indicated how wicking and spreading depend directly upon the surface tension of the liquid and the contact angle of the liquid against the solid (or alternatively, on the three surface free energies). Changes in the surface tension may also influence the structure of the porous solid, particularly when it is composed of flexible fibers. Liquid bridges form between these fibers, as shown in Fig. 19. These bridges draw the wetted fibers together so that the entry of the liquid is often self-blocking. The "adhesive" force between the fibers is directly proportional to the surface tension, so that with respect to the self-blocking effect, wicking flow is favored by low surface tension. Stiff-fibered structures, resistant to collapse, would be less subject to this effect.

4.8. Summary of Role of Contact Angle, Surface Tension and Interfacial Free Energies in Absorbency

In summary, penetration absorption of liquids into randomly-structured, non-collapsible solid porous media is favored by the smallest possible contact angle of the liquid against the solid and the largest possible value of the liquid surface tension. When

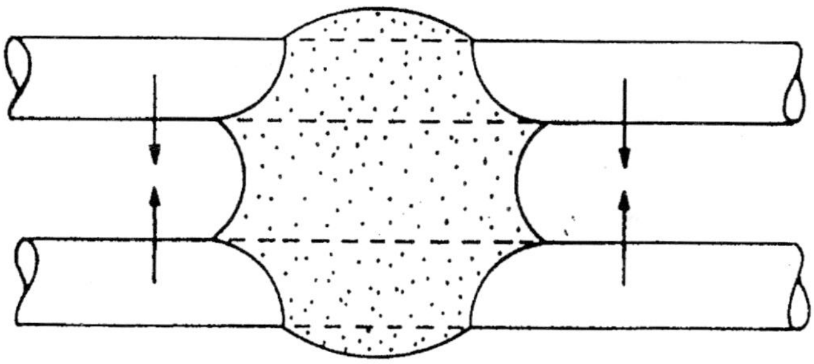

Fig. 19. Liquid bridge between fibers, drawing them together.

surfactants are added which absorb significantly only at the liquid-gas interface, the effect of the resulting reduction in the advancing contact angle is canceled by the simultaneous reduction in surface tension, in accord with eq. 19. Net increases in the rate of penetration are thus effected principally when (σ_{SG} - σ_{SL}) is increased. The presence of physically adsorbed* surfactants at interfaces at equilibrium will always reduce the corresponding surface energies. Thus an increase in (σ_{SG} - σ_{SL}) caused by such adsorption must be traced to the solid-liquid interface and the reduction in σ_{SL} that it can effect. In certain other structured or textured porous media the geometry of the system is such that a liquid surface tension reduction by itself, caused by surfactants, may be sufficient to promote penetration absorption. Also, when the porous solid consists of flexible fibers, low values of surface tension will tend to prevent the self-blocking effort of liquid bridges, and therefore may be favorable to wicking in such cases.

The motion of liquid threads into the interior of a porous solid and surface wicking in grooves along the surface are also favored by low contact angle and high surface tension. The movement of finite liquid masses is retarded by large contact angle hysteresis for a given advancing angle.

For Immersional absorption, where trapped pockets of air are to be removed from a porous mass, the process is favored by low surface tension, low contact angle and low contact angle hysteresis. The spreading of a liquid over a smooth surface requires a zero advancing contact angle and is favored by the lowest values for both σ_{SL} and σ.

The dependence of absorbency on surface properties thus depends on the nature of the wicking flow involved, the flexibility of the solid fibers or walls and the geometry of the porous matrix. In most cases of penetration absorption, however, absorbency is favored by low advancing contact angle and high liquid surface tension, so that surfactants used to promote such absorbency should be absorbed effectively at the solid-liquid interface and not so effectively (from the standpoint of surface tension reduction) at the gas-liquid interface. For immersional absorption, in which the removal of trapped air pockets is critical,

---

* Chemisorbed fibers may lead to an increase in surface free energy if the chemical anchoring of the absorbate molecule causes its lyophobic moiety to be exposed to the solvent medium.

absorbency is favored by both low advancing contact angle (which usually also means lower hysteresis) and low surface tension. Thus a surfactant chosen to promote such absorbency should adsorb effectively at the solid-liquid and significantly reduce the surface tension.

5. THE USE OF SURFACTANTS TO PROMOTE ABSORBENCY

There are two principal modes by which surfactants are used to promote absorbency. In one mode, the surfactant is added to the liquid prior to its contact with the solid porous mass. Used in this manner to promote absorbency by promoting wetting, surfactants are termed wetting agents. Wetting agents are used as aids in a wide variety of processes including the cleaning of natural fiber textile materials of greases and other dirt, paper making materials or rosins and pitches, and both natural and synthetic fibrous materials of lubricants, debonding agents and other chemicals used in various process steps. They are used in many processes whose object is the impregnation of porous solids such as textiles or papers with various materials such as sizing, wet- and dry-strength additives, antistatic agents, dyes, fillers, softeners, water-proofers, soil release additives, etc. In every case, the object of the wetting step is the rapid and thorough displacement of air from the surface of the solid so that a uniform and complete removal from the solid surface or deposition upon it of various other chemicals may be effected. The wetting agent is usually just a vehicle for the accomplishment of the necessary liquid-solid contact, but in some cases in which it is desired to deposit a chemical on the solid surface, the chemical itself may also be a wetting agent.

The second major mode in which surfactants are used to promote absorbency is as pre-treatments of the solid surface with an adsorbent which will promote subsequent wetting by a given liquid. Such a surfactant is termed a rewetting agent, and it acts by yielding a minimum value for the solid-liquid interfacial tension while reducing the gas-solid interfacial tension as little as possible. No significant effect on the liquid surface tension occurs unless the rewetting agent is rapidly desorbed. Rewetting agents are often important in the manufacture of absorbent products of the type listed in Table 1 and are often coincident with materials used as soil-release, wearing comfort and antistatic agents in the manufacture of textiles, particularly those composed of fibers which are hydrophobic in the absence of such surface treatment.

While it is sometimes true that good wetting agents for a given solid-liquid system are also good rewetting agents for the same system, the usual case is just the reverse. The criteria for choosing surfactants for either purpose and the optimal conditions for their application are examined below in light of the foregoing considerations of surfactant properties, adsorption equilibria and mechanisms of absorbency.

5.1. Surfactants as Wetting Agents

5.1.1. The dynamics of wetting

Whether the mode of absorption is one of penetration, immersion, a combination of these modes or one involving other types of wetting, the process by which solid is brought into contact with liquid is a dynamic one. This is especially important where the process is aided by surfactants used as wetting agents. The properties of the interfaces which are important are those existing at the advancing solid-liquid-gas interline. Since absorption of liquid is rapid (usually occurring in a few seconds to minutes) and involves the movement of the interline over large areas of solid in a short time, it is very unlikely that surfactant adsorption equilibria will exist near the interline [65]. The surfactant will be much depleted from its bulk concentration by absorption onto the solid-liquid interface (and also onto the gas-liquid interface if spreading or surface wicking is involved). The concentration of surfactant at the advancing front will depend on both the efficiency of surfactant mass transfer to the interfaces and the adsorption equilibrium, which determines the driving force for adsorption. The use of equilibrium values (corresponding to the bulk surfactant concentration) for the surface tension and contact angle in the Washburn equation and related expressions for various types of absorption promoted by wetting agents is likely to be seriously in error.

The mechanisms of mass transfer of dissolved surfactants to the interface are complex and depend on the fiber structure and the flow patterns in the liquid. Despite the complexities, the transport rate in systems below the CMC is generally limited by diffusion through a liquid boundary layer over the solid and therefore varies approximately as the square root of the diffusivity, D. In situations in which micelles exist, the transport rates may not follow a Fick's law dependence on the concentration gradient, and the effective diffusivity may be much reduced. There are some cases in which the adsorption step itself may be the rate determining step, with overall transport rates being as much as several orders of magnitude lower than expected from Fick's law diffusion [66]. The reason for such slow adsorption are many and varied but often can be traced to the processes of diffusion into micropores in the solid or into the solid itself. These processes are often not relevant to the relatively brief wetting process, although they are often of vital importance to other steps in the overall process (such as detergency) of which wetting is just the first step. Wetting agents have no primary effect in promoting the penetration of such micropores. A surfactant solution, for example, will not penetrate a sheet of cellophane any more rapidly than pure water [67]. For present purposes, it is reasonable to assume that the molecular diffusivity is a good measure of the efficiency of surfactant transport in wetting. The overall effectiveness of a surfactant in promoting the dynamic wetting process thus depends on a balance between effective adsorption equilibrium and effective bulk transport.

5.1.2. General criteria for good wetting agents

Adsorption equilibria and the interfacial property changes which promote wetting have been considered earlier. In order to improve adsorption, the adsorption must generally reduce the value of σ_{SL}. For aqueous media, this requires that the solid surface becomes more hydrophilic by amphipathic adsorption onto hydrophobic surfaces or hydrophobic patches on the solid surfaces. Such adsorption obeys Traube's rule, according to which the extent of equilibrium adsorption increases exponentially with the size of the hydrophilic portion of the surfactant for a given hydrophobe structure and a given hydrophilic group. In general, the denser the amphipathic monolayer coverage, the greater will be the reduction in the solid-liquid interfacial tension. The denser the film at the gas-liquid interface, the greater will be the reduction in the surface tension, also favoring immersional absorption. For a given hydrophobe size, denser adsorption is achieved for straight-chain compounds as opposed to those possessing significant branching. For a given hydrophobe, increasing hydrophilicity of the hydrophilic group leads to decreased adsorption. In particular, the equilibrium adsorption of anionic surfactants of a given hydrophobe on a negatively charged or uncharged textile fiber will be one-half to one-third of that corresponding to adsorption of nonionic surfactant of the same hydrophobe [68]. Cationic surfactants will adsorb with reversed orientation on negatively charged surfaces (amphiphilic adsorption), rendering them more hydrophobic.

The amount of adsorption at the advancing interline depends not only on the equilibrium adsorption for a given bulk surfactant concentration but also on the effective mass transfer coefficient, which is generally proportional to $\sqrt{D}$. The structural features of a surfactant leading to high diffusivity are often opposite to those leading to high equilibrium amphipathic adsorption so that choosing a wetting agent generally involves compromises.

In general, the diffusivity of a surfactant of a given hydrophilic group will decrease with the size of the hydrophobe. For example, for both sodium n-alkyl sulfates and unionized n-alkyl carboxylic acids, the effective diffusivity obeys the simple relationship [51]:

$$D = D_o - an \quad , \qquad (40)$$

where D_0 and a are constants, and n is the number of carbon atoms in the chain. Concentrations for which eq. 40 holds are well below the CMC, and as the chain length increases beyond n = 7 for the fatty acids and n = 14 for the alkyl sulfates, the dependence of the diffusivity on chain length is less pronounced. The constant "a" is approximately 0.7 for both types of compounds, but for a given chain length, the ionic sulfate has a diffusivity about five times as great as the unionized fatty acid. Data for amino acids [69], however, suggest that in general the diffusion coefficient in aqueous solution is decreased by any increase in the degree of hydration of the molecule. Diffusivities for polyoxyethylene nonionics are significantly lower than those of the corresponding anionics with the same hydrophobe [51]. Thus with regard to both linear hydrophobic chain length and hydrophilic group type, the requirements for rapid diffusion and high equilibrium adsorption are opposed. At bulk concentrations near the CMC (0.1%) the compromise for optimum wetting of negatively charged or neutral absorbents by n-alkyl anionic surfactants at 25°C is a chain length of 12-14 carbon atoms [69]. At very low surfactant concentrations (<0.01%) or at higher temperatures, the optimum corresponds to longer-chain compounds in both cases because of changes in the adsorption equilibria. As the concentration is raised, the longer-chain compounds are first to form micelles, and in general any factor which tends to promote micellization leads to a decreased effective diffusivity and hence decreased effectiveness as a wetting agent. Best results for wetting are usually obtained for bulk concentrations near the CMC of the particular surfactant. This leads to concentrations less than the CMC at the advancing front, but not so low that the extent of adsorption is small there. Under these conditions, the rate of wetting (immersional), as expressed, for example, by the wetting time, t_W, in the Draves test is given for $C_2 > 0$ by:

$$\log t_w = A' - B' \log C_2 , \qquad (41)$$

where C_2 is the surfactant concentration and A' and B' are empirical constants [69].

Interestingly, the same type of relationship as eq. 36 has been observed for penetration wetting [70] (although more data are needed to firmly establish it), despite the predicted non-dependence of that wetting mode on the liquid surface tension.

For nonionic surfactants, optimum wetting behavior at 250°C appears to be attained for compounds with 6-8 oxyethylenes units and an effective hydrophobic chain length of 10-11 carbon atoms [69]. Highest wetting power is generally achieved by surfactants whose

Table 3.
The effect of Hydrophobe chain Branching on Immersional Absorbency [58]. Results of wetting test DIN (German Bureau of Standards) 53901 with sodium hexadecylsulfates: $(C_{n_1}H_{2n_1+1})(C_{n_2}H_{2n_2+1})$ $CHCH_2OSO_3^-Na^+$

n_1/n_2	Wetting time (sec)
13/1	42
12/2	35
11/3	35
10/4	29
9/5	29
8/6	27
7/7	25

cloud points are just above the temperature at which the test is conducted, and for nonionic block copolymer surfactants, optimum wetting is achieved by materials with just enough polyoxyethene units to confer a minimal solubility.

For both ionic and nonionic [71] surfactants a considerable improvement in wetting behavior can be achieved by employing branched chain hydrophobes rather than straight chain compounds. In a study of a series of branched C_{14} sodium sulfates as wetting agents for cotton, the results shown in Table 3 have been reported [51]. Optimum wetting is achieved when the hydrophilic group is centrally located in the molecule. Equilibrium adsorption of the branched chain compounds (on activated carbon, which is a good representation of hydrophobic surfaces) is reduced progressively up to a factor of nearly two as the hydrophilic group moves from the end of the molecule to the center [51]. This unfavorable effect upon adsorption equilibrium is more than compensated by an increase in diffusivity. The diffusivity of the C_7-C_7 sulfate is more than twice that of the C_{14} straight-chain sulfate. For similar reasons, *o*-sulfonated alkylbenzenes are better wetting agents than the corresponding *p*-sulfonates. Much information on the behavior of specific surfactants as wetting agents can be found in surfactant manufacturer brochures and technical bulletins.

Little attention has been given to the ability of wetting agents to reduce the liquid surface tension as a criterion for choosing them. According to relationships developed earlier, penetration absorption will be negligibly (or sometimes adversely) affected by reductions in surface tension while other forms of wetting appear to be favored by such reductions. The situation is often not clear in a complex practical system. Generally, strong reductions in surface tension are desirable, but such reductions usually accompany the use of surfactants chosen on the basis of more critical criteria. The exception to the above are most polymeric surfactants, which in any event are poor wetting agents due to their slow diffusion to and adsorption at the solid-liquid interface.

It is interesting that the properties of surfactants leading to optimum wetting and optimum absorbency are usually not the same as those leading to overall optimum performance in many of the processes in which wetting is the first step. Thus good wetting agents are often poor candidates for detergency. Paraffin chain salts of 16 carbon atoms give good detergency but are poor wetting agents, whereas the C_{12} compounds are good wetting agents but poor detergents. Good detergent action depends on the highest possible equilibrium adsorption while wetting is more strongly dependent on kinetics. In choosing a wetting agent for use in conjunction with an overall process (such as detergency) involving several steps, other consideration must be taken into account. Powerful detergent action, for example, may be achieved by using mixtures of strongly adsorbing nonionics in mixture with branched-chain anionics [51].

5.2. Surfactants as Rewetting Agents

5.2.1. The purpose of rewetting agents

In many situations where high absorbency is desired it is not possible to add a wetting agent to the liquid prior to its contact with the solid. This is the case for most of the materials of Table 1 which are used for the absorption of physiological fluids, or the cleanup of liquid spills. The uptake of body moisture by textiles for wearing comfort cannot be controlled using wetting agents. Absorbency in such cases may be enhanced through the use of rewetting agents, i.e., surfactants which are absorbed to the surface of the solid matrix (generally from a liquid solution) and remain adsorbed with an orientation which enhances hydrophilicity (or lyophilicity) when the solvent medium is removed by evaporation. The properties required for a good rewetting agent are quite different from those for a good wetting agent, and many good wetting agents are poor rewetters, and vice versa.

Rewetting agents may be applied to a wide variety of solid substrates. Most of the synthetic textile fibers (polythenes, polyesters, polyamides, polyacrylonitriles, polyurethanes, etc.) and some natural fibers (e.g., wool, silk) are quite hydrophobic, giving products made from them poor absorbency and poor properties related to absorbency, viz, antistatic and soil release characteristics.

It has been suggested [72] that rewetting agents may also be used to upgrade relatively inexpensive, but insufficiently hydrophilic, wood pulps such as mechanical, thermomechanical or refined pulp so that they may be used in absorbent products.

Most of the absorbent products of the type listed in Table 1 as well as many textiles are presently made from cellulosic fibers, which are naturally hydrophilic. Cotton which has been thoroughly scoured and bleached to remove the natural fats, waxes and proteins that the raw material contains is generally in excess of 99% cellulose and exhibits instant and complete wetting and absorption upon contact with water [73]. It is also instantly wet by most organic liquids. The profusion of hydroxyl groups along the cellulose chain provide ample opportunity for hydrogen bonding. The heat evolved when dry cellulose is wetted with water amounts to 10 cal/g, indicating a strong interaction [74]. The advancing contact angle of water against pure cellulose fibers has been measured at 0° [75]. Cotton, particularly when oxidized, also contains isolated carboxyl groups and carbonyl groups, which do not detract significantly from its hydrophilicity. The use of rewetting agents (or wetting agents) does not improve upon the absorbency of pure cellulose materials. In fact, good anionic wetting agents for other materials show little tendency to adsorb on pure cellulose. The use of rewetting nd wetting agents for cotton is aimed at countering the effect of sometimes localized hydrophobicity due to incompletely-removed natural hydrophobic impurities or more often the hydrophobicity resulting from lubricants and sizes added at various stages in the preparation of the fabric. An unused cotton towel is often found to be poorly absorbent, but after one or two launderings is highly absorbent. The initial poor absorbency is caused by residual sizing and other finishing materials. A rewetting agent may be used to provide good initial absorbency. Many surfactants, as defined in this chapter, are good rewetters for this purpose. They provide good wetting only for the initial contacting of the material with water but do not provide a permanent enhancement of hydrophilicity.

Similar considerations apply in the use of rewetting agents on cellulosic materials prepared from wood. Many of the high absorbency products such as disposable diapers, kitchen towels, sanitary napkins, etc. are prepared from desized, dry-formed wood "fluff pulp". Here again, any behavior suggesting less than total hydrophilicity is traceable to hydrophobic impurities. The nature of these contaminants and the problems they present are somewhat different from those of cotton materials. The naturally-occurring "pulp extractives" consist of a mixture of resin acids, fatty acids and "unsaponifiables" (fatty alcohols, esters, etc.) Hardwood and softwood pulps also differ significantly from one another with respect to their extractives. Almost half the resin in softwood consists of resin acids, whereas hardwoods have none. Hardwoods have a lower total extractive content but a higher fraction of fatty acids and unsaponifiables. The latter are not generally removed in alkaline cooking, so that despite their lower overall content of hydrophobic impurities, hardwoods are less desirable for fluff pulp. (Another reason for preferring solfwood pulps is their generally greater fiber lengths.) Mechanical pulps of either kind are significantly less absorbent than chemical pulps due to their greater residual resin content. Solvent extraction of the resins from mechanical softwood pulp has been shown to increase absorbency (as measured by sink time) more than ten fold [76]. Rewetting agents thus can have an important role to play in the use of mechanical pulps for absorbent products. Even "fully deresinated" pulps contain residual pockets of resin which offer little deterrent to absorbency in the fresh product. It has long been known, however, that after aging for a period of weeks or months (particularly in hot weather), the product may become much less absorbent or even water-repellent. This problem of "self-sizing" has been traced to the slow vaporization and subsequent deposition of the unremoved resin pockets to form patches of hydrophobic surface on the cellulose [77]. Some rewetting agents have proven effective in stabilizing the pulps against self-sizing [78,79].

5.2.2. General criteria for good rewetting agents

In all cases, the role of rewetting agents is to cover either an entire hydrophobic surface or selectively cover hydrophobic patches of otherwise hydrophilic surfaces. There are a number of general criteria for good performance of a rewetting agent (for aqueous systems). First, it must adsorb strongly and evenly on the hydrophobic surface from aqueous solution exposing its hydrophilic moiety to the solution phase. This usually implies amphipathic adsorption, with all the same criteria for such adsorption discussed earlier. The kinetics of the adsorption are relatively unimportant in selecting rewetting agents as adequate time can be made available for equilibrium to be achieved. The relevant measure of the extent of adsorption is thus the equilibrium isotherm. Adsorption onto negatively-charged (the usual case for moderate to high pH) or neutral surfaces with the requisite orientation dictates the use of anionic or nonionic surfactants, with the latter generally being preferred due to the greater density of their equilibrium adsorption. Some examples of compounds commonly used as rewetting agents are those listed in Table 4. Many more are listed elsewhere [20]. Some of the specific example compounds are also good wetting agents, but most are not. In the latter case, other members of the same general type can usually be found which are good wetting agents, but poor rewetting agents. In addition to ordinary surfactants, a variety of

Table 4.
Examples of Nonionic Surfactants Used as Rewetting Agents.

Type	Examples	Application	Ref.
1. PEO alkylphenols	IGEPAL CO-630, CO-710	Paper towels	91
	TRITION N-57 X-14; X-155	Polyurethane sponge	20
2. PEO alcohols	STEROX AJ	Paper towels	92
	NEODOL 25-12, 91-8	Soil rewetting	93
	TRICOL DA-6	pad bath penetrant	94
3. PEO osters of fatty acids	STEROX CD	Synthetic and natural fabrics	95
	TRYDENT OA-10	Paper	94
4. Polyalkeylene oxide block copolymers	PLURONIC L31, L72, L81, L92, & L101	Absorbent papers, anti-self-sizing agents	78 96
5. PEO triglycerides	TRYLOX CO30	Wet-strength papers	94
6. Nonionic blends	WITCOSPERSE 201	Wettable-powder pesticides	97
7. Alkyl sulfosuccinates	PROTOWET D-75	Textile rewetting	98
8. Sulfonated alkyl	REXOWET RW	General rewetting	99
9. PEO cationics	BEROCELL 564	Soft paper, cellulose fluff, anti-self-sizing	79
10. Organic siloxane surfactants		Polyurethane foams as sponges, absorbent pads, tampons	100

hydrophilic polymers may be used as rewetters [80], but these are outside the scope of the present chapter.

A second general criterion for rewetting surfactants is that their anchorage to the solid surface be at least sufficiently strong that they resist reorientation and desorption over the period of time of the storage of the treated product or during the wetting period. Poorly anchored surfactants amphipathically adsorbed from aqueous solution may quickly overturn when the aqueous phase is removed and the surface is exposed to air. Thermodynamics favors such reorientation since it results in a lowering of surface free energy. Lower energy hydrocarbon groups, for example, are therefore exposed, rendering the surface hydrophobic. This may be an explanation for self-sizing in some cases, and is likely to be the cause of the observed loss in hydrophilicity with use of certain ultrafiltration membranes [81]. The overturning of monolayers in the opposite sense when the gas-solid interface is covered with water is also well-known [82,83] and may be rapid enough to be an explanation in some cases for observed contact angle hysteresis. In extreme cases, the anchorage may be so weak that the adsorbed surfactant desorbs into the solution. Such materials are poor rewetting agents. Strong amphipathic anchoring is favored by relatively large hydrophobes and hydrophilic groups of just sufficient size and hydrophilicity to confer the necessary solubility to the surfactant for purposes of its application. The head group hydrophilicity must also be sufficient to impart the necessary wetting characteristics to the surface. The latter consideration is seldom a limiting factor, however. These considerations again generally favor nonionics over ionics, and they are most effective when applied at a temperature just below the cloud point.

As discussed above, most non-polymeric rewetting agents are designed for single-use absorption ("non-durable finishes"), whereas textile finishes for soil release, antistatic and wearing comfort characteristics as well as absorbency are preferably permanent. If a high degree of permanence or substantively is required, the surfactant must generally be covalently bonded to the surface, or polymerized in situ, or cross-linked, etc.

One way of achieving an amphipathic anchor of required strength to synthetic textile fibers is to use a crystallizable hydrophobe containing sufficient repeat units identical with those of the synthetic fiber being treated that modest heating (curing) of the treated material leads to co-crystallization [20]. The hydrophilicity so imparted may have such permanence with repeated washings that the finish may be regarded as "durable." Another approach not requiring a chemical bonding between the surfactant and the surface is to design compounds capable of lateral chemical linkage (polymerization) once they are adsorbed. The resultant intertwining provides the anchor. A recent example of this type to the durable modification of various hydrophobic or partially hydrophobic fibrous materials employs a particular type of polyoxyethylene containing saline [84] capable of polymerizing in this way upon being cured after being amphipathically adsorbed on the substrate.

The more common approach to assuring anchorage of the surfactant sufficient in strength to prevent overturning in contact with air and desorption when rewetted with water is to employ an amphiphilic linkage with the substrate. One recent example [72] concerns the treatment of surfaces containing accessible hydroxyl groups. The surfactant used is a sulfosuccinate wherein the ester groups are substituted with amine and/or hydroxyl groups. It is fixed to the surfaces using a diepoxide as a bridging or cross-linking agent which forms ether linkages with the hydroxyl groups of surface and either ether or amide linkages with the functional end groups of the substituted surfactant. A similar use of diepoxide as a bridging agent for attaching surfactant to surfaces containing amine groups has also been made [85].

A very promising approach to obtaining significant permanence without chemical modification is to use surfactants capable of an electrostatic interaction with the surface but at the same time containing additional functional groups which interact strongly with water. The most important types of such surfactants are the polyethoxylated cationics, for use with negatively-charged substrates [70,79,86,87]. One example is shown in Fig. 20, but many others have been proposed. Some results obtained with the use of compounds of this type are rewetting agents for a mechanical fluff pulp are shown in Table 5. Ethoxylated citionics are especially effective as debonding agents in that they display the usual ability of citionics to reduce the interfiber bonds of the cellulose without the constant loss of hydrophilicity. The preferred structure of the surfactant does, however, involve a trade-off between debonding ability and wettability. Greater reduction in mechanical strength (i.e., debonding) is achieved as the number of carbon atoms in the alkyl chains, R, is increased, but wettability is decreased. The reverse occurs as the number of oxyethylene units increase, other parts of the molecule remaining unchanged. A typical compromise for compounds of the type of Fig. 20 has alkyl chains of 14 to 20 carbon atoms and n and m in the range from 2 to about 6 [86]. Improved performance with respect to both properties has been claimed for compounds in which in place of the R alkoxy groups, alkyl-substituted phenolic groups having from about

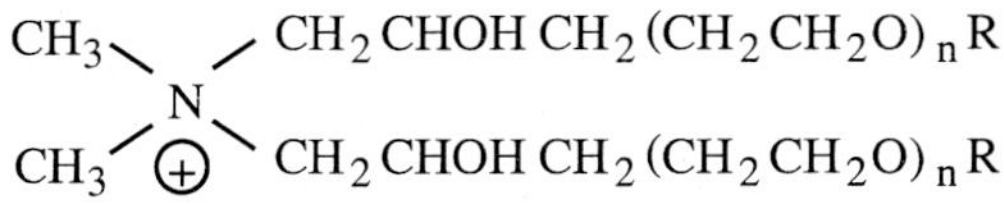

(a)

$$(CH_2)_nO-(CH_2CH_2O)_mH \quad + \quad \oplus NR_4$$

(b)

Fig. 20. Examples of surfactants of surfactant combinations used as rewetting agents and debonding agents for fluff pulp [70]. (a) Quaternary ammonium compound with two long chairs consisting of a PEO chain terminated with a hydrocarbon group. (b) PEO alcohol used in combination with a quaternary ammonium compound.

20 to 38 carbon atoms are used [87]. Additional improvement of the rewettability can be achieved by using nonionic surfactants such as PEO alcohols or alkylphenols in combination with the ethoxylated cationics [87]. Another approach to improving the strength of the amphipathic anchorage without chemical modification of the surface is through cooperative adsorption. This is accomplished by using a combination of a cationic and nonionic surfactant, as illustrated in Fig.. 20. The cationic establishes the initial adsorption by ion pairing, after which the nonionic possibilities for designing rewetting agents of this type are clearly evident. The substantivity of the cationic surfactant CTAB for cellulose has been studied extensively [25], for example, although a systematic study of the dependence of such adsorption on the cation structure appears not to have been made. The possibilities for designing rewetting surfactants for specific adsorption while conferring hydrophilicity to the surface would appear limitless. Any one of the various amphiphilic mechanisms discussed earlier could be employed. For example, polyvalent cation bridging could be used to fix polyoxyethylated anionic surfactants to negatively charged substrates, or polyoxyethylenated fatty acid may be bound by hydrogen bonding to the carbonyl oxygen of polyester fibers under acidic conditions.

Table 5.
Influence of Rewetting Agents of the Type shown in Fig. 20a on the Penetration Wetting of a Mechanical Fluff Pulp [70] of Initially Poor Absorbency. Results obtained before and after aging (self-sizing) are shown.

Surfactant CMC (%)	Absorption time (sec) Before aging	After aging
0	13	66
0.10	11	32
0.25	9	21
0.50	8	17

A third general criterion for choosing rewetting agents concerns the ability of the material to adsorb evenly over the solid. The surfactant concentration in the initial treatment bath should not be so high that when adsorption is complete a significant excess of surfactant is left in the solution. Such surfactant becomes deposited as bulk phase in the fiber pores and junctions where the water is last to evaporate upon drying. Most crystalline surfactants show poor wettability when first contacted with water, as the formation of the crystal in air exposes the lower-energy hydrophobic moieties of the individual molecules at the crystal surface. Wettability is eventually restored as the surface molecules reorient themselves upon prolonged contact with water, but the process is not rapid enough to be consistent with good rewetting characteristics for a dry porous solid containing such crystals. It is observed that surfactants which are liquids or amorphous solids when dry make better rewetting agents than those which are crystalline [88].

Beyond the general requirements for good rewetting characteristics, each situation usually presents its own requirements. If rewetting is to occur under strongly acid or alkaline conditions, for example, the anchoring "bond" cannot be one which is easily destroyed under such conditions. High temperature applications generally require larger hydrophobes for adequate amphipathic anchoring, etc. Occasionally, rewetting and wetting agents are both used in a given process. One example is the shrink-proofing of cotton by Sanforizing. The fabric is subjected to the mechanical rubbing action of a series of friction elements or "shoes" in a thoroughly wetted condition. Prior to initial drying, the material is treated with a rewetting agent. Then before entering the zone of mechanical action, the fabric is sprayed with water containing a wetting agent. Under such conditions, the wetting agent and rewetting agent must be compatible, i.e., there must not be a specific interaction between the hydrophilic groups of the two surfactants (as between an anionic and a cationic) producing an adsorbed bilayer with a hydrophobic exterior.

Frequently rewetting agents must be used in situations where other surfactants have been added for purposes other than to facilitate wetting. Often the surfactants act in opposition in achieving their specific objectives. Good compatibility should be sought, but it cannot always be achieved. One example of a situation of the above type, referred to earlier, occurs in the production of fluff pulp. Often a "debonding agent' must be added to the pulp prior to its drying to assist in the subsequent mechanical process of separating the fibers from one another (dry defibration) to product a low density fluff. The process is carried out in hammer mills, tooth roll defibrators or combinations of shredders and disc-refiners. The wood fibers are tightly bonded to one another through the multiple hydrogen bonds which form between the hydroxyl groups of the cellulose fiber surfaces. Debonding agents are surfactants_which adsorb to the surface and partially block the opportunity for such interfiber bonding. This is usually accomplished with amphiphilically adsorbed surfactants which reduce the fiber hydrophilicity. A solution to this problem is achieved by using surfactants of the type shown in Fig. 20 which yield both debonding and rewetting characteristics.

Similar problems are encountered in textile finishing [89]. Resiliency, dimensional stability and wrinkle resistance are associated with hydrophobicity while soil release, moisture absorbency and antistatic properties are associated with hydrophilicity. Finishes with start as surfactants but are often rendered "durable" through in situ bonding to the fiber or lateral polymerization, must provide acceptable compromises with respect to these properties. This often requires the use of fairly complex mixtures of surfactants, each having a different function. As an example, some quantitative results obtained using a mixture of butyl stearate (as a lubricant: hydrophobic), sorbitan mono-oleate (for hydrophilicity) and ethoxylated stearamine (for antistatic properties) for coating Nylon fibers have been reported [90].

6. SUMMARY

Surfactants provide a powerful means for promoting absorbency, acting either as wetting agents or rewetting agents. In either case, the adsorption of the surfactant to the solid surface exposing lyophilic functional groups is essential. Anchorage to the solid is usually amphipathic, but new compounds are being devised which are absorbed amphilically to the solid while exposing other lyophilic groups to the liquid. Substances used as wetting agents must diffuse quickly to the interface and therefore branched-chain, medium molecular weight materials are preferred. Governing considerations in the choice of rewetting agents, in addition to those of strong equilibrium adsorption, are the strength of the anchor against monolayer overturning and desorption. They must furthermore not accumulate as residual hydrophobic crystals in the dried porous solid. The use of surfactants as either wetting agents or rewetting agents to improve absorbency must be compatible with the presence of chemicals added for other purposes or with other process steps subsequent to the absorption itself.

7. ACKNOWLEDGEMENT

The author is grateful to Ross Rieke and Thomas Daugherty of the Procter and Gamble Company, Mark Bowns of the Weyerhauser Company and Jared Austin of the Kimberly-Clark Company for helpful discussions.

8. REFERENCES

1. M. J. Rosen, Surfactants and Interfacial Phenomena, 2nd edit.,Wiley-Interscience, New York, 1989.
2. J. L. Molliet, B. Collie and W. Black, Surface Activity, 2nd edit., Van Nostrand, Princeton, 1961.
3. A. M. Schwartz, W. Perry and J. Berch, Surface Active Agents and Detergents, Wiley-Interscience. New York, Vol. I, 1949, Vol. II, 1958.
4. R. Defay, I. Prigogine, A. Bellemans and D. H. Everett, Surface Tension and Adsorption, Longmans, London, 1966, Ch. 2, p. 21.
5. K. B. Blodgett, J. Am. Chem. Soc., 57 (1935) 1007-1022.
6. K. B. Blodgett and I. Langmuir, Phys. Rev., 51 (1937) 964-982.
7. A. Weissberger and B. W. Rossiter (Eds.), Physical Methods of Chemistry, Vol. I, Part V, Wiley-Interscience, New York, 1971, Ch. 9, p. 501.
8. R. Defay et al., loc. cit. [4], Ch. 7, p. 85.
9. S. Voyutsky, Colloid Chemistry, Mir. Moscow, 1978, pp. 155 ff.
10. W. J. Popiel, Indtroduction to Colloid Science, Exposition Press, Hicksville, New York, 1978, Ch. 7, p. 114.
11. L. I. Osipow, Surface chemistry, Reinhold, New York, 1962, pp. 185-188, 217-220.
12. J. L. Moilliet et al., loc. cit. [2], pp. 98 ff.
13. R. Defay et al., loc. cit. [4], pp. 110 ff.
14. A. W. Adamson, Physical Chemistry of Surfaces, 5th edit., Interscience, New York, 1990, p. 427.
15. P. Mukerjee and K. J. Mysels, Critical Micelle Concentrations of Aqueous Surfactant Systems. NSRDS-NBS 36, U.S. Dept. of Commerce, Washington, D.C. 1971.
16. M. J. Rosen, loc. cit. [1], Ch. 3, p. 108.
17. M. J. Rosen, loc. cit. [1], Ch. 5, p. 207.
18. J. L. Moilliet et al., loc. cit. [2], p. 79.
19. J. J. Kipling, Adsorption from Solution of Non-Electrolytes, Academic Press, London, 1965, pp. 179 ff.

20. J. E. McIntyre and M. M. Robertson (to ICI, Ltd.) U.S. Patent 3, 416, 952, 1968.
21. F. M. Fowkes and M. A. Mostafa, Ind. Eng. Chem. Prod. Dev., 17 (1978) 3-7.
22. G. J. Kahan, J. Coll. Sci., 6 (1951) 571-575.
23. M. E. Ginn, in E. Jungermann (Ed.), Cationic Surfactants, Surfactant Science Series, Vol. 4, Dekker, New York, 1970, Ch. 10, p. 341.
24. T. Wakamatsu and D. W. Fuerstenau, in W. J. Weber, Jr. and E. Matijevic (Eds.), Adsorption from Aqueous Solution, Advance in Chemistry Series, Vol. 79, ACS, Washington, D.C., 1968, Ch. 13, p. 161.
25. H. J. White, Jr., in E. Jungermann (Ed.), Cationic Surfactants, Surfactant Science Series, Vol. 4, Dekker, New York, 1970, Ch. 9, p. 311.
26. M. J. Rosen, Surfactants and Interfacial phenomena, Wiley interscience, New York, 1978, p. 51.
27. L. R. Snyder, J. Phys. Chem., 72 (1968) 489-494.
28. F. M. Fowkes, in L. H. Lee (Ed.), Adhesion and Adsorption of Polymers, Polymer Science and Technology, Vol. 12A, Plenum, New York, 1980, pp. 43-65.
29. R. S. Drago, G. C. Vogel and T. E. Needham, J. Am. Chem. Soc., 93 (1971) 6014-6026.
30. R. S. Drago, L. B. Parr, and C. S. Chamberlain, J. Am. Chem. Soc., 99 (1977) 3203-3209.
31. J. J. Keavney and R. J. Kulick, in J. P. Casey (Ed.), Pulp and Paper Chemistry and Chemical Technology, 3rd edn., Vol. III, Wiley-Interscience, New York, 1981, Ch. 16, p. 1547.
32. H. Schott, Kolloid-Z.Z. Polym., 219 (1967) 42-48.
33. R. H. Peters, Textile Chemistry, Vol. 3, The Physical chemistry of Dyeing, Elsevier, Amsterdam, 1975.
34. J. J. Kipling, loc. cit. [19], p. 185-197.
35. R. Defay et al., loc. cit. [4], Ch. 15, p. 217.
36. N. R. S. Hollies and R. F. Goldman, Clothing Comfort, Ann Arbor Science, Ann Arbor, 1977, pp. 45 ff.
37. C. M. Aberson, in D. Pge (Ed.), TAPPI STAP, 8 (1970) 282-307.
38. J. L. Moilliet et al., loc. cit. [2], pp. 157 ff.
39. N. R. S. Hollies, M. K. Kaessinger and H. Bogarty, Text. Res. J., 26 (1956) 829-835.
40. F. W. Minor, A. M. Schwartz, E. A. Wilkow and L. C. Buckles, Text. Res. J., 29 (1959) 931-939.
41. S. Martinis, J. L. Ferris, P. J. Balousek and M. P. Beetham, Preprint No. 7-3, TAPPI Ann. Mtg. Chicago, IL, March 2-5, 1981, pp. 1-8.
42. S. Levine and G. Neale, in J. F. Padday (Ed.), Wetting, Spreading and Adhesion, Academic Press, London, 1978, Ch. 11, p. 241.
43. D. H. Everett, J. M. Haynes and R. J. L. Miller in Fibre-Water Interactions in Paper-Making, Vol. 2, The British Paper and Board Industry Federation, William Clowes & Sons, London, 1978, pp. 519-534.
44. E. T. Hodgson and J. C. Berg, Wood fiber Sci., 20 (1), 3 (1988).
45. D. D. Fley and D. C. Pepper, Trans. Far. Soc., 42 (1946) 697-702.
46. K. T. Hodgson and J. C. Berg, J. Colloid Interface Sci., 121, 22 (1988).
47. R. Defay et al., loc. cit. [4], pp. 14 ff.
48. R. A. Pyter, G. Zografi and P. Mukerjee, J. Colloid Interface Sci., 89, 144 (1982).
49. B. Lindman, M. C. Puyal, R. Rymden and P. Stilbs, J. Phys. Chem., 88, 5089 (1984).
50. R. M. Wenheimer, D. F. Evans and E. L. Cussler, J. Colloid Interface Sci., 80, 357 (1981).
51. M. J. Schwuger, in E. H. Lucassen-Reynders (Ed.), anionic surfactants: Physical Chemistry of Surfactant Action, Surfactant Science Series, Vol 11, Dekker, New York, 1981, Ch. 7, p. 267.
52. L. C Gionotti, M. S. thesis, University of Washington, Seattle, 1982.
53. Y. Okamura, K. Gotoh, M. Kosaka and N. Tagawa, J. Adhesion Sci. Tech., 12, 639 (1998).
54. R. E. Johnson, Jr. and R. H. Dettre, in E. Matijevic (Ed.), Surface and Colloid Science, Vol. 2, Wiley-Interscience, New York, 1969, pp. 85-153.
55. R. J. Good, in R. J. Good and R. R. Stromberg (Eds.), Surface and Colloid Science, Vol. 11, Plenum, New York, 1979, pp. 1-29.
56. A. M. Schwartz and S. B. Tejada, J. Coll. Interface Sci., 38 (1972) 359-375.
57. E. B. Dussan V., Ann. Rev. Fluid Mech., 11 (1979) 371-400.
58. T. D. Blake, in J. C. Berg (Ed.), Wettability, Surfactant Science Series, Vol. 49, Dekker, New York, 1993, Ch. 5, p. 251.
59. S. F. Kistler, in J. C. Berg (Ed.), Wettability, Surfactant Science Series, Vol. 49, Dekker, New York, 1993, Ch. 6, p. 311.
60. N. K. Adam, Disc. Far. Soc., 3 (1948) 5-11.
61. S. Baxter and A. B. D. Cassie, J. Textile Inst., 36 (1945) T67-T90.
62. M. J. Rosen, loc. cit. [1], p. 260.
63. A. W. Adamson, loc. cit. [14], p. 110.

64. A. M. Schwartz, Ind. Eng. Chem., 61 (1969) 10-21.
65. D. Durham, Surface Activity and Detergency, MacMillan, London, 1961, pp. 206 ff.
66. J. L. Moilliet, et al., loc. cit. [2], pp. 126 ff.
67. A. W. Schwartz, W. Perry and J. Berch, Vol. I, loc. cit. [3], p. 322.
68. M. J. Schwuger and H. G. Smolka, Colloid Polym. Sci., 256 (1978) 1014-1020.
69. M. J. Rosen, loc. cit. [26], p. 190 ff.
70. H. Kolmodin, Research-Svensk Papperstidning, No. 12 (1981) R73-78.
71. M. J. Schick, Text. Chem. Col., 9 (1977) 31-37.
72. L. P. James (to Personal products Co.) U.S. Patent 4,368,323, 1983.
73. J. J. de Boer, Text. Tes. J., 10 (1980), 624-631.
74. D. Durham, loc. cit. [54], p. 84.
75. T. H. Daugherty, M. S. Thesis, University of Washington, Seattle, WA, 1981, p. 68.
76. J. H. Field, Preprint No. 6-2, TAPPI Ann. Mtg., Chicago, IL, March 2-5, 1981, pp. 125-131.
77. J. W. Swanson and S. Cordingly, TAPPI, 42 (1959) 812-819.
78. I. R. Smolka, in M. J. Schick (Ed.), Nonionic surfactants, Surfactant Science Series, Vol. 1, Dekker, New York, 1967, Ch. 10, p. 328.
79. Technical Bulletin on Berocell 564, Berol Kemi AB, 1982.
80. F. W. Marco (to Deering Milliken Res. Corp.) U.S. Patent 3,535,141, 1970.
81. L. A. Errede and P. D. Martinucci, Ind. Eng. Chem. Prod. Res. Dev., 19 (1980) 573-580.
82. D. Durham, loc. cit. [54], pp. 61-62.
83. J. A. Finch and G. W. Smith, in E. H. Lucassen-Reynders (Ed.), Anionic Surfactants: Physical Chemistry or Surfactant Action, Surfactant Science Series, Vol. 11, Dekker, New York, 1981, Ch. 8, p. 317.
84. J. Lipowitz and R. E. Kalinowski (to Dow Corning Corp.) U.S. Patent 4,207,071, 1980.
85. T. S. Keller, A. S. Hoffman, B. D. Ratner and B. J. McElroy, in K. L. Mittal (Ed.), Physicochemical Aspects of Polymer Surfaces, Vol. 2, Plenum, New York, 1983, p. 861.
86. J. G. Emmanuelsson and S. L. Wahlen (to Berol Kemi AB), U.S. Patent 4,144,122, 1979.
87. S. L. Wahlen and J. G. Emmanuelsson (to Berol Kemi AB), Can. Patent 218,551, 1978.
88. A. M. Schwartz, W. Perry, and J. Berch, Vol. I, loc. cit. [3], p. 324.
89. G. C. Tesoro, J. Am. Oil Chem. Soc., 45 (1968) 351-353.
90. G. E. P. Elliott, T. A. Elliott, S. M. Rowan and I. D. Severn, in J. F. Padday (Ed.), Wetting, Spreading and Adhesion, Academic Press, London, 19978, Ch. 17, p. 391.
91. C. R. Enyeart, in M. J. Schick (Ed.), Nonionic Surfactants, Surfactant Science Series, Vol. 1, Dekker, New York, 1967, Ch. 3, p. 44.
92. W. B. Satkowski, S. K. Huang and R. L. Liss, in M. J. Schick (Ed.), Nonionic Surfactants, Surfactant Science Series, Vol. 1, Dekker, New York, 1967, Ch. 4, p. 86.
93. Technical Bulletin SC:7-81, Shell Chemical Co., 1981.
94. Technical Bulletin S-45, Emery Industries, Inc., 1981.
95. W. B. Satkowski, S. K. Huang and R. L. Liss, in M. J. Schick (Ed.), Nonionic Surfactants, Surfactant Science Series, Vol. 1, Dekker, New York, 1967, Ch. 5, p. 142.
96. Technical Data on Pluronic Polyds, BASF Wyandotte, 1981.
97. Technical Bulletin No. 256, Witco Chemical Co., 1981.
98. McCutcheon's Emulsifiers and Detergents, North American Edition, MC Publ. Co., Glen Rock, NJ, 1981, p. 217.
99. McCutcheon's, ibid., p. 223.
100. K. G. Spitler and D. A. Radovich (to Mobay Chemical Corp.), U.S. Patent 4,128,512, 1978.

Absorbent Technology.
P.K. Chatterjee and B.S. Gupta, editors.

CHAPTER VI

FIBERS AND FIBROUS MATERIALS

LUDWIG REBENFELD
TRI/Princeton, P.O. Box 625, Princeton, New Jersey 08542, USA

Contents

1. INTRODUCTION

This chapter provides an overview of the structure and properties of the natural and synthetic fibers used in the production of textile materials, including woven, knitted and nonwoven products. The principal fiber types, including the cellulosics, acrylics, polyamides, polyesters, and polyolefins are discussed in some detail. Special attention is given to numerous specialty fibers that are used in absorption applications. These include alloy rayon fibers, modified cellulose fibers, acrylic and polyester fibers, and superabsorbent fibers. The methods of formation and the important structural characteristics of fibrous materials produced from fibers, including yarns, woven and knitted fabrics, and nonwoven products are also discussed.

The sorption of fluids by fibrous materials is of importance both in the production and in the utilization of a wide variety of products, ranging from classical woven and knitted textiles to the numerous types of nonwovens that are gaining increasing commercial utilization and consumer acceptance. Simply defined, fibrous materials are substrates that are composed of fibers. By a variety of processes, fibers are arranged into desired configurations and assemblies, and bound, by either chemical or mechanical forces, into reasonably cohesive substances that we shall refer to as fibrous materials. The salient structural characteristic of these materials is that they are discontinuous or heterogeneous, so that the spaces between the component fibers, referred to as pores, play a key role in their physical properties. Fibers themselves are frequently described as fibrous materials, but in our use of the term we shall take it to refer to the macroscopic structure that is composed of fibers. Many fibrous materials are specifically designed to serve as absorbent products, and the importance of sorption processes in these cases is quite obvious. Other fibrous products, whose primary function may not be absorbency, nevertheless frequently interact with various fluids, and an understanding of sorption processes in these cases is of importance also.

The sorption process is by no means simple. Concepts and ideas that may at first appear to be obvious and predictable are quite often not confirmed by observation. The complexity of the sorption process reflects to a large extent the rather complex nature of fibrous materials themselves. The purpose of this chapter is first to describe the various types of fibrous materials, and then to describe the chemical structure and physical properties of the major types of textile fibers used in the production of these materials. While some of the fibers may not be widely used in absorption applications, an understanding of their structure and properties will aid in achieving a better perspective of fibrous materials. Several specialty fibers specifically designed for absorbency applications are also discussed.

The physical properties of fibrous materials depend on two separate but somewhat interdependent factors. The first involves the individual component fibers constituting the fibrous material in question. Fibers have a wide range of chemical, electrical, mechanical, optical, and other properties of matter, and the properties of a fibrous product naturally depend on the properties of the fibers of which it is composed. A typical classification of such fiber properties is given in Table 1.

The second factor which influences the physical properties of fibrous materials is the geometric arrangement of the component fibers. The fibers comprising the fibrous material can be fully extended and straight, or curled but more or less aligned in a preferred orientation, as they are in typical textile yarns and fabrics. Alternatively, they can be randomly arranged as they are in certain nonwoven products. Manufacturing technologies are available to produce fibrous

Table 1.
Classification of Fiber Properties.

Geometric	Physical	Chemical
length	density	response to
average value	linear	acids
distribution	bulk	alkalies
		oxidation
cross section	thermal	reduction
average value	melting point	heat
distribution	transitions	
shape	conductivity	sorption
		moisture
crimp	optical	dyes
frequency	birefringence	
form	refractive index	swelling
	luster and color	anisotropy
	electrical	
	resistivity	
	dielectric constant	
	surface	
	roughness	
	friction	
	mechanical	
	tension	
	compression	
	torsion	
	bending	
	shear	

materials that lie anywhere between the two extremes of perfect alignment or orientation and complete isotropy or randomness. The important point is that the physical properties of fibrous materials depend both on the physical properties of the fibers themselves and on their geometric arrangement. This dual dependence is true for all physical properties of fibrous materials, including strength, abrasion resistance, resilience, and bulk density, and is particularly true for sorption characteristics.

In attempting to understand sorption characteristics in terms of fiber properties, one must distinguish to a certain extent between the sorption of liquids and that of vapors and gases. Adsorption and absorption of vapors and gases take place by the fibers themselves. The extent of this sorption depends on the chemical nature of the sorbing species and on the chemical and physical structure of the fiber. This type of sorption also takes place in the case of liquids; however, another form of sorption by fibrous materials is possible in the case of liquids, namely, that of capillary sorption. Interfiber spaces or pores in fibrous materials can have such dimensions that liquids can be held or retained by means of capillary forces. Fibrous materials are typical porous materials with porosity values ranging from about 0.6 to 0.9 and even higher.

For this type of liquid sorption, the surface properties of the fibers become controlling factors, because quite evidently the fiber surfaces constitute the pore walls. Thus, such physical quantities as fiber wettability and fiber surface energy become important. In general, vapors and gases are not held by capillary forces in interfiber spaces, although vapor condensation can occur under certain conditions. In such cases capillary sorption and fiber surface properties again become important. Furthermore, many fibers, particularly naturally occurring cellulosic fibers but also certain man-made fibers, have intricate (micro)pore structures themselves which can serve as capillaries for liquid (and vapor) sorption.

2. TYPES OF FIBROUS MATERIALS

2.1. General Considerations

One can distinguish among several types of fibrous materials primarily on the basis of fiber organization. On one end of the scale, there are isotropic assemblies where the fiber arrangement is completely random with no preferred orientation in any of the three principal spatial axes. By definition, the physical properties of an isotropic fiber assembly would be independent of the test direction, and any point in a unit volume space of such an assembly would be indistinguishable from another. Isotropic fiber assemblies are quite rare in most end-products and in fact are quite difficult to achieve because of the high length to width ratio of fibers. This high aspect ratio causes fibers to align themselves in stress fields, thereby creating a preferred orientation of the fibers in the processing direction. Thus, considerable effort must be exerted to overcome the tendency for preferred orientation when an isotropic assembly is desired. On the other end of the fiber organization scale, there is a wide variety of fibrous materials where the fibers are arranged in well defined spatial patterns.

2.2. Textile Yarns

Textile yarns and several preliminary linear structures, such as roving and sliver, are typical of those structures where there is a high degree of fiber orientation with respect to the principal axis of the material. The degree of fiber parallelization and orientation can be controlled by variation in fiber geometric properties and in the processing conditions used in yarn manufacture. The high degree of structural anisotropy of textile yarns is reflected in large differences between axial and transverse physical properties, and is the basis of their unique combination of high strength and low bending rigidity.

Textile yarns are produced from staple (finite length) fibers by a combination of processing steps referred to collectively as yarn spinning. Such yarns are referred to as spun yarns. Depending on the specific processing conditions, the degree of fiber parallelization and surface hairiness of spun yarns can vary over a considerable range. These structural characteristics strongly influence physical properties. Sorption properties in particular are affected by the fiber organization in a spun yarn. The staple fibers may be either natural fibers, such as cotton or wool, or any of a number of man-made fibers. The cohesion of spun yarns is achieved through interfiber friction effects and by means of a twist that is introduced during spinning. Schematic representations of typical yarns are given in Fig. 1.

Textile yarns are also produced from continuous filament man-made fibers. In these multifilament yarns, there is nearly complete filament alignment and parallelization with respect to the yarn axis. The degree of twist introduced into a multifilament yarn is usually quite low

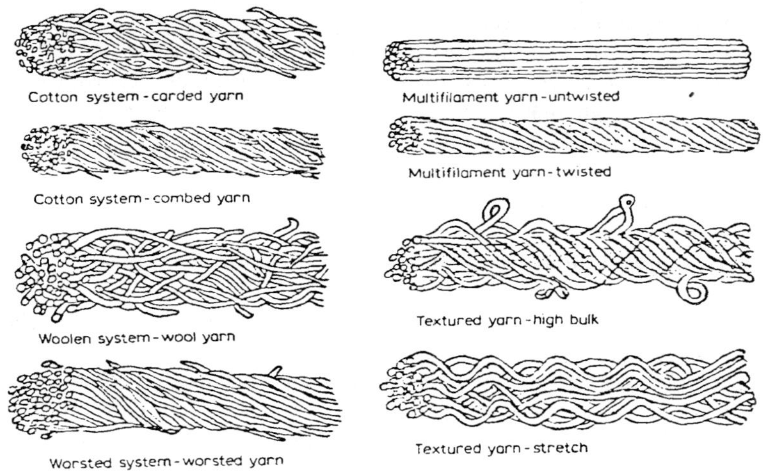

Figure 1. Schematic description of spun and multifilament yarns; from Goswami, Martindale and Scardino [1].

and just adequate to produce some level of interfilament cohesion. Such yarns are quite compact and smooth in appearance. A variety of processes have been developed to introduce bulk and texture to multifilament yarns [1].

2.3. Textile Fabrics

Yarns are used principally in the formation of textile fabrics either by weaving or knitting processes. Fabrics are a form of planar fibrous assembly where the high degree of structural anisotropy characteristic of yarns is minimized but not totally eliminated. In a woven fabric, two systems of yarns, known as the warp and the filling, are interlaced at right angles to each other in various patterns. The woven fabric can be viewed as a planar sheet-like material with pores or holes created by the yarn interlacing pattern. The dimensions of the fabric pores are determined by the yarn structure and dimensions, and by the weaving pattern. Within a given fabric, the pores are reasonably uniform in size and shape. The physical properties of woven fabrics are strongly dependent on the test direction.

Generally strength is higher and extensibility is lower in the warp direction than in the filling direction. These properties are usually intermediate in the non-orthogonal directions, but specific values depend on the weaving pattern [2]. Schematics of some simple weave patters are shown in Fig. 2.

Knitted fabrics are produced from one set of yarns by looping and interlocking processes to form a planar structure. The pores in knitted fabrics are usually not uniform in size and shape and again depend largely on yarn dimensions and structure and on the numerous variables of the knitting process. Knitted fabrics are normally quite deformable, and again physical properties are strongly dependent on the test direction.

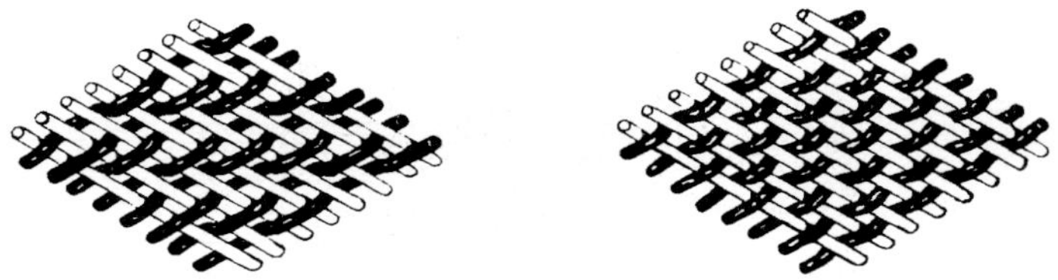

Figure 2. Schematics of typical simple weave patterns; from Berry [2].

2.4. Nonwovens

The term "nonwoven" simply suggests a textile material that has been produced by means other than weaving, yet these materials really represent a rather unique class of fibrous structures. In nonwovens the fibers are processed directly into a planar sheet-like fabric structure, bypassing the intermediate one-dimensional yarn state. The fibers are then either bonded chemically or interlocked mechanically (or both) to achieve a cohesive fabric such as is shown in Fig. 3.

Typically, fibers are dispersed in a fluid (liquid in the wet-laid process of manufacture or air in the dry-laid process of manufacture) and deposited in sheet-like planar form on a support base prior to bonding or interlocking. The paper-making process is a well known example of a wet-laid nonwoven process which utilizes short paper (wood) fibers. Within the plane of a nonwoven material, the fibers may be either completely isotropic or there may be a preferred fiber orientation or alignment usually with respect to a machine or processing direction. Fiber orientation may be randomized into the third dimension, perpendicular to the plane of the fabric, by a process known as needlepunching. The type of nonwoven material that is produced depends largely on the fiber type used and on the method of manufacture [3]. Typically air-laid nonwovens are less dense and compact and would tend to be softer, more deformable, and somewhat weaker. Wet-laid or paper-like nonwovens would be more dense, stronger, more brittle and less permeable to fluids. It is dangerous, however, to generalize. Nonwovens by either process can be produced to achieve a wide variety of products with a broad range of physical properties. The methods used for bonding and for interlocking the constituent fibers to form the cohesive nonwoven material may be of greater importance than the method used for producing the initial nonwoven web. The degree of bonding or interlocking also has a critical influence on the final properties of the nonwoven material. In general, strength increases, extensibility decreases, softness decreases and sorptive capacity decreases with increasing degree of bonding or interlocking [4-7]. In view of the importance of fiber orientation in nonwovens in terms of their properties, there has been much research on quantifying their structure by image analysis techniques [8].

Figure 3. Scanning electron photomicrograph of a polyester nonwoven fabric; courtesy of S.B. Ruetsch, TRI/Princeton.

Table 2.
Classification of Textile Fibers

A. Naturally occurring fibers

1. Vegetable: (based on cellulose) cotton, linen, hemp, jute, ramie, kenaf
2. Animal: (based on proteins) wool, mohair, vicuna, other animal hairs, silk
3. Mineral; asbestos

B. Man-made fibers

1. Based on natural organic polymers
   - a. rayon: regenerated cellulose (viscose and cuprammonium processes)
   - b. lyocell: regenerated cellulose (solvent process)
   - c. acetate: partially acetylated cellulose derivative
   - c. triacetate: fully acetylated cellulose derivative
   - e. azlon: regenerated protein
2. Based on synthetic organic polymers
   - a. acrylic: based on polyacrylonitrile (also modacrylic)
   - b. aramid: based on aromatic polyamides
   - c. nylon: based on aliphatic polyamides
   - d. olefin: based on polyolefins (polypropylene)
   - e. polyester: based on a polyester of an aromatic dicarboxylic acid and a dihydric alcohol
   - f. spandex: based on segmented polyurethane
   - g. vinyon: based on polyvinyl chloride
   - h. vinal: based on polyvinyl alcohol
   - i. carbon: based on polyacrylonitrile, rayon or pitch
3. Based on inorganic substances
   - a. glass
   - b. metallic
   - c. ceramic

3. STRUCTURE AND MORPHOLOGY OF FIBERS

3.1. Classification of Fibers

Textile fibers used in the production of fibrous materials may be conveniently divided into two major categories and into several subcategories as indicated in Table 2. Naturally occurring fibers are derived from the vegetable, animal and mineral kingdoms. Among the natural fibers indicated in Table 2, cotton and wool are by far the most important from a commercial point of view. In 1996 the worldwide production of cotton amounted to 19.4 million metric tons, and that of wool to 1.5 million metric tons [9]. The worldwide production of other vegetable fibers, including linen, jute, hemp, ramie, and kenaf, referred to as bast fibers, is estimated at about 5.0 million metric tons.

Man-made fibers may be categorized according to those based on naturally occurring polymers, those based on synthetic organic polymers, and those based on inorganic substances. The consumption of man-made fibers based on synthetic polymers, frequently referred to as synthetic fibers, has grown rapidly since their introduction into various markets in the 1940s and 1950s. Today in the United States, more than 50% of the fibers used in textile applications are man-made. The 1996 worldwide production of man-made fibers amounted to about 27 million metric tons [9].

3.2. Concepts of Fiber Structure

With the exception of glass fiber, asbestos, and a variety of specialty fibers based on inorganic substances, fibers are a class of solid organic polymers that are distinguishable from other polymers by their physical properties and by their characteristic geometric dimensions. A fiber is readily identifiable as a substance that is extremely long with respect to its width (or diameter), is flexible, and has highly anisotropic physical properties. However, from a more fundamental point of view, one wishes to differentiate between fibers and other solid organic polymers on the basis of molecular structure. It is self-evident that the chemical and physical properties of fibers are a reflection of their molecular structure and also of the intermolecular organization [10,11].

For a complete description of fiber structure, it is useful to consider three levels of molecular organization, each relating to certain aspects of fiber behavior and properties. The organochemical structure defines the structure of the repeating unit in the base polymer, and the nature of the polymeric link. This level of molecular structure is directly related to chemical properties, dyeability, moisture sorption, swelling characteristics, and indirectly to all physical properties. The macromolecular structure describes the family of polymer molecules in terms of chain length, chain-length distribution, chain stiffness, molecular size, and molecular shape. The supermolecular structure provides a description of the arrangement of the polymer chains, primarily in terms of such factors as orientation, crystallinity, and fibrillar character.

As a broad generalization, all fibers that are used in textile applications are semicrystalline, irreversibly oriented polymers. This means that fibers have certain regions in which the molecular chains are highly oriented, closely packed, and arranged in near-perfect register. These regions are usually referred to as crystalline regions or as crystallites. The degree of orientation and the degree of crystallinity are important quantities that strongly influence fiber physical properties. In other regions of the fiber the molecular chains are not as well ordered, tending to a random-coil configuration, and these are usually referred to as amorphous regions. In the case of natural fibers, there are also various identifiable aggregates of polymer chains

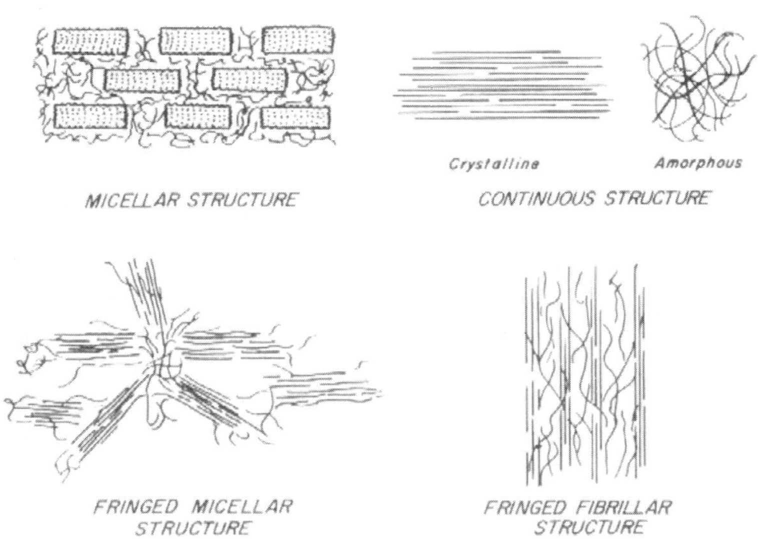

Figure 4. Schematic representation of several fiber structure models; from Rebenfeld [10].

which have been referred to at various times as micelles, fibrils, microfibrils, and macrofibrils. These morphological characteristics play an important role in fiber chemical and physical properties, and are of special importance with regard to absorbency behavior. Schematic views of several fiber structure models are shown in Fig. 4.

4. DESCRIPTION OF GENERIC FIBER TYPES

4.1. Cellulosic Fibers

As is evident from the classification of textile fibers shown in Table 2, cellulosic fibers bridge the gap between natural and man-made fibers. The principal cellulosic fibers are cotton, rayon, acetate, and triacetate. While wood pulp fiber is strictly speaking not a textile fiber because of its relatively short length (2-3 mm), it is an important cellulosic fiber that is also used extensively in fibrous materials associated with absorption applications. The chemical structures descriptive of the principal cellulosic fibers are shown in Fig. 5.

4.1.1. Natural Cellulosic Fibers. The supermolecular structure of natural cellulosic fibers is quite complex and has an important bearing on their physical properties. Some of the unique supermolecular and morphological characteristics of cotton have been well documented by microscopy techniques [12]. The structural features of a cotton fiber are illustrated schematically in Fig. 6 and a photomicrograph of cotton is shown in Fig. 7. The secondary wall of the cotton

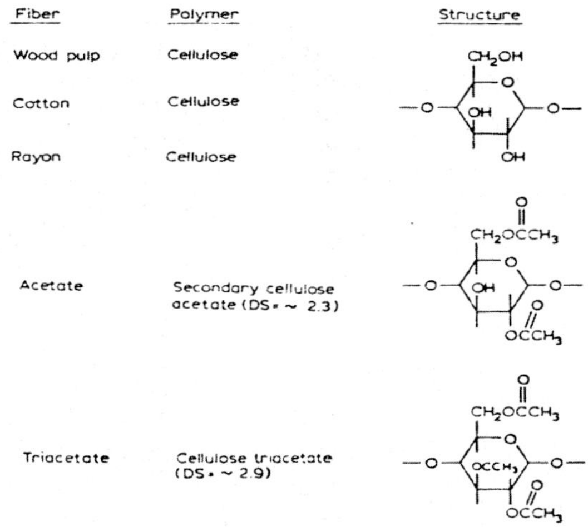

Figure 5. Chemical structures of cellulosic fibers.

fiber, which accounts for the bulk of the fiber, is composed of uniformly sized fibrils in which the cellulose chains are well aligned and closely packed in crystalline units [13]. These fibrils, oriented at a defined angle (ca. 23°) with respect to the fiber axis, are packed in the fiber in concentric sheets or layers. These layers are referred to as growth layers, because their deposition within the fiber is associated with the diurnal cycle during fiber development. The rate of secondary wall development is a varietal characteristic, but it is affected by environmental

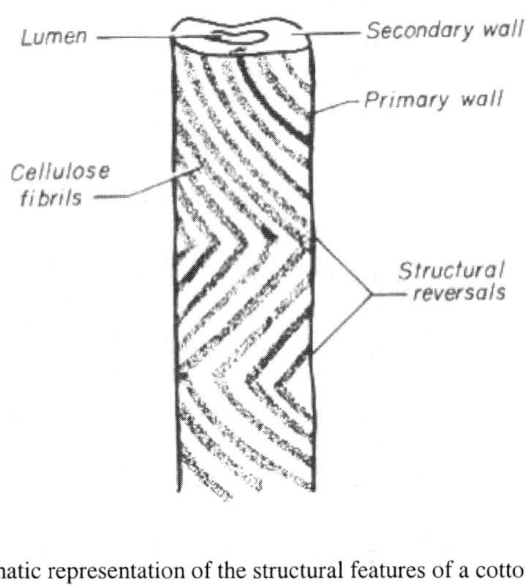

Figure 6. . Schematic representation of the structural features of a cotton fiber; from Rebenfeld [10].

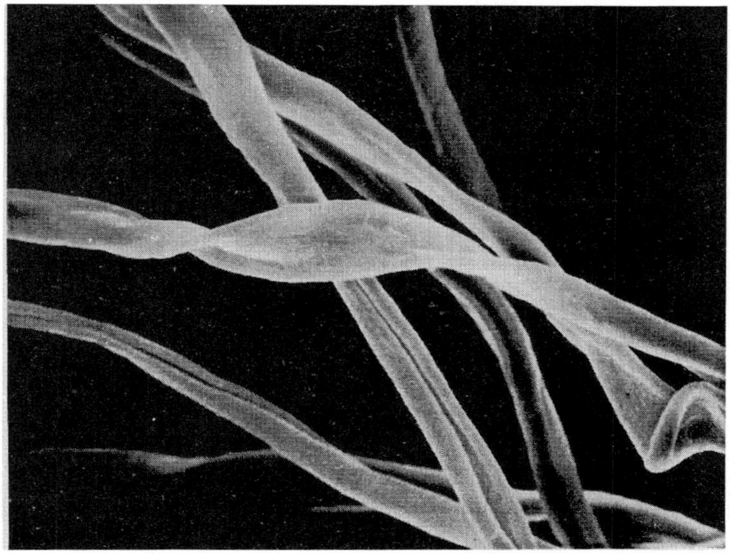

Figure 7. Scanning electron micrograph of cotton fibers; courtesy of S.B. Ruetsch, TRI/Princeton.

conditions such as light intensity and temperature [14].

The outer skin of a cotton fiber, referred to as the primary wall, is also largely cellulosic and fibrillar but the fibrils are overlapping and crossed, providing a tougher surface structure which serves to protect the inner secondary wall of the fiber. It has been shown that the primary wall is remarkably strong, probably reflecting its basket-weave type structure [15]. However, it has also been established that the cellulose in the primary wall is quite low in molecular weight [16]. The outermost layer of the primary wall is essentially, hydrophobic because of the presence of waxy substances and other non-cellulosic compounds. In the secondary wall, the fibrillar deposition is in the form of a helix with the sense of the helix reversing frequently along the length of the fiber [17]. These points of reversal provide the fiber with considerable extensibility and resilience. These helical reversals are unique for cotton.Wood pulp and other natural cellulosic fibers also have fibrillar structures and some type of growth layer features. However, these fibers must be considered as composites since the cellulosic fibrils are effectively dispersed in a lignin matrix [18]. In general, wood pulp and many of the other natural cellulosic fibers are stronger but less extensible than cotton [19]. Rayon and acetate fibers are not fibrillar in character because the natural cellulose morphology is eliminated during the manufacturing process. These fibers have distinctly different cross-sectional shapes as is illustrated in Fig. 8.

4.1.2. Man-Made Cellulosic Fibers. The cellulose polymer is highly reactive because of the three free hydroxyl groups on each anhydroglucose unit, the basic monomeric repeating unit of cellulose. Two of these hydroxyl groups on cellulose are secondary while one is primary. The chemical reactivities for the three hydroxyl groups vary as might be expected from well known reaction rates for aliphatic alcohols. Cellulose can undergo the typical reactions of alcohols to

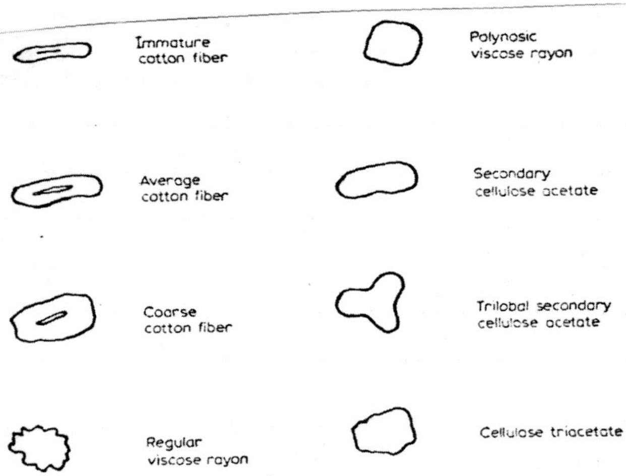

Figure 8. Cross-sectional shapes of some cellulosic fibers.

form cellulose derivatives such as the cellulose acetates and butyrates, methylcellulose, carboxymethylcellulose, hydroxyethylcellulose, and a nearly inexhaustible list of other organic and inorganic cellulose ethers and esters [20].

The three free hydroxyl groups in cellulose serve as principal sorption sites for water molecules. Directly sorbed water is firmly chemisorbed on the cellulosic hydroxyl group by hydrogen bonding, while additional water is held somewhat less firmly through secondary polar interactions. Presumably water is sorbed only in the noncrystalline or amorphous domains and does not penetrate the cellulose crystallites. It is important, however, to recognize that the crystallites are not necessarily structurally perfect and therefore may be somewhat penetrable by water and other solutes. Also one must include the surfaces of cellulose crystallites as accessible sorption sites, and therefore the crystallite dimensions (and shape), and thus their surface to volume ratio, is important in determining the sorptive capacity of cellulosic fibers. Primarily, however, it is the degree of crystallinity that controls the moisture sorption capacity of cellulosic fibers, as is evident from the data in Table 3. The equilibrium moisture regain of the secondary cellulose acetate fibers, and even of the triacetate fibers, is quite high despite the fact that many of the hydroxyl groups are acetylated and therefore no longer can serve as sorption sites for water. This exemplifies the importance of accessibility and of the three-dimensional superstructure of fibers. The acetyl groups because of their size relative to free hydroxyl groups open the structure of the fiber to such an extent that the remaining free hydroxyl groups are easily accessible to water and can serve effectively as primary and secondary sorption sites.

Table 3.
Crystallinity and Moisture Regain of Cellulosic Fibers

Fiber	Degree of Crystallinity (%)	Moisture Regain at 65% R.H. (%)
Cotton	70-90	7-8
Mercerized cotton	50-70	9-10
Rayon and lyocell	30-50	11-13
Acetate(secondary)	10-20	6.0-6.5
Triacetate	40-60	4.0-4.5

The hydroxyl groups on the anhydroglucose units of cellulose are also capable of strong intermolecular and intramolecular hydrogen bonding. Such bonding accounts for the high cohesive energy density of cellulosic fibers and for the high degree of stability and impenetrability of the cellulose crystallites. The inherent stiffness of the cellulose chain also contributes to these characteristics. These factors are also relatable to the mechanical properties of these fibers. The natural cellulosic fibers, such as cotton, are strong, modestly extensible, and have a high modulus of elasticity. Cotton fibers are quite flexible reflecting the fact that their cross-sectional shape is highly elliptical or even ribbon-like. The regenerated cellulosic fibers (rayon) are much weaker due to their lower degree of structure regularity and perfection, and also the lower degree of polymerization (molecular weight) of the cellulose chains. The derivative fibers, cellulose acetate and triacetate, are also weaker and more extensible than cotton and more nearly equivalent to rayon. Cellulose triacetate fibers, because of their high degree of acetylation, are thermoplastic. This means that they can be stabilized or heat-set at 170 to 220°C for 1 to 2 minutes [21].

The major obstacle in the production of regenerated man-made cellulosic fibers from naturally occurring cellulose is the lack of solubility of this polymer in common non-degrading solvents. It is necessary to use complexing solvents, such as aqueous cuprammonium hydroxide or cupriethylene diamine, or to form partial derivatives, such as the acetates or nitrates, which are then soluble in organic solvents. Several processes for the manufacture of regenerated (and derivative) cellulosic fibers are based on these solubilization methods, but none has achieved the success of the well known viscose process. This process for the production of rayon from wood pulp and/or cotton linters was the first commercially successful process for the manufacture of man-made fibers. The viscose process involves the formation of cellulose xanthate by reaction of cellulose impregnated with sodium hydroxide with carbon disulfide, with subsequent extrusion of the solubilized cellulose xanthate in a wet spinning process to form the regenerated cellulosic fiber [22]. It is an enormously versatile process that allows the production of a wide range of diverse fibers with applications not only in traditional textile products but also in industrial markets, notably tire cord. As environmental issues became of increasing concern, and despite many important technological improvements in process control, the viscose process did not meet the air and water pollution standards established in the United States and in many other industrialized nations. The viscose process continues to be the major process for the

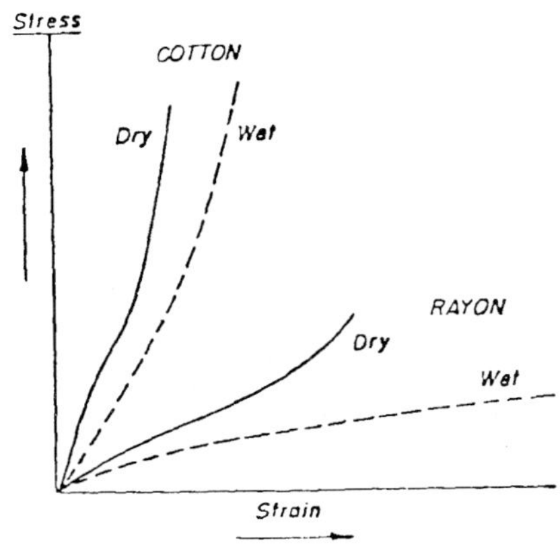

Figure 9. Typical stress-strain curves of cotton and rayon fibers in the dry and wet states.

production of rayon, but in view of environmental regulations it is practiced only in South America, Eastern Europe, Asia, and in the developing regions of the world.

With the obvious environmental limitations of the viscose process, and in view of the desirable properties of man-made cellulosic fibers, a major research effort was undertaken for alternate processes that would be environmentally acceptable. The search focused on means of dissolving cellulose without derivitization directly in a solvent from which the fiber could be formed by solvent or dry spinning. The intensive research efforts culminated in a new type of cellulosic fiber, generically designated as lyocell, which is produced by direct solvent or dry spinning of cellulose dissolved in aqueous N-methyl-morpholine oxide [23]. With complete solvent recovery, the lyocell process is an environmentally acceptable technology for the production of man-made cellulosic fibers, and represents an important innovation in fiber technology. Several major lyocell production plants are now operative in the United States, Canada, and Europe.

4.1.3. Properties of Cellulosic Fibers. The physical properties of cellulosic fibers reflect their highly hydrophilic chemical structure and their complex fine structure. Typical properties of the textile cellulosic fibers are given in Table 4. Of particular interest in the case of cellulosic fibers is the response of their mechanical properties to variations in moisture content. Generally, in the case of regenerated and derivative cellulosic fibers, strength decreases and extensibility increases with increasing moisture content. In contrast, the strength of cotton is either unaffected by variations in moisture content or may even increase with increasing moisture. Typical stress-strain curves of cotton and rayon at 65% RH and in the wet state are shown in Fig. 9. The contrast between these fibers in their response to moisture is generally explained in terms of intermolecular bonding between cellulose chains and their relative degrees of crystallinity [24,25]. Cotton with its high crystallinity and high degree of intermolecular bonding is effectively a highly cross-linked polymer system. This implies polymer rigidity and a lack of chain mobility. A cotton fiber is strong because of this effective cross-linking, but under an

Table 4.
Typical Physical Properties of Cellulosic Fibers

	Cotton	Rayon and Lyocell	Polynosic Rayon	Acetate	Triacetate
Tenacity at Break (gpd)					
65% RH	3.0-5.0	1.5-3.5	3.0-5.5	1.0-1.5	1.1-1.4
Wet	3.0-6.0	0.8-2.5	2.5-5.0	0.7-1.1	0.8-1.0
Exten. at Break (%)					
65% RH	5.0-9.0	10-30	7-10	25-40	25-35
Wet	6.0-12.0	20-40	10-15	30-45	30-40
Elastic Mod (gpd)					
65% RH	60-80	40-70	45-70	25-40	35-40
Wet	50-70	15-35	35-55	20-35	25-35
Moisture Regain (%) at 65% RH	7.0-8.0	11-13	10-12	6.0-6.5	3.5-4.5
Water Vol. Swell. (%)	40	40-80	30-40	10-30	5-15
Specific Gravity	1.54	1.52	1.52-1.54	1.32	1.32-1.35

applied load, the molecular system in the dry state is unable to rearrange to assume an optimal stress distribution. At high relative humidities and in the wet state some of the effective cross-links in cotton are broken, allowing a certain degree of chain mobility and an optimal stress distribution, resulting in higher fiber strength than in the dry state. In rayon fibers the crystallinity and intermolecular bonding are low, or at least well below optimum, so that fiber strength in the dry state is low compared to that of cotton. Here fiber strength is determined or limited primarily by the low degree of effective cross-linking whereas in cotton it is limited primarily by a lack of chain mobility. At high humidities or in the wet state, intermolecular hydrogen bonds are broken and cross-linking in rayon is decreased further so that strength is lower than in the dry state.

Important innovations in rayon technology in recent years have produced various high wet-strength and high wet-modulus fibers, with some achieving nearly the properties of cotton. These rayon fibers, frequently referred to as polynosic rayons, are generally used in textile apparel applications, while the more standard types of rayon are used in products where absorbency is of primary interest.

4.2. Wool

The hair covering of certain animals provides several fibers that are extensively used for textile purposes. The fiber that is obtained from the fleece of sheep and lambs is referred to as wool, although the term is also used to describe the hair covering of related animals. In addition to wool, the following animal fibers are of interest: mohair (from the Angora goat), cashmere wool, common goat hair, camel hair, llama hair, alpaca hair, and vicuna wool. All these fibers, as well as human hair, are structurally very similar and are principally composed of a class of proteins known as keratins. Keratins are natural cellular systems of fibrous proteins which have evolved to serve as a protective outer barrier for the higher vertebrates. Keratin proteins are also the basis of the outer layer of skin and of appendages such as hooves, scales, hair and feathers [26].

Wool and other keratin fibers are structurally very complex and hierarchical. There are three main components - the cuticle, the cortex, and the medulla. The cuticle is a multilayered system of overlapping cells that forms the protective surface covering of the cortex which constitutes the bulk of the fiber mass (80 to 90%). The cortex is also multicellular with closely packed spindle-like cells that are aligned along the fiber axis. The cortical cells are 50 to 100 μm in length and 3 to 6 μm in diameter. Many animal fibers also have an inner canal known as the medulla, a series of dried cells which may contain granular proteins and inorganic pigments. An intercellular cement, also referred to as the cell membrane complex, is a non-keratin proteinwhich serves to bind the cuticular and cortical cells and to provide a diffusion path into the fiber. The cortical cells are composed of crystalline microfibrils (now referred to as intermediate filaments) embedded in an amorphous matrix. In turn, the microfibrils are composed of several protofibrils which are two- or three- stranded ropes of α-helices. The polypeptide chains constituting the α-helices are composed of at least nineteen amino acids. Each α-helix is inter- and intramolecularly stabilized by hydrogen bonds, ionic bonds or salt linkages, van der Waals forces, hydrophobic bonds, and disulfide bonds arising from the amino acid cystine [27].

Wool fibers vary from 5 to 30 cm in length, with diameters of about 18 to 25 μm. The fibers are generally circular or slightly elliptical in cross-section. Many wool fibers, particularly the fine (small diameter) wools used in apparel, are highly crimped, with a regular sinusoidal wave pattern. The crimp in these fibers is structurally based, with ortho and para cortical cells forming a bilateral structure in which the para cortex is on the inside and the ortho cortex on the outside of each crimp curvature [28, 29]. Another key structural characteristic of wool and other animal fibers is that the outermost layer of cuticular cells forms a system of overlapping scales on the surface of the fiber, as is shown in Fig. 10. These scales serve to protect the fiber and the host animal from the external environment. They also cause the fiber's coefficient of friction to be higher in one direction than in the other, giving rise to the differential friction effect which is responsible for the felting shrinkage of wool products [30].

An interesting feature of wool and other natural animal fibers is that their surface is quite hydrophobic and not readily water wettable, yet these fibers are able to absorb water vapor to a

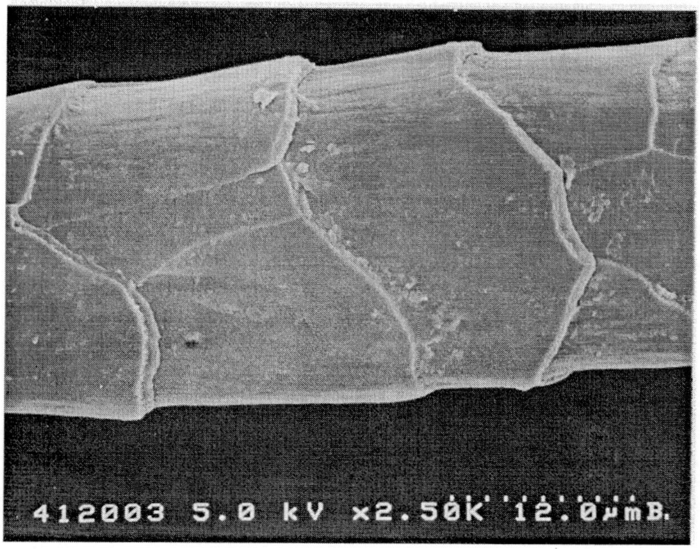

Figure 10. Surface structure of a typical keratin fiber (human hair); courtesy of S.B. Ruetsch, TRI/Princeton.

very high extent. Their equilibrium moisture regain under standard conditions (65% RH and 21°C) is about 16%. The hydrophobic surface is due to the chemical structure of the epicuticle, and the high moisture absorption capacity reflects the hydrophilic components of the proteins in the internal fiber structure. A diffusion pathway for the moisture into the interior fiber is obviously necessary and is provided by the cell membrane complex discussed above.

The mechanical properties of wool reflect the complex morphology of these fibers. The stress-strain curve reveals several distinct regions which are associated with the response of the crystalline microfibrils to the deformation [31]. After an initial Hookean region, the stress-strain becomes essentially flat during which the α-helices unfold into a β pleated sheet extended chain structure. This α to β transformation takes place against very little resistance and would result in fiber extensions of about 100% if it were to go to completion. However, at a strain of about 30%, the stress-strain curve enters the post-yield region where there is a sudden increase of stress with increasing strain, resulting in fiber extensions at break of about 50% (in the wet state). Some typical physical properties of wool fibers are summarized in Table 5.

4.3. Polyamide Fibers

While polyamide fibers are not extensively used in absorbency applications, they constitute an important category of fibrous materials. Aliphatic polyamides, now generically referred to as nylons, were the first truly synthetic fibers introduced into the marketplace in the 1940's primarily as replacements for silk filaments. The two most common nylons are nylon 6 and nylon 66, the numerals designating the nature of the monomer from which the condensation polymer is derived. In nylon 66 the monomer units are hexamethylenediamine and adipic acid, the numerals indicating the number of carbon atoms in the diamine and dibasic acid, respectively [32]; in nylon 6 the monomer is caprolactam which is the lactam formed from the six carbon e-aminocaproic acid [33]. The nominal structures are shown below:

$$-\overset{\overset{O}{\|}}{C}-CH_2-CH_2-CH_2-CH_2-\overset{\overset{O}{\|}}{C}-\overset{H}{N}-(CH_2)_6-\overset{H}{N}-$$

Nylon 66

$$-CH_2-CH_2-CH_2-CH_2-CH_2-\overset{\overset{O}{\|}}{C}-\overset{H}{N}-$$

Nylon 6

Table 5.
Typical Physical Properties of Wool and Synthetic Polyamide Fibers

	Wool	Nylon 66	Nylon 6	Kevlar7	Nomex7
Tenacity at Break (gpd)					
65% RH	1.5	4.0-8.0	4.0-8.0	20-23	4.0-5.5
Wet	1.0	3.0-7.0	3.0-7.0	20-23	3.0-4.0
Extension at Break (%)					
65% RH	30	15-35	20-40	2.5-4.0	20-30
Wet	40	20-40	25-45	2.5-4.5	20-30
Elastic Modulus (gpd)					
65% RH	30	30-50	30-45	500-800	70-90
Wet	25	25-45	25-40	500-800	50-70
Equil. Moisture Regain at 65% RH (%)	16	4.0-4.5	4.5-5.0	4.5	8.0
Melting Point (^{0}C)	~300	260	225	decomp. at 460	380
Specific Gravity	1.30	1.14	1.14	1.44	1.38

The important polymeric linkage for all nylons is the amide group which provides the strong intermolecular hydrogen bonding network characteristic of these polymers. It is the amide linkage and the resulting intermolecular hydrogen bonding that is responsible for the high melting points, high strength, and the exceptionally high elastic properties of these filaments. Otherwise, the reasonably long chain methylene units $-(CH_2)_6-$ could not provide desirable properties because of free rotation around single carbon bonds and because only relatively weak intermolecular Van der Waals forces would be operative between adjacent chains.

The amide group also serves as the site for moisture sorption in these fibers which makes them reasonably hydrophilic. As a result, the physical properties of nylon fibers are quite strongly dependent on relative humidity or moisture content, particularly as far as their elastic properties are concerned. Mechanical properties of nylon fibers are also quite strongly temperature dependent. This reflects the thermal stability of the critical hydrogen bonds in nylon and also the increased flexibility of methylene chains at elevated temperatures. Nylon fibers are also subject to oxidative and UV degradation especially at elevated temperatures; however, anti-oxidants and UV stabilizers are usually added to the nylon melt prior to filament extrusion by melt spinning processes to overcome this problem. Typical physical properties are summarized in Table 5.

Nylon 6 and 66 are by far the most extensively used aliphatic polyamides, but just about any combination of diamine and dibasic acid can be polymerized by condensation reactions to form a polymer from which fibers might be produced. However, only a few of the possible nylons have the proper balance between physical properties and processability to make them useful as fiber-forming polymers.

Aromatic polyamide fibers, generically referred to as aramids, are a rapidly growing category of fibers that are being used increasingly in a wide range of applications, although never as absorption media. Because of their high strength, inherent flame resistance, and thermal stability, aramids, are used as thermal and electrical insulation, as flame resistant clothing and upholstery, and in a wide range of high performance applications such as in fiber-reinforced composites [34]. The condensation polymer of 1,4-diaminobenzene and terephthalic acid is the basis of one of the most important commercial aramid fibers, Kevlar7 [35]. The corresponding *meta* isomer, the condensation product of 1,3-diaminobenzene and isophthalic acid , is the basis of the commercial fiber Nomex7 [36]. The nominal structures are shown below:

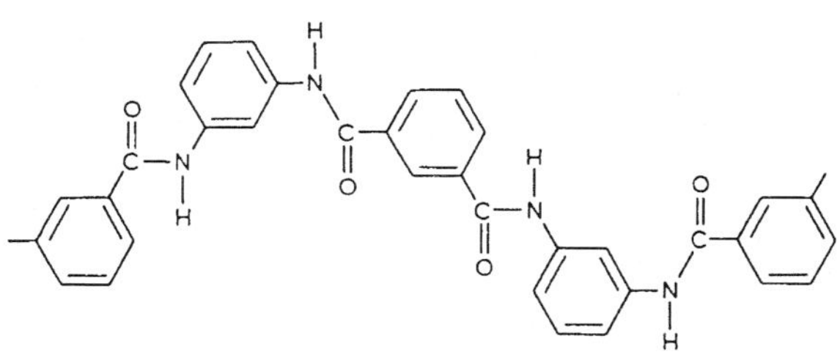

Nomex®

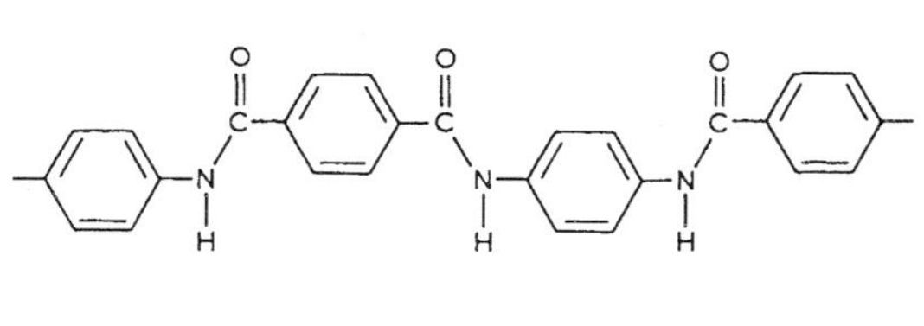

·Kevlar®

While the structures of these fibers are quite similar, the isomeric difference is largely responsible for the differences in the physical properties that are shown in Table 5. Because of the linear *para* structure, the molecular chains in Kevlar7 are extremely stiff and can be fully aligned in drawing operations. The resulting highly crystalline fiber is correspondingly strong and stiff. The *meta* structure of Nomex7 does not allow such orientation and crystallinity, and the important characteristics of this fiber are its thermal stability and flame resistance.

4.4. Acrylic Fibers

Acrylic fibers are an important category of man-made fibers based on polyacrylonitrile produced by addition polymerization of acrylonitrile (vinyl cyanide). Most commercial acrylic fibers are in fact copolymers of acrylonitrile and other vinyl monomers, such as vinyl chloride, vinyl acetate, vinyl alcohol, vinylidene chloride, acrylic acid, methacrylic acid, and methacrylate esters [37,38]. By definition, however, an acrylic fiber must be composed of a polymer based upon at least 85% by weight of acrylonitrile. If less than 85% by weight but at least 35% by weight is based on acrylonitrile, the fiber is referred to as a modacrylic.

Acrylic fibers are produced either by dry or wet spinning extrusion processes. The fibers have reasonable chemical stability, but they undergo several transitions in chemical and physical structure at elevated temperatures. Among these transitions is the formation of ladder type polymers containing six-membered rings, the exact structures of which depend upon whether the pyrolysis is carried out in oxidative or inert atmospheres. Acrylic fibers are used extensively as precursors for the manufacture of carbon and graphite fibers.

The physical properties of acrylic fibers (Table 6) are strongly dependent on the comonomer used with acrylonitrile to produce the polymer. Sorption properties in particular vary with chemical composition, not only because of the sorption sites that may be introduced with a particular monomer, such as with vinyl alcohol, but also because of the structural irregularity that is brought about by copolymerization which allows greater accessibility to moisture and other absorbates. There are many specialty acrylic fibers for absorption applications, as will be discussed later in this chapter.

4.5. Polyester Fibers

Fibers in this category constitute the most rapidly growing segment of man-made fibers. The fibers are extremely versatile and are used extensively in a variety of apparel, household and industrial applications [39, 40]. Polyester fibers, produced by melt spinning operations, are based on condensation polymers formed from terephthalic acid and dihydric alcohols. The most widely used polymer is that formed from terephthalic acid and ethylene glycol to produce poly(ethyleneterephthalate). Polyester fibers based on terephthalic acid and 1,4-cyclohexanedimethanol and also 1,3-dihydroxy propane are also used commercially.

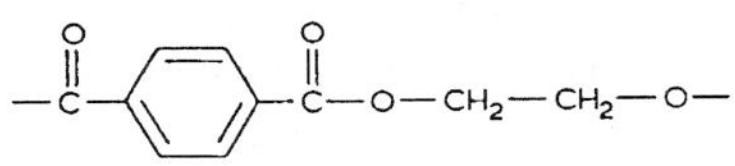

Poly (ethylene terephthalate)

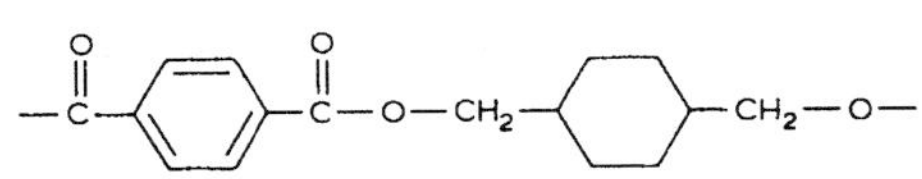

Poly (1,4 - cyclohexylenedimethylene terephthalate)

Table 5.
Typical Physical Properties of Several Man-Made Fibers

	Polyester	Acrylic	Polyethylene	Poly-propylene
Tenacity at Break (gpd)				
65% RH	3.5-7.0	2.5-4.5	2.5-4.0	4.0-6.5
Wet	3.5-7.0	2.0-4.0	2.5-4.0	4.0-6.5
Extension at Break (%)				
65% RH	15-40	15-30	10-50	15-30
Wet	15-40	20-35	10-50	15-30
Elastic Modulus (gpd)				
65% RH	85-90	40-70	20-30	30-45
Wet	85-90	30-60	20-30	30-45
Equil. Moisture Regain at 65% RH (%)	0.4	1.5-2.0	< 0.1	< 0.1
Melting Point (°C)	260	softens at 235	130	170
Specific Gravity	1.38	1.17	0.95	0.90

Copolymerization with variations in either the dibasic acid component or the dihydric alcohol component or with p-hydroxybenzoic acid is used extensively to produce special combinations of fiber physical properties. Intermolecular cohesive forces in polyester fibers are relatively weak, and the good fiber forming characteristics are due primarily to the inherent chain stiffness arising from the partial aromatic structure. Completely aliphatic polyesters, such as might be formed from adipic acid and ethylene glycol, have low melting points (below 100°C) and correspondingly inadequate mechanical properties. On the other hand, polyester fibers where the dihydric alcohol component is also at least partially aromatic are of considerable interest in view of their high strength and stiffness. Such aromatic polyester polymers are able to form liquid crystalline domains, and the properties of the corresponding fibers reflect the highly ordered and oriented structures.

Polyester fibers can be highly drawn to produce fully oriented and highly crystalline structures. The degree of orientation can be varied over a wide range to achieve fibers with various combinations of physical properties. Typical values of important fiber properties are shown in Table 6. The low moisture sorption characteristics of polyester fibers reflect the lack of sorption sites and the rigid backbone polymer which restricts chain mobility and accessibility. Nevertheless, polyester fibers are used in many liquid absorbency applications. This is because it is possible to produce small diameter (fine) fibers and also to produce fibers with irregular or noncircular cross-sectional shapes [41]. Both these fiber characteristics contribute to the formation of effective interfiber capillaries in a fibrous network that account for the high liquid absorbency.

4.5.1. Microfibers. It has always been well known that fiber diameter, frequently referred to as fiber fineness, has an important influence on the properties of textile structures such as yarns and fabrics. Prior to the advent of man-made fibers, the approximately 10-20 μm diameter of a cotton fiber and the 10-15 μm diameter of silk filaments were considered as the finest fibers achievable. When the man-made fibers were introduced, it was necessary to match the geometric properties of these fibers (diameter, length, crimp curvature) to that of either cotton or wool so that these new fibers could be processed on textile equipment that evolved over centuries for the processing of natural fibers. Thus, for many years the lowest linear densities of the man-made fibers were in the 2 to 3 denier range. With increased understanding of how fiber linear density and fiber diameter influence the properties of yarns and fabrics, particularly softness and other tactile characteristics, it became of interest to develop small diameter fibers, referred to as microfibers, for certain applications.

Pioneering work in this area of fiber technology was done with polyester fibers in Japan in the 1980s. The first approach taken was to reduce the fiber diameter by controlled surface dissolution of the polyester by treatment in aqueous NaOH solutions at elevated temperatures. Typically weight losses achieved were in the 20 to 30% range producing fibers with linear densities as low as 1 denier. Even finer polyester fibers may be achieved by this approach, but the economics of controlled fiber dissolution are not particularly attractive. Depending on treatment conditions, the reduction in fiber linear density and diameter may be accompanied by changes in fiber surface texture, such as the fiber pitting shown in Figure 11.

More innovative approaches have been taken to the development of microfibers involving the extrusion of bicomponent fibers of the island-in-the-sea and citrus structure type. After extrusion, the lesser component can be removed either by dissolution or by mechanical splitting. These methods have produced polyester, nylon, and polypropylene microfibers not only with linear densities as low as 0.1 denier, but also with controlled cross-sectional shapes [42]. The advent of microfibers has enormously broadened the range of fibrous products for textile consumers. Polyester fabrics with the softness, drapeability and tactile properties previously associated only with the finest silks are now routinely available [43].

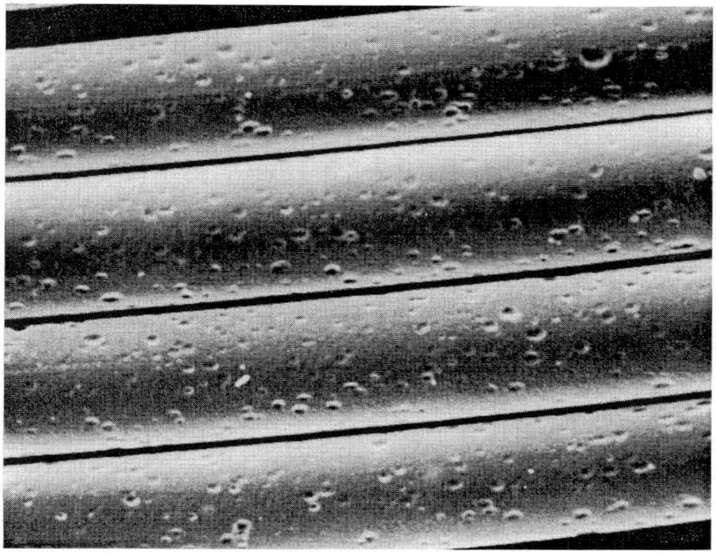

Figure 11. Photomicrograph of polyester fibers treated in aqueous NaOH solution; courtesy of S. B. Ruetsch, TRI/Princeton.

4.6. Polyolefin Fibers

While a number of polyolefin polymers may be extruded into fibers, only polyethylene and polypropylene have found widespread application in fibrous form. Fibers formed *from* high molecular weight isotactic (stereoregular) polypropylene have good physical properties, adequate thermal stability, as well as the necessary versatility for a variety of applications. As in the case of polyester, polyolefin fibers have no specific sorption sites for moisture which accounts for the low equilibrium moisture regain of these fibers. In the case of polypropylene attempts have been made to modify the basic chemical structure either by copolymerization with other monomers during the polymer formation process or by grafting on other polymers after fiber formation by melt spinning [44, 45, 46]. Sorption as well as other physical properties can be enhanced by such chemical modifications, but none of these processes has found widespread acceptance. The economic advantage of polypropylene over other typical fiber forming polymers would be significantly impaired by these modification treatments. Metallocene-based catalysis for the polymerization of both isotactic and syndiotactic polypropylene has recently been developed, which is able to produce more uniform polymer in terms of molecular weight, tacticity, and chemical composition. Finer fibers of higher strength allow improved processability and product properties.

4.7. Vinal and Vinyon Fibers

Vinal fibers are formed from poly(vinyl alcohol) polymers by either wet or dry spinning methods [47]. The fiber is not manufactured in the United States, but is produced and used in Japan. It is a reasonably strong fiber with considerable hydrophilicity. The response of the fiber to water is strongly dependent on its degree of crystallinity. The degree of swelling of the fiber and its solubility in water decrease with crystallinity which can be controlled by heat treatments and processing conditions after fiber formation. These properties can also be controlled by reaction with formaldehyde which renders the fiber more hydrophobic and may also cause some degree of polymer crosslinking.

Vinyon fibers are formed from poly(vinyl chloride) and are used extensively as bonding fibers in nonwovens [48]. These fibers have low shrinkage and softening temperatures (−80°C), and are quite weak with a tenacity at break of approximately 2.0 gpd. They have an equilibrium moisture regain at 65% RH of less than 1.0% and do not swell in water. An important feature of these fibers is their non-flammability.

4.8. Bicomponent Fibers

The utilization of two different polymers in the production of man-made fibers has led to a variety of bicomponent or biconstituent fibers. Such polymer blend fibers can be either homogeneous, where there is no separation of the two polymers, or heterogeneous where the two polymers are segregated into spatial regions. Homogeneous bicomponent fibers are not very common because it is difficult to obtain polymers where some degree of phase separation would not occur and because phase separation is usually necessary to achieve the desired effects. Heterogeneous bicomponent fibers are being used more extensively as new blend systems are developed and as new production technologies become available. Schematic representations of the three principal types of heterogeneous bicomponent fibers are shown in Fig. 12. Various other configurations are possible, but they all fall into one of the three categories shown. Side-by-side and sheath-core bicomponent fibers are by far the most common and are produced by special co-extrusion and mixed stream technologies [49].

One of the first side-by-side fibers was a viscose rayon produced by co-extrusion of two different cellulose xanthate solutions to produce a chemically crimped fiber [50]. Similarly, bicomponent helically crimped acrylic fibers have been produced by co-extrusion [51], simulating natural bicomponent structure of crimped wool. Bicomponent nylons and polyesters have also been produced, as have asymmetric sheath-core polyethylene/polypropylene fibers.

4.9. Hollow Fibers

The production of man-made fibers with one or more internal channels running along the fiber axis is a logical extension of the underlying technology for the production of bicomponent fibers [52]. Several rayon fibers with a hollow lumen-like channel to simulate cotton have been produced and are used extensively in certain products. These fibers are of low nominal density and are useful in absorbency applications. Hollow polyester fibers with four channels are produced to provide high bulk and resilience, and are also useful in high absorbency applications. A cross-sectional view of a multichannel polyester fiber is shown in Fig. 13. Similar fibers are also produced from polyamides.

5. SPECIALTY FIBERS FOR ABSORPTION PRODUCTS

5.1. General Considerations

While many of the fibers described in the previous section can be used in absorbency applications, several specialty superabsorbent fibers have been developed exclusively for this purpose. In many cases these fibers reflect rather minor structural modifications from one of the

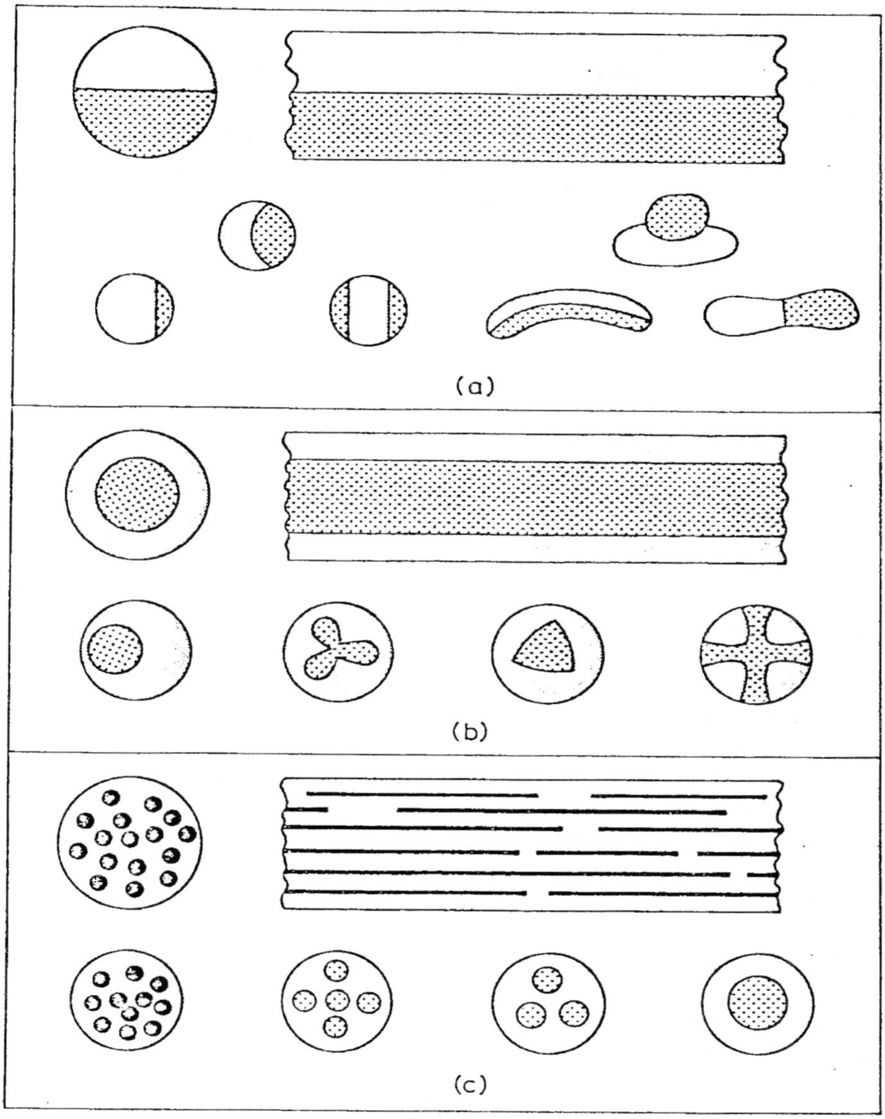

Figure 12. Typical phase arrangements in bicomponent fibers: (a) side-by-side, (b) sheath-core, and (c) matrix-fibril; from Paul and Newman [49].

generic fiber types, while others are quite unique in chemical composition usually incorporating one or more highly hydrophilic polymers. Also, in some cases highly hydrophilic polymers are extruded directly into filaments without the support of a generic fiber.

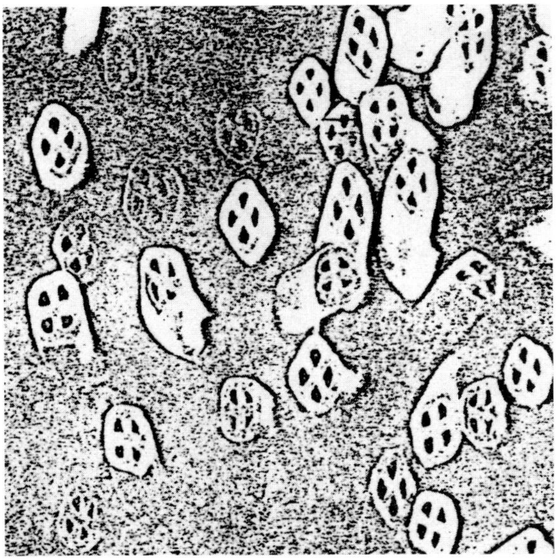

Figure 13. Cross-sectional view of a Holofil7 polyester fiber; courtesy of Joel Brostin Personal Products Co.

The properties of these hydrophilic polymeric absorbates depend not only on chemical structure but also on polymer chain mobility and cross-linking density. While cross-linking is important for decreasing solubility of hydrophilic polymers, it also directly affects absorption rate, absorption capacity and gel strength. Under free-swell conditions increasing the cross-link density of an absorbate increases the absorption rate and gel strength while decreasing the absorption capacity. In many end-uses, however, fibers are under compression, which restricts swelling. In these cases a higher cross-link density increases the absorption capacity.

Although molecular considerations are important for producing a highly hydrophilic material, optimal utilization of any material is dependent on physical form. The geometric shape and the porous structure of fibers enhance their value as absorbent materials. Since the principal end-uses of superabsorbent materials are usually in sheet form (e.g., paper towels and other fabrics), fibers are convenient because they are readily fabricated into woven and nonwoven products. The same hydrophilic polymer in the form of powder or granules would need a support system for an even distribution in a sheet. These supports are difficult to produce and retard fluid flow to the absorbate.

5.2. Alloy Rayon Fibers

Alloy rayon fibers are homogeneous multicomponent polymer blends. In the production of these fibers, the cellulose xanthate and one or more superabsorbent polymers are mixed and spun in a manner analogous to normal viscose rayon processing. Some typical absorbent polymers blended with rayon include sodium carboxymethyl-cellulose, polyvinylpyrrolidone, alkali metal salts of alginic acid, and polyacrylic acid salts. The water retention and fluid holding capacity of typical alloy rayon fibers, as measured by standard industrial testing procedures, are approximately double those of standard rayon. Equilibrium moisture regain at 65 % RH may be

15% for the alloy rayon and 11% for a standard rayon, while the other physical properties are essentially equivalent. The large increase in water retention and fluid holding capacity for these fibers reflect not only the higher hydrophilicity as indicated by the equilibrium moisture regain, but also the increased liquid retention in interfiber capillaries.

The absorption properties of a polyvinylpyrrolidone/rayon fiber are listed in Table 7. The data indicate clearly the increased absorption properties with increasing polyvinylpyrrolidone content. The absorption capacity of this matrixed cellulose can be further enhanced by including a modified cellulose, cyanoethylcellulose, in the alloy mixture. Fluid holding capacities for fibers of a regenerated cellulose containing various quantities of cyanoethylcellulose and polyvinylpyrrolidone are given in Table 8. In contrast to the increased absorption capacity obtained when polyvinylpyrrolidone is the sole polymer additive, cyanoethylcellulose decreases the fluid holding capacity of rayon. A mixture of the two polymers, however, produces a superior superabsorbent alloy fiber.

The water retention capacity for an alloy rayon containing acrylic acid/methacrylic acid copolymers is shown in Table 9. It is clear that the water retention increases significantly by incorporating the copolymers in the rayon structure. Various samples were examined to determine the effect of the acrylic acid to methacrylic acid ratio. As the data indicate, water retention increases slightly with increasing acrylic acid content in the copolymer. The purpose of the methacrylic acid is to retain cohesiveness of the viscose material which is necessary if the fiber is to be carded and otherwise processed into end products. In another example, an alloy rayon containing alkali metal salts of alginic acid may be produced by dispersing sodium alginate in the cellulose xanthate [56]. An equal blend of the alloy fiber with regular rayon has a fluid holding capacity of 4.28 cm^3/g whereas the rayon control has a fluid holding capacity of only 2.7 cm^3/g.

Table 7.
Absorption properties of a polyvinylpyrrolidone/rayon fiber [53].

Cellulose (%)	Polyvinyl-pyrrolidone (%)	Fluid-holding capacity (cm^3/g)	Water Retention (%)
100	0	3.06	105
95	5	3.16	112
90	10	3.52	121
80	20	4.15	145
70	30	4.69	186
65	35	4.68	178
60	40	4.65	190

Table 8.
Absorption properties of a polyvinylpyrrolidone/cyanethylcellulose/rayon [54].

Cellulose (%)	Cyanoethyl-cellulose (%)	Polyvinyl-pyrrolidone (%)	Fluid-Holding Capacity (cm^3/g)
100	0	0	3.11
90.9	9.1	0	2.53
83.3	16.7	0	3.10
71.4	28.6	0	3.42
90.0	0	9.1	3.53
76.9	0	23.1	4.69
80.0	10.0	10.0	5.04
74.0	13.0	13.0	5.38

5.3. Coated Fibers

Superabsorbent properties of fibers may be achieved by coating methods. Applicable superabsorbent materials for this process include sodium carboxymethylcellulose (CMC), cross-linked hydroxyethylcellulose, cross-linked partial free acid CMC, and starch. As with other fibers containing superabsorbent polymers, the enhancement of properties depends on the ratio of the two polymers [57].

5.4. Modified Cellulose Fibers

Modified cellulose fibers have the basic backbone of a cellulose fiber but some or all of the hydroxyl groups have been substituted. Modified cellulose fibers produced from insoluble cross-linked carboxymethylcellulose (CMC) are an example of hydrophilic polymers being spun directly into filaments [58]. The cross-linking is achieved by esterification of the carboxyl groups in the free acid form after filament extrusion. Typically the cross-linked CMC has a carboxymethyl degree of substitution of 0.7 with about 65-70% of the carboxyl groups being in the sodium salt form. These conditions make the polymer highly hydrophilic and swellable but yet insoluble. Typical data for two types of absorbent CMC fibers are shown in Table 10 and

Table 9.
Water retention of acrylic acid (AA)/methacrylic acid (MAA) copolymer/rayon alloy fibers [55].

Polymer Added (%)	Water Retention (%)
None	97
AA/MAA 90/10	136
AA/MAA 80/20	130
AA/MAA 50/50	127

Table 10.
Absorbency of simulated urine of modified cellulose (carboxymethylcellulose) fibers [58].

Material	Liquid Retention (%) after draining	Liquid retention (%) after centrifuging
Cellulose fibers	32	1.0
Absorbent CMC #1	19	16.0
Absorbent CMC #2	24	21.0

compared with standard cellulose fibers. The data indicate that cellulosic fibers have a greater capacity for simulated body fluids than absorbent CMC under swell conditions, however, ~97% of the liquid is expelled by centrifugation from the cellulose fiber in comparison with only 15% for the absorbent CMC products.

Another means of modifying the properties of cellulose involves polymer grafting. A cellulose graft copolymer fiber with a natural or regenerated cellulose backbone and alternating ionic and nonionic side chains grafted onto it is highly water absorbent. Typical ionic polymer moieties may include poly(acrylic acid), sodium poly(acrylate), poly(methacrylic acid), potassium poly(methacrylate), poly(vinyl alcohol sulfate), poly(phosphoric acid), poly(vinyl amine), poly(4-vinylpyridine), hydrolyzed poly(acrylonitrile) and several others. Examples of nonionic polymer moieties are poly(methylmethacrylate), poly(ethylmethacrylate), poly(ethylacrylate), poly(butylacrylate), poly(vinylacetate), poly(styrene), poly(butadiene) and poly(isoprene) [59].

Absorption data for various graft compositions of cellulose, sodium poly(acrylate), and poly(ethylacrylate) are given in Table 11. Addition of poly(ethylacrylate) alone increases fluid retention only slightly. With the addition of the ionic polymer moiety, sodium poly(acrylate), the fluid holding capacity increases significantly. On the other hand, the ionic polymer alone forms a gel on the fiber surface which inhibits further fluid absorption and retention. These data illustrate the sensitivity of fluid absorption not only to the chemical composition of the fiber, but also to the physical structure.

5.5. Acrylic Fibers

An acrylic fiber in which the increased absorptive properties are attributed primarily to its physical form has been described [60]. The fiber has a very thin capillary structure which enables it to absorb and retain large quantities of liquids. Since the material is intended primarily for apparel applications, the high liquid retention characteristic is important. For example, the material does not feel damp until it attains a moisture content of 15% in comparison to 3% for conventional acrylics. Mechanical properties of this fiber are also affected by its porous structure. The tenacity at break and the elastic modulus are lower than for conventional acrylics, and the extensibility is somewhat higher.

A process for the production of an absorptive acrylic fiber utilizes retention of selective solvents in the fiber structure during extrusion to create a high tensile strength, water retentive, moisture absorptive fiber from a polyacrylonitrile copolymer [61]. The fiber has a unique core

Table 11.
Fluid retention of sodium polyacrylate-ethylacrylate copolymer grafts on cellulose [59].

Cellulose (%)	Polyethylacrylate (%)	Sodium polyacrylate (%)	Fluid Retention (cm^3/g)
100	0	0	1.5
20.7	79.3	0	2.9
20.2	57.3	22.5	12.0
20.3	49.2	30.5	14.0
19.8	33.9	46.3	9.5
22.6	27.3	50.1	9.3
24.1	19.3	56.6	9.5
33.2	12.4	54.3	10.8
34.4	0	65.6	Gelled

and sheath structure. The sheath is highly porous and has a high substitution of carboxyl salt groups to provide the hydrophilic character. Liquid retention is high due to the capillary forces that pull liquid into the core of the fiber. This transport of liquid to the core makes the fiber feel outwardly dry while it has higher levels of water retention than cotton. The fiber can contain up to 19% water before being perceptibly damp, compared to 13% for wool, 11% for cotton, 8% for nylon, 5% for standard acrylics, and 2% for polyester [62].

5.6. Polyester Fibers

While polyester fibers are inherently hydrophobic, the absorptive capacity of a poly(ethylene terephthalate) fiber can be increased by chemical modifications and treatment. For example, treatment in aqueous hydroxyamine solutions provides hydroxy sorption sites on the fiber [63]. After twelve clear water washings of materials treated with ethanolamine, diethanolamine and 2-diethylaminoethanol, the time for absorption of a water droplet was one to two orders of magnitude less than the absorption time of an untreated sample. In another approach to increased hydrophilicity in polyester, water soluble aliphatic polyamide polymers are coextruded with the polyester [64]. After fiber formation the polyamide is extracted with water to leave a more porous absorptive polyester substrate.

5.6.1 Deep Grooved Fibers. A particularly interesting development among specialty polyester fibers for absorbency applications are deep-grooved fibers. These are fibers produced by melt spinning from poly(ethylene terephthalate) polymer with novel cross-sectional shapes that create deep grooves or channels along the longitudinal axis of the fiber [65]. A cross-sectional view of a deep groove polyester fiber is shown in Figure 14. The grooves provide ducts to move liquid

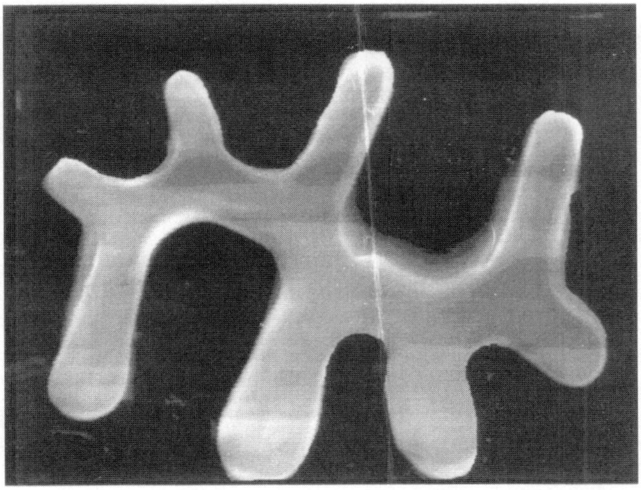

Figure 14. Cross-sectional view of a deep groove polyester fiber (4DG Fiber7); courtesy of L.R. Dean, Eastman Chemical Company.

spontaneously by capillary forces, and also spaces to trap and store liquid for increased effective absorbency [66].

5.7. Superabsorbent Fibers

Superabsorbent polymers in powder form have high fluid holding capacity due to their hydrophilic chemical structure and osmotic pressure swelling effects. The essential element of superabsorptivity is that the polymer swells upon absorbing water to form a hydrogel, which can then retain even larger quantities of fluid. A further feature of superabsorbents is that fluid can be retained even when external pressure is applied. Such powders are usually dispersed in multilayer nonwoven webs of cellulosic and other textile fibers to provide absorbent pads. Even greater fluid holding capacity can be achieved when the superabsorbent polymer is in fiber form. The modified cellulosic fibers discussed in Section 5.4 are examples of superabsorbent fibers. The cellulose is modified by esterification or etherification, and in many products by controlled grafting of other hydrophilic polymers.

A recent example of superabsorbent fibers is a cross-linked acrylate copolymer, partially neutralized to the sodium salt, which is now commercially available [67]. Nonwoven webs andpads containing superabsorbent fiber have faster fluid uptake than when the superabsorbent is in powder form due to wicking effects created by the small diameter fibers.

5.8. Corn Fibers

A recent development and still in the early experimental stages is a fiber based on polylactic acid derived from corn [68]. Its physical properties approach those of the more common man-made fibers, but its principal feature is its complete biodegradability. It is a highly crystalline fiber with a melting point of 175°C.

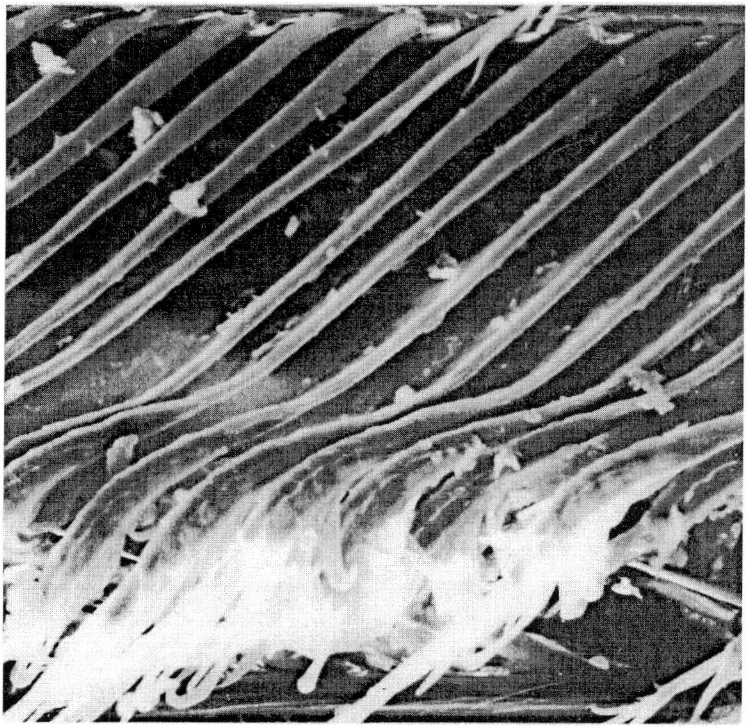

Figure 15. An SEM photograph of purified chicken feather fibers; courtesy of E. Erbe, G. Gassner, and W.F. Schmidt, U.S. Department of Agriculture.

5.9. Chicken Feather Fibers

Fibers produced from waste chicken feathers have been developed by the U. S. Department of Agriculture, and are being commercially marketed for various end uses, including absorption products [69, 70]. It is estimated that there are 2 to 4 billion pounds of feathers produced annually as a byproduct of poultry production. A key step in the preparation of useful fibers is the separation of the fine feather fibers from the quill, which is accomplished by mechanical means. The fibers are microcrystalline keratin protein. They have a diameter of about 6 μm and an average length of 8 mm. An SEM photograph of purified feather fibers is shown in Figure 15. The fibers are hollow which aids in liquid transport. However, the major advantage in terms of absorption properties of products made from these fibers is their wet collapse resistance. The products retain high bulk even when saturated with liquid.

6. REFERENCES

1. B. C. Goswami, J. G. Martindale and F. L. Scardino, *Textile Yarns: Technology, Structure and Applications*, John Wiley & Sons, Inc., New York, 1977.
2. E. B. Berry, in *The American Cotton Handbook Volume* 2, 3rd Edition, pp. 573-669, D. S. Hamby, editor, John Wiley & Sons, Inc., New York, 1966.

3. J. Lunenschloss and W. Albrecht, editors, *Nonwoven Bonded Fabrics*, Halsted Press, John Wiley & Sons, Inc., New York, 1985; and O. Jirsak and L. C. Wadsworth, *Nonwoven Textiles*, Carolina Academic Press, Durham, N.C., 1999.
4. S. C. Winchester and J. C. Whitwell, Textile Res. J., 40 (1970) 458-471.
5. D. D. Hubbell and J.C. Whitwell, Textile Res. J., 40 (1970) 521-529.
6. J. W. S. Hearle and A. Newton, Textile Res. J., 37 (1967) 778-797.
7. M. M. Besso, G. E. Gillberg and D. E. Stuetz, Textile Res. J., 52 (1982) 587-597.
8. B. Pourdeyhimi, R. Ramanathan and R. Dent, Textile Res. J., 67 (1997) 181-187.
9. *Fiber Organon*, Fiber Economics Bureau, Washington, D.C.
10. L. Rebenfeld, J. Polymer Science, Part C., 9 (1965) 91-112.
11. J. W. S. Hearle, *Polymers and Their Properties, Volume 1: Fundamentals of Structure and Mechanics*, , Elsherwood Ltd. and Halsted Press, John Wiley & Sons, Inc., New York, 1982.
12. I. V. de Gruy, J.H. Carra and W.R. Goynes, *The Fine Structure of Cotton*, Marcel Dekker, Inc., New York, 1973.
13. N. Morosoff and P. Ingram, Textile Res. J., 40 (1970) 250-255.
14. W. R. Goynes, B.F. Ingber and B.A. Triplett, Textile Res. J., 65 (1995) 400-408.
15. J. J. Hebert, Textile Res. J., 63 (1993) 695.
16. E. K. Boylston and J. J. Hebert, Textile Res. J., 65 (1995) 429-431.
17. G. Raes, T. Fransen and L. Verschraege, Textile Res. J., 38 (1968) 182-195.
18. J. A. Bristow and P. Kolseth, editors, *Paper - Structure and Properties*, Marcel Dekker, Inc., New York, 1986.
19. K. W. Britt, editor, *Handbook of Pulp and Paper Technology*, 2nd Edition, Van Nostrand Reinhold Co., New York, 1970.
20. N. M. Bikales and L. Segal, editors, Cellulose *and Cellulose Derivatives*, John Wiley & Sons, Inc., New York, 1971.
21. B. S. Sprague, J. L. Riley and H. D. Noether, Textile Res. J., 28 (1958) 275-287.
22. W. A. Sisson, Textile Res. J., 30 (1960) 153-170.
23. H. Chanzy, M. Paillet and R. Hagege, Polymer, 31 (1990) 400-405.
24. J. L. Gardon and R. Steele, Textile Res. J., 31 (1961) 160-171.
25. L. Rebenfeld and H.D. Weigmann, SIRTEC Proceedings, pp. 595-611, Institut Textile de France, Paris, 1969.
26. H. P. Lundgren and W. H. Ward in *Ultrastructure of Protein Fibers*, R. Borasky, editor, Academic Press, New York, 1963.
27. P. Jolles, H. Zahn and H. Hoecker, editors, *Formation and Structure of Human Hair*, Birkhaeuser Verlag, Basel, Switzerland, 1997.
28. M. Horio and T. Kondo, Textile Res. J. 23 (1953) 373-386; and E. H. Mercer, Textile Res. J. 23 (1953) 383-397.
29. J. H. Bradbury, in *Advances in Protein Chemistry, Volume 27*, pp. 111-211, Academic Press, New York, 1973.
30. K. R. Makinson, *Shrinkproofing of Wool*, Marcel Dekker, Inc., New York, 1979.
31. M. Feughelman, *Mechanical Properties and Structure of Alpha-Keratin Fibers, Wool, Human Hair and Related Fibers*, UNSW Press, Sydney, 1997.
32. J. H. Saunders, in *Polymers: Fibers and Textiles, A Compendium*, pp. 634-669, J. I. Kroschwitz, editor, John Wiley & Sons, Inc., New York, 1990.
33. W. Sbrolli in *Man-Made Fibers: Science and Technology, Volume II*, H. F. Mark, S. M. Atlas and E. Cernia, editors, John Wiley & Sons, Inc., New York, 1968.
34. T. W. Chou and F. K. Ko, editors, *Textile Structural Composites*, Elsevier Publishing Co., New York, 1988.
35. H. H. Yang, *Kevlar7 Aramid Fiber*, John Wiley & Sons, Inc., New York, 1993.
36. H. H. Yang in *High Technology Fibers, Part C*, M. Lewin and J. Preston, editors, Marcel Dekker, Inc., New York, 1993.
37. C. W. Davis and P. Shapiro in *Encyclopedia of Polymer Science and Technology, Volume 1*, pp. 342-373, John Wiley & Sons, Inc., New York, 1964.

38. J. Minor in *Encyclopedia of Chemical Technology, Volume 19*, pp.379-419, John Wiley & Sons, Inc., New York, 1982.
39. G. W. Davis and J. R. Talbot, in *Polymers: Fibers and Textiles, A Compendium*, pp. 670-745, J. I. Kroschwitz, editor, John Wiley & Sons, Inc., New York, 1990.
40. I. Goodman, in *Encyclopedia of Polymer Science and Engineering, Volume 12,* pp. 1-75, H.F. Mark et al. editors, John Wiley & Sons, Inc., New York, 1988.
41. E. M. Hicks, et al., *Textile Progress, Volume 3, Number 1,* The Textile Institute, Manchester, 1971.
42. T. F. Cooke, *Bicomponent Fibers - A Review of the Literature*, Technical Information Center Report #44, TRI/Princeton, 1993.
43. M. Fukuhara, Textile Res. J., 63 (1993) 387-391.
44. M. Ahmed, *Polypropylene Fibers - Science and Technology*, Elsevier Scientific Publishing Co., New York, 1982.
45. L. M. Landoll, in *Polymers: Fibers and Textiles, A Compendium*, pp. 611-633, J. I. Kroschwitz, editor, John Wiley & Son, Inc., New York, 1990.
46. L. M. Landoll, in *Encyclopedia of Polymer Science and Engineering, Volue 10*, pp. 373-395, H. F. Mark et al. editors, John Wiley & Sons, Inc., New York, 1987.
47. F. L. Marten, in *Encyclopedia of Polymer Science and Engineering, Volume 17*, pp. 167-198, H.F. Mark et al. editors, John Wiley & Sons, Inc., New York, 1989.
48. L. Gord in *Man-Made Fibers: Science and Technology, Volume III*, H.F. Mark, S.M. Atlas and E. Cernia, editors, John Wiley & Sons, Inc., New York, 1968.
49. D. R. Paul and S. Newman, editors, *Polymer Blends*, Academic Press, New York, 1978.
50. W. A. Sisson and F. F. Morehead, Textile Res. J., 23 (1953) 152-157.
51. E. M. Hicks, Jr., J. F. Ryan, Jr., R. B. Taylor, Jr. and R. L. Tichenor, Textile Res. J., 30 (1960) 675-697.
52. A. J. Hughes and J. E. McIntyre, *Textile Progress, Volume 8, Number 1*, The Textile Institute, Manchester, 1976.
53. F. R. Smith, USP 4,041,121 (August 9, 1977) and USP 4,136,697 (January 30, 1979), assigned to Avtex Fibers, Inc.
54. 54. F. R. Smith, USP 3,919,385 (November 11, 1975) and USP 3,951,889 (April 20, 1976), assigned to Avtex Fibers, Inc.
55. T. C. Allen and D. B. Denning, USP 4,066,584 (January 3, 1978), assigned to Akzona, Inc.
56. F. R. Smith, USP 4,063,558 (December 20, 1977), assigned to Avtex Fibers, Inc.
57. A. R. Reid, USP 4,128,692 (December 5, 1978), assigned to Hercules, Inc.
58. T. J. Podlas, in INDA Technical Symposium Proceedings ANonwoven Technology Challenges and Achievements@, pp 25-39 (March 1976).
59. P. K. Chatterjee and R. F. Schwenker, Jr., USP 3,889,678 (June 17, 1975), assigned to Personal Products Co.
60. P. L. Mazzucho and L. Patron, Selezioni Tessile (1980) 67-69; also P. Lennox-Kerr, Textile Institute and Industry, 19 (1981) 83-84.
61. E. Radlmann, E. Reinehr and G. Nischk, USP 4,143,200 (March 6, 1979), assigned to Bayer AG, Germany.
62. P. Hoffmann, Chemische Rundschau, 29 (1978) 15-18.
63. A. S. Forschirm, USP 4,063,887 (December 20, 1977), assigned to Celanese Corp.
64. British Patent No. 1,585,399 (March 4, 1981), assigned to Bayer AG, Germany.
65. B. M. Phillips and S. Bagrodia, USP 5,611,981 (March 18, 1997), assigned to Eastman Chemical Company.
66. B. M. Phillips, INDA Fundamental Research Conference, Auburn University, July 28-29, 1994.
67. Oasis Superabsorbent Fibers, *Medical Textiles*, International Newsletters, England, December 1997.
68. Corn Fiber, Medical *Textiles*, International Newsletters, England, December 1996.
69. W. F. Schmidt and M. J. Line, *Proceedings TAPPI Nonwovens Conference*, pp. 135-140 (1996).
70. W. F. Schmidt, *Proceedings Poultry Waste Management Conference,* Springdale, AR, October 19-22, 1998.

Absorbent Technology.
P.K. Chatterjee and B.S. Gupta, editors.

CHAPTER VII

CROSS-LINKED CELLULOSE AND CELLULOSE DERIVATIVES

RAYMOND A. YOUNG
Department of Forest Ecology & Management, University of Wisconsin, Madison, Wisconsin, U.S.A.

Contents

1. INTRODUCTION

Although cotton cellulose has been utilized in textile applications for thousands of years, it wasn't until 1839 that Anselme Payen isolated and named the same polysaccharide from wood [1]. The hydrophilic and absorbent characteristics of cellulose have been known for an equivalent length of time. With the advent of absorbent disposable products in recent years, the demand for cellulose fibers in absorbent applications has increased dramatically. The importance of cellulose-water interactions is also reflected in the voluminous amount of literature devoted to the subject [2-4].

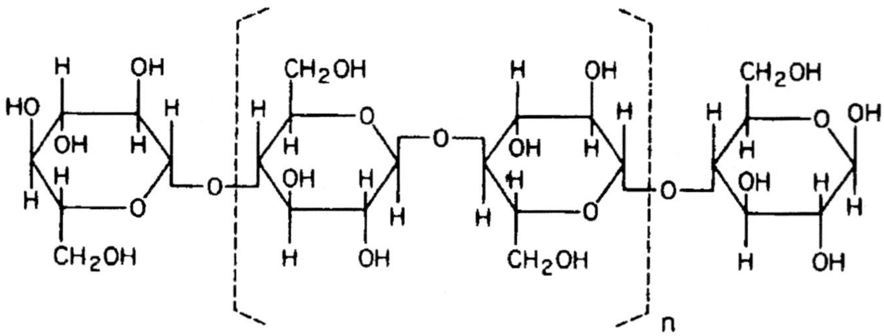

Fig. 1. Chemical structure of cellulose [11].

There are a variety of products which utilize absorbent cellulose. The list includes diapers, catamenial napkins, tampons, nursing pads, wound dressings and other incontinence products. Practically all forms of cellulose can function in absorbent products, including wood pulps, rayon and cotton. However, the wood pulps are often desirable in commercial applications for economic reasons. Chemical wood pulps have been typically employed although a significant effort has been expended on absorbent applications of mechanical pulps, especially chemithermomechanical pulps [5].

The high bulk of fluffed pulp absorbent products and the skyrocketing costs of packaging and shipping, which are very dependent on product bulk, have spurred developments of compact materials with extraordinary absorbent characteristics, the superabsorbents. Superabsorbent napkins are typically 67% lower in bulk volume than napkins constructed from fluffed pulp and can absorb up to 70 times their weight of liquid [6-8]. In addition to the high fluid absorption capacity and low bulk, superabsorbent polymers have the ability to hold fluids under pressure; another important advantage over fluffed cellulose pulp.

However, superabsorbents are significantly more expensive per pound than fluff pulp [7]. Apparently the improved absorbent properties and the savings in packaging and transportation costs compensate for the higher material costs since superabsorbent polymers are now present in major brands of baby diapers, sanitary napkins and incontinence product.

The creation of cellulosic superabsorbents attests to the ingenuity of cellulose chemists around the world. The synthesis of these materials was based on previous knowledge of both the morphology and chemical reactions of the cellulose macromolecule. The pertinent physical and chemical characteristics of cellulose and various cellulose derivatives which led to the development of superabsorbence technology are briefly reviewed in this chapter.

2. CHEMISTRY AND MORPHOLOGY OF CELLULOSE

Cellulose is a polymer of anhydroglucose linked by glycosidic bonds as depicted in Fig. 1. In the native state cellulose has a degree of polymerization (DP) of about 14000. When isolated from wood via chemical pulping methods, the DP is considerably reduced, to

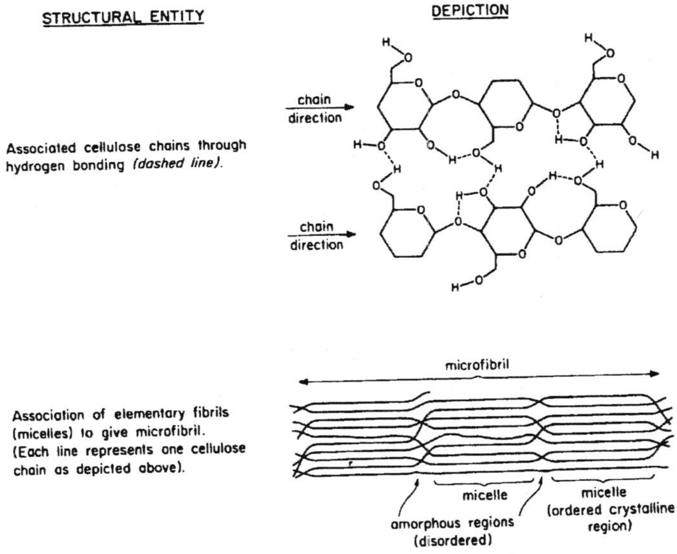

Fig. 2. Fine structure of cellulose (not all hydroxyl groups in the anhydroglucose units are shown) [9].

about 2000. This is the result of either acid or base cleavage of the glucosidic bonds, depending on the pulping method [9].

In the plant, cellulose chains are laid down in a parallel arrangement and the long-chain molecules strongly associate through secondary forces (hydrogen bonds) (10). This association between the cellulose chains results in a very uniform crystalline structure known as micelles or microcrystallites (Fig. 2). The micelles are also associated in the plant to give long threadlike structures known as microfibrils also shown in Fig. 2. Spaces or dislocations (amorphous regions) of about 15-20Å have been reported between the micelles in the microfibrils, thus there is ample room for solvents and other reactive molecules to enter the fiber structure for modification reactions [11,12].

Two distinctive modes for cellulose modification can be described. The first type is fiber modification reactions which do not destroy the original crystalline/amorphous morphology of the fiber. If the micellar structure is broken down and there is loss of the fibrous integrity then we have the second class of reactions, termed conversion processes. Both fiber modification reactions and conversion processes begin as heterogeneous reactions. However, the conversion processes ultimately result in a homogeneous reaction in which a soluble derivative is usually converted to a different useable form such as sheets, foils, films, rods and filaments or it is molded with plasticizers to plastics of other useful shapes. In the formation of superabsorbent fibers from cellulose, the crystalline-amorphous morphology is considerably altered although the fiber integrity can be maintained. This is generally achieved through cross-linking of the modified cellulose fiber, further discussed in the following sections.

Practically all cellulose modification reactions occur at the available free hydroxyl groups on the cellulose macromolecule (Fig. 1) and the modification of cellulose to form superabsorbents is no exception. The extent of derivatization is characterized by the degree of substitution (DS) which is the number of substituted hydroxyl groups per anhydroglucose unit. Thus the maximum DS is 3, with a DS less than or equal to 1 more common.

Both esterification and etherification reactions at the cellulose hydroxyl groups have been utilized for the formation of superabsorbent celluloses. After the reagents have penetrated the pores and dislocations in the cellulose as described above, the relative reactivities of the primary (C6) and two secondary (C2 and C3) hydroxyl groups (per anhydroglucose unit) become the critical factors. In general, reversible equilibrium controlled reactions (i.e., esterification) tend to favor 6-O-substitution while irreversible, rate controlled reactions (i.e., etherification) tend to favor 2-O-substitution [13].

However, many other factors can influence the final substitution pattern. For example, the relative substitution in the 2-O-position in irreversible reactions decreases as the size of the incoming substituent increases. Etherification with the bulky trityl chloride results in most of the substitution occurring at the 6-O-position with only 5-12% of the substituent at the 2-O-position [13].

2.1. Esterification

As a polyhydroxy macromolecule, cellulose readily undergoes esterification reactions. The equilibrium shown below lies partly to the right in the presence of a strong acid and can be shifted further to the right for more complete ester formation by removal of the water [14-18].

$$\text{acid} + \text{alcohol} = \text{ester} + \text{water} \qquad (1)$$

The most common cellulose organic ester is probably cellulose acetate. This cellulose ester was discovered in 1865 and has been prepared in essentially the same way since the late 1800's; by treating cellulose with sulfuric acid and acetic anhydride as shown below [14].

$$\text{Cell-(OH)}_3 + H_2SO_4 + 3\,(CH_3CO)_2O \longrightarrow \text{Cell}\begin{cases} OSO_2OH \\ (OCOCH_3)_2 \end{cases} + 4CH_3COOH \qquad (2)$$

As depicted the initial product is a mixed ester of sulfuric and acetic acid. The sulfuric acid groups can be removed in two ways; by transesterification in which magnesium oxide is added to neutralize the catalyst and continuing acetylation to the triacetate or by hydrolysis in which the dilute acetic acid is added and the reaction continued to give the secondary acetate (DS about 2.5) [17,19]. The dramatic affect of the DS on the properties of cellulose derivatives is shown for cellulose acetate. At a DS above 2.8 (triacetate) the cellulose acetate is soluble in chlorinated solvents, at a DS of 2.2-2.8 in acetone, ketones, etc., and at DS 1.2-1.8 in methyl cellosolve. The cellulose acetate is water soluble at a low DS, in the range of 0.6-0.8. Products of lower DS are soluble only in cellulose solvents [15,17]. Cellulose acetate has been widely used in conversion processes. The extruded fiber products are used in both textiles and cigarette filters.

Acetylation has also been used for fiber modification to improve the affinity of pulp fibers for moisture. The acetyl content must be below 10% or about 0.4 DS; above this level the moisture take-up is less than the untreated sample [20].

Two inorganic esters of cellulose have been utilized for formation of superabsorbents, namely cellulose sulfate and phosphorylated cellulose. Generally cellulose sulfate is prepared with sulfuric acid in the presence of an aliphatic alcohol (see eq. 2). The reaction is dependent on such variables as the chain length and structure of the alcohol, the sulfuric acid:alcohol molar ratio, and the temperature and time of reaction [14,19]. Petropavlovskii and Krunchak [21] found that for a sulfuric acid-propanol system the sulfur content of the product was directly proportional to the amount of acid in the mixture, up to 3.5 to 1 (acid to alcohol) with no further increase in DS at higher ratios. The maximum DS achieved was 1. A number of other methods have been employed to obtain higher DS values [19].

Phosphorylated cellulose has been extensively studied because of flame retardant applications for textiles [19]. Urea and phosphoric acid were utilized by Katsumura and Nonaka [22] to prepare phosphorylated cellulose. Degradation of the cellulose was minimized by using higher temperatures and shorter reaction times with a constant molar ratio of urea to phosphoric acid. More stable, essentially undegraded, water soluble cellulosic acid phosphates were prepared by Touey [23] with an anhydrous solution of orthophosphoric acid, phosphorus pentoxide and an alcohol diluent. Phosphorylation of cellulose for superabsorbent applications will be further discussed in a later section.

2.2. Etherification

The use of cellulose ethers for formation of superabsorbents far exceeds the esterified celluloses. Etherification is almost always carried out with cellulose which has been "activated" by alkali (alkali cellulose). Mercer in 1844 was the first to point out the considerable changes in the properties of cellulose treated with alkali and later work has verified that important structural, physiochemical and chemical changes take place with the lye treatment [14-17,24,25].

The alkali cellulose is then typically treated with alkyl chlorides or oxizarnes to form a variety of important cellulose ethers including methyl-, ethyl- and carboxymethyl-celluloses. A generalized reaction is shown in eq. 3.

$$\text{Cell-(OH)} + \text{NaOH} + \text{RCl} \quad \rightarrow \quad \text{Cell-(OR)} + \text{NaCl} + H_2O \qquad (3)$$

The alkali cellulose attacks the alkyl halide by an S_N2 nucleophilic bimolecular mechanism. Solvent effects are also important particularly if the substituent is bulky. Methyl- and ethyl-cellulose react by the S_N2 mechanism but chloracetic acid approaches an S_N1 mechanism [26]. Isopropyl alcohol is preferred for carboxymethylation while benzene is more suitable for ethylation of cellulose [14,25,26]. Salts are also formed as shown in eq. 3 and must be washed from the product in many applications.

Methyl- and ethyl-cellulose have similar structures but quite different properties. Although the methyl- and ethyl-substituents are hydrophobic, at low degrees of substitution water solubility is conferred to the cellulose. This fact not only attests to the profound influence of physical structure on the hydrophilic properties of cellulose but also demonstrates the utility of cellulose modification reactions for control of these properties (also cf. cellulose acetate).

Commercial methylcellulose is produced in the range of DS = 1.6 to 2.0 and is soluble in organic solvents when completely substituted. The remarkable property of methylcellulose is its inverse viscosity versus temperature reaction. When the "gel point" is reached the viscosity suddenly increases tremendously [27]. Ethylcellulose is used in different applications such as durable films, paper coatings, lacquers and adhesives.

Carboxymethylcellulose (CMC) is probably the most important cellulose derivative for preparation of cellulosic superabsorbent products. This derivative has the largest commercial volume of all the cellulose ethers mainly because it provides controlled properties at prices unmatched by other materials with similar solution properties [27].

Sodium carboxymethylcellulose is made by treating alkali cellulose with monochloracetic acid. In addition to the main reaction shown below (eq. 4), a principal side reaction occurs to form sodium glycolate, also shown in eq. 5:

$$\text{Cell-(OH)} + NaOH + ClCH_2COONa \rightarrow \underset{\text{CMC}}{\text{Cell-}(OCH_2COONa)} + NaCl \qquad (4)$$

$$ClCH_2COONa + 2NaOH \rightarrow HO\text{-}CH_2COONa + NaCl + H_2O \qquad (5)$$

It can be seen from the formula of CMC that it is an ether acid; it can also be considered as the ether of cellulose and glycolic acid [17,25].

CMC is utilized in a wide variety of products including detergents, foods, drilling muds, textiles, paper sizes, pharmaceuticals, paints and others. Industrial CMC is a white or yellowish odorless powder which is very hygroscopic. Under ambient conditions the product has a moisture content of about 11-12% [24,25,27].

Carboxymethylation has also been used for cellulose fiber modification [20]. The carboxymethylated fibers have enhanced hydrophilicity and the modified fibers fibrillate well on beating and develop high bonding strength. The bursting strength of carboxymethylated pulps can be increased by over 150% by raising the DS from 0.002 to 0.062. With the enhanced affinity of the CMC fibers for water they apparently have greater plasticity and hydrodynamic specific volume resulting in increased bond area and bonding strength [20].

Partially carboxymethylated cotton also exhibits high moisture regain, increased absorbency and greater swellability [24,28]. The enhanced swelling and absorbency properties can be attributed in part to the increased hydrophilicity of the carboxyl groups. However, the bulky carboxyl group also undoubtedly contributes to opening up the cellulose structure which allows diffusion of water into the microscopic interstices. Since alkali is utilized in the preparation of CMC, the carboxymethyl substituents are substituted on previously inaccessible regions and the bulky groups keep the cellulose structure from collapsing and reforming the tightly hydrogen bonded structure after drying. Thus previously inaccessible regions become available for penetration by liquids [28]. Similar factors are important in the development of superabsorbent CMC polymers.

Hydroxyethylcellulose (HEC) is another cellulose ether which has been utilized in the preparation of superabsorbent cellulose products. The derivative is formed by reaction of ethylene oxide with cellulose in the presence of sodium hydroxide which is not consumed. Alkali cellulose is formed as an intermediate which reacts with ethylene oxide regenerating sodium hydroxide and producing hydroxyethylcellulose.

$$\text{Cell-OH} + \underset{\diagdown\;O\;\diagup}{CH_2\text{-}CH_2} \xrightarrow{NaOH} \text{Cell-OCH}_2\text{CH}_2\text{OH}$$

$$\text{Cell-OCH}_2\text{CH}_2\text{OH} + \underset{\diagdown\;O\;\diagup}{CH_2\text{-}CH_2} \xrightarrow{NaOH} \underset{\text{HEC}}{\text{Cell-OCH}_2\text{CH}_2\text{OCH}_2\text{CH}_2\text{OH}} \tag{6}$$

Since the sodium hydroxide is regenerated in the reaction and the attached hydroxyethyl group has a free hydroxyl group, the substituted group can further react with ethylene oxide to add another hydroxyethyl group as shown in the equation above. The addition can be repeated several times until a chain of polyoxyethylene is attached to the cellulose [17,24,26]. The molar substitution (MS) is thus, usually much greater than the DS. The minimum MS value for water solubility is 1.0, which represents a DS of about 0.4. The ratio of MS to DS is the average length of the oxymethylene pendant chain. Because the anhydroglucose units are not uniformly substituted, a single cellulose chain can contain both water soluble HEC and unsubstituted cellulose segments. Byproduct glycols are also formed in the reaction since water competes with cellulose for the reagent as shown below [26].

$$\text{HOH} + {}^{-}\text{OCH}_2\text{CH}_2\text{OH} \rightarrow \text{HOCH}_2\text{CH}_2\text{OH} + \text{OH}^{-} \tag{7}$$

The increased hydrophilicity of HEC is conferred in part by the long side chains which inhibit the cellulose from reforming a tightly bonded structure as discussed for CMC. HEC is chiefly used as a component of latex paints [27].

3. SUPERABSORBENTS

Absorbent polymers occur in nature as both agarose and pectin. Both of these substances are polysaccharides which contain additional hydrophilic functional groups. Agarose, from the red algae, contains sulfuric acid monoesters and pectin, a fruit carbohydrate, contains a high proportion of carboxyl groups. Fucoidan, a sulphated polysaccharide from brown algae, has also been reported to be a very hygroscopic polymer and may serve to prevent dehydration of the seaweed plant upon long exposure [29,30].

Based on similar principles man has been able to engineer polymers with superabsorbent characteristics. Basic research in the field of superabsorbents was initiated in the late 1960's and has continued through to the present. Table 1 summarizes the presently available highly absorbent polymers into two groups; the modified natural polymers (cellulose and starch) and the synthetic polyacrylamide and polyacrylate polymers. Several reviews have been published on methods for producing highly absorbent cellulose products [31-35]. The emphasis in this chapter is on the cross-linked and derivatized celluloses. The properties of superabsorbents are often characterized according to the water or salt water retention value (WRV or SRV) [32,36,37]. Saline solutions of about 1% concentration are utilized to simulate body fluids. WRV (SRV) is defined as the amount or percent of water (salt water) retained by the absorbent polymer after centrifugation, per dry weight of the sample. Common wood pulp or paper products have a WRV of only about 50%, uncoated

Table 1.
Highly absorbent polymers

Modified natural polymers
Cross-linked cellulose
Cross-linked cellulose derivatives
Grafted cellulose (hydrolyzed)
Grafted starch (hydrolyzed)
Synthetic polymers
Cross-linked polyacrylamides and polyacrylates

cellophane (regenerated cellulose) has a WRV of 130%, while superabsorbents can have a WRV in the range of 200-7000% (centrifuged at 160OXg) [32].

The phenomenon of water absorption by superabsorbents is dramatically depicted in Fig. 3 and 4 (38). Figure 3 is a photomicrograph of dry superabsorbent fibers which have never been wet with water and are similar in appearance to the original wood pulp fibers. Figure 4 is a photomicrograph of the same single fiber wet with water. The fibers have absorbed water equal to 32 times their dry weight and have increased in diameter about 5 times [38]. It should be noted that the fiber integrity is intact with the fiber lumen still visible in Fig. 4. A detailed description of the preparation of highly absorbent cellulose products follows.

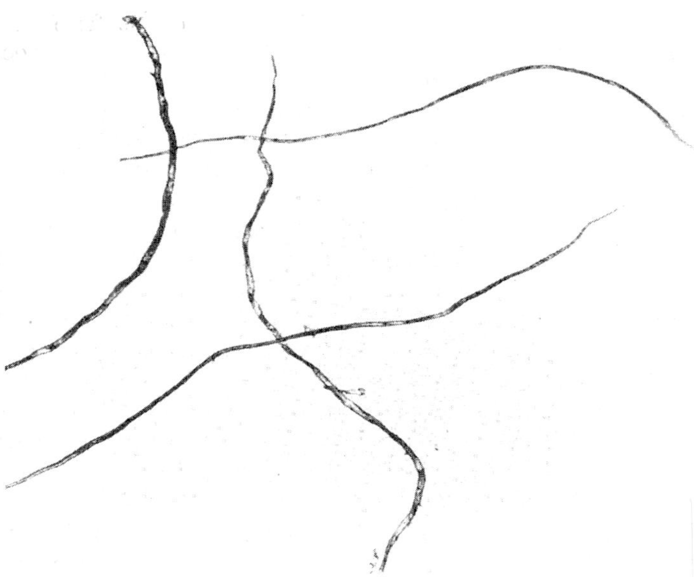

Fig. 3. Photomicrograph of cross-linked CMC fiber (CLD) never wet with water (74X, in 2-propanol) (Courtesy of W. L. Dean).

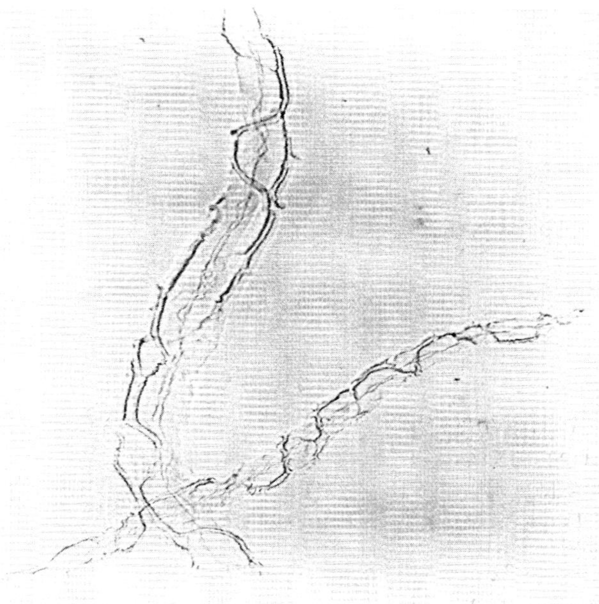

Fig. 4 Photomicrograph of cross-linked CMC fiber (CLD) wet with water (74x) (Courtesy of W.L. Dean).

3.1. Cross-linked Cellulose

Wood pulp has been used for making absorbent products for many years with satisfactory results. The pulp fluff is usually prepared by grinding sheets of wood pulpboard in hammer mills or similar devices. The pulp is then delivered in fluff form to the production lines for incorporation into absorbent products. For reasons already discussed, improvements in the absorbent properties of wood pulp fluff have been sought in recent years.

A number of patents have been issued for cross-linking of cellulosic fibrous material for improvement of fluid absorbency, fluid retention, dry and wet resilience and others. The cross-linked cellulose material is obtained by combining at least two hydroxyl groups in a cellulose molecule or in adjacent cellulose molecules. A variety of cross-linking agents are available to achieve the connection, some of which are listed in Table 2 [31,39-41]. The reactive groups which combine with the hydroxyl groups may be present prior to the reaction, as with glyoxal, or they may be generated during the reaction as in the case of the sodium thiosulfate derivative of divinyl sulfone [37]. Obviously, to cross-link the cellulose, the agent must be difunctional with respect to cellulose for reaction with two hydroxyl groups. Although formaldehyde is monofunctional in many reactions, it is difunctional with respect to cellulose. Formaldehyde is often the preferred cross-linking agent because it is generally less expensive and fully effective at low levels of addition [39,41,42].

The cross-linking of cellulose with formaldehyde has been extensively investigated in the past because of applications in the textile industry for crease resistance and wrinkle recovery of cotton fabrics [41,42]. The reaction consists essentially of two chemical steps; the formation of the hemiacetal (or 0-methylolcellulose) (eq. 8) and the cross-linking per se by formation of a formal (or oxymethylene) bridge (eq. 9) [41].

Table 2.
Cellulose cross-linking agents[a]

Formaldehyde
Methylolated nitrogen compounds
Dimethylolurea
Dimethylolethyleneurea
Dimethylolimidazolidone
Dicarboxylic acids
Maleic acid
Dialdehydes
Glyoxal
Diepoxides
Diisocyanates
Divinyl compounds
Divinyl sulfone
Dihalogen containing compounds
Dichloroacetone
1,3-Dichloropropan-2-ol
Halohydrins
Epichlorohydrin

[a]For a more complete listing and discussion of cross-linking agents see ref. 39.

$$\text{Cell-OH} + \text{HCHO} \leftrightarrows \text{Cell-OCH}_2\text{OH} \tag{8}$$

$$\text{Cell-OCH}_2\text{OH} + \text{Cell-OH} \leftrightarrows \text{Cell-OCH}_2\text{O-Cell} \tag{9}$$

There is evidence to suggest that the formaldehyde condenses with cellulose to form intermediate cross-links between the macromolecular chains.

According to Smith [43], the formation of the cross-links depends on resonance stabilization of the carbonium ion centered on the methylene carbon in the acid catalyzed reaction. Thus the rate of formaldehyde add-on is increased by increasing the hydrogen ion activity and by including an alcohol in the formaldehyde-acid solution. Some hydrolysis of the cellulose also occurs in the cross-linking reaction. Meyer et al. [44] carried out a simultaneous analysis of the rates of cross-linking and hydrolysis for pad-cure cross-linking of cotton cellulose and determined that the apparent activation energy of the cross-linking reaction (119.8 kJ/mol) was about 14.3 kJ/mol smaller than that of the hydrolysis rate (128.2 kJ/mol). A decreased reaction temperature should therefore favor the cross-linking reaction.

Many cellulose cross-linking agents are prepared by methylolation of amino groups of parent substances such as ureas, cyclic ureas, carbamates, acid amides, and aminotriazines, to their methylol compounds (Table 2). All methylolations can be accelerated by alkalis or acids [41,45]. A generalized acid catalyzed reaction is shown in eq. 10.

$$\equiv NH - HCHO \overset{H^+}{\leftrightarrow} \equiv N\text{-}CH_2OH \qquad (10)$$

The dimethylol compound is then used to cross-link cellulose as shown in eq. 11 with dimethylolurea (DMU).

$$2Cell\text{-}OH + HOCH_2\text{-}NH\text{-}C(=O)\text{-}NH\text{-}CH_2OH \xrightarrow{H^+} Cell\text{-}OCH_2\text{-}NH\text{-}C(=O)\text{-}NH\text{-}CH_2O\text{-}Cell \qquad (11)$$

Epichlorohydrin has also been used to cross-link cellulose. This cross-linking agent is considered to be both mono- and di-functional because it contains a reactive chlorine substituent, but can also undergo a ring opening reaction. There is some evidence that the reaction proceeds through the formation of the chlorohydrin ether which is capable of further reaction to form the cross-linked cellulose as depicted below [41,46].

$$2\ Cell\text{-}OH + ClCH_2\underset{\diagdown O \diagup}{CHCH_2} \rightarrow Cell\text{-}OCH_2CH(OH)CH_2O\text{-}Cell \qquad (12)$$

As shown, cellulose can be cross-linked in a number of ways depending on the reactive agent. There are also two general methods for cross-linking cellulose; wet-cross-linking and dry-cross-linking. A few examples are given below for each method.

3.1.1. Wet cross-linking. Wet cross-linked cellulose is obtained when the crosslinking agent is treated with the cellulose fibers while they are in the swollen state. Usually water is the swelling agent although other inert solvents can be utilized [39]. Cross-linking tends to stabilize a fiber in the particular state of swelling it has attained at the time of reaction. Thus the extent of swelling at the time of cross-linking has a dramatic effect on the subsequent physical properties [42]. Fibers reacted in the swollen state tend to maintain their distended structure and exhibit high moisture regain, high water inhibition and good wet resilience. A disadvantage is that the dry resilience is not improved to the same extent.

Steiger [39] described a process for wet cross-linking of wood pulpboard by immersion of the sample at room temperature in a solution containing (by volume) 20% formalin (37% HCHO), 50% hydrochloric acid (37% HCl) and 30% water. The wood pulpboard was then neutralized with sodium bicarbonate, washed and air dried. The wet cross-linked pulp was then ground or shredded to produce a pulp fluff with improved absorbency.

However, during the grinding operation undesirable clumps and knots are produced to various degrees. When the fiber clump and knot content rises above 50-75% the product is unsuitable for many purposes. The wet cross-linked product described by Steiger was reported to have a 35% lower knot content than a comparable dry cross-linked product (described in the next section). In a different disclosure [47] an improved tampon product from rayon was produced by subjecting the continuous fibrous elements (i.e., rayon filaments) to bulking, crimping or texturizing prior to the wet cross-linking reaction. The crimp in the fibrous elements is set by the cross-linking reaction. The crimped filaments are then stretched and folded to provide a tampon of improved absorbency. The fluid capacity was 36% better than the untreated product with the wet cross-linking alone; but with crimping, wet cross-linking and stretching the product had a fluid capacity far greater than the untreated product [47].

Graef (48) described a wet cross-linking method for making a hydrophilic cellulose pulp for absorbent products by applying the cross-linking agent directly on a wet or partially dried sheet while it is being formed on a conventional paper machine. Glyoxal and propylene glycol were the preferred cross-linking agents and the best results were obtained when the agent was added at a point about midway through the press section of the paper machine. It was claimed that the products had greatly increased absorbency rate and a somewhat higher water holding capacity. In a later patent Graef and Hunter (49,50) found that it was useful to also add a de-bonding agent directly into the pulp slurry in the papermaking process to aid re-dispersion and re-pulping of the absorbent product. The de-bonding agent is preferably added before the latent cross-linking agent, prior to the headbox of the paper machine, and the cross-linking reactant is applied near the end of the forming section or at the press section.

Any of a variety of cationic, anionic and non-ionic de-bonding agents were suggested by Graef and Hunter (49,50) for improving the ease of disintegration of the absorbent pulp. Cationic de-bonders are usually based on quaternary ammonium salts while the anionic de-bonding agents include a large class of materials, many of which possess surfactant properties, i.e., sulfated fats, fatty esters and fatty alcohols. Nonionic de-bonding agents also include a large class of materials such as aliphatic acids with ethylene oxide, propylene oxide or mixtures of these two materials. When anionic or nonionic de-bonders are used, cationic retention aids are also necessary to improve substantivity to the cellulose fibers. It was found that although a nonionic softener significantly improved the ease of dispersibility and bulk of the absorbent product, the cationic de-bonding agents were superior and when a nonionic de-bonder was used in combination with the cationic material, the effectiveness of the cationic de-bonder was reduced.

Bernardin et al. (51) also disclosed a process for creation of an absorbent product from wet cross-linked rayon fibers. Two features distinguish the method developed by Bernardin et al., that: (1) the surfaces of the fibers are free of any furnishes which are applied during processing to the rayon and (2) the use of "inflated cellulose fibers". The fiber surface finishes are removed through simple solvent extraction or by scouring procedures resulting in improved vertical wicking properties of absorbent webs from the fibers. The "inflated cellulose fibers" are rayon fibers that contain a lumen as a result of incorporation of a blowing agent in the rayon extrusion process. Such fibers are commercially available from Courtaulds, Ltd., under the trade name Viloft™ or Courcel™. Bartholomew et al. (52) have described the manufacture, properties and uses of inflated viscose fibers. The inflated cellulose fibers must be able to at least partially collapse to form ribbon-like fibers to give the maximum improvement in absorbency characteristics. Wet cross-linking of the rayon was typically carried out on carded webs with 1,3-dichloro-2-propanol in a caustic solution.

In 1990, Bernardin and Heimbach (53) described the importance of obtaining a critical density of cellulose fiber webs to obtain the optimum properties from the wet cross-linked fibers. They indicated that it was necessary that the webs have a density from about 0.10 to about 0.35 g/cc and preferably, from about 0.15 to 0.3 g/cc for air-laid webs and from about 0.08 to about 0.35 g/cc, preferably from about 0.08 to 0.30 g/cc, for water-laid webs of the wet cross-linked fibers. Either formaldehyde (formalin in an acidic solution) or 1,3-dichloro-2-propanol in a caustic solution were the preferred cross-linking agents. Their results indicated a vertical wicking for the wet cross-linked products that was superior to similar dry cross-linked materials. For a 1,3-dichloro-2-propanol wet cross-linked spruce pulp it was found that the vertical wicking capacity increased as the density increased up to a maximum at 0.20 g/cc and then began to fall off as the density increased. It was also noted that the vertical wicking rates were initially lower at the higher densities. However it was found that the vertical wicking rates for higher density mats crossed over those of the lower density webs, resulting in higher subsequent vertical wicking rates for the higher density webs.

3.1.2. Dry cross-linking. Dry cross-linked cellulose is obtained when the cellulose fibers are in a collapsed state at the time of cross-linking. The fibers collapse when the water which causes them to swell is removed. Dry cross-linking is primarily used to impart dry and wet resilience to the final product. Wet cross-linking does not give good resilience properties as already discussed and for this reason dry cross-linking is preferred in many applications.

One method of dry cross-linking is to pass the cellulosic fibers through a boric acid solution, dry and then heat in a sealed tube in the presence of paraformaldedhyde. An aqueous technique is more commonly utilized, however, and involves application of the cross-linking agent and catalyst to the cellulose fiber in a water bath. The fibers are then dried, with the cross-linking reaction occurring in a subsequent curing step [39].

A disadvantage of dry cross-linking is that considerable problems arise in the subsequent shredding or grinding step. The dried product is difficult to disintegrate which results in severe fiber breakage and a high content of fiber clumps or knots.

Bernardin [54] patented a process whereby the air dried fibers containing the cross-linking agent (formalin) and catalyst (aluminum sulfate), are first disintegrated to substantially individual or a loosely associated state before the cross-linking reaction. The post cure at 110-115°C initiates the cross-linking. In this way cross-linking between the fibers is inhibited and cross-linking between the macromolecules within the fiber are promoted; avoiding hard clump formation. In effect, the fibers are set by the cross-linking action in a fluffed condition while maintaining a minimum of cohesive forces between the fibers. The water holding capacity and wet bulk were reported to be approximately double that of the untreated pulp.

The fibers produced by the method described by Bernardin [54] can also be rapidly dewatered. In the original untreated cellulose the water was retained within the pores of the cell wall; while in the cross-linked fluffed product, water is retained mainly within the capillary voids between the stiffened fibers. Although the fibers lose their bonding ability, they can be wet laid with unmodified pulp fibers to form an absorbent product or for filter applications.

Chatterjee [55] described an alternate method for avoiding the large quantities of fiber clumps and knots which arise in disintegration subsequent to dry cross-linking. With this method the wood pulp is pretreated to remove or decrease the hemicellulose content. Hemicelluloses are a group of gummy, amorphous polysaccharides with a DP no greater than 150. They are predominantly composed of xylose (xylan) in hardwood pulps and glucose and mannose (glucomannan) in softwood pulps. A considerable amount of these hemicelluloses are distributed on the surface of the pulp fibers.

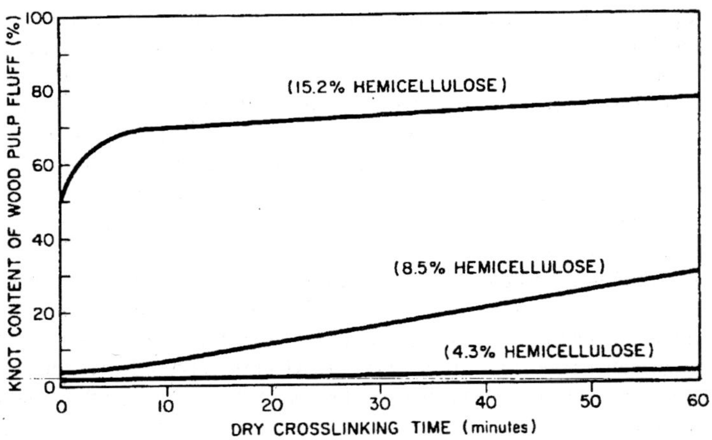

Fig. 5. Effect of fiber hemicellulose content and dry cross-linking time on knot content of wood pulp fluff [55].

The hemicelluloses are very important to fiber bonding in paper and similarly cause intercross-linking between adjacent cellulose fibers in the dry cross-linking process. This leads to excessive formation of fiber clumps and knots in the disintegration operation. However, if the hemicelluloses are preferentially extracted with about 5-12% cold caustic at temperatures from 15-35°C then the "knot problem" can be substantially reduced. Figure 5 shows the percent knot content of a wood pulp fluff versus dry cross-linking time at three different hemicellulose contents. These results demonstrate that the cellulose fibers should contain no more than 7% hemicellulose content for optimum low levels of knots and fiber clumps [55].

Investigators at the Centre Technique de l'Industrie de Papier in France have been issued several patents for cross-linking pulp. In French and German patent descriptions [56,57] a dried pulp was cross-linked with formaldehyde in the presence of a hydrochloric acid catalyst in aqueous acetone. Thus bleached softwood kraft pulp was rinsed with acetone and treated with a mixture containing formaldehyde, HCl water, and acetone (86.9%) for 5-6 minutes at about 50°C to give a cross-linked product containing 1.3% HCHO and an absorptive capacity significantly higher than the wood pulp.

Formaldehyde cross-linked cellulose fibers were also described by Lesas and Pierre in U.S.A. and German patents [58,59]. According to the invention fluffed paper pulp is exposed to the reagents in the vapor phase or as finely divided droplets which contain 1-6% (by weight of pulp) of formaldehyde and hydrochloric formic acid catalysts. The reaction time is extremely short (1-10 seconds) such that the temperature of the fibers do not reach more than 50°C in a hot air stream (180-200°C). In this way fiber degradation is minimized since the reaction occurs mainly at the surface of the fibers. This method contrasts with other procedures in which the cellulosic fibers are submitted to prolonged immersion and contact with the cross-linking reagents. Lesas and Pierre claim that the fibers are therefore more flexible and provide better touch and feel (hand) characteristics. The cross-linked fibers had excellent water absorption capacity.

Schoggen et al. [60] suggested that improved performance absorbent structures can be produced by a process that involves several defibration steps to produce individualized cross-linked cellulose fibers which impart improved wet resiliency. Thus, moist cellulose fibers, at or above 30% moisture content (MC), are defibrated and dried to 18-30% MC under substantially unrestrained conditions. The fibers are then cross-linked with a solution of one of C2-C8 dialdehydes, preferably glutaraldehyde, to form interfiber cross-link bonds and are again defibrated to individual fibers and dryed under unrestrained conditions.

Dean et al. [61,62] also described a process for dry cross-linking of cellulose fibers with C2-C8 dialdehydes that, when formed into a web, have dry densities greater than their corresponding equilibrium wet densities. The cross-linking agent was typically glutaraldehyde, at 0.75-2.5 mole%, to give intrafiber cross-link bonds. The process produces twisted and curled cellulosic fibers which, when formed into compressed absorbent structures, expand on wetting or into lower density absorbent structures which essentially maintain constant volume upon wetting. Dean et al. also reported that the water retention value, absorbency properties and the wet resiliency of these absorbent structures could be further improved by reducing the level of unstable and unreacted cross-linking agent on the individualized fibers, by washing the fibers with an alkaline solution.

Jewell [63] also described a system to produce cross-linked individualized fibers for absorbent applications. The cellulose fibers were treated with one of a variety of different cross-linking agents, including cyclic N-sulfatoimide, a mixture of glyoxal and imidazolidone, a periodate or dimethoxyethanal and a dihydroxy compound containing sulforyl or phosphorous. The cross-linking agent was sprayed onto a mat of the cellulose fibers and the mat was disintegrated in a hammer mill to yield substantially unbroken individual cellulose fibers. The fibers were then dryed and cured to give a product with a nit level of no more than about 3.

Bourbon and Ryan [64] described a process for both dry and wet cross-linking utilizing a glutaraldehyde cross-linking agent. In one example they described a method to produce wet-laid structures containing the individualized, stiffened fibers. Thus the cross-linked fibers were formed into sheets on a modified paper machine where immediately after deposition on the Fourdinier forming wire, a series of five streams of water of sequentially decreasing flow rates were directed upon the fibers. The showers, spaced in the machine direction, were oriented in the cross machine direction, such that fiber flocculations were removed and further formation of flocculations were substantially inhibited. The fibers were set in place from the pressure of the Dandy roll (cylindrical screen roll on top of Fourdrinier forming wire) on the paper machine.

Herron et al. [65-70] in a series of patents described a processes for preparing cross-linked cellulose fibers using C2-C9 polycarboxylic acids. The cross-linking acids included, for example, citric acid, 1,2,3,4 butane tetracarboxylic acid, 1,2,3, propane tricarboxylic acid, oxydisuccinic acid and tartrate disuccinic, with citric acid the preferred agent. The cross-linking agent is applied in solution followed by dewatering and mechanical separation (disk refiner) of the fibers . The fibers are then dried and and the cross-linking agent reacted to form intra-fiber cellulose bonds with the fibers in highly twisted configurations. Preferably 1.5-6 0 mole % of cross-linking agent are utilized per cellulose anhydroglucose unit. The resulting fibers have a WRV of 28-60 and provide good absorbent performance characteristics in absorbent structure applications. A very similar type of approach for producing high bulking resilient fibers with polycarboxylic acid cross-linking agents was described by Kokko [71].

Norlander [72] described a cellulose cross-linking system for fluff pulp based on agents used to impart crease resistance to cotton fabrics. Dihydroxy ethylene urea and its derivatives are commercially available and can be used for both dry and wet cross-linking cellulose fibers. After addition of the cross-linking agent the curing mechanism is activated at temperatures of 150-200 °C for 10 minutes, preferably on dry defibrated pulp. The absorption capacity of the modified pulp was 11 g/g.

Several inventors described an alternate approach for dry cross-linking of cellulose fibers through use of polymeric cross-linking agents. Herron and Dean [73] cross-linked the fibers with polyacrylic acid types macromolecules. Thus to a slurry of softwood kraft pulp fibers, an acrylic acid/maleic acid copolymer (65%/35%, M.W. about 9,000) and sodium hydroxide was added and then soaked for 60 minutes. The slurry was de-watered by centrifugation and de-fibrated in a Sprout-Waldron disk refiner. As the individualized fibers exited the refiner they were flash dried with hot air in vertical tubes which caused the fibers to twist and curl. The fibers were then cured at about 188°C for about 8 minutes. The washed and screened cross-linked cellulose fibers had a WRV of 43 and contained 4.6% by weight of the acrylic acid/maleic acid copolymer.

Norlander [74] prepared water soluble polymers containing free acid and aldehyde groups from, for example citric acid/glycerol and maleic acid/glycerol (50/50 mixture, M.W. 450-10,000), for use as cross-linking agents for preparation of absorbent cellulosic fibers. The absorbent fibers were produced by dipping kraft pulp sheets in an aqueous solution of the cross-linking polymers pressing out excess liquid and drying at 85°C. The sheets were then de-fibered in a hammer mill before cross-linking at180°C for six minutes.

Grafting of polyacrylic polymers has also been utilized to increase the absorbency of cellulosic fibers [75,76]. Gatenholm [76] treated the fibers with an ozone gas mixture in water for up to 90 minutes at temperatures up to 50°C. Ionizable monomers such as acrylic acid and methacrylic acid containing a cross-binder such as diethylene glycol methacrylate were then graft copolymerized to the cellulose fibers in a solution, which also contained Mohr's salt (ferric sulfate/ammonium sulfate), at pH 1-3, 60°C, for up to 3 hours. Poccia et al. [77] polymerized and partially grafted acrylic type monomers to cellulose fibers in the form of a web to give a composite type structure. The polymerization was initiated with an electron beam or chemical initiator. The composite web was then ground to yield particulate superaborbent materials and fibers containing bound and cross-linked superabsorbent particles. The inventors indicated that the fibers with the bound superabsorbent were stronger and retained their bulk to a greater extent when wetted by liquids. They also indicated that absorbent structures containing these fibers had less tendency for gel blocking due to superior distribution of the superabsorbent in combination with the fibers.

3.2. Modified Carboxymethylcellulose

Carboxymethylcellulose is the most widely utilized cellulose ether for superabsorbent products and this section is therefore devoted mainly to applications of modified CMC. Other ethers such as HEC will also be treated in this section where appropriate. The chemical reactions and structure of CMC and the other important cellulose ethers have already been presented in a previous section.

A number of patents have been issued for improving the absorbent properties of cellulose fibers by low degree substitution (less than 0.35) of carboxymethyl substituents

[20,78] as previously discussed. The absorbent properties of CMC are greatly enhanced when the DS of the CMC is above about 0.35. However, at this DS and above, CMC is soluble in water [26]. Utilization of water soluble CMC and other soluble cellulose ethers in absorbent applications requires that the polymer be insolubilized in some fashion to avoid dissolution. A variety of methods have been disclosed in the patent literature for insolubilization of CMC. The various methods can be roughly categorized into three groups; insolubilization can be achieved by: (1) Addition of a cross-linking agent; (2) Heat treatment and (3) Conversion of ionic state (i.e., conversion of CMC salt to free acid). Each of these methods will be discussed in more detail in the following subsections.

In a very recent patent, Bahia and James [79] were able to avoid the additional cross-linking steps, as described below, by using solvent spun cellulosic fibers, such as commercially available Tencel™ fibers, instead of regenerated cellulose (cellulose xanthate) rayon-type fibers. They indicated that the greater uniformity and crystallinity of the solvent spun fibers made this approach possible. Thus the solvent spun fibers were directly carboxymethylated with a caustic solution of monochloroacetate to give a high absorbency fibrous material. Regenerated cellulose and cotton fibers both have a relatively dense skin type layer at the surface, whereas the solvent spun fibers lack this skin type structure. According to Bahia and James, this allows the production of high absorbency fibers without weakening of the fiber during the carboxymethylation process to such an extent that the fiber structure is lost. The degree of substitution was preferably in the range of 0.2-0.5. It was also noted that the presence of polyvalent metal ions, such as calcium, barium and magnesium, in the reagent solution helped to avoid the formation of modified fibers which were soluble in distilled or demineralized water. Bahia and James indicated that the polyvalent metal ions may form cross-links between carboxylic acid groups on different cellulose chains. Although the fibers did show some partial solublity in distilled water, in most applications the products would be in contact with body fluids which contain salts. The absorbency of the CMC solvent spun fibers was measured at 30-40 g/g of 0.9% saline solution and the tenacity of the modified fibers was 15-25 cN/tex.

3.2.1. Addition of cross-linking agent. Many of the cross-linking agents already described in the section on "Cross-linked Cellulose" can be utilized to cross-link CMC for creation of superabsorbent products (see Table 2). The important physical properties such as water absorption capacity, WRV (SRV), wicking rate, etc. of cellulose ether superabsorbents are dependent on a large number of parameters which include the following [32,33,80]: (1) Type of substituent (carboxymethyl for CMC); (2) Degree of substitution (DS); (3) Distribution of substituents; (4) Degree of cross-linking; (5) Type of cross-linking agent; (6) Reaction procedure; and (7) Physical form of product.

The type of substituent can have a dramatic influence on the WRV or SRV of a superabsorbent polymer. For example, at an equal level of cross-linking an ionic ether can retain up to four times as much deionized water as that of a non-ionic ether (2000% versus 8000% WRV) [32,33]. However, the advantage of the ionic ether is considerably reduced in saline solution. In a 1% sodium chloride solution the difference is only 2.5 times in favor of the ionic ether and there is no advantage in an 8% sodium chloride solution. Body fluids such as blood and urine contain the equivalent of about 1% sodium chloride, thus ionic ethers such as CMC (salt) should have an advantage over non-ionic counterparts in many absorbent characteristics.

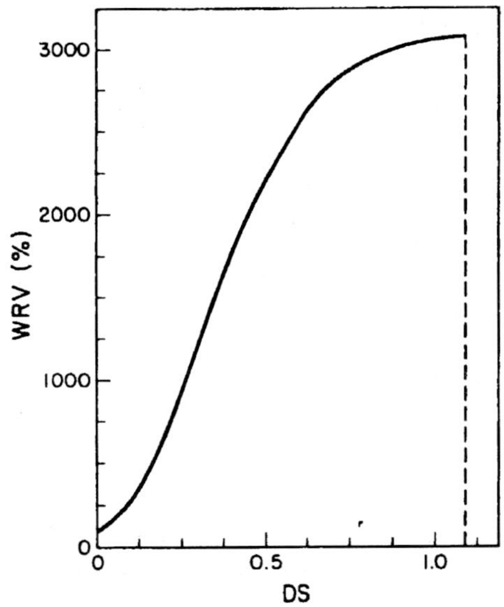

Fig. 6. Effect of degree of substitution (DS) of cross-linked CMC on water retention value (WRV). Constant degree of cross-linking at 0.02 mol/mol of cellulose [32].

The alteration of the ionic state of cross-linked CMC can also serve as a control mechanism of the shrinking and swelling behavior of the absorbent product. By partial or complete ion exchange of the sodium ions with hydrogen ions (or other cations) in the cross-linked CMC, the swelling behavior of the superabsorbent can be reduced to various degrees. Dean [38] reported that the WRV of a cross-linked CMC in sodium salts could be 16 times the WRV of carboxymethylcellulose in the acid form. Obviously the ionic state of the CMC is important to the absorbent properties.

The degree of substitution is another important parameter affecting water retention capacity of cross-linked CMC absorbents. Holst [32,33] described the effect of DS on WRV as shown in Fig. 6. The amount of cross-linking was kept constant at 0.02 mol/mol of cellulose. The water retention capacity (WRV) increases almost linearly with DS up to about a DS of 0.5 and then increases at a decreasing rate up to about a DS of 1.0. The use of DS values greater than one offered no advantage in water retention capacity for samples at this cross-linking density. It is important to note that the rate of absorption shows an opposite effect, decreasing almost exponentially at a DS greater than 0.70 [80].

The large increase in WRV at low DS is probably related mainly to the opening-up of the tightly hydrogen bonded cellulose structure by the bulky carboxyl substituents. Of course if the cross-links were not present in the sample, a water soluble product would be obtained at about DS = 0.35. The limiting water retention value at DS = 1.0 is dependent on the cross-linking density, discussed below.

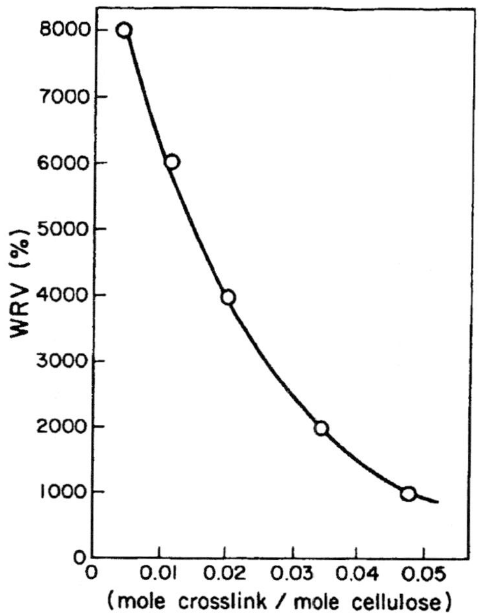

Fig. 7. Effect of the amount of cross-linking on the water retention value (DS = 1.0) [32].

The distribution of the substituents along the cellulose backbone will influence the properties of the polymer. Uneven distribution normally gives a tendency of high "gel-blocking" of the superabsorbent and consequently low rate of absorption. The gel-blocking phenomenon occurs when an absorbent polymer swells and effectively prohibits fluid distribution in an absorptive pad. This effect is still an open field of research, since little is really known about gel-blocking. However, in end-product uses this effect might have a significant influence on the spreading (transport) characteristics of absorbed fluids and largely influence the over-all absorption capacity of the product [80].

The effect of the degree of cross-linking on the WRV is shown in Fig. 7 at a constant DS = 1.0 [32]. The water absorption capacity (WRV) is enhanced by decreasing the level of cross-linking and ranges from about 1000 to 8000% at 0.01 and 0.05 mol of cross-links/mol of cellulose, respectively [32,33]. Obviously the amount of water retention desired of a cross-linked CMC superabsorbent can be conveniently manipulated by proper control of the DS and degree of cross-linking.

Direct comparative studies on the effect of the type of cross-linking agent and the reaction procedures have not been published, however, a number of variations have been disclosed in the patent literature. For example, Dean and Ferguson [81] described a method for production of "bibulous cellulose fibers" by cross-linking CMC with epichlorohydrin, similar to equation 12 for cellulose. At a 5% level of application of epichlorohydrin the WRV was about 1300%; but at 3% epichlorohydrin a much higher WRV (about 3200%) was

Table 3.
Cellulose ether cross-linking agents

Dean and Ferguson [81]
Bis(epoxypropyl)ether
Dichloroethane
Divinylsulfone
Epichlorohydrin
Ethylene glycol-bis(epoxypropyl)ether
Formaldehyde
Vinylcyclohexene dioxide
1,3-Dichloro-2-propanol
1,3-Bis((-hydroxy-(-chloropropoxy)-2-propanol
1,3-Bis((-hydroxy-(-chloropropoxy)ethane
1,2: 3,4-Diepoxybutane
1,2: 5,6-Diepoxyhexane
2,3-Dibromo-1-propanol
2,3-Dichloro-1-propanol
2,2-Dichloroethyl ether
Holst et al. [82]
Methylenebis(acrylamide)
N, N'-Dimethylol(methylenebis(acrylamide))
Trisacrylolhexahydrotriazine
Acrylamidomethylene chloroacetamide
2,4,6-Trichloropyrimidine
2,4,5,6-Tetrachloropyrimidine
Cyanuric chloride
Triallyl cyanurate
Dichloroacetic acid
Phosphorus oxychloride
Bis(acrylamido)acetic acid

reported; which again illustrates that lower levels of cross-linking give higher water absorbency.

Many other cross-linking agents for insolubilization of CMC were suggested by Dean and Ferguson and are listed in Table 3 [81]. A number of the cross-linking agents in Table 3 were also suggested for use in cross-linking cellulose pulp (Table 2). Included in the list are several additional cross-linking agents suggested by Holst et al. [32,33,82-86]. With the exception of formaldehyde which requires acidic conditions, all the cross-linking agents listed by Dean and Ferguson [81] will cross-link CMC under alkaline conditions.

Epichlorohydrin was preferred for the wet cross-linking of CMC because both the wet cross-linking and etherification could be accomplished in a single alkaline medium [81]. Reactions involving epoxides may also be preferred because the reactions are easily controlled and the rate can be adjusted by varying the concentration of additive and the pH of the environment [80].

For preparation of the cross-linked CMC, the CMC sodium salt (DS = 0.6-1.2) was treated with a 3-7% solution of epichlorohydrin under alkaline conditions for about 24 h at ambient temperature to yield the superabsorbent product. At a DS = 1.0 and a 3% epichlorohydrin addition level the product had an extremely high WRV [81]. The Buckeye Cellulose Corporation (Proctor and Gamble) marketed an epichlorohydrin cross-linked CMC superabsorbent product. The cellulose raw material was a softwood pulp and the modified absorbent material exhibited an extremely high swelling in water [38]. A cross-linked CMC superabsorbent is also distributed by Billerud Uddeholm AB of Sweden [6].

Holst et al. [32,33,82-86] in a series of patents described a large number of additional cross-linking agents some of which are listed in Table 3. Additional modifications for production of cellulose ether superabsorbents were also suggested by the German inventors and included the use of a reaction medium other than isopropyl alcohol in the etherification and use of decreased percentages of water in the modification reaction. As a result of procedural changes Holst et al. reported that they were able to achieve much shorter reaction times for the formation of cellulose based superabsorbents.

In addition to formation of CMC products, etherification reactions were described by Holst et al. [82-86] for HEC, carboxymethylhydroxyethylcellulose (CMHEC) and methylhydroxyethylcellulose (M-HEC). The cellulose ethers were crosslinked with several types of compounds as depicted below [32,33].

Functional group	Cross-linking agent (examples)
CH_2=CH-C(=O)-N- Acrylamide	CH_2-NH-C(=O)-CH CH_2
CH_2 =CH-CH_2-OC=N- Allyloxyazomethine	2,4,6-Triallyloxy-1,3,5-triazine
-N= C(Cl)- Chloroazomethine	2,4, 6-Triallyloxy-1,3,5-triazine
Activated Chlorine compound	Dichloroacetic acid Phosphoroxychloride Epichlorohydrin

Depending on the specific cross-linking agent, from 0.001 to 0.2 parts by weight were used by weight of cellulose. An exception was dichloroacetic acid of which at least 0.1 parts by weight were necessary. In one method for production of CMC or CM-HEC absorbents, the alkylation, etherification and cross-linking were preferably carried out in the presence of 0.85-3.0 parts by weight of isopropanol per one part by weight of cellulose. To provide the necessary alkali, aqueous lye was generally utilized. Water was thus introduced into the reaction by aqueous addition of lye, isopropanol (87%) and the cross-linking agent. However, according to Holst et al. the water introduced into the reaction mixture should never be more than the quantity by weight of isopropanol in the mixture; and the water content should preferably be less than two thirds of the isopropanol. For products with very high WRV it was found necessary to use organic solvents such as dioxane, methyl ethyl ketone, ethanol, acetone, or t-butyl alcohol as the liquid reaction medium. Holst et al. [82-86] reported that the quantity of water retained by the modified cellulose ethers was up to 60 times the weight of the original sample.In the reactions to form the superabsorbent materials, low temperatures (80°) and short reaction times (1 h) were claimed as illustrated in the following example: 125 g of cellulose was sprayed with 228 g of an aqueous solution of sodium hydroxide (28 %) and alkalized with continuous mixing for 45 minutes at 20°C. Then 169 g of sodium monochloroacetate was added together with 1.2 g of methylenebis(acrylamide) and simultaneous etherification and cross-linking took place for 1 h at 80°C. The cleaned and dried product had an exceedingly high water retention value.

Surprisingly Holst et al. [87-89] were also able to obtain swellable cellulose ethers with reactive agents which were only monofunctional with respect to cellulose. The reaction conditions were similar to those described above and the specific compounds utilized are listed in Table 4. Apparently strong secondary intermolecular associations are important since the cellulose ethers are not covalently cross-linked. The agents capable of incorporating enhanced absorption in the cellulose through monofunctional reactions have the general formulas shown below [32,33]:

Table 4.
Reagents used as monofunctional reactants [86-88].

N-Methylol acrylamide
N-(Acrylamidomethylene)acetamide
N-(Acrylamidomethylene)formamide
N-(Acrylamidomethylene)amylurethane
N-(Acrylamidomethylene)methylurethane
N-(Acrylamidocarboxymethylene)ethylurethane
N-(Acrylamidomethylene)methoxyethylurethane
Vinylsulfonamide

$$CH_2{=}CH\text{-}\overset{\displaystyle O}{\overset{\|}{C}}\text{-}NH\text{-}\underset{\displaystyle R^2}{\underset{|}{CH}}\text{-}R^1 \quad \text{or} \quad CH_2{=}CH\text{-}\overset{\displaystyle O}{\overset{\|}{\underset{\displaystyle O}{\underset{\|}{S}}}}\text{-}\underset{\displaystyle R^4}{\underset{|}{N}}\text{-}R^3$$

where R^1 = H, CH_3 or CH_2OH, R^2 = CH_3, CH_2OH or N-methyleneacrylamide and R^3 and R^4 = H or CH_3.

Chatterjee [90] offered a unique alternative to producing a cross-linked, modified cellulose superabsorbent. In this invention the cross-linking agent not only contains reactive functional groups for linking adjacent cellulose chains, but also has bulky, hydrophilic carboxyl groups present in the structure which improve the absorbency properties of the product. Thus a one-step reaction is possible instead of two separate reactions; one for derivatizing the cellulose and the second for cross-linking the modified cellulose.

The invention was based on cellulose bis-alkane derivatives formed by etherification reactions with cross-linking agents of the following general formula:

$$\begin{array}{l} \;\;| \\ \text{-}C\text{-}(CH_2)_k\text{-}X \\ \;\;| \\ (CH2)_n \\ \;\;| \\ \text{-}C\text{-}(CH_2)_m\text{-}COOH \\ \;\;| \end{array}$$

where X = H or COOH and k, n and m are integers from 0 to 4. This structure comprises dihalogenated (CI or Br) mono- or di-carboxylic acids (and salts). Typical monocarboxylic acids include dibromopropionic, dichlorobutyric, dichlorotrimethylacetic, etc., and typical dicarboxylic acids include dibromosuccinic, dibromomalonic, dibromoadipic, dichlorosuccinic, etc.

Thus according to Chatterjee [90], wood pulp is treated with aqueous sodium hydroxide for 30 minutes at room temperature. The cross-linking agent 2,3-dibromosuccinic acid is then added, the mixture is stirred for 30 minutes and then heated in an oven at 55°C for 3.5 h. The sample is purified and the modified cellulose absorbent product has the structure shown below (eq. 13).

$$2\ Cell\text{-}OH + HOOC\overset{\displaystyle H}{\overset{|}{\underset{\displaystyle Br}{\underset{|}{C}}}}H\overset{\displaystyle H}{\overset{|}{\underset{\displaystyle Br}{\underset{|}{C}}}}HCOOH \rightarrow Cell\text{-}O\text{-}\overset{\displaystyle COONa}{\overset{|}{CH}}\text{-}\underset{\displaystyle COONa}{\underset{|}{CH}}\text{-}O\text{-}Cell \qquad (13)$$

Tajiri et al. [91,92] in Japan described a process for simultaneous carboxylation and cross-linking of cellulose pulp fibers. Bleached softwood kraft pulp was reacted in an aqueous solution of sodium hydroxide and monochloroacetic acid for carboxymethylation and the solution also contained ethylene glycol diglycidylether for simultaneous cross-linking. In several examples the pulp was laminated by spun-bonding or intertwined with

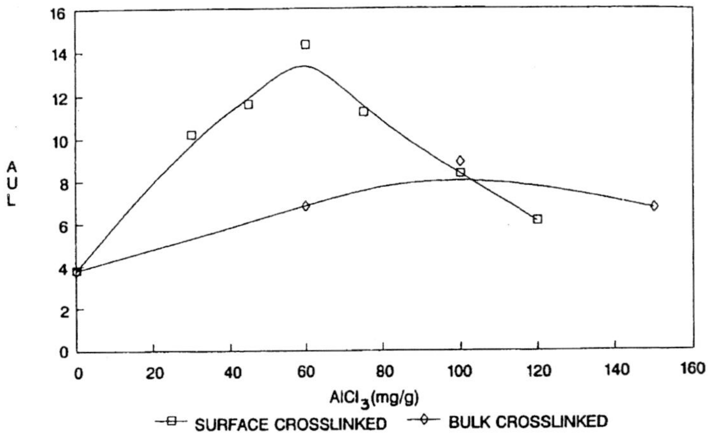

Fig. 8. The effect of cross-linking of CMC with aluminum chloride on the absorbency under load (AUL) [94].

polypropylene filaments. The initial water content was 50-90% by weight with the carboxylating agent present in a molar ratio of 0.7-2.0 per glucose glucose unit. The water content was then adjusted to 20-60% by weight, and to at least 5% by weight below the initial water content, by evaporation. The carboxylation/cross-linking reactions were then initiated by heating to 50-110°C while maintaining the water content in the impregnated sample at 20-60% by weight and at least 5% below the initial water content. The resultant cross-linked, carboxylated cellulose had a degree of substitution of carboxylalkyl groups of 0.35-0.8 and 1.5-15% cross-linking agent based on cellulose weight. The water retention of the modified product was approximately double that of the unmodified pulp.

Several inventors have recently disclosed the use of metal salts and metal-based compounds for cross-linking of CMC materials for superabsorbent applications. Cottrell et al. [93] described the use of titanium and zirconium cross-linking agents for cross-linking of polysaccharides, mainly illustrated with carboxymethyl guar and sodium zirconium lactate. However, the method was also suggested to be applicable to CMC. Qin [94,95] disclosed the use of aluminum chloride for surface cross-linking of CMC as opposed to typical bulk cross-linking. He found that it was desirable to have a metal cation having a valency of at least 3 to encourage intermolecular bonding between adjacent polymeric chains, since divalent metal cations favored intramolecular bonding within the polymer chain. The effect of the surface cross-linking of Aqualon™ CMC with aluminum chloride versus bulk cross-linking is shown in Fig. 8. As shown the absorbency under load (AUL) is improved for the surface cross-linked CMC up to about 100 mg/g of applied aluminum chloride.

Improved absorbency for cross-linked CMC products were described by a group of Japanese inventors [96,97] by reaction in solvents and/or further treatment of the cross-linked CMC with different swelling solvents. Thus CMC sodium salt was treated with epichlorohydrin in an aqueous isopropanol solution in the presence of NaOH to give a fibrous cross-linked cellulose derivative. This material was then swelled in an aqueous epichlorohydrin in an aqueous isopropanol solution in the presence of NaOH to give a fibrous

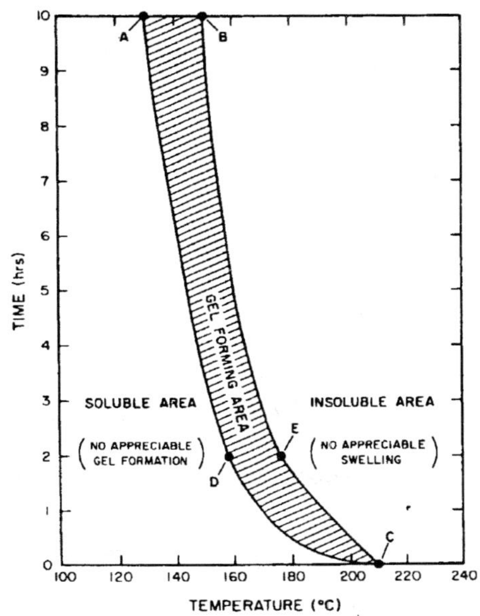

Fig. 9. Time- temperature relationship for heat treatment of CMC [99].

cross-linked cellulose derivative. This material was then swelled in an aqueous methanol solution to a swelling degree of 50 g/g, de-swelled with Me_2CO, filtered, dispersed in Me_2CO, left at room temperature for 6 hours and dried to give an absorbent product which absorbed urine to 86 g/g. Several other variations on this type of approach were described in their patent disclosures. Carboxymethylcellulose can also be insolubilized by heat treatment as discussed in the next section.

3.2.2 Heat treatment. It was noted over 30 years ago that water soluble CMC polymers could be insolubilized by heating at moderately high temperatures without losing the moisture absorption capacity. In a 1949 disclosure [98] a dilute aqueous solution of CMC (2%) was applied to a fibrous sheet and the coated fibers were heated to 150-230°C. A water absorbent fibrous mat was obtained. If the sample was heated to only 120°C insolubilization of the CMC did not take place. It was suggested that water soluble HEC could be similarly modified, but in this case, it was found advantageous to dry the fibrous sheet impregnated with HEC at 105°C before subjecting it to the elevated temperature (above 150°C).

In a latter disclosure, Elliott [99] defined the effect of heat on CMC more succinctly as shown in Fig. 9. The object of Elliott's invention was to provide a alkali-metal CMC which was swellable in water to highly-swollen, discrete gel particles which were dispersible in water to form a uniform, stable, aqueous suspension of highly swollen gels. Thus a CMC salt with a DS = 0.5 - 1.0 was heated to 130-210°C in a dry and finely divided particulate form. The time and temperature conditions were selected so that the product had the gel-forming properties defined by ABC in Fig. 9. When suspended in water the heat treated CMC particles were reported to swell from 2.5 to 10 times their original diameter [99].

Further refinements on heat cross-linked cellulose ethers have resulted in a number of commercial superabsorbent CMC products. In 1968, Reid [100] patented a method for preparation of absorbent but water insoluble ethers and esters of cellulose. It was required that the polymers contain both hydroxyl and carboxyl groups and be water soluble. A portion of the carboxyl groups in the polymer were converted to the free acid form by acidifying (70% nitric acid) until the ratio of the free acid to the salt form of the carboxyl groups was in the range of 0.007/1 to 3/1, but usually 0.2/1 to 2/1. It was necessary to operate within this range so that the polymer could be insolubilized by curing but remain sufficiently water soluble prior to curing. After drying the acidified polymer at room temperature, it was cured within a broad range of time and temperature conditions, the shortest of which was 1-2 minutes at 200°C.

A wide variety of polysaccharides were suggested by Reid [100] as starting materials for preparation of the absorbent products. The specific cellulose carboxyalkyl ethers prepared in aqueous isopropanol were CMC, hydroxypropylcellulose(HPC) and hydroxypropyl-CMC. Several cellulose esters from dibasic acids, such as cellulose succinate, were also suggested as potential absorbents.

By mildly esterifying several cellulose ethers Desmarais and Reid [101] obtained improvements in the water dispersibility and interfacial activity of the cellulose derivatives. The ethers listed above are very hydrophilic and aggregate and lump when mixed with water. The addition of acyl groups (0.1-18%) increased the hydrophobic character of the samples resulting in delayed hydration and better absorbent characteristics.

Based on Reid's earlier work [100], Hercules Incorporated marketed a self cross-linked CMC absorbent fibrous product. The CMC is cross-linked by heating (drying) the polymer at low pH where many of the carboxyl groups are in the free acid form, as previously described. The cross-linking is believed to be due to esterification reactions as shown in Fig. 10. The absorbent CMC has a DS of about 0.7 with 65-70% of the carboxyl groups in the sodium salt form. Thus the polymer has a high degree of hydrophilicity but does not dissolve in water because of the heat-induced cross-links. The product has good water absorption characteristics and a high degree of maximum swelling.

Often times CMC products harden and hornify in the drying stages which make it difficult to subsequently disintegrate the material. Schoggen et al. [103] disclosed a method to avoid hardening and hornification of CMC products and which improved absorbency of the final product. Thus, the cellulose derivative was first slurried in an aqueous isopropyl alcohol solution (44% by weight) followed by washing with aqueous methyl alcohol rather than directly drying. These investigators found that it was advantageous to reduce the water content gradually in successive stages through alcoholic displacement washing. The first stage must contain at least 10% by weight of water so that the free water content of the product is gradually reduced. Typically three wash stages were utilized in a sequence of 80% methyl alcohol, 90% methyl alcohol, and finally 90-100% methyl alcohol. Better communition quality was achieved if the last stage contained some water, that is, less than 100% methyl alcohol. Chatterjee and Kwok disclosed in a series of patents [104,105] that it was not necessary to use such extensive and elaborate purification and post-reaction procedures; and they reported additional advantages with their method. In the invention the

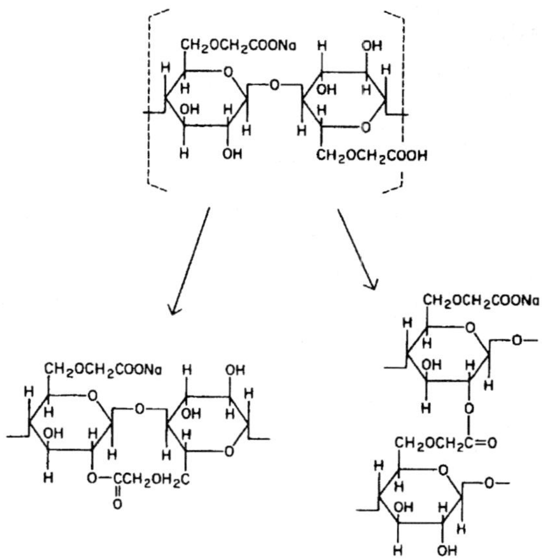

Fig. 10. Possible esterification (cross-linking) reaction for heat treated CMC [102].

carboxyalkylated product was simply drained and filtered to remove only a portion of the carboxyalkylating reactants, residues, impurities and byproducts formed during the reaction. Chatterjee and Kwok found that if at least 3-4% by weight (less than 50%) of these materials were present during the subsequent heat treatment, catalytic reactions took place to insolubilize the carboxyalkylated cellulose in drastically reduced time. It was also observed that increased degrees of carboxyalkyl substitution and cross-linking were attained along with improved color and brightness of the product. These investigators suggested that their invention resulted in a process which was much simpler and potentially less expensive than previous methods primarily because fibers are cross-linked prior to the washing step and do not require a washing with a non-aqueous solvent. The absorbent fibers also show less tendency to harden or hornify and absorb water without the formation of a surface gel which inhibits further water absorption.

There are several possible mechanisms for cross-linking in the process designed by Chatterjee and Kwok. One is esterification between the introduced carboxyalkyl groups and the unreacted hydroxyl groups of the cellulose similar to the reactions depicted in Fig. 10. Another possibility is the formation of cross-links containing glycolide and polyglycolide structures as shown in Fig. 11. The glycols arise from side reactions in the etherification (see eq. 5 and 7). If chloropropionic acid instead of chloroacetic acid is used as the carboxyalkylating agent then the cross-links are bridged by C_2H_4 radicals compared to CH_2 radicals, respectively [104,105].

Shinohara and Field [106] utilized both carbon dioxide gas and heat to form an absorbent cross-linked CMC product. These investigators claimed that the $C0_2$ catalyzed the transformation of the soluble CMC salt to an insoluble product such that milder reaction

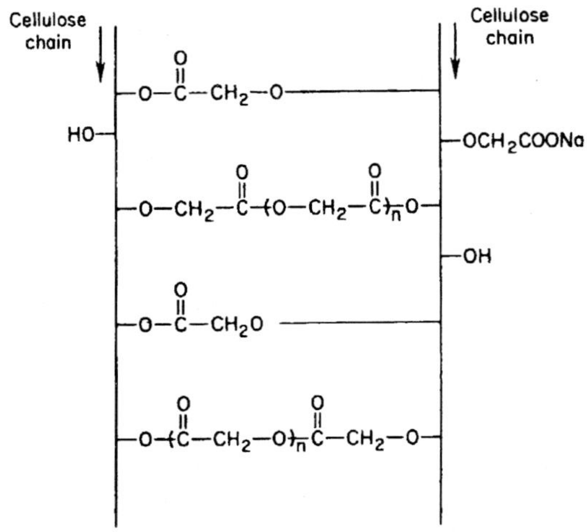

Fig. 11. Possible structures of glycolide and polyglycolide cross-links for heat treated CMC [104,105].

time/temperature cycles could be used and which provided products with less degradation and discoloration. Equipment costs would also be reduced. These inventors suggested that the CMC cross-linking occurs, in part, by reaction 14.

$$\text{Cell-OH} + \text{Cell-COONa} \rightarrow \text{Cell-O-}\overset{\overset{\text{O}}{\|}}{\text{C}}\text{-Cell} + \text{NaOH} \tag{14}$$

The CO_2 gas may react with the sodium hydroxide formed in the reaction to give sodium bicarbonate, resulting in a shift of the equilibrium towards completion. It is also possible that the CO_2 acts as a precipitant in the neutralization of the CMC product [106].

A product with similar properties was also prepared by treating dry, granular CMC sodium salt with HCl gas (degree of acidification 15%) and heating in a forced air oven for 60 minutes to give a cross-linked partially acid CMC with a high swell-ratio [107].

The Buckeye Cellulose Corporation marketed an ester cross-linked CMC product [103,108,109]. The cross-linked product is produced directly in a slurry process. The additional step involved with insolubilization of the water soluble CMC is avoided with this invention and product coloration engendered by post preparative heat treatments does not occur. The method is based on two important modifications, (1) the amounts of etherifying agent (monochloroacetic acid) and neutralizing agent (sodium hydroxide) are regulated so that a molar excess of sodium chloroacetate is present after neutralization. Also a sufficient number of moles of caustic are present per mole of cellulose after neutralization (0.5 mol caustic/mol cellulose) to insure attainment of a DS (carboxyalkyl substituents) of about 0.4 or above. As the excess etherifying agent is increased the time for insolubilization of the product is reduced, up to 8 mol of monochloroacetic acid per mol of cellulose. Also, the combined time and temperature conditions in the etherification reaction is sufficient to result in substantial insolubilization of the resulting CMC product. Equivalent products were produced at 60°C for 10 h or 80°C for 3 h. An increase in the excess moles of acid to moles of unreacted sodium hydroxide up to 1.5 was found to reduce the time necessary for insolubilization.

The critical factor in the above process is that an acidic condition be engendered during etherification. Ways to achieve acidic conditions described in the patent [108] included addition of hydrochloric or nitric acid so that an excess of monochloroacetate is present relative to sodium hydroxide; or by direct addition of monochloroacetate prior to completion of the reaction. The product has good absorption characteristics, similar to epichlorohydrin cross-linked CMC, and has good brightness. Reflectance brightness values of 75-80 were reported.

A number of variations of the basic processes already described have been patented. For example, in applications of cross-linked carboxyalkylcelluloses to tampon products, the absorbency rate may not be sufficient. The desirable fluid capacity and retention are sometimes achieved at the expense of a reduced absorbency rate as previously discussed. Thus the tampon product, although highly absorbent, has a reduced ability to immediately take up fluid from the time of insertion. This problem can be partially alleviated by the use of layered structures comprised of both the more economical conventional cellulose fibers which accept fluid readily and the superabsorbent cross-linked CMC fibers, for example. Karmazyk et al. [110] described a tampon product which had the two types of fibers (conventional and superabsorbent) intermixed within the product. The superabsorbent fibers were composed of CMC with a DS = 0.4-2.0 and 5-40% of the carboxyl groups in the free acid form as earlier described by Reid [100]. The tampon product was claimed to have an improved absorbency rate.

The absorbency time was defined as the number of minutes it takes one gram of absorbent material compressed to 5 cm^3 in a glass syringe to absorb a 0.9% sodium chloride solution up to the 5 cm^3 mark. Peak performance for tampons containing the described fibers was attained when the absorbency time was about 25 minutes [110]. Ahlgren and Kloow [6] have also reported that an absorbent product with both high fluid capacity and high rate of absorption could be produced by mixing cross-linked CMC with fluffed pulp.

An additional unique approach to production of absorbent cellulose ether products was developed by Karlsson et al. [111] in Sweden. The method is based on CMC or other cellulose ethers with low DS (0.35 or at most 0.40), the limit of water insolubility. Generally CMC fibers with this low DS are so similar in water absorption properties to those of the non-substituted product that the cost of substitution is not justified, although patents have been issues for absorbent low DS CMC fibers [20,78]. However, the Swedish inventors subsequently swelled the derivatized fibers to the maximum state and then "fixed" the fibers in the highly swollen state. According to the invention, the swollen fibers are fixed by either forming into a paste and heating on a roll dryer (the fibers become attached to the roller) or freeze drying. With the fibers mechanically fixed in this state, the resulting drying forces cause the fibers to burst. The bursting takes place in such a manner that the fibers burst open along one line or major break and, in so doing, the surface of the burst fiber available for water absorption becomes several times greater than a fiber dried in the normal manner without swelling and bursting. In addition to better absorption properties the burst fibers did not exhibit the gel blocking effect; which arises with highly substituted CMC that swells so greatly at the surface that further transportation of liquid through the material is prevented.

Hercules Incorporated marketed a sheeted product for a time which was produced by the method of Karlsson et al. [111].

3.2.3. Conversion of ionic state. Most of the methodology described in this section overlaps to some extent that already described in the previous section on heat treatment. CMC can be insolubilized by acidifying a dilute aqueous solution of the polymer and casting a film [112-114]. The film is then soaked in a strongly acidic solution, washed with a dilute acid solution and with water to remove salts and excess acid and finally cured. Obviously the majority of the carboxyl groups in the CMC are converted to the free acid form in contrast to the invention previously described [100]; where only a portion of the carboxyl groups are converted to the free acid, the rest remaining in the salt form.

The salt ions in CMC can also be replaced with hydrogen ions by contacting an aqueous solution of CMC with a cation exchange resin in the free acid form. Essentially all the cations present in the solution become attached to the resin and the hydrogen ions supplied by the resin become attached to the CMC to form in situ the free acid form of CMC as a stable aqueous colloidal dispersion. Films are cast from this dispersion and dried [100].

Surface insolubilization of CMC has been achieved by treating CMC coatings or films, deposited from water solutions, with aluminum sulfate [114]. Due to the dense nature of the deposited coating or film, such insolubilization is confined mainly to the surface layer and does not extend uniformly throughout the deposited coating or film. The films are water sensitive and tend to disintegrate in water. The films are also contaminated with salt impurities. This method of insolubilization is not utilized in superabsorbent applications.

3.3. Cellulose Derivatives Containing Phosphorus and Sulfur

A number of patents have been issued for superabsorbents based on cellulose derivatives which contain phosphorus or sulfur. Most previous investigations on derivatization of cellulose with phosphorus containing reagents have been carried out to impart flame retardancy to textile fibers and fabrics. The previous knowledge of etherification and esterification reactions of cellulose with phosphorus and sulfur containing compounds has been utilized to produce unique superabsorbent products.

It has already been shown that cellulose is rendered more hydrophilic by substitution of the hydroxyl groups with ionizable groups such as the carboxyl functional groups in CMC. The presence of such groups makes cellulose more absorbent, but the ionic groups also tend to make the modified cellulose more water soluble than unmodified cellulose. The greater the degree of substitution, the greater the degree of water solubility. Cross-linking of the water soluble polymer is therefore necessary for superabsorbent applications [115].

Similar principles were utilized by Chatterjee to prepare phosphorus containing cellulose superabsorbents [115]. Thus wood pulp cellulose was etherified with a phosphonoalkylating agent such as disodium chloromethyl phosphonate to form hydrophilic disodium phosphonomethylcellulose. The cellulose derivative is crosslinked simultaneously during the etherification in a one-step process, or prior to the etherification reaction either by wet or dry cross-linking methods such as those previously described. If formaldehyde is used as the cross-linking agent, the product has the general structure shown in Fig. 12.

The optimum absorbency consonant with water insolubility was obtained through adjustment of the degree of phosphonoalkylation and the degree of cross-linking. The

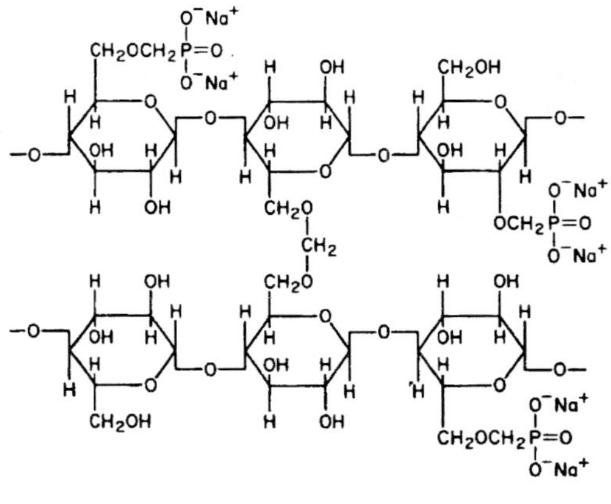

Fig. 12. Possible structure of formaldehyde cross-linked phosphonoalkylcellulose [115].

preferred phosphorus content was about 2% by weight and the preferred range of cross-linking density was 0.015-0.110 mole of cross-link units per mole of cellulose [115].

The preferred cross-linking agent for the two-step wet cross-linking process was formaldehyde (see eq. 8 and 9) while epichlorohydrin or dichloro-2-propanol was preferred for the one-step dry cross-linking process (see eq. 12). Additional cross-linking agents are listed in Tables 2 and 3.

Superabsorbent cellulose materials containing phosphorus have also been prepared by esterification reactions. Bernardin [116] of the Kimberly-Clark Corporation phosphorylated cellulose by the urea phosphate method, wherein the fibers were saturated with a solution of urea and orthophosphoric acid and reacted at near 160°C for about 20 minutes. Phosphorus oxychloride alone or in combination with other reagents such as pyridine was a less satisfactory phosphorylating agent.

To obtain the superabsorbent properties a number of additional treatments were necessary. After the phosphate groups were substituted on the cellulose hydroxyls, the cellulose fiber walls were hydrolyzed with a hot acid solution (3% HCl, 70°C) for 30 minutes. This step removes the constriction of the primary wall which then allows more efficient ballooning and gelling of the fiber and fosters enhanced capillary suction in the final product. Ballooning of pulp fibers has been previously observed when the fiber walls are damaged during mechanical treatment as shown in Fig. 13 [116].

The acid treatment of the phosphorylated fibers also converts the phosphate to the free acid form. If the fibers are converted back to the salt form with dilute alkali the absorbency is improved about five times over that in the acid form. Even greater absorbency is developed in the phosphorylated cellulose (salt form) if the fibers are refined for at least one minute before drying. The refining increases the absorbent properties threefold over the unrefined fibers [116].

Bernardin [116] also observed, unexpectedly, that if the refined fibers are reconverted to the acid form, they retain much of the increased absorbency developed by refining and exhibited by salt form fibers. This discovery means that the product can have both absorbency and ion-exchange properties which complement one another. The treated fibers

Fig. 13. Ballooning of secondary wall of radiata pine fiber. Primary wall has burst but is still attached to the fiber at several points (Courtesy of John Wiley & Sons).

are finally solvent dried instead of by water evaporation to avoid appreciable hydrogen bonding and collapse of the water swollen fibers. Drying with such solvents as acetone, alcohol, alcohol followed by hexane, etc. avoids the horny, hard mass which is obtained when water is directly dried from the fibers. According to Bernardin [116] freeze drying can be substituted for solvent drying in the final step to accomplish the same objective. Mats of the freeze dried fibers were also found to have more resilience which may be desirable for some end uses.

The significance of the ion-exchange capability of the phosphorylated cellulose absorbent is for pH control in tampon products. Normally the vagina is acidic, having a pH in the range 3.8-4.5. At this pH beneficial microorganisms are present which provide protection and resistance to infection and inflammation. During menstruation, however, a slightly alkaline pH is established which inhibits growth of the beneficial microflora. The vagina is then more susceptible to infection and inflammation during menses [116].

Obviously it is desirable to use the tampon product for pH control during menses. Kellet [117] incorporated glyceryl triacetate as a physiological blostat and automatic pH controller; while Burgeni [118] used CMC in the hydrogen form as an acidifying polymer in combination with regenerated cellulose fibers. The phosphorylated cellulose product devised by Bernardin combines both the pH control and absorbency functions in one fiber.

Fewer absorbent cellulosic products containing sulfur are described in the literature. Arai and Goda [119] described an absorbent product prepared by cross-linking cellulose sulfate with glutaraldehyde in dimethylformamide and neutralizing with aqueous NaOH. The DS of the cellulose sulfate decreased markedly with increased degree of cross-linking. However, even at lower values of DS (0.2) the absorbency was good in de-ionized water. But it was also noted that, in aqueous electrolytic solutions, higher DS values gave higher absorbency.

Tajiri and Tsukamoto [120] utilized sulfonic acid compounds bearing halogenated C1-C3 aliphatic and/or epoxy groups to modify cellulose materials in aqueous alcoholic solutions containing an alkali hydroxide catalyst. Thus, cellulose pulp was etherified with sodium chloromethanesulfonate in an aqueous butanol solution containing NaOH and cross-linked with epichlorohydrin to give a fibrous product with very good absorbent properties for urine.

In another variation for the introduction of sulfur onto cellulosic materials for improved absorbency, Shet and Wallajapet [121] first oxidized the cellulose, which opened anhydroglucose rings in the polysaccharide chain, and then they sulfonated the oxidized cellulose to give a product with an absorbency of 8 g/g. The oxidation was preferably carried out with sodium metaperiodate which preferentially cleaves the between carbons C2 and C3 in the anhydroglucose structure to yield a cellulose chain with aldehyde groups at these two positions. The oxycellulose is then reacted with sodium bisulfite to yield a hydroxysulfonic acid with a DS of 0.2-0.5 which is water insoluble and has improved absorbency characteristics.

3.4. Extruded and Regenerated Filaments

A number of new and novel approaches for preparation of highly absorbent products have been patented. In this section two such approaches will be briefly discussed. The first is filament and web formation from modified cellulose fibers; and the second, is coextrusion of absorbent cellulose derivatives with rayon viscose.

A unique development was disclosed by Lassen and coworkers [122-124]. The invention provides filaments and webs formed by extrusion of chemically modified, highly swollen cellulose fibers. In the filaments the individual fibers are identifiable and are predominantly aligned parallel to the filament axis. Many channels or capillaries are present within the filament which greatly enhance the transport of liquids. The filaments can also be formed into webs by random deposition onto a moving wire prior to drying. Contact bonds or fused bonds are formed at the cross-over points. A flow diagram illustrating the process for producing the filaments is shown in Fig. 14.

Although filaments can be prepared from several types of highly absorbent fibers such as cross-linked CMC, Lassen [122] preferred phosphorylated pulp fibers (3% substitution), the preparation of which was described in the previous section [116]. It was found that curing the phosphorylated pulp in a microwave oven instead of a standard oven resulted in softer filaments and webs with very little discoloration. The modified pulp was acid hydrolyzed to break down the outer cell wall as previously described and then refined. After the refining step the individual fibers retain their identity but are in a highly swollen, gel-like form. At this point, particularly if physiological effects will be important in the product, the pH of the fibers is adjusted to 5.5-8.

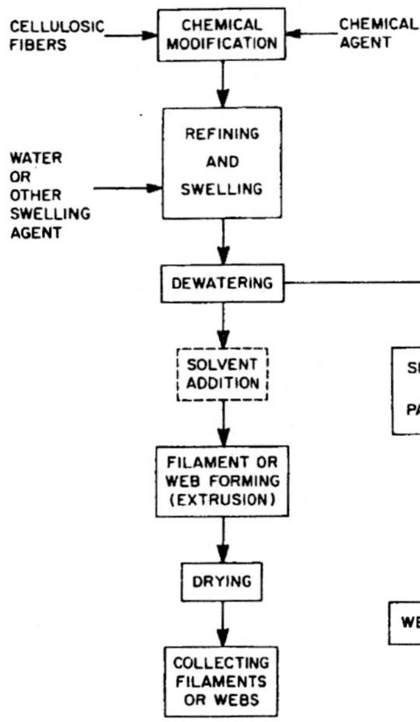

Fig. 14. Flow diagram of process for producing filaments and webs from modified and swollen cellulose fibers [116,122,123].

Further tailoring of the product can be done in the filament forming or spinning step. The orifice through which the sample is extruded has a significant effect on the final filament properties. Lassen [122] found that generally for batches of equally refined fibers, the larger the orifice (up to 0.089 cm) the better the wicking properties of the resulting filaments in terms of liquid transport. However, finer filaments tended to exhibit a faster rate of absorption probably due to the larger surface area per unit weight. The structure of two air-dried filaments prepared from cross-linked CMC are shown in Fig. 15. The large microporous character of the filaments is clearly evident [125].

To obtain the best properties, the filaments are preferably extruded directly into a solvent exchange bath. However, the selection of the solvent substantially affects the softness and wicking properties of the resulting filaments. An acetone bath containing less than 10% water appeared to give the best combination of softness and wicking properties. Addition of some solvent (2-50%) to the highly swollen fiber mass before extrusion also improved absorbency of the final filaments [123].

As already mentioned, filaments of cross-linked-CMC fibers can also be produced by this method. However, the CMC fibers should not be so extensively cross-linked that the

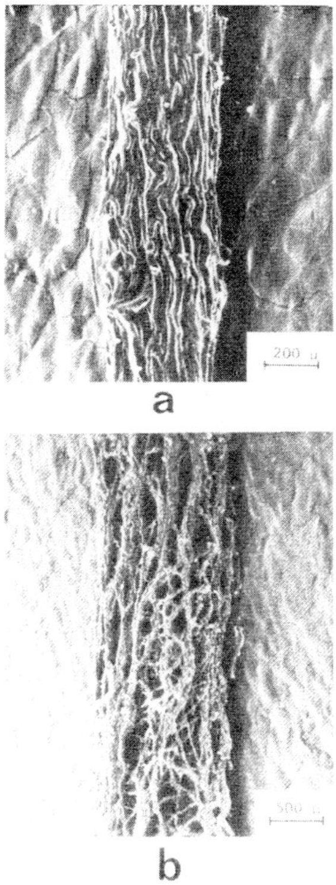

Fig. 15 Scanning electron micrograph of air dried superfibers from cross-linked CMC, (a) CLD, (b) Aqualon [125] (Courtesy of R.H. Marchessault).

fibril structure is destroyed and the modified product hydrogen bonds extensively during drying. Several commercially available cross-linked CMC absorbent fibers can be modified by refining which improves the interfiber bonding properties and also enhances the absorbency of the final filament.

A web is formed by directing the filaments onto a moving screen which, in turn, carries the web into the solvent bath. This results in interfilamentary bonds at crossover points, particularly when the bath is acetone. Fused bonds are formed when the filaments are highly swollen and contact bonds when the filaments are swollen less. Softer feeling webs have also been prepared by first partially dewatering the swollen filaments and depositing them from air via a pneumatic sprayer [124]. Dean [38] has also prepared webs directly from highly swollen cross-linked CMC fibers (not filaments).

Another method for producing highly absorbent filaments is incorporation of an absorbent cellulose derivative into a cellulose xanthate (viscose) solution and coextrusion of the polymers. A variety of modified cellulose polymers have been suggested for this application in the patent literature including cellulose sulfate [126] (see eq. 2), CMC [127-129], cross-linked CMC [130], and even starch [131].

Takebe and Yamazaki [132] disclosed a method for directly forming absorbent filaments from lower alkyl or hydroxyalkyl substituted cellulose ethers. The absorbent product was water dispersible and when thrown away (in water) it could be readily flushed without plugging drains. Thus staple fibers were spun from methylcellulose, ethylcellulose, HEC, HPC, and hydroxypropylmethylcellulose. Standard spinning methods were utilized with variations depending on the cellulose ether. For example, methylcellulose was spun from water into an acetone bath at room temperature by a wet spinning process. Methylcellulose could also be dry spun after it was dispersed and dissolved in ethylene dichloride and methyl alcohol and extruded into hot air (100-110°C). HPC was melt spun by first softening at 150°C, kneading at 190°C and extruding at 210°C. Five parts by weight of octanol was added to the melt for melt spinning hydroxypropylmethylcellulose under similar temperature conditions. The cut staples from the cellulose ethers were mixed with cotton fibers to form tampon products.

The physical form of the regenerated fiber can also affect the absorbency properties. It has been reported that rayon fibers with symmetrical cross-sectional shapes such as Y, X, H or T, give improved absorbency (Galaxy rayon fibers) [133]. Nguyen et al. [133] found that the absorbency of these fibers could be further improved by mixing in approximately 30% of standard rayon fibers such as Danufil rayon fibers. The carded webs with the mixture of fibers had approximately 15% greater extent of absorption.

3.5. Microfibrillated Cellulose

Microfibrillar cellulose has been utilized in a variety of ways to create highly absorbent, retentive cellulose pulp by a relatively simple process. Cellulose fibers are composed of layers of associated microfibrils, the walls are denotated from the outside of the fiber to the lumen as, primary, secondary and tertiary walls. The secondary wall contains over 90% of the mass of the fiber and is further delineated into S1, S2 and S3 layers. When a fiber is mechanically beaten in a Valley beater or disk refiner, the bonding of the fiber in a paper sheet is dramatically improved. This is due in part to the loosening and disrupting of the microfibrils in the outermost layers of the fiber [11]. However care must be taken to avoid excessive damage to the layers. To prepare microfibrillated cellulose the fibers are beaten to a much greater degree to where at least the outermost secondary walls are essentially completed disintegrated to microfibrillar form. This involves beating to a Canadian Standard Freeness (CSF) of less than 50 at a solids content of 1.5-6.0% for the slurry. The beaten dispersion is then freeze dried which results in sublimation of the water from the ice form. The freeze dried product contains many cellular voids giving it a sponge-like appearance under the microscope. The cellulose also appears to have a unique structure with the freed fibrils in the compressed sheet forming discrete platelets or sheets around the cellular voids. The resulting product from the beaten pulp had a porous plate capacity of at least 10 g/g with a 1% sodium chloride test solution under a hydraulic head of 35.5 cm of liquid [134].

In another disclosure for preparation of absorbent materials from microfibrillated cellulose, Makoui and Chatterjee [135] described a more complicated process utilizing a cross-linking agent. The microfibrillated cellulose was prepared in a similar way as described above through beating, however, in the example given, the beaten slurry was placed in a reaction vessel and heated to 80°C and then the beaten slurry was poured into a container and frozen to -25°C for 24-72 hours. The frozen cake was then solvent exchanged by placing in an acetone bath until the ice was melted and then washed three more times with acetone. Zinc chloride and a glutaraldehyde cross-linking agent were added in sufficient acetone to cover the product and, after evaporation of the acetone, the product was cured at 100°C for one hour. Finally the low density cross-linked product was washed in water for 15 minutes, excess water removed and dried in an oven at 105°C for 30 minutes. The cross-linked refrigerated microfibrillated product (XRMFC) had superior absorbent properties to both the uncross-linked freeze dried material (FMFC) and a cross-linked freeze dried pulp (XFMFC). For example, when the products were densified to 0.4 g/cc, the XRMFC had an absorbency of 16 g/cc and a retention of 8 g/g (porous plate test, 1% NaCl solution) while the XFMFC had an absorbency of 10 g/cc and a retention of 3 g/g and the FMFC had an absorbency of only 6 g/cc and a retention of 5 g/g [135].

Pores in microfibrillated cellulose (MFC) products were created through freeze-dry approaches in the products discussed above. However pores can also be generated in these materials by incorporation of certain chemicals in aqueous or non-aqueous systems [136,137]. In one approach MFC was dispersed in a 2% acetone slurry, followed by addition of zinc chloride and glutaraldehyde for cross-linking purposes. Then 25 g of sodium sulfate salt was added for each gram of fiber as a pore generator. After mixing the slurry, the acetone was evaporated and the product cured at 100°C for one hour. The cured cake was washed for 30 minutes in water to remove the salt leaving a pores in the structure where the salt previously occupied. In another non-aqueous approach, ammonium bicarbonate was utilized as the pore generator instead of sodium sulfate. In this case the ammonium bicarbonate was added first and then decomposed by heating in an oven before cross-linking since this chemical interfered with the cross-linking reaction.

A porous MFC was also produced by incorporating polystyrene [136,137]. In this non-aqueous approach, shredded polystyrene was dispersed with the MFC in an alcohol solution on a one to one basis. The same zinc chloride/glutaraldehye cross-linking agent was incorporated and the resultant stabilized and cured MFC cake was subsequently washed with methylene choride to remove the polystyrene and generate the porous structure. Polystyrene, MFC and the same cross-linking agent were also combined in water to give a cured cake which was extracted with methylene chloride to give a porous MFC by an aqueous based system. The porous MFC products had an absorbency of 15 g/g (porous plate test, 1% NaCl solution) and a retentive capacity of 6-7 g/g.

3.6. Lignin-Containing Cellulose Fibers

It would be very desirable to be able to use high lignin content fibers in superabsorbent products since this would provide greater use of the natural fibrous resource and avoid pollution problems associated with chemical delignification and bleaching. Several approaches have been explored for use of high lignin content fibers in superaborbent applications.

Pilon et al. [138] found that when several strains of fungi were allowed to grow in mechanical pulps the WRV of the pulps increased up to 88% over the controls. Samples treated with white rot fungi showed the highest increase in WRV. This increase in WRV could be due to partial lignin removal and/or modification of the lignin (oxidation) and random cleavage of cellulose chains at the surface of the crystalline regions. Another, less probable reason for the increase in WRV is the adsorption of highly hydrated fungal polysaccharides onto the fibers.

Naieni [139] described a method for improving the absorbent characteristics of lignin containing fibers at least 10% through heat treatment. Thus a high yield pulp, such as northern softwood chemithermomechanical pulp (CTMP), is dispersed in water at 10% consistency and the pH is adjusted to below 7 with citric or sulfuric acid. The pulp is then dewatered by contrifugation to provide a cake of 50% consistency. The dewatered cake is air dried to about 60% consistency, fluffed in a disk refiner and flash dried at high temperature to about 90% consistency and then heated at an air temperature of 165°C for 60 minutes. The absorbent pads are prepared by compressing to a density of 0.1 g/cc with a basis weight of about 0.13 g/sq. inch.

Improved absorbent characteristics for lignin containing fibers were also realized by esterification of CTMP with C2-C9 polycarboxylic acids [140]. These acids must contain two or more carboxyl groups, where 1,2,3-propane carboxylic acid would be considered a C3 carboxylic acid. Specific carboxyl groups in an aliphatic or alicyclic polycarboxylic acid must be separated from a second carboxylic acid group by no less than 2 carbon atoms and no more than three carbon atoms. Examples of suitable polycarboxylic acids include citric, maleic, 1,2,3-propane carboxylic, oxydisuccinic, etc. In a typical example, northern softwood CTMP, with a lignin content of 20%, is dispersed for 30 minutes in a solution at 10% consistency with citric acid at a pH of 3. The pulp is dewatered to 50% consistency and air dried to 60% consistency. The pulp is then flash dried at high temperature to 90% consistency. The product had a WRV of 131, a drip capacity of 5.9 g/g and a wet compressibility of 7.0 cc/g.

CTMP fibers of increased specific surfaced area and improved absorption properties were prepared by Eriksson et al. [141] by precipitation of a porous layer of hydrophilic chemicals onto the fiber surface. Thus, CTMP was impregnated with an aqueous solution of sodium metasilicate for 30 minutes and after dewatering the pulp was impregnated with poly(aluminum chloride). The pH was then adjusted to 9 and the pulp allowed to absorb for 30 minutes. The modified dewatered pulp exhibited improved absorption rate dependent on the percent of aluminum in the reaction. Absorption rates in the range of 3-4 ml/s were reported for the modified pulp. It was also found that anchoring of the porous layer could be improved by addition of cationic starch derivatives or synthetic polymers such as modified polyacrylates. The inventors noted that the exchange between the aluminum ions and the hydroxyl ions results in the formation of electrically charged hydroxyl aluminum polymers. The high electrical charge of these complex aluminum compounds promotes the coagulation of the negative particles. The presence of silicate amplifies the surface deposition and improves the absorption properties of the pulp.

3.7. Additional Methods for Cellulosic Absorbents

Many additional variations and methods for creating modified cellulosic superabsorbents have been patented. Absorbent cross-linked CMC and HEC have been applied as a coating to fibrous mats to improve absorbency by several different inventors [142-145]. Pietsch and Horn [145] found that treatment of cross-linked and modified cellulose with polyethylene glycol (0.5%) enhanced absorbency characteristics. Pierre et al. [146] heated and bonded cellulose fiber mats with alginate and aqueous polyvinyl alcohol to form an absorbent pad.

An absorbent sodium cellulose carboxylate was prepared by Godsay et al. [147] by oxidation of softwood pulp with gaseous nitrogen dioxide. The absorbent product was formed without loss of fibrous character or gel formation and with a high water absorptivity, only if an efficient buffer was used at pH 6.5-7.0. Higher pH levels lead to gel formation and at lower pH levels the absorptivity was impaired. A $NaHSO_3/Na_2SO_3$ buffer system gave the best results. Use of sodium hydroxide in place of the buffer resulted in gelation rather than yielding a highly absorbent salt. The diameter of the 6-carboxycellulose fibers increased about 90% during the buffer treatment (pH = 7).

Mun et al. [148] and Kim et al. [149] produced a superabsorbent product by grafting of cyanoethylated wood pulp. Pulps of various degrees of cyanoethylation (DS - 0.13, 0.35 & 0.63) were grafted with polyacrylamide, which was subsequently hydrolyzed to improve absorbency. The best absorbencies were from the grafted (hydrolyzed) pulp at the lower DS levels, with the pulp at DS 0.35 exhibiting the best absorbency characteristics. The absorbency was about twice that shown for an unmodified wood pulp.

In a series of papers, Miyata and Sakata [150-152] described the formation and properties of superaborbents prepared from hydroxyethylcellulose (HEC) and other substrates through grafting of polyacrylamide, polymethylmethacrylate, poly-2-(dimethylamino)ethylmethacrylate (PDA) or polyacrylic acid sodium salts. It was found that a partially hydrolyzed polyacrylamide (p-PAM) grafted HEC gave the best absorbency properties of the above modified materials. The water absorbency was 3000 g/g and for a saline solution (0.85% NaCl) the absorbency was 270 g/g, both of which are superior to values for many commercial absorbents. As the degree of grafting increased, the water absorbency increased to a maximum value of 200-400%, and lower molecular weight HEC (p-PAM grafted) gave superior absorbency compared to a high molecular weight HEC backbone. However, the compressive strength of the gels showed a reverse trend with the higher molecular weight HEC (p-PAM grafted) showing greater compressive strength. The surfaces of the p-PAM grafted HEC had a granular structure compared to (PDA) grafted HEC which had a network type structure.

4. MECHANISM OF SWELLING AND WATER RETENTION

The interaction of water with cellulose and modified cellulose is only briefly reviewed in this section; further discussions are given in other chapters in this book and in several references [2,3,125,153-160]. Most neutral polysaccharides in nature are hydrated, generally as an element of structural adaptation. Conceptually, the water of hydration can be in the form of sheets or columns of water molecules depending on the structure of the polysaccharide [153]. Native cellulose, a structural polysaccharide, exhibits a comparatively low degree of hydration and cellulose fibers are only mildly swollen by water. Thus the term "hydrophilic" more aptly describes the character of unmodified cellulose fibers as they are utilized in absorbent applications [125].

Marchessault et al. [125] have pointed out that "the so-called hydrated state, familiar to the fine paper specialist; wherein the pulp slurry exhibits a slimy feeling and slow drainage is related to colloid characteristics and not to a high degree of swelling in the macromolecular sense of the word". These investigators presented five different ways in which free water can be immobilized in gel-type systems as shown in Fig. 12 of Chapter 11 under the section on "Equilibrium Swelling".

Although cellulose can display all the states described in Chapter 11, it usually is found in an intermediate form. In addition, modified cellulose in superabsorbent fibers can exhibit microparticulate gel characteristics limited by the fiber dimensions [125]. X-Ray diffraction studies of modified CMC superabsorbent fibers indicated the structure to be an oriented non-crystalline partial derivative of cellulose, although the pattern showed traces of cellulose II. This compares with a grafted cellulosic superabsorbent which exhibited the characteristic oriented pattern of native cellulose in the X-ray diffraction diagram. Marchessault et al. [125] concluded that model (e) in Fig. 12 of Chapter 11 best described the grafted cellulose superabsorbent fibers, which they termed "constrained paracrystalline gels". Weak mechanical forces are responsible for holding the fiber together and, indeed, a short vigorous mechanical action dispersed the grafted absorbent fiber into microfibrillar elements which formed a thixotropic gel [125]. Based on the X-ray diffraction diagrams of the modified CMC superabsorbent fibers, it appears that the appropriate model for these fibers is further down the continuum and models (a) and (b) may more appropriately describe this system.

Westman and Lindström [159,160] explained the swelling behavior of cross-linked cellulose in terms of gel swelling theory. They found that swollen cross-linked cellulose samples exhibited a swelling maximum in sodium hydroxide solutions of different concentrations and suggested that when the concentration of NaOH is high enough to cause de-protonization of the cellulose hydroxyls, the swelling increases due to the electrostatic repulsion between the charged groups in the network. Beyond the swelling maximum, no more charges are introduced in the cellulose chains and further addition of NaOH then only leads to an increase in the ionic strength, which results in a de-swelling of the gels due to shielding of the charges. Similar explanations have been offered by Lindström et al. [157,158] for factors affecting hornification of cellulose fibers.

Grignon and Scallan [155] considered the swelling of cellulose gels to be caused by an osmotic pressure differential which results from a difference in concentration of mobile ions between the interior of the gel and the exterior solution. These investigators also felt that the presence of ionizing groups in the gel (i.e., carboxyl groups in CMC) were the predominant factor creating the unequal distribution of mobile ions. Their postulation was based on earlier work of Proctor [161,162] and Neale [163,164].

Proctor [161,162], in an investigation of acid-gel equilibria, suggested that when a gel, with ionizable groups attached, is placed in an ionic solution an exchange of mobile ions takes place; but the gel attains a higher concentration due to the presence of bound groups. Because of this difference in concentration, water enters the get to reduce the osmotic pressure differential and swells the gel. Proctor proposed that the swelling would continue until the osmotic pressure differential was equal to the resistance from the cohesive forces of the polymer network [155]. Neale et al. [163,164] presented a similar hypothesis for swelling of cellulose and confirmed that, at least in dilute solutions, the behavior was in accord with the Donnan theory [165,166].

According to the Donnan theory the presence of a nondiffusible ion on one side of a membrane affects the partition of all diffusible ions. This is essentially the situation with an electrolytic gel and an exterior solution [155].

Grignon and Scallan [155] based their theoretical treatment of swelling on the methods of Donnan, Proctor and Neale and derived an expression for the concentration of all diffusible ions, E, between the gel and the external solution:

$$E = ((\beta - 1) / (\beta + 1))\, C \tag{15}$$

where C is the concentration of ionized groups and ß is the distribution coefficient of the various ions in solution. Since E is directly proportional to the osmotic pressure differential (Van't Hoff equation), it was also considered proportional to the degree of swelling. Thus

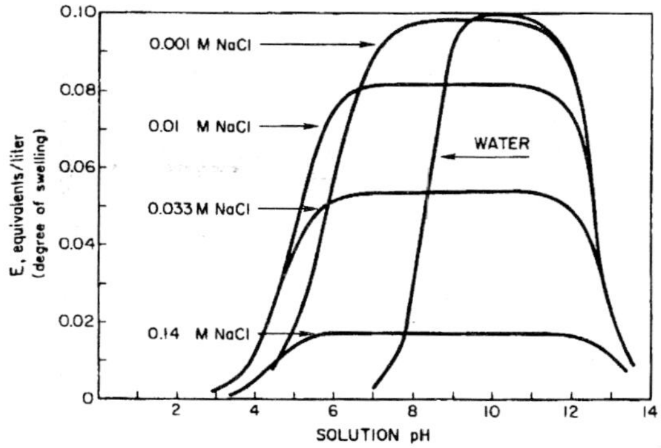

Fig. 16. Theoretical plot of degree of swelling as a function of pH for a 'gel containing weak acid groups [155].

swelling from electrolytic effects is brought about by a concentration of nondiffusible ions within the gel which is a function of the internal pH.

Figure 16 shows a theoretical curve developed by Grignon and Scallan [155] for the variation in swelling with pH for gels containing the same concentration of acid groups but different dissociation constants. Maximum swelling is achieved upon complete ionization. As the salt concentration increases the maximum is decreased. Qualitatively, the salt permits the transport of hydrogen ions out of the gel and at a certain level of salt addition, the chloride ions have more control over swelling than pH [155].

Empirically determined plots of WRV versus pH for actual superabsorbent products follow the predicted trends (Fig. 17) [155]. The swelling curve rises at pH values characteristic of weak acidic groups, reaches a maximum near neutrality, and decreases at a pH beyond 10. At the higher pH, appreciable quantities of ions accumulate outside the gel and swelling is reduced. A lowering of the whole curve occurs with the added salt. Grignon and Scallan felt that the swelling behavior of superabsorbent gels is almost completely governed by this mechanism [155].

A marked difference in the swelling behavior of cross-linked cellulose in LiOH, NaOH and KOH solutions was also noted by Westman and Lindström [159,160]. Sodium hydroxide caused the greatest swelling and LiOH the least. Their explanation for these differences was based on differences in the hydration of the alkali metal ion and in the tendency to form ion pairs, that is, to form MOH in solution instead of dissociating into M^+ and OH^-. The tendency to form ion pairs is greatest in LiOH followed by NaOH; while no experimental evidence exists for ion pair formation in KOH.

Apparently KOH, in contrast to LiOH and NaOH, causes no intracrystalline swelling of the cellulose. This may be related to size differences of the unhydrated ions. In the unhydrated state, K^+ is the largest alkali metal ion based on crystallographic radius. This may also indicate that alkali metal hydroxides do not penetrate into the crystallites in the form of hydrates [159,160].

Typically superabsorbent materials are added to a fibrous matrix in many superabsorbent applications such as disposable diapers and feminine hygiene products. Thus the water of imbibition in these composite structures is affected by the relative proportion and

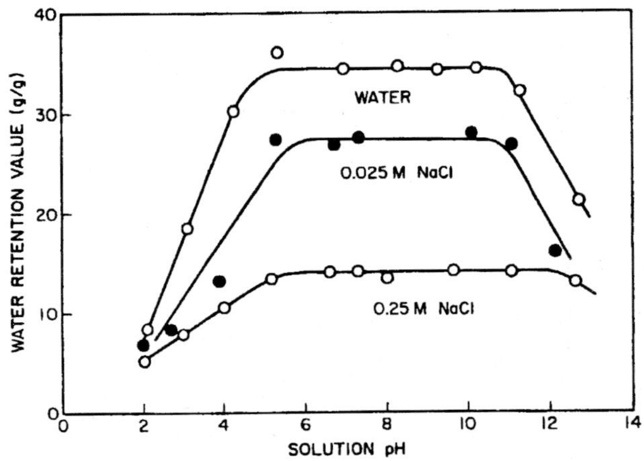

Fig. 17. Empirical plot of degree of swelling of superabsorbent pulp as a function of pH and salt concentration of the external solution [155].

distribution of the component materials. Many other factors are important to the perfomance of the composite such as fiber and superabsorbent surface properties and swelling characteristics, gel particle size and distribution, etc. [167-169].

Liquid uptake in fibrous structures is driven by capillary action and liquid retention occurs primarily in the interfiber pores. Superabsorbents, of course, will increase substantially the water holding capacity of the composites and in their presence the swelling either results in overall expansion of the structure or in the partial filling of the pore spaces between the fibers. Thus the transport of liquid is governed by capillary action in the interfiber pore space and by diffusion of the liquid into the superabsorbent. However diffusion of the liquid into the superabsorbent reduces the pore space and decreases the rate of absorption, such that the two mechanisms are tied together. Gel blocking occurs when extensive swelling of the superabsorbent causes closure of the wicking channels such that further liquid transport is prevented. Thus product design must be carefully considered for maximum performance [167,168].

Schuchardt and Berg [167] evaluated the wicking flow of water in composite networks of cellulose and CMC fibers to obtain a more fundamental understanding of the relationship between fiber swelling and liquid transport in fiber networks. They evaluated the imbibition of water in paper strips and fluff pads with various proportions of the two components and interpreted the data according to the Lucas-Washburn equation. The wettability and swelling of the fibers was determined by Wilhelmy method [11,170]. The cellulose fibers swelled instantly to a small degree while the CMC fibers doubled in diameter over several minutes. When n-octane was used as a nonswelling reference liquid, the liquid imbibition was adequately described by the Lucas-Washburn equation for all the fiber networks. However, when water was imbibed, the results deviated substantially from the Lucas-Washburn equation due to the swelling of the CMC fibers. The deviations were attributed to the long-term swelling effects of the CMC fibers which caused a dynamic

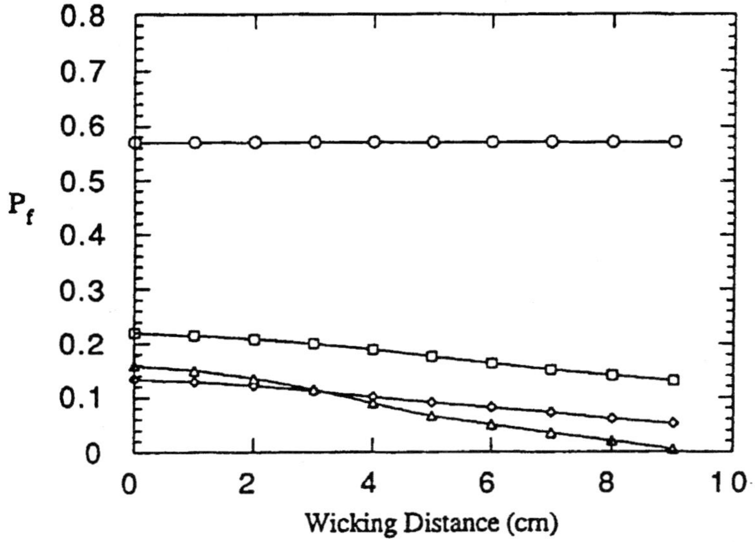

Fig. 18. Permeability factor for water in cellulose-CMC composite strips as a function of wicking distance. (O) 0% CMC; ( ) 10% CMC; ( ) 20% CMC; (Δ) 30% CMC.

reduction in the effective interfiber pore dimensions in the structure. Schuchardt and Berg proposed a modified capillary model to describe the rate of advance of the liquid front for these systems in which the hydrodynamic radius of the capillary decreases linearly with time.

Schuchardt and Berg [167] quantified the net pore restriction caused by fiber swelling with a "permeability factor", Pf, which was defined as the ratio of the wicking-equivalent radius in the swollen state to that in the unswollen state. Values of Pf were computed for each centimeter of wicking distance and plotted as shown in Fig.18 for a series cellulose-CMC composites [168]. Quite extensive pore blocking was evident even at only 10% superabsorbent content with slight increases in the effect as the content is increased to 30%. The expected greater pore blocking effects at longer wicking distances is also shown in Fig. 18.

The permeability factor was also used by Wiryana and Berg [168] for comparison of the behavior of powdered versus fibrous CMC superabsorbents in composite structures. As shown in Fig 19, at 0 wicking distance the composite with the fibrous CMC shows an unexpected increase in permeability which was attributed to open channels created in the structure when dried. The powdered superabsorbent did not show such an enhancement in permeability. The difference was attributed to the differences in the shapes of the open spaces created when the composites were dried. The channel type open spaces created by the dried superabsorbent fibers were more effective in promoting rapid wicking and, as shown in Fig. 20, these channels were more effective in compensating for the ultimate effects of the re-swelling compared with the isolated cell-like opening created by dried superabsorbent in powder form [167,168].

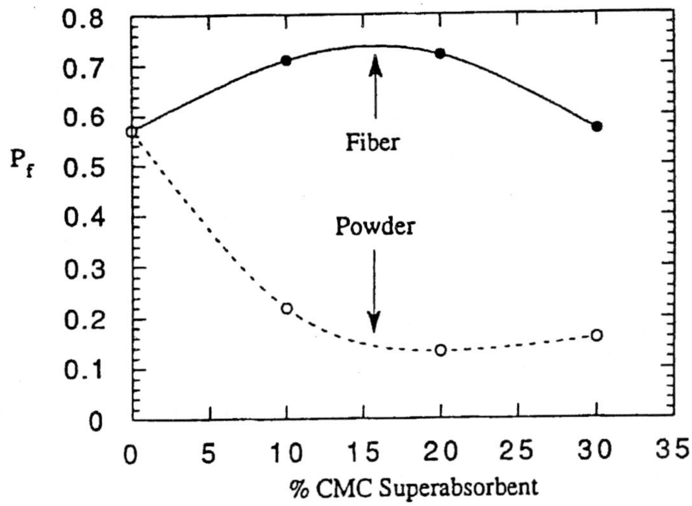

Fig. 19. Comparison of permeability factors (Pf) at zero wicking distance for water in cellulose-CMC superabsorbent composite strips with the superabsorbent in powder versus fibrous form.

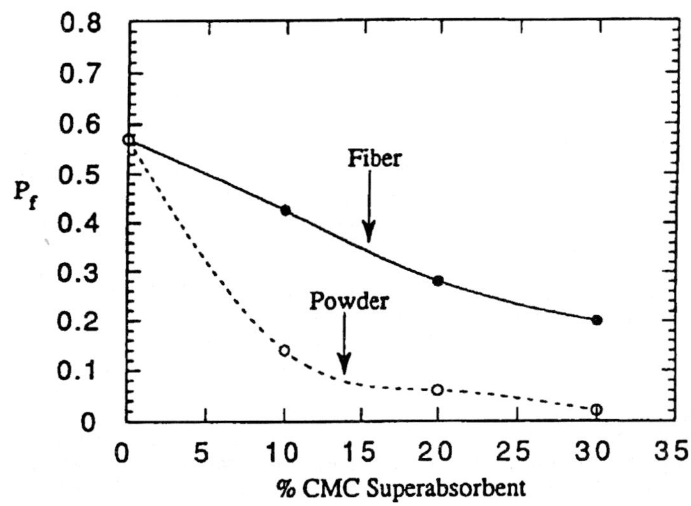

Fig. 20. Comparison of permeability factors (Pf) at a wicking distance of 8 cm for water in cellulose-CMC superabsorbent composite strips with the superabsorbent in powder versus fibrous form.

5. USES AND APPLICATIONS

Superabsorbent celluloses can be produced in a variety of physical forms such as powders, granules, flakes, fibers, laminates and sheets. Generally absorption capacity and rate are increased at lower density of material. The final application will dictate to some extent the form of the superabsorbent material. In absorbent products, the uses of cellulosic derivatives are similar to that described in other chapters in this book.

There has also been considerable interest in utilizing cross-linked-CMC and other superabsorbents as tablet disintegrants [31,171-176]. The superabsorbent is mixed in powder form with the other components and pressed to form a tablet which, upon contact with water or acidic solution, rapidly expands to aid disintegration of the tablet. Compared to traditional starch and starch derivatives, the cross-linked CMC reduces the disintegration rate and increases the dissolution rate [31,171].

A disadvantage of the superabsorbents for tablet disintegration is that they are highly hydrophilic and absorb moisture which causes the tablets to soften when exposed to high humidities. It is also possible that moisture-labile drugs may not be as stable in formulations containing superabsorbents [31,171].

6. ACKNOWLEDGEMENT

Appreciation is expressed to Robert Sitaru for searching out the literature and patents for this second edition of the chapter.

7. REFERENCES

1. A. Payen, Troisieme memoire sur les Developpernents des Vègètaux. Extrait des Mèmoires de l'Academie Royale des Sciences. Tome VIII des savants Étrangers Imprimerie Royale, Paris, 1842.
2. H. Corte, Cellulose-Water Interactions, in H. F. Rance, (Ed.), Handbook of Paper Science Vol. 1. The Raw Materials and Processing of Papermaking, Elsevier, New York, 1980, pp. 1-91.
3. Fund. Res. Comm.: Fiber-Water Interactions in Paper-Making, Trans. of the 6th Fundamental Res. Symp., Oxford, 1977, Tech. Div. Brit. Paper Board Ind. Fed., London, 1978.
4. S. H. Zeronian, Intercrystalline swelling of cellulose, in T. P. Nevell and S. H. Zeronian, (eds.), Cellulose Chemistry and its Applications, Ellis-Horwood Ltd., Chichester, England, 1985, p. 138.
5. S. A. Naieni, U.S. Patent 5,709,774, January 20, 1998.
6. P. A. V. Ahlgren and G. Kloow, The Billerud fluff pulp concept and cekosorb superabsorbent, Proc. Insight 81 Absorbent Products Conf., San Antonio, Texas, September 16-18, 1981.
7. E. Richman, Permasorb in a new, convenient sheet laminate, Proc. of Insight 81 Absorbent Products Conf., San Antonio, Texas, September 16-18. 1981.
8. G. Goldstein and M. Pierre, Superabsorbent, how a Europeanuser sees the development of new products, Proc. of Insight 81 Absorbent Products Conf., San Antonio, Texas, September 16-18. 1981.
9. E. Sjöström, Wood Chemistry, Fundamentals and Applications, 2nd edition, Academic Press, NY, 1993.
10. T. Kondo, hydrogen bonds in cellulose and cellulose derivatives, in S. Dumitriu, (ed.), Polysaccharides, Structural, Diversity and Functional Versatility, Marcel Dekker Inc., NY, 1998, p. 131.
11. R. A. Young, Structure, swelling and bonding of cellulose fibers, in R. A. Young and R. M. Rowell, (eds.), Cellulose, Structure, Modification and Hydrolysis, John Wiley & Sons, NY, 1986, p.91.
12. H. A. Krässig, Cellulose: Structure, Accessiblity and Reactivity, Gordon Breach Science Pub., NY, 1993.
13. S. P. Rowland, Hydroxyl reactivity and availability in cellulose, in R. M. Rowell and R. A. Young (eds.), Modified Cellulosics, Academic Press, New York, 1978.

14. A. Ishizu, Chemical modification of cellulose, in D. S. Hon and N. Shirashi, (eds.), Wood and Cellulosic Chemistry, Marcel Dekker Inc., NY, 1991, p. 757.
15. K. Kamide, Characterization of chemically modified cellulose, , in D.S. Hon and N. Shirashi, (eds.), Wood and Cellulosic Chemistry, Marcel Dekker Inc., NY, 1991, p. 801.
16. R. A. Young and R. M. Rowell, (eds.), Cellulose, Structure, Modification and Hydrolysis, John Wiley & Sons, NY, 1986.
17. R. M. Rowell and R. A. Young, (eds.), Modified Cellulosics, Academic Press, NY, 1978.
18. L. C. Wadsworth and D. Daponte, Cellulose esters, in T. P. Nevell and S. H. Zeronian, (eds.), Cellulose Chemistry and its Applications, Ellis-Horwood Ltd., Chichester, England, 1985, p. 344.
19. G. D. Hiatt and W. J. Rebel, Esters, in N. M. Bikales and L. Segal (eds.), Cellulose and Cellulose Derivatives, Part V, Wiley-Interscience, New York, 1971, p. 741.
20. K. Ward, Jr., Chemical Modification of Papermaking Fibers, Marcel Dekker, New York, 1973.
21. G. A. Petropavlovskii and M. M. Krunchak, Kh. Pukl. Khim., 36 (1963) 2506.
22. K. Katsumura and S. Nonaka, Sen-i Gakkaishi, 13 (1957) 24.
23. G. P. Touey, U.S. Patent, 2,759,924, March 14, 1956.
24. M. D. Nicholson and F. M. Merritt, Cellulose ethers, in T. P. Nevell and S. H. Zeronian, (eds.), Cellulose Chemistry and its Applications, Ellis-Horwood Ltd., Chichester, England, 1985, p. 363.
25. N. I. Nikitin, The Chemistry of Cellulose and Wood, Academy of Sciences of the USSR, Israel Program of Scientific Translations, Jerusalem, 1966.
26. A. B. Savage, Ethers, in N. M. Bikales and L. Segal (eds.), Cellulose and Cellulose Derivatives, Part V, Wiley-Interscience, New York, 1971, p. 785.
27. D. F. Durso, Chemical modification of cellulose: A historical review, in: R. M. Rowell and R. A. Young (eds.), Modified Cellulosics, Academic Press, 1978, p. 23.
28. T. C. Allen and J. A. Cuculo, J. Polym. Sci.: Macromolecular Rev., 7 (1973) 189.
29. S. Dumitriu, (Ed.), Polysaccharides, Structural, Diversity and Functional Versatility, Marcel Dekker Inc., NY, 1998.
30. E. Percival and R. H. McDowell, Chemistry and Enzymology of Marine Algal Polysaccharides, Academic Press, New York, 1967.
31. F. L. Buchholz and N. A. Peppas, (eds.), Superabsorbent Polymers, Science and Technology, ACS Symp. Ser. No. 573, Am. Chem. Soc., Washington, D.C., 1994.
32. A. Holst, Die Angewandte Makrol. Chem., 76/77 (1979) 177.
33. A. Holst, Das Papier, 32 (1978) V7.
34. G. Goldstein, Revue A.T.I.P., 35 (1981) 115.
35. H. Kubota, Kagaku Kogyo, 30 (1979) 928.
36. J. G. Penniman, Paper Trade J., May, 1980, p. 44,
37. D. Abson and R. D. Gilbert, Tappi, 63 (1980) 146.
38. W. L. Dean, Cross-linked derivative (CLD) fibers as binders for nonwovens, Proc. of the INDA Second Technical Symp., Washington, D.C., 1974.
39. F. H. Steiger, U.S. Patent 3,658,613, April 25, 1972,
40. L. Roth and J. Weiner, Chemical Modification of Cellulose. 1. Cross-linking, Bibliographic Set. No. 228, Inst. Paper Chem., 1966.
41. S. L. Vail, Cross-linking of cellulose, in T. P. Nevell and S. H. Zeronian, (Eds.), Cellulose Chemistry and its Applications, Ellis-Horwood Ltd., Chichester, England, 1985, p. 384.
42. A. G. Pierce, Textile Bull., 94 (1968) 41.
43. F. R. Smith, Text. Res. J., 42 (1972) 301.
44. V. Meyer, K. Miller, and H. Zollinger, Text. Res. J., 46 (1976) 756,880.
45. H. Petersen, Text. Res. J., 38 (1968) 156.
46. G. C. Tesoro and J. J. Willard, Cross-linked Cellulose, in N. M. Bikales and L. Segal (eds.), Cellulose and Cellulose Derivatives, Part V, Wiley- Interscience, New York, 1971, p. 835.
47. F. H. Steiger, U.S. Patent 3,320,956, May 23, 196.
48. A. Graef, U.S. Patent 4,853,086, August 1, 1989.
49. P. A. Graef and F. R. Hunter, U.S. Patent 5,225,047, July 6, 1993.
50. P. A. Graef and F. R. Hunter, Patent Cooperation Treaty, WO 88/04704, June 30, 1988.
51. L. Bernardin, P. Rhode and C. Heimbach, European Patent 0410480A2, January 1, 1991.
52. J. Bartholomew, J. Industrial Fabrics, 4(1) (1985) 4.
53. L. Bernardin and C. Heimbach, European Patent 0399564A2, November 28, 1990.
54. L. J. Bernardin, U.S. Patent 3,224,926, December 21, 1965.

55. P. K. Chatterjee, U.S. Patent 3,932,209, January 13, 1976.
56. J. Querre and G. Guiroy, Fr. Dernande 2,224,486, October 31, 1974; 2,265,763, October 24, 1975.
57. S. Sangenic, G. Guiroy and J. Querre, Ger. Offen. 2,416,532, October 24, 1974.
58. C. H. Lesas and M. Pierre, U.S. Patent 4,204,055, May 20, 1980; 4,204,054, May 20, 1980.
59. C. H. Lesas and M. Pierre, Get. Offen. 2, 646, 317, June 23, 1977.
60. H. L. Schoggen, J. T. Cook and R. M. Bourbon, U.S. Patent 4,889,596, December 26, 1989.
61. W. L. Dean, D.R. Moore, J. W. Owens and H. L. Schoggen, U.S. Patent 4,822,453, April 18, 1989.
62. W. L. Dean, D. R. Moore, J. W. Owens, H. L. Schoggen R. M. Bourbon and J. T. Cook, European Patent 0251676A2, January 7, 1988.
63. R. A. Jewell, Patent Cooperation Treaty, WO 93/14261, July 22, 1993.
64. R. M. Bourbon and J. J. Ryan, Jr., U.S. Patent 4,889,597, December 26, 1989.
65. C. M. Herron, W. L. Dean, D. R. Moore, J. W. Owens and H. L. Schoggen, U.S. Patent 4,889,595, December 26, 1989.
66. C. M. Herron, W. L. Dean, D.R. Moore, J. W. Owens and H. L. Schoggen, European Patent 0252649A2, January 13, 1988.
67. C. M. Herron and J. D. Cooper, European Patent 0429112A2, May 29, 1991.
68. C. M. Herron and J. D. Cooper, European Patent 0427316A2, May 15, 1991.
69. C. M. Herron, T. R. Hanser, J. D. Cooper and B. S. Steven, European Patent 0427317A2, May 15, 1991.
70. C. M. Herron and J. D. Cooper, U.S. Patent 5,137,537, August 11, 1992.
71. B. J. Kokko, European Patent 0440472A1, August 7, 1991.
72. L. Norlander, Patent Cooperation Treaty, WO 93/16228, August 19, 1993.
73. C. M. Herron and W. L. Dean, Patent Cooperation Treaty, WO 95/34710, December 21, 1995.
74. L. Norlander, Patent Cooperation Treaty, WO 97/02844, January 30, 1997.
75. H. T. Lokhande and P. V. Varadarajan, Bioresource Technol., 45(3), (1993) 161.
76. P. Gatenholm, Patent Cooperation Treaty,WO 97/22744, June 26, 1997.
77. J. P. Poccia, K. J. Ovans and H. Pieniak, European Patent 0402650A2, December 19, 1990.
78. G. L. Graham, Jr., U.S. Patent 3,005,456, October 24, 1961.
79. H. S. Bahia and J. R. James, U.S. Patent 5,731,083, March 24, 1998.
80. G. Kloow, Billerud Uddeholm AB, Skoghall, Sweden, personal communication.
81. W. L. Dean and G. N. Ferguson, U.S. Patent 3,589,364, June 29, 1971.
82. A. Holst, H. Lask and M. Kostrzewa, U.S. Patent 4,066,828, January 3, 1978; 4,068,068, January 10, 1978.
83. A. Holst, M. Kostrzewa and H. Lask, U.S. Patent 3,936,441, February 3, 1976.
84. Holst and H. Lask, Get. Offen. 2,519,927, November 18, 1976.
85. A Holst, H. Lask and M. Kostrzewa, Ger. Offen. 2,520,336, November 18, 1976; 2,520,337, November 18, 1976; 2,357,079, May 28, 1975.
86. A. Holst and M. Kostrzewa, Ger. Offen 2,543,187, March 31, 1977.
87. A. Holst, H. Lask and M. Kostrzewa, U.S. Patent 4,008,067, January 10, 1978.
88. A. Holst and H. Lask, U.S. Patent 3,965, 091, January 22, 1976.
89. A. Holst, M. Kostrzewa and G. Bachberger, U.S. Patent 4,117,222, September 26, 1978.
90. P. K. Chatterjee, U.S. Patent 3,971,379, July 27, 1976.
91. K. Tajiri, M. Maeda and H. Tsukamoto, U.S. Patent 5,725,601, March 10, 1998.
92. K. Tajiri, K. Ren and H. Tsukamoto, Japanese Patent 05049925A2, March 2, 1993.
93. I. W. Cottrell, A. Goswami and M. S. Chowdhary, U.S. Patent 5,532,350, july 2, 1996.
94. J. Qin, U.S. Patent 5,470,964, November 28, 1995.
95. J. Qin, U.S. Patent 5,498,705, March 12, 1996.
96. K. Ren, K. Tajiri and H. Tsukamoto, Japanese Patent 08041103A2, February 13, 1996.
97. K. Ren, K. Tajiri and H. Tsukamoto, Japanese Patent 06172401A2, June 21, 1994.
98. R. B. Seymour and G. H. Schroder, U.S. Patent 2,486,805, November 1, 1949.
99. E. S. Elliott, U.S. Patent 2,639,239, May 19, 1953.
100. A. R. Reid. U.S. Patent 3,379,720, April 23, 1968.
101. A. J. Desmarais and A. R. Reid, U.S. Patent 3,435,027, March 25, 1969.
102. T. J. Podlas, Internally cross-linked carboxymethyleellulose fibers for absorbent nonwoven applications, Proc. of INDA Symp., March, 1976, p. 25.
103. H. L. Schoggen, J. R, Holmgren and T. S. Harris, U.S. Patent 3,826,711, July 30, 1974; British Patent 1,424,311, February 11, 1970.

104. P. K. Chatterjee and M. C. Kwok, U.S. Patent 3,723,413, March 27, 1973; British Patent 1,407,132, September 24, 1975.
105. P. K. Chatterjee, U.S. Patent 3,731,686, May 8, 1973.
106. M. Shinohara and N. D. Field, U.S. Patent 4,200,736, April 26, 1980.
107. H. L. Marder, N. D. Field, and M. Shinohara, British Patent Appl. 2,021,116, November 28, 1979.
108. H. L. Schoggen, U.S. Patent 3,678,031, July 18, 1972.
109. H. L. Schoggen and J. W. Smith, U.S. Patent 4,252,761, February 24, 1981.
110. L. M. Kacmarzyk, J. J. Hlaban and L. J. Bernardin, U.S. Patent 4,044,766, August 30, 1977.
111. A. J. Karlsson, H. E. Lundstrom and L. H. Alsson, U.S. Patent 4,256,877, March 117, 1981.
112. L. Lilienfeld, U.S. Patent 1,682,293, August 28, 1928.
113. A. R. Reid, U.S. Patent 2,495,767, January 31, 1950.
114. R. W. Butler, U.S. Patent 3,064,313, November 20, 1962.
115. P. K. Chatterjee, U.S. Patent 4,020,271, April 26, 1977.
116. L. J. Bernardin, U.S. Patent 3,658,790, April 25, 1972; 3,691,154, September 12, 1972; 3,739,782, June 19, 1973.
117. L. Kellet, U.S. Patent 3,091,241, May 5, 1963.
118. A. Burgeni, U.S. Patent 3,187,747, June 5, 1965.
119. K. Arai and H. Goda, Sen'i Gakkaishi, 49(9), (1993) 482.
120. K. Tajiri and H. Tsukamoto, Japanese Patent 04146902A2, May 20, 1992.
121. R. T. Shet and P. Wallajapet, U.S. Patent 5,703,225, December 30, 1997.
122. F. O. Lassen, U.S. Patent 3,997,647, December 14, 1976; 4,256,111, March 17, 1981.
123. W. C. Sigl and F. O. Lassen, U.S. Patent 4,049,764, September 20, 1977.
124. T. B. Rydell, U.S. Patent 4,174,417, November 13, 1979.
125. R. H, Marchessault, M. Dube, J. St. Pierre and J. F. Revol, New insights into fibre swelling, interfibre bonding and wet strength, in: Trans. Symp. on Fibre-Water Interactions in Paper-Making, Oxford, 1977, Tech. Div. Brit. Board Ind. Fed., London, p. 795.
126. Autex Fibers, Jap. Patent 80,163,205, December 19, 1980.
127. H. Pietsch and W. Horn, U.S. Patent 4,169,121, September 25, 1979.
128. H. Kraessig, F. Gatschy and H. Teichmann, Ger. Offen. 2,809,312, September 28, 1978.
129. I. Marini, J. Lenz, J. Schmidtbauer and K. Geppert, Get. Offen. 3,303,415, April 30, 1981.
130. A. Holst, M. Kostrzewa and H. Lask, Ger. Offen. 2,364,628, June 28, 1975.
131. F. R. Smith, U.S. Patent 4,144,079, March 13, 1979.
132. T. Takebe and T. Yamazaki, U.S. Patent 4,084,591, April 18, 1978.
133. H. Nguyen, N. Martens and G. Garbolino, Patent Cooperation Treaty WO 97/23248, July 3, 1997.
134. P. K. Chatterjee and K. B. Makoui, U.S. Patent 4,474,949, October 2, 1984.
135. K. B. Makoui and P. K. Chatterjee, U.S. Patent 4,689,118, August 25, 1987.
136. K. B. Makoui and P. K. Chatterjee, U.S. Patent 4,911,700, March 27, 1990.
137. K. B. Makoui and P. K. Chatterjee, U.S. Patent 4,767,848, August 30, 1988.
138. L. Pilon, M. Desrochers, L. Jurasek and P. J. Neumann, Proc. Tappi Ann. Mtg. March 2-5, 1981, Chicago, Tappi Press, Atlanta, p. 293.
139. S. A. Naieni, U.S. Patent 5,709,774, January 20, 1998.
140. S. A. Naieni, C. M. Herron and T. R. Hanser, U.S. Patent 5,484,896, January 16, 1996.
141. I. V. Eriksson, G. E. Annergren and l. Wagberg, U.S. Patent 5,492,759, February 20, 1996.
142. A. Holst and H. Lask, Get. Offen. 2,534,358, May 26, 1977.
143. L. Grosse and H. Doerr, Ger. Offen. 2,554,558, June 8, 1977.
144. A. R. Reid, Ger. Offen. 2,821,968, November 30,1978.
145. H. Pietsch and W. Horn, U.S. Patent 4,169,121, September 25, 1979.
146. M. Pierre, G. Goldstein and C. Lesas, Ger. Offen. 2,831,211, March 22, 1979.
147. M. P. Godsay, V. N. Gupta, R. L. Lafond and D. M. MacDonald, J. Polym. Sci., Pt. C, 36 (1971) 529.
148. S. P. Mun, J. P. Kim, M. Yoshinobu and M. Morita, J. Korean Fiber Society, 33(8) (1996) 715.
149. J. P. Kim, S. P. Mun, S. B. Park and J. M. Rhee, , J. Korean Fiber Society, 33(8) (1996) 722.
150. N. Miyata and I. Sakata, Sen-i Gakkaishi, 47(8) (1991) 428.
151. N. Miyata and I. Sakata, hydroxyethylcellulose graft copolymers as super water-absorbents, in: J.F. Kennedy, G. O. Phillips and P. A. Williams (eds.), Cellulosics: Chemical, Biochemical and Material Aspects, Ellis Horwood, NY, 1993.

152. N. Miyata and I. Sakata, Physicochemical properties of hydroxymethyl cellulose and other polysaccharide graft copolymers as super water absorbents, in: : J. F. Kennedy, G. O. Phillips and P. A. Williams (eds.), Wood and Cellulosics, Ellis Horwood, NY, 1987
153. T. Bluhm, Y. Deslandes, R. H. Marchessault, and P. R. Sundararajan, New insights into the crystal structure hydration of polysaccharides, Chap. 15 in: S. D. Rowland (ed.), Water in Polymers, ACS Symp. Ser. No. 127, Am. Chem. Soc., Washington, D.C., 1980
154. B. Nystrom, M. E. Moselet, W. Brown and J. Roots, J. Appl. Polym. Sci., 26 (1981) 3385.
155. J. Grignon and A. M. Scallan, J. Appl. Polym. Sci., 25 (1980) 2829.
156. A. J. Pennings and W. Prins, J. Appl. Polym. Sci., 58 (1962) 229.
157. T. Lindström and M. Kolman, Svensk Papperstidn., 85 (1982) R140.
158. T. Lindström and G. Carlsson, Svensk Papperstidn., 85 (1982) R146.
159. L. Westman and T. Lindström The Swelling of Cellulose Hydrogels in Concentrated Alkaline Solutions, paper presented at Ninth Cellulose Research Conf., Syracuse, N.Y., May, 1982.
160. L. Westman and T. Lindstrom, Swelling and Mechanical Properties of Cellulose Hydrogels. V. Swelling in Concentrated Alkaline Solutions, in press.
161. H. R. Proctor, J. Chem. Soc., 105 (1914) 313.
162. H. R. Proctor and J. A. Wilson, J. Chem. Soc., 109 (1916) 307.
163. S. M. Neale, J. Text. Inst., 20 (1929) 373; 21 (1930) 225.
164. J. Farrar and S. M. Neale, J. Colloid Sci., 7 (1952) 186.
165. F. G. Dorman and A. B. Harris, J. Chem. Soc., 99 (1911) 1554.
166. F. G. Dorman, Z. Electrochem., 17 (1911) 572.
167. D. R. Schuchardt and J. C. Berg, Wood Fiber Sci., 23(3), (1991) 342.
168. S. Wiryana and J. C. Berg, Wood Fiber Sci., 23(3), (1991) 457.
169. S. A. Goldman, J. C. Horney and H.L. Retzsch, U.S. Patent 5,061,259, October 29, 1991.
170. R. A. Young, Wood Fiber Sci., 8(2), (1976) 120.
171. R. Shangraw, A. Mitrevej, and M. Shah, Pharm. Techn., 4 (1980) 48.
172. R. A. Miller, G. R. B. Down, C. H. Yates, and J. F. Millar, Canad. J. Pharm. Sci., 15 (1980) 55.
173. E. M. Rudnic, C. T. Rhodes, J. F. Bavitz and J. B. Schwartz, Drug Devel. Ind. Pharm., 7 (1981) 347.
174. R. P. Bhatia, K. J. Desai, and B. B. Sheth, Drug and Cosmetic Ind., 122 (1978) 38.
175. . F. Bavitz, N. R. Bohidar, and F. A. Restaino, Drug Devel. Commun., 1 (1974/1975) 331.
176. . Esteve, P. J. Solanas, R. Salazar, A. del Pozo and C. Fauli, paper presented to the Congreso National de Farmaceuticos En La Industria, Sevilla, Spain), October 13-15, 1977.

Absorbent Technology.
P.K. Chatterjee and B.S. Gupta, editors.

CHAPTER VIII

SYNTHETIC SUPERABSORBENTS

THOMAS L. STAPLES
The Dow Chemical Company, Midland, MI 48674

PRONOY K. CHATTERJEE
Nutech International Co., 331 McDowell Dr.,East Brunswick, NJ 08816

Contents

1. INTRODUCTION

Absorbent fibers have been employed for years in articles ranging from diapers to surgical sponges. Initially, these products made use of cellulosic fibers solely as their absorbent material. The transport of fluids by these absorbents was accomplished primarily by capillarity, without any significant swelling of the fibers.

Several important developments occurred in this field in the early 1970's with the disclosure of a new class of materials called "superabsorbents". The superabsorbent material is able to retain a substantial quantity of fluid in its polymer network, significantly more than what could be held in conventional absorbent fibers.

Initially, superabsorbents were made either from chemical substitution on cellulose or by grafting polyelectrolytes onto starch or cellulose [1]. Many products made use of these two types of superabsorbents due to their tremendous water absorption capacity. Although modified natural products were the original examples, today's marketplace is dominated by synthetic superabsorbents. Consistency and greater ability to design in performance favor synthetics, which are the subject of this chapter.

The impact of superabsorbent polymers on the performance of high volume personal care products has accelerated in the last 15 years. Baby diapers exhibiting a reduced tendency to leak represent the most common application of superabsorbents. Global production of synthetic superabsorbents is over 1 million metric tonnes or 2.2 billion pounds (Fig. 1) [2].

In this chapter we will first briefly review the physical chemistry of aqueous fluid absorption and superabsorbents. We will then discuss the manufacture of the major commercial synthetic materials, how their performance is evaluated, and finally we will review the primary applications. Throughout this chapter the emphasis will be on crosslinked sodium polyacrylate gels because this is the dominant chemistry for commercial superabsorbents.

Several reviews and monographs have appeared in recent years. Brannon-Peppas and Harland [3] made one of the earliest reviews. Harland and Prud'homme [4] and Buchholz and Peppas [5] are compilations of papers from symposia sponsored by the American Chemical Society. In 1998 Buchholz and Graham [6] published an entire monograph devoted

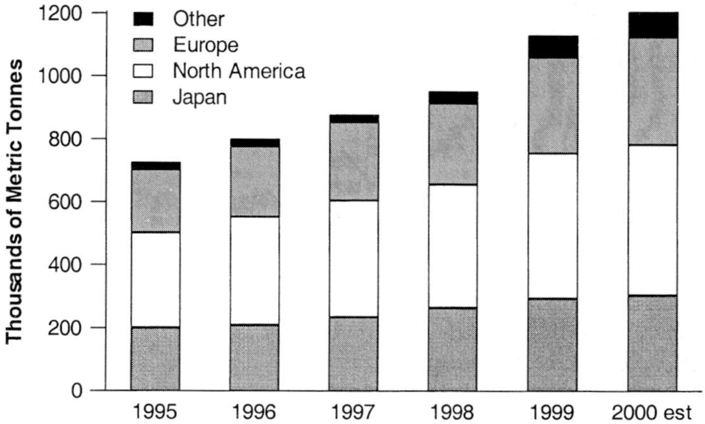

Figure 1: Manufacturing capacity of superabsorbents (data compiled by The Dow Chemical Company)

to polyacrylate superabsorbents, the primary subject of this chapter. One of the current authors (Staples) was a contributor, and this book is recommended as an expanded source for this area. For a rigorous description of the theory of polyelectrolyte gels, the Thesis of Yin [7] is recommended. Recently, a substantial collection of synthetic references has been published [8].

2. PHYSICAL CHEMISTRY OF POLYMERIC ABSORBENTS

2.1. Capillarity vs. Osmosis

Two quite different mechanisms are employed to retain fluids in absorbent products, capillarity and osmosis. In another chapter of this monograph one can find a detailed discussion of capillary absorption. The essential characteristic of this phenomenon is illustrated by the spontaneous rise of fluid in a small tube. The total surface energy of a system is lowered by wetting of the tube wall. For aqueous fluids this means polar surfaces are more easily wetted than nonpolar ones. Available surface is also important, and superior capillary absorbents generally have a large proportion of wettable surfaces; this high specific surface area is usually associated with small pores. A familiar example of a good capillary absorbent for aqueous fluids is cellulose fluff. The capillary suction pressure draws fluid into the porous structure. It is more or less completely determined by the pore dimensions and surface energy of the absorbent substance and the surface tension of the fluid.

Osmosis, the other mechanism for fluid transport and storage, is driven by the same thermodynamics that cause dissolution. The free energy of mixing drives fluid into a polymeric sample as the polymer chains dissipate into the surrounding solvent. With a linear polymer this process leads to the formation of a polymer solution. For a gel, crosslinks eventually constrain this dissipation of chains, and the swelling equilibrium is reached. The most striking characteristic of superabsorbents, whether synthetic or made from natural products, is their ability to swell and contain fluid by osmosis.

Contrasting the general features of these two phenomena is informative, but as with all generalizations, caution must be used. As a rule, however, capillary transport is a much faster process than molecular diffusion. A drop of water touched by the corner of a paper towel disappears almost instantly into the sheet. If a drop is placed instead on a piece of gelatin, the absorption process is much slower. Unfortunately, the term "diffusion" is frequently used to describe capillary transport, but in this chapter we will restrict its use to mean molecular diffusion. It should be noted, however, that the mathematical descriptions of both capillary and molecular "diffusion" are analogous [9].

It is more difficult to generalize about how effectively fluids are immobilized by either capillary attraction or osmosis. A liquid held by either mechanism can be removed by evaporation, pressure, or extraction. Of particular interest in many personal care applications is the ability of fluids to be contained while under an external load. A classic situation is a baby sitting on a wet diaper. Squeezing out capillary liquid requires deforming the pores in which it is held; this is in part a question of the compressive strength of the porous structure rather than its capillary attraction. Resisting compression while remaining soft, often a requirement in personal care, is a challenge for materials that depend on capillarity to retain fluids.

On the other hand, osmotic pressures can be quite high for superabsorbent gels, and this allows high fluid retention under load. However, for a saturated gel, one that has reached its equilibrium swelling capacity, *any* incremental pressure will cause fluid to exude from the gel, by definition of the equilibrium state. One cannot make a blanket statement as to which mechanism is the better method to hold fluid.

Indeed, both phenomena can and do operate simultaneously. For example, capillary transport, while not the earmark of superabsorbents, can be an important ancillary feature in their performance. When gel particles swell together forming a monolith, all capillary transport (the faster process) ceases and fluid moves only by molecular diffusion (the much slower process). This phenomenon is called by the descriptive term "gel blocking" and is generally undesirable in personal care applications, where one desires rapid fluid movement during transient events such as urination. Correspondingly, the swelling, and consequent softening, of a cellulose fiber by osmosis has important ramifications for its ability to transport and store additional water between itself and other fibers in a fluff mat or pad.

2.2. Polymer Solutions and the Swelling of Gels

Billmeyer [10] in his *Textbook of Polymer Science* describes this situation as follows:

Dissolving a polymer is a slow process occurring in two stages. First, solvent is slowly imbibed into the polymer to produce a swollen gel. In the second stage, the gel gradually disintegrates into a true solution. Only the second stage is materially speeded by agitation. If the polymer is crosslinked by primary valence bonds or strong hydrogen bonds, or is highly crystalline, only swelling may take place.

Our interest in this chapter is polymer networks or gels, crosslinked systems that do not form true solutions but stop at the swollen state noted by Billmeyer. It is important to recognize, however, that the swelling process begins much as the formation of a true polymer solution. The thermodynamic description of swelling can be summarized as

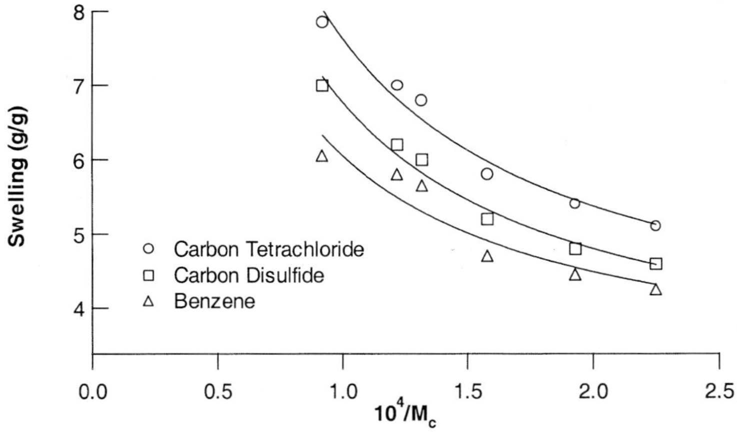

Figure 2: Swelling of a series of crosslinked rubbers in three solvents [11]

$$\Delta F = \Delta F_{mix} + \Delta F_{elas} \tag{1}$$

The free energy is composed of two portions, the ΔF_{mix}, which is analogous to that for polymer solutions, and ΔF_{elas}, which reflects the elastic restraint created by the crosslinks. For the systems of interest here, mixing will be spontaneous and hence ΔF_{mix} less than zero. Expanding the network by swelling requires work and thus ΔF_{elas} will be positive. When the overall ΔF is at a minimum and equilibrium is reached, the size of the polymer network at that point represents its swollen capacity in that solvent. We will follow common practice and use the term *solvent* to describe a compatible liquid in a system that only swells and does not dissolve. In this section we will first illustrate the behavior of gels in a phenomenological way and then summarize some of the current theoretical treatments.

The most important generalization concerning the swelling of gels is that more crosslinks mean less swelling—the higher the crosslink density of the network the lower its equilibrium swelling capacity. Another factor is the quality of the solvent. Figure 2 demonstrates both these effects with plots of swelling capacity of multiple samples of natural rubber with varying crosslink densities, each in three solvents, carbon tetrachloride being a better solvent for this system than benzene [11].

Polarity, high dielectric constant, and hydrogen bonding all make water an unusual swelling solvent; the term "hydrogel" is often used to describe aqueous crosslinked systems. Despite the unique properties of water, *nonionic* polymer gels do not behave much differently from, *e.g.*, swollen rubbers. Significant differences arise, however, with *polyelectrolyte* gels.

A polyelectrolyte has either a cation or an anion distributed along the polymer chain. The oppositely charged species, the counterion, must remain in the vicinity of the polymer chain to maintain charge neutrality. In the case of a swollen polyelectrolyte gel in an excess of solvent (water), the counterions must remain within the bounds of the gel and cannot

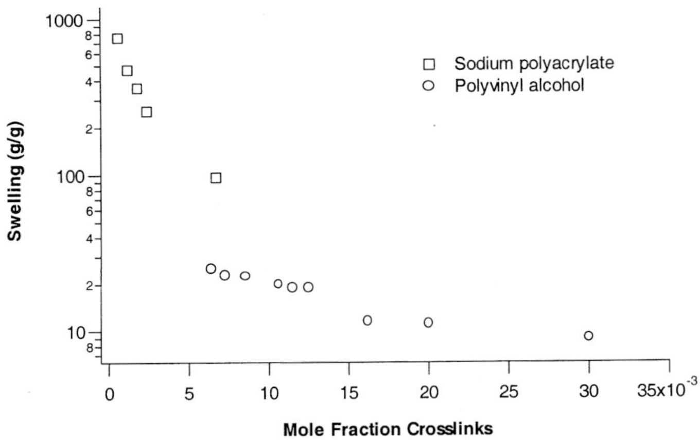

Figure 3: Swelling of nonionic and ionic gels in water [7,12]

migrate into the surrounding fluid. The osmotic potential of these many counterions greatly increases the tendency for water to enter the network and swell the gel. For this reason, it is much easier to prepare polyelectrolyte gels that can absorb large amounts of water, as compared to nonionic gels. Figure 3 illustrates this; data for a nonionic polyvinyl alcohol is included for comparison.

If a polyelectrolyte gel is immersed not in pure water but in an aqueous salt solution, a more complex equilibrium is established. The Donnan equilibrium refers to the state in which the water is in thermodynamic balance between the salt solution outside the gel and the polyelectrolyte solution inside the gel. Some salt can move into the gel, but the ions forced to remain in the gel, the polyion network and its counterions, will exclude a portion of the salt. Because the driving force for water into the gel is reduced by the presence of ions already in the free fluid, the swelling of the gel can be significantly reduced in salt solutions, as compared to its behavior in fresh water (see Figure 4).

Another effect is that of pH. Polyelectrolytes are the salts of either weak of strong acids or bases. Consider a weak polyacid such polyacrylic acid. In a low pH environment the polymer is protonated and essentially ceases to be a polyelectrolyte, resembling polyvinyl alcohol. As the system is neutralized the enhanced osmotic pressure from the counterions causes considerable expansion of the gel. Figure 4 illustrates this as well.

Polyelectrolytes are also chelators. This leads to a significant impact on swelling by multivalent ions. Figure 5 shows the effect of added calcium ions to the extent of swelling of sodium polyacrylate. One caution in interpreting such data is that these chelates may have long, but not infinite, lifetimes. Slow changes (over hours or days) may continue as the ionic networks continue to readjust to the most favorable configurations and the multivalent ions attempt to distribute themselves uniformly through the gel. Indeed, for some polysaccharide systems, calcium can be used as a crosslinker (Section 3.2.5.). The even more complex

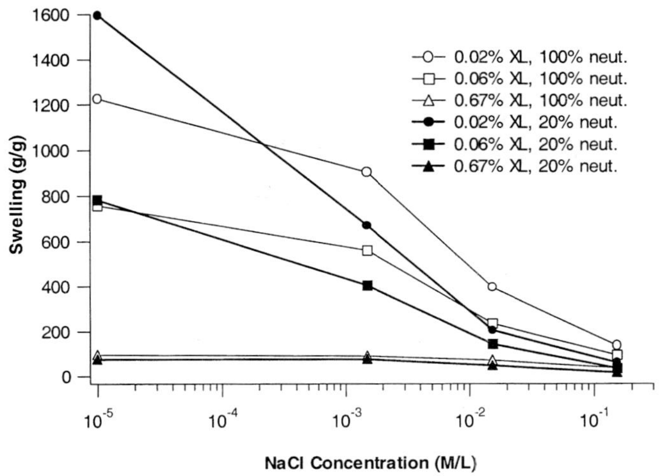

Figure 4: Effect of salt and pH on swelling [7]

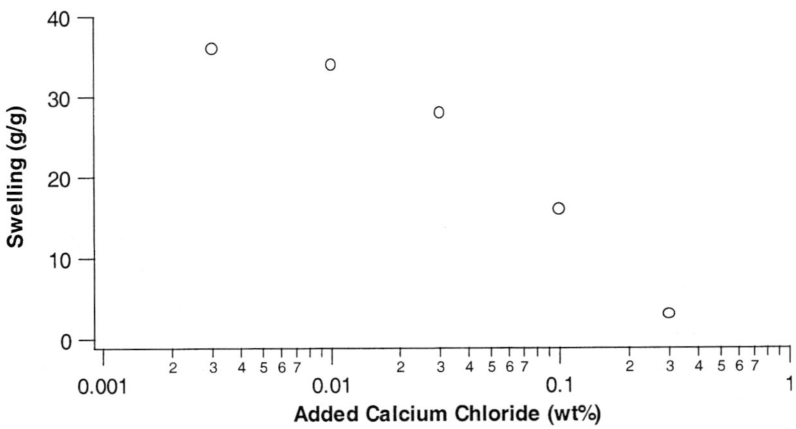

Figure 5: Impact of added calcium ions on swelling of crosslinked sodium polyacrylate in 0.9% NaCl

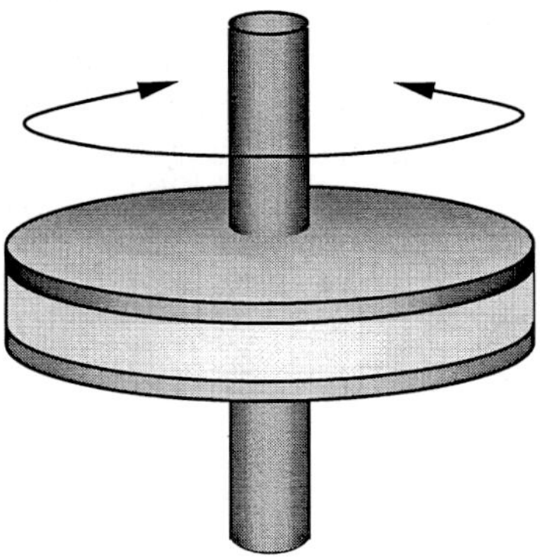

Figure 6: Schematic of gel modulus measurement

swelling behavior of polyacrylate and polyacrylamide gels in aqueous solutions of non-solvents has been reviewed by Tanaka [13], who pioneered this work.

2.3. Modulus of Gels

Aside from swelling capacity, another property characteristic of a gel is its toughness or resiliency. Consider the preparation of a gelatin dessert. The mixture begins as liquid solution, little different in consistency from water. As it "gels", the mixture first thickens then stiffens into an elastic solid, eventually holding its shape sufficiently to be cut into cubes. Quantification of this stiffness constitutes measuring the elastic modulus of the gel.

Figure 6 is an illustration of a general measurement technique for shear modulus [14]. Other modes of deformation may be used, but for polyelectrolyte gels, this is the most commonly sought value. As the mobile surface of the sample moves there is a resistance. Ideally this resistance is elastic; the displacement fully recovers once the force is released. In the case of polymeric systems, there may be a significant time for full recovery as the chains rearrange. This time delay can be related to the viscous drag of the system. In swollen superabsorbent gels, the polymer density is often rather low and hence this effect is of less interest than for elastomers.

The major determinant of modulus in a gel is the number of elastic chains per unit volume. To be an elastic chain, a segment of polymer must be attached to the bulk of the gel at each of its ends. These attachments are the crosslinks. Modulus and swelling capacity are related. As the volume of a single gel sample increases with swelling, the number of chains, or crosslinks, per unit volume naturally decreases. One can qualitatively observe the extent of swelling of a gel and can simultaneously sense its stiffness tactilely. Experience tells us that more highly swollen gels tend to be softer. Indeed, one cannot speak of the modulus of a gel sample without defining the extent of swelling at which the measurement was made. When considering only systems swollen to equilibrium, gels with higher crosslink densities will swell less and hence have higher moduli at equilibrium.

If only one end of a polymer segment is connected to the bulk of the gel, this is referred to as a dangling end; it cannot contribute to the modulus of the gel. If neither end is connected the polymer is soluble, not formally part of the network at all. In an excess of solvent these soluble chains can be leached into the surrounding fluid when the gel is swollen, forming what is frequently termed the "extractable" portion of a particular sample. The probability of forming either unattached or singly attached chains would be expected to go down as the backbone molecular weight of the polymer increases.

2.4. Summary of Theoretical Descriptions of Polyelectrolyte Gels

Referring to Eq 1, our task now is to develop expressions for the various terms to create understanding from experimental results.

$$\Delta F = \Delta F_{mix} + \Delta F_{elas} \tag{1}$$

Ideal solutions composed of small molecules behave in predictable ways because the thermodynamics of solutions are well understood [15].

$$\Delta F_{mix} = \Delta H_{mix} - T\,\Delta S_{mix} \tag{2}$$

For ideal solutions, there is no heat of mixing; enthalpic changes are zero. For small molecules, the entropy can be determined from the mole fractions,

$$\Delta S_{mix} = -k\,(n_1 \ln x_1 + n_2 \ln x_2) \tag{3}$$

$$\Delta F_{mix} = kT\,(n_1 \ln x_1 + n_2 \ln x_2) \tag{4}$$

where n_i and x_i represent the number of moles and the mole fraction, respectively, of each component in the system; k is Boltzmann's constant.

By definition, polymers have a high molecular weight, and hence there are relatively few *moles* of polymer, compared to solvent, in a typical system, and early workers found that the equations above no longer adequately described the results. It was found that the *volume* fraction (v_i) rather than the mole fraction was a better measure of entropy for these systems.

$$\Delta S_{mix} = -k\,(n_1 \ln v_1 + n_2 \ln v_2) \tag{5}$$

Nonideality in solutions of small molecules arises when ΔH is no longer zero due to interaction of the solute with the solvent. For a great many polymer/solvent systems, it is also the case that the heat of mixing is not negligible, and this deviation from ideality for polymer solutions is handled with the well-known Flory-Huggins interaction parameter χ("chi"). This essentially represents the heat of mixing on a monomeric unit basis. Consolidating these conclusions the full expression for the free energy of mixing for polymeric solutions becomes then,

$$\Delta F_{mix} = kT\,(n_1 \ln v_1 + n_2 \ln v_2 + \chi n_1 v_2) \tag{6}$$

As just noted, for high molecular weight polymers, the number of *polymer* moles, n_2, is small relative to n_1, the moles of *solvent*. In the case of a crosslinked gel, there is only *one* molecule of polymer, and hence the second term in parentheses above is completely negligible, and the mixing contribution to free energy becomes

$$\Delta F_{mix} = kT\,(n_1 \ln v_1 + \chi n_1 v_2) \qquad (7)$$

The derivation of this equation and its limitations are presented in many polymer texts, particularly the classic original by Flory [16], *Principles of Polymer Chemistry.*

Developing the second term of Eq. 1, ΔF_{elas}, begins with the recognition that, as in mixing, there is also no enthalpic change in the system due to *elastic* deformation. Hence, ΔH_{elas} is taken as zero, and is just - $T\Delta S_{elas}$; ΔS_{elas} is the entropy associated with the configurational change in the network as its volume increases with swelling from a volume V_0 at which the gel is crosslinked to V. In developing the theory of rubber elasticity, Flory[16] derives the following expression,

$$\Delta F_{elas} = (kT\nu_e/2)\,(3\,\alpha^2 - 3 - \ln \alpha^3) \qquad (8)$$

where ν_e is the number of elastic chains, proportional to crosslinks, and $\alpha^3 = V/V_0$, termed the linear deformation factor. By deriving the expression for the chemical potential of the solvent from the total free energy of the system (Eq 1) and setting it to zero, Flory obtained the following approximate relation for equilibrium swelling,

$$q^{5/3} \approx (V_0/v_{sol}\nu_e)\,(1/2 - \chi) \qquad (9)$$

The substitution of swollen volume q for $1/v_2$ has been made; the term v_{sol} is the molar volume of the solvent. This expression is derived only for moderately swollen systems; extreme swelling will be discussed below. This result is consistent with the earlier qualitative statement indicating a hyperbolic relation between capacity q and crosslink density, proportional to ν_e. Equation 9 can be written in what turns out to be a somewhat more convenient terminology,

$$q^{5/3} \approx (v_{sp}\,M_c/v_{sol}\rho_0)\,(1/2 - \chi)/(1-2M_c/M) \qquad (10)$$

in which M_c is the molecular weight between crosslinks and M is the backbone molecular weight, *i.e.*, that of the uncrosslinked system; the term v_{sp} is the specific volume of the polymer. These two molecular weight terms allow prediction of the effect of the dangling ends noted in the previous section; if there were no dangling ends, M would be infinite and the denominator would be unity. One caution is that these expressions are developed largely for the swelling of rubbers, which are generally crosslinked neat. As we will see in later sections, the networks of interest in this chapter are generally crosslinked with significant solvent (water) in the system. For systems crosslinked without solvent, $V_0/V = v_2$, but $V_0/V > v_2$ for most superabsorbents.

Moving from nonionic to polyelectrolyte gels, it is common practice[7] to restate the free energy equation as

$$\Delta F = \Delta F_{mix} + \Delta F_{os} + \Delta F_{elas} \qquad (11)$$

The additional term, ΔF_{os}, indicates the free energy associated with the osmotic effects. In a sense it is a mixing phenomenon but is so significant that it is developed separately from ΔF_{mix}. Simply adding the ΔF_{os} to the equation, however, is insufficient. This approach breaks down because at the high degrees of swelling seen with polyelectrolytes, especially in pure water, the other terms must be modified as well. The network configuration can no longer be accurately represented by the Gaussian chain statistics on which the Flory-Huggins approach is based. An improvement in the model was made by Hasa et al. [17], who proposed use of the inverse Langevin function to describe chains at high extension. Unlike Gaussian statistics, which allow consideration of chains of infinite length, the Langevin approach requires a fixed limit for chain extension. Oppermann [18] has presented data illustrating the situation where chains approach their limit in Reference 4.

Two additional related considerations for polyelectrolytes are 1) chain stiffening resulting from charge repulsion and 2) charge condensation. In the first situation, like charges along the backbone of a polyelectrolyte repel one another; this gives rise to an additional stiffening of the backbone beyond that expected from entropic considerations. The result is a higher extension, and thus greater swelling, than would be the case for a nonionic polymer. An opposing effect, first proposed by Manning [19], is termed counterion condensation. According to this description, bare charges cannot be sustained on, for example, every monomer unit of a polyacrylate because of dielectric considerations. A significant number of cations will condense onto the backbone charges. In the case of sodium polyacrylate, for example, only between one third and one half of the potential groups will actually be ionized at any time, even when the system is 100% neutralized. This reduces the amount of chain extension, and hence swelling, otherwise expected from charge repulsion.

Oppermann [18] developed a model accounting for the limited extensibility of the chains and Manning charge condensation. One of the most comprehensive attempts to handle this is contained in Yin's Thesis [7] In addition to the theoretical treatment, an extensive collection of equilibrium swelling data for polyacrylic acid gels is presented as a function of crosslink density, degree of neutralization, and salinity; much of the data was also presented in a chapter [20] of reference 4.

However, for a great many applications, including most of those for personal care, polyelectrolyte networks are not fully extended, and the assumption of Gaussian statistics is satisfactory. Buchholz [21] has discussed this situation in considerable detail in Chapter 5 of Reference 6 and derives the following equation,

$$q^{5/3} = (2M_c/v_{sol}\rho_0 V_0^{2/3})\,(1/2 - \chi)/(1 - 3M_c/M) \qquad (12)$$

All the parameters retain their meaning, and ρ_0 is the density of the polymer at preparation.

3. SYNTHESIS AND MANUFACTURE

One could conclude from the discussion in the previous section that a simple, yet reasonably accurate, chemical definition of superabsorbents would be crosslinked polyelectrolytes. The greater swelling capacity of ionic systems adds value, and hence practically no commercial *superabsorbents* (as opposed to hydrogels, *e.g.*, those used in ophthalmology) are nonionic. Extending this thought leads to maximizing charge-to-mass ratio for the greatest absorbent efficiency. Among the many polymers tested over the years, no other polymer can provide a high ratio as economically as poly(sodium acrylate). The resultant dominance of polyacrylates in the marketplace is why in this section we will primarily discuss these materials. Also treated are the chemically similar polyacrylamides, copolymers of maleic anhydride which are particularly useful for fibers, and polyaspartic acid, of interest for its biodegradability. In addition, a very brief survey of important nonionic aqueous polymers and modified natural gums is included to direct the interested reader to sources for these closely related topics.

One shortcoming of the definition above is that it includes another commercially important class of materials that are not used as superabsorbents, ion exchange resins. Although these are indeed crosslinked polyelectrolytes, they are highly crosslinked in order to swell as *little* as possible. Most ion exchange resins are made by the substitution of ionic groups, either anionic or cationic, onto a polystyrene backbone crosslinked with divinyl benzene, usually prepared in the form of a bead. References abound on these materials and no further discussion of them will be presented here.

3.1. Polyacrylates

Greater than 95% of the superabsorbents in the marketplace are crosslinked, partially neutralized poly(acrylic acids), generally with sodium as the counterion. Manufacturing comprises an aqueous free-radical polymerization of acrylic acid, either preneutralized or unneutralized, usually with a soluble vinyl crosslinker. Two general polymerization processes are used, solution (gel) or water-in-oil suspension, the former accounting for the great majority of production. Despite considerably different process equipment, the underlying chemistry of the two methods is similar. In this section we will first discuss the monomer and its polymerization chemistry, and then describe the two types of process in some detail. Graham and Wilson [22] should be considered as a detailed reference for both processes.

3.1.1. Monomers and Crosslinkers. Acrylic acid [23] has grown to dominate as a feedstock for the personal care markets, largely because of its rapid polymerization and efficiency in delivering carboxylate functionality. Worldwide capacity exceeds 2 million tonnes per year, distributed primarily in North America, Europe and Japan. The most commonly used manufacturing process is a catalytic oxidation of propylene in two steps, first to acrolein and then to the acid.

Acrylic acid is a liquid that freezes somewhat below room temperature, as does the more common acetic acid, the “active” component of vinegar. General properties of acrylic acid are given in Table 1. Examination of this table shows a high heat of polymerization, similar on a molar basis to other vinyl monomers; but with the relatively low molecular weight of the monomer, the result is considerable energy stored per unit mass. Acrylic acid is a rapidly polymerizing monomer, and all these characteristics combine to make storing and handling of this material challenging. The monomethyl ether of hydroquinone (MEHQ) is typically used as an inhibitor, but to be effective the inhibited monomer must be stored in contact with air because the interaction of MEHQ with oxygen is required to maintain inhibition.

Table 1.
Properties of Acrylic Acid.

Property	Units	Value
CAS registry number		[79-10-7]
Molecular formula		CH_2=CHCOOH
Molecular weight	g/mol	72.06
Melting point	°C	13.5
Boiling point	°C	141
Density at 20°C	g/cm^3	1.040
$\Delta H_{polymer}$	kJ/mol	77.4

An unusual feature of acrylic acid is its tendency to form a dimer, 3-acryloxypropionic acid (CAS registry number [24615-84-7]). This reaction is unrelated to its polymerization, the reaction is a Michael addition of one monomer molecule to a second.

$$2\ CH_2{=}CHCOOH \rightarrow CH_2{=}CHCOOC_2H_4COOH$$

The product is itself a monomer and can copolymerize into the polyacrylic acid chain. When subsequently heated, as during drying, this moiety can thermally crack leading to a new molecule of acrylic acid. Because formation of this dimer is not a radical process, inhibition with MEHQ has no effect on the dimer content of acrylic acid. Instead, distillation during production generally results in a low level; and dimer increases at a predictable rate during storage [24,25].

All superabsorbents are crosslinked gels and controlling the incorporation of these crosslinks represents a major objective of the study of these materials. Two approaches to crosslinking are used in the production of commercial superabsorbents, 1) copolymerization of monomers with multiple vinyl groups or 2) post-reaction, *i.e.,* after the polymerization, of already formed chains, analogous to a "curing" process. Based on the total number of crosslinks, the former strategy predominates; post-reactions are generally utilized to add a few strategic crosslinks to improve properties. Several copolymerizable crosslinkers frequently cited in both the scientific and the patent literature are illustrated in Figure 7.

Post-reaction crosslinking of acrylic acid polymers is generally effected by esterification of the many available acid groups. Both glycerin and polyethylene glycol have been used [26],as have alkylene carbonates [27]. The enhanced reactivity of epoxy compounds has been utilized, ethylene glycol diglycidyl ether (EGDGE) in particular has been cited [28]. Another method of post-polymerization crosslinking of polycarboxylates is through the use of multivalent cations [29]. A review of metal ion crosslinking can be found

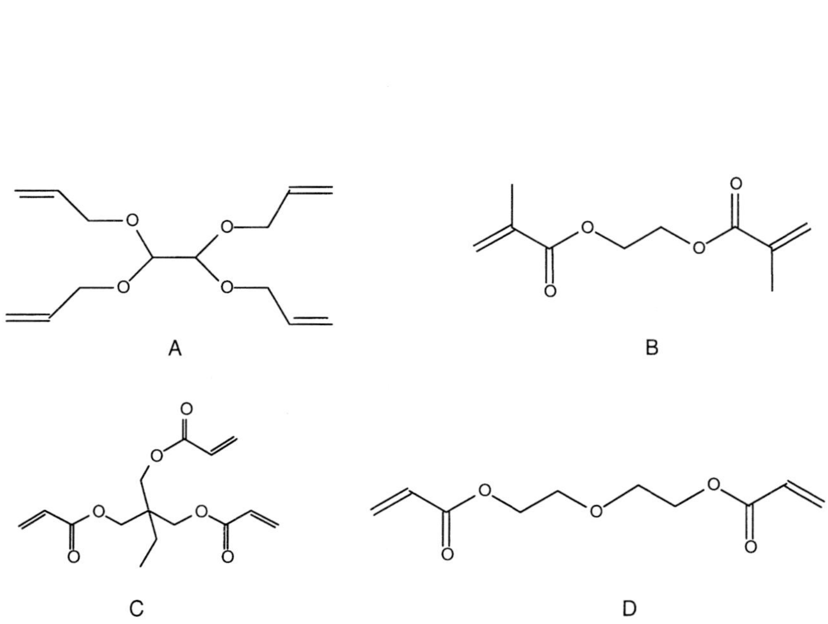

Figure 7: Selected vinyl crosslinkers for superabsorbents. A: Tetrallyloxyethane. B: Ethyleneglycol dimethacrylate. C: 1,1,1-trimethylolpropane triacrylate. D: Diethyleneglycol diacrylate.

in the Thesis of Uhl [30]. Organic polycations have also been used [31]. These post polymerization crosslinking techniques can be used for other carboxylated polymers besides sodium polyacrylate, and specific mention will be made in later sections.

3.1.2. Synthesis of Polyacrylate Gels. A typical preparation of a polyacrylate gel begins with an aqueous monomer mix, generally containing between 20 and 40% by weight acrylic acid or its salts with less than 1% by weight of appropriate vinyl crosslinkers. Oxygen is a powerful inhibitor of the polymerization, and sparging with an inert gas such as nitrogen or carbon dioxide is important for reproducible results. Polymerization is initiated by generating free radicals through thermal decomposition of peroxides, persulfates, diazenes, redox systems, or with photoinitiators. The reaction releases considerable heat; viscosity rapidly builds and the gel point is reached. Convection quickly slows and only by careful control and a large heat transfer surface can temperature be maintained. Two standard laboratory preparations are described in considerable detail in Chapter 2 of Reference 6.

Numerous kinetic studies of this polymerization have been made [32,33,34,35,36], but mechanistic questions remain. Nevertheless, experimental results from several sources show the clear impact of pH. Figure 8, with data taken from Ito et al. [32] and Kabanov et al. [33], shows relatively high rate of polymerization at the low and moderately high pH values with a trough in the middle and another fall off at very high pH. This is important because the major products of commerce are around neutral pH because skin contact with personal care applications demands it. To get to this common point from the acid monomer, two process approaches have evolved, either preneutralization (neutralization of the monomer) or

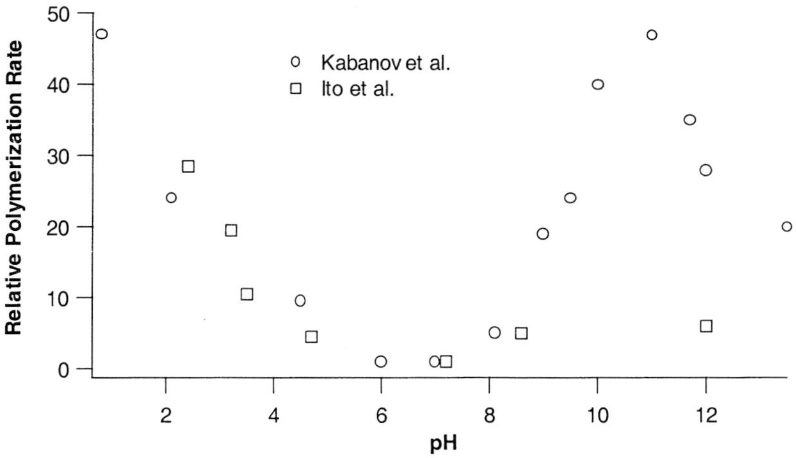

Figure 8: Effect of pH on polymerization rate of acrylic acid.

postneutralization (neutralization of the polymer). The relative advantages of each will be discussed in Section 3.1.3.

Another unusual feature of acrylic acid polymerization is the higher than first power dependence on monomer concentration. Figure 9 shows three polymerizations at different monomer concentrations, normalized to fractional conversion [36].

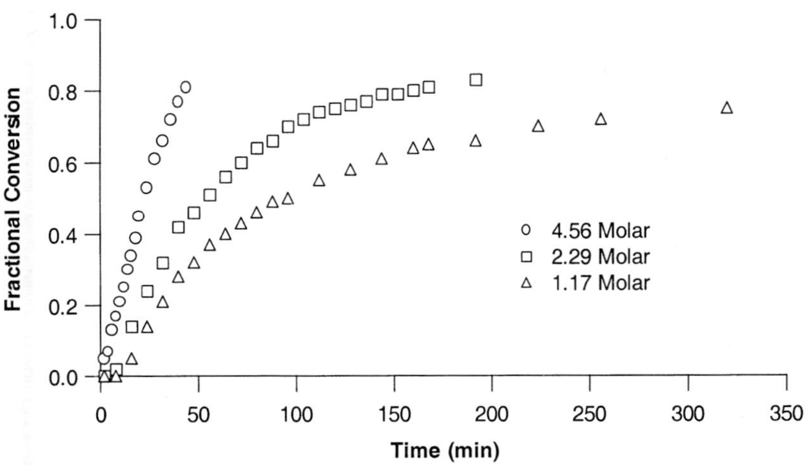

Figure 9: Dependence of polymerization rate of acrylic acid on monomer concentration.

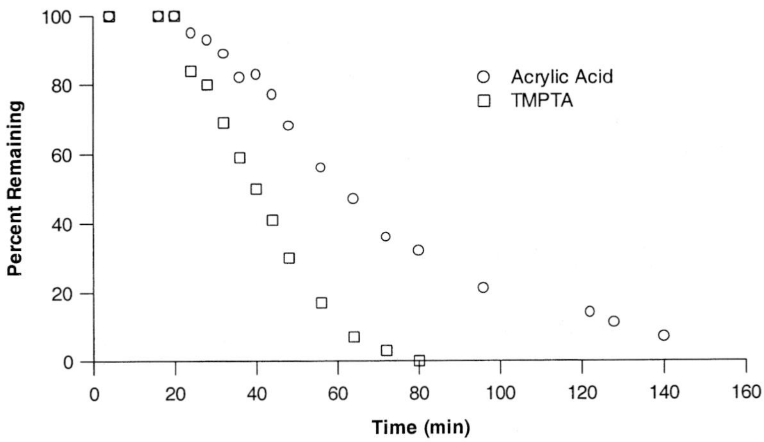

Figure 10: Copolymerization of TMPTA crosslinker with acrylic acid. Conditions were 42% solids, 65% neutralized, and isothermal at 55°C

Incorporation of vinyl crosslinkers into the polymerization of a gel largely determines its properties. An interesting result reported by Arriola et al. [37] shows considerable nonuniformity during the course of polymerization for an ester crosslinker (Figure 10). NMR techniques developed to follow the polymerization of acrylic acid and cited above [36] were used with an isotopically enhanced sample of crosslinker to follow the reaction. These data would suggest that a relatively large fraction of the polymer would not be connected to the network because of depletion of crosslinker by 70% conversion. This is not the practical experience with these polymerizations; analyses of commercial samples typically show less than 10 weight percent soluble polymer. Therefore, some other grafting or crosslinking events must occur. These have not been elucidated at this time.

Gel produced by these polymerizations can be studied without further treatment, but commercial products are typically dry powders. Laboratory drying of these gels can be done in a conventional oven. One can also recover the polymer by precipitation with, for example, methanol. Grinding can be accomplished in a laboratory blender, followed by appropriate screening.

3.1.3. Gel Polymerization Processes. As noted, superabsorbent manufacturers choose to polymerize either acrylic acid solution and neutralize the resulting acidic gel or neutralize the monomer before polymerization. A variety of specific considerations influence these choices, ownership of the various technologies naturally being a major factor. In general, "post-neutralization" (after polymerization) requires the use of crosslinkers stable to hydrolysis but allows one to polymerize the acid, which reacts more rapidly than the salt, especially in more dilute solutions. It is somewhat more difficult to efficiently mix neutralent with a gelled reaction mixture than with a monomer solution. "Pre-neutralization" (as a monomer, before polymerization) is itself a simpler process but requires polymerization of sodium (or potassium) acrylate solution, which is in general a more sluggish monomer than the acid. Solubility of organic crosslinkers in the high ionic strength monomer mix is also a

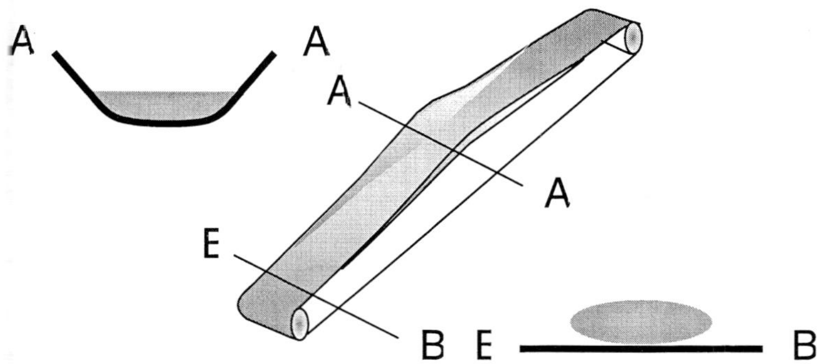

Figure 11: Schematic of belt polymerization equipment.

consideration; Siddall and Johnson [38] used poly(vinyl alcohol) as a dispersant to mitigate this situation.

For cost considerations, sodium is the typical counterion. Solutions of either sodium hydroxide or sodium carbonate are used as the neutralent. The heat of neutralization is significant, and management of this heat is important for controlling the temperature during addition of base. A too rapid addition can result in a temperature rise in a monomer solution, which could lead to polymerization, with even greater evolution of heat, leading ultimately to a catastrophic situation. With sodium carbonate, release of carbon dioxide must be managed, but this gas evolution can provide considerable cooling.

Both batch and continuous reactors have been patented for commercial scale polymerization of sodium polyacrylate gels. Key elements of this polymerization that must be considered by any process are 1) managing the large amount of heat released during this rapid polymerization, 2) comminution, somewhere in the process, of the tough gel to facilitate drying. Optimizing the solutions to these two of problems has resulted in a variety of reactor configurations including a moving belt, agitated vessels, a boiling reactor, and suspension systems. The latter are sufficiently different from gel processes that they are discussed separately in Section 3.1.4.

Among continuous processes, one commonly cited device is a belt polymerizer in which a flexible belt carries a continuous mass of gel until polymerization is completed; the material then goes to a grinder. One such device used for both acrylamide (*v.i.*) and acrylic acid was described in a 1973 patent[39]. The liquid monomer mix added to a portion of the belt that is cupped by several rollers into a trough. As polymerization ensues the thickened mass will hold its shape and the belt is allowed to flatten. Figure 11 illustrates the concept.

The continuous gel slab thus formed may be 6 to 12 cm thick, and the belt speed may vary from 5 to 12.5 cm/min. To minimize air intrusion into the polymerizing gel, the device may be encased and blanketed with an inert atmosphere. Temperature control is somewhat limited after the gel forms because there is limited surface area. Spraying the bottom side of the belt to remove some of the heat of polymerization is taught. The recipes contain 30-43% partially (50-70%) neutralized acrylic acid.

One strategy around the temperature control problem is to devise a recipe that allows an adiabatic reaction. In one such case a relatively dilute solution of 17-25% (unneutralized) acrylic acid is used [40]. Reflecting an approximate 50°C temperature rise and a starting temperature of 20°C, this concentration would allow the reaction mass to stay below 70°C. It is generally easier to maintain product quality when polymerization temperatures do not exceed 90°C. Neutralization slows the reaction, as noted in the previous section, and hence the more dilute reaction mixture was apparently found to proceed more favorably in the acid form. Of course post-neutralization of the polymer is required because the products of commerce must yield an approximately neutral pH for skin contact applications. The crosslinker tetraallyloxyethane is claimed; the ether is less likely to hydrolyze during the neutralization process than an ester. Grinding of the gel to facilitate mixing of the base is required. The choice of initiator system is constrained by the conditions of temperature range and pH as well; examples can be found in the patents cited.

Some superabsorbents are produced by grafting onto a substrate loaded to the reactor. An early patent to Masuda et al. [41] is a graft polymer prepared by copolymerizing acrylic acid onto starch. Additional crosslinking was used and the polymer was post-neutralized.

Agitation during the reaction presents a significant challenge as the tough gel forms, but the advantages are twofold: 1) Breakup of the gel can be accomplished when the it is weaker, minimizing the power required to grind the fully polymerized gel logs. 2) Certain temperature control strategies can be used to manage the high exotherm of polymerization. Several configurations of agitated reactors have been illustrated. Tsubakimoto and Shimomura [42] describe a continuous horizontally arranged device, which they note produces a finely divided gel with a minimum of large particles. An anonymous disclosure [43] of a vertical reactor with a slow moving agitator for “gelatinous and pasty” materials discusses its use for aqueous sodium polyacrylate gels.

Vaporization of a portion of the water charged to the reactor can be used to control the temperature if there is enough surface area in the divided gel. Such a procedure has been described by Siddall and Johnson [38]. An even more extensive use of this is strategy is described for both a batch reactor [44] and a continuous system [45]. High concentrations (>70%) of *potassium* acrylate, which has a higher water solubility as a monomer than the sodium salt, are polymerized in a boiling reactor, resulting in a product substantially dry to the touch.

Considering the concentration ranges of from 17 to 43% cited for most of the processes (boiling reactors [44,45] being an exception), two to four parts of water will need to be removed from the reaction mass for each part of polymer. After reduction of the aqueous gel to a crumb, either through agitation during polymerization or a separate gel mincing step, drying can be accomplished in more or less any conventional dryer. Two methods specifically cited in the literature are through circulation belts [42,46] and rotating drums [40]. The drying process may be continuous, even if the polymerization has been batchwise.

As noted, for most superabsorbent applications the item of commerce is granular, resembling table sugar. The product resulting from drying even after breakup of the gel is generally not properly sized, and some grinding and classifying process will be required. When dry, sodium polyacrylate is rather brittle and grinds reasonably well; unfortunately, this also leads to overgrinding which may form a fine fraction that is often separated from the material as sold. Methods to recycle these fines back into the polymerization [47,48] or blend them into the gel [49,50] have been patented.

Throughout the discussion thus far, the gel particle has been assumed to be monolithic, *i.e.*, crosslink density and other morphological features are uniform throughout the particle. As will be discussed in greater detail in Section 5 (Applications), current superabsorbent products are typically crosslinked to a greater extent near the surface. Direct evidence of this morphology in "second generation" products has been presented in two recent reports using sophisticated techniques [51,52]. The post-polymerization crosslinking techniques mentioned in the previous section are often used to obtain this gradient of crosslink density.

Manufacture of these improved polymers generally requires blending the dried and sized polymer with an additive, followed by an additional reaction step. For example, from 0.001 to 10% polyhydric alcohols were added in a ribbon mixer and heated in disc dryer or similar device [26]. Extensions of this technology to high speed paddle mixers have been made [53]. Because these reactants must penetrate the gel particle, a small amount of rehydration is necessary; addition of any water to a superabsorbent often leads to stickiness, and thus typically another agent, such as an alcohol [26] or an aqueous salt solution [54], is added with the crosslinking agent. These materials suppress or at least slow the swelling to facilitate even distribution of the reactant. Glycidyl compounds [28] and alkylene carbonates [27] have already been cited as surface crosslinkers. Burgert et al., have shown a similar significant improvement in properties without adding a surface treatment [55].

A useful measure of improved performance is the Absorbency Under Load (AUL) test, described in Section 4.1.3. The data in Table 2 are taken from Reference 6 (p101) and illustrate AUL values of second generation products.

Table 2.
Data from Various Processes to Obtain Improved Properties.

	Initial Swelling Cap (g/g)	Swelling Cap after Treat (g/g)	0.6 AUL after Treat (g/g)
Aluminum acetate	39	35	18
Glycerin	39	35	21
EGDGE	40	35	31
Ethylene carbonate	50	36	25
Process of Burgert et al.	39	36	26

3.1.4. Suspension Processes. Reactions in which droplets of aqueous monomer are suspended in another fluid and polymerized look quite different from gel processes, but the chemistry occurring is very similar. These processes can be considered as an approach to solving the "torque problem" of breaking up tough gels; sizing the monomer mixture into small droplets while it is still a liquid requires much less energy and lighter duty equipment than does grinding the fully polymerized gel. Furthermore, the relative ease of agitation of the suspension allows much improved heat transfer over what is possible in gel reactors. An isothermal polymerization is in fact possible.

To maintain the discrete drops in a second fluid, typically a hydrocarbon oil, requires sophisticated suspending agents, the subject of several patents [22]. These suspending aids, though often surface active, should not be confused with emulsifiers, and suspension polymerization is not the more common emulsion polymerization. In the latter the dispersed phase containing the polymer is much more finely divided, and the monomer migrates through the continuous phase during polymerization. In suspension systems, each droplet of aqueous monomer mixture remains discrete during the polymerization process. Suspension droplets are generally several microns in diameter and will settle spontaneously when agitation ceases. Because most industrial polymers are plastics, the more common suspension polymerization processes have a continuous phase of water and polymerizing droplets of organic monomers. For that reason, the systems of interest here, in which the discrete phase is aqueous, are commonly referred to as "inverse" suspensions.

The aqueous monomer phase is similar to the preneutralized reaction mixture for a gel polymerization [22]. Unneutralized acrylic acid will dissolve in the continuous organic phase, so only the neutralized recipes (>50%) will work in a suspension process. Suspension failure describes the situation, to be avoided at all costs, in which the droplets stick to one another, perhaps because of some adventitious polymerization in the oil phase, causing a catastrophic build of viscosity and possibly runaway polymerization.

After mixing, the aqueous sodium acrylate solution is dispersed in the oil phase using fairly rapid agitation with the addition of suspending aids. The phase ratio, the relative volumes of the oil and the aqueous phases, can vary. A high ratio in favor of the aqueous monomer makes efficient use of equipment and hence is more economical, but at the risk of premature suspension failure caused by the close contact and more frequent collisions of the droplets. This is the inducement to develop robust suspending aids. Initiation is typically effected by heating a thermally labile compound. The exotherm can be more easily handled in the agitated system and evaporative cooling through azeotroping the water and oil is possible.

Isolating the product is relative straightforward. If agitation is stopped the typical suspension will settle, allowing decantation and filtering. Essentially all of the water in the aqueous phase may be removed by azeotropic distillation if desired [56]. Some washing of the hydrophobic surface resulting from oily residuals may be necessary. It is important to recognize that no grinding or screening steps would typically be required, in vivid contrast to the gel polymerization processes discussed in the preceding section.

The typical product of a suspension polymerization process is a bead. As a commercial superabsorbent for personal care there are two potential shortcomings to this form: 1) The round shape is difficult to keep in place in a diaper or other construct, it tends to roll about [57]. 2) A sphere offers the *minimum* surface area for wetting, and this can limit the rate of absorption. One tactic to ameliorate both these problems is to develop porosity, and often roughness at the same time. Boiling on the surface of the bead [58], fugitive additives [59], and controlled agglomeration [60] have been described. Another approach is a macroscopic change in shape; for example, sausage-shaped particles have been obtained by controlling both the shear field and the rheology of the droplet during polymerization [57]. One novel strategy is to build convolutions into the surface [61]. By carefully controlling both the location and the timing of initiation, extensive folding of the particle surface can be built in; these products show quite high absorption rates, among other characteristics. This technology has also been used to agglomerate polymer particles obtained from gel polymerization processes [62].

A somewhat related method of polymerizing gels is a precipitating polymerization. In this system the unneutralized acrylic acid is used, which is soluble in many nonaqueous solvents (unlike the salt). The polymer (crosslinked or not) is, however, insoluble in these systems. The intent of these polymerization schemes is to provide for an orderly separation of the precipitating polymer from the reacting monomer solution. These processes are not yet commercialized for superabsorbent gels, but the BFGoodrich Company has produced Carbopol™ polymers, useful as thickeners, in such a process [63]. A particularly interesting solvent for precipitating polymerization of acrylic acid is supercritical carbon dioxide [64].

3.1.5. Miscellaneous Acrylate Polymers. In addition to sodium polyacrylate, certain other acrylate based polymer systems have been developed as superabsorbents. An ethyl acrylate/methacrylic acid latex, prepared as a conventional polymer-in-water emulsion, subsequently hydrolyzed, has been coated and crosslinked on tissue to form a laminate [65]. The same polymer has been used to print a superabsorbent pattern on a substrate or to spin fibers [66].

Another fiber product is spun from a copolymer of acrylic acid, methyl acrylate, and either hydroxyethyl acrylate or hydroxypropyl methacrylate [67,68]. This product has been marketed by Technical Absorbents Limited, a joint venture of Allied Colloids and Courtaulds Fibers Limited, under the tradename Oasis.

Gels made from methacrylic acid are well known in the scientific literature [69], but only polymers in which the monomer is part of a copolymer have been commercialized. As a homopolymer, the polymerization rate is slower [33], and the hydrophilicity is less. The much smaller scale of production makes methacrylic acid more expensive than acrylic [70]; it is a minority co-product of methyl methacrylate manufacture. On the positive side, the use of methacrylic acid and esters such as ethyl acrylate allow polymerization in a non-aqueous phase, which often facilitates recovery by minimizing the magnitude of the drying process [65].

3.2. Other Polymeric Gels

3.2.1. Polyacrylamides. Polyacrylamides are widely used in an uncrosslinked form as thickeners and flocculants in the oil industry, ore recovery, water and sewerage treatment, and a variety of industrial and consumer products [71]. Crosslinked polyacrylamides, particularly when partially hydrolyzed to yield acrylic acid moieties, are superabsorbent gels and have been used in selected applications over the years. Nonhydrolyzed, and hence nonionic, polyacrylamide gels are used for electrophoresis media in biotechnology laboratories [72,73]. Particulate polyacrylamide gels are useful in horticulture [74].

Hydrolysis of polyacrylonitrile can yield first polyacrylamide and eventually some acrylic acid groups. Because polyacrylonitrile can be readily formed into fibers, it has been possible to form absorbent fibers by hydrolyzing the surface of such materials [75,76].

Acrylamide monomer is a solid, but the typical commercial product is a 50% solution in water. The monomer is a neurotoxin, and attention to safe handling is essential; the liquid form allows less opportunity for physical contact and is therefore safer.

3.2.2. Maleic Anhydride Copolymers. Another major category of superabsorbents are those derived from maleic anhydride. Although this monomer will not homopolymerize, it can form alternating copolymers with several vinyl monomers. Hydrolysis yields a structure reminiscent of polyacrylic acid.

At the present time, most commercial superabsorbents are sold as granular solids. Fibrous forms are a much smaller fraction of the market but are of interest because of their physical similarity to the cellulose fibers blended with superabsorbents to form composites [77]. Making a polymeric fiber is a drawing process and is not possible with a fully crosslinked polymer. What is required is a high molecular weight, non-crosslinked polymer that can be "set" in some way after being drawn into its fibrous form, either from solution or a melt. Indeed, a fibrous superabsorbent can be produced by the copolymerization of maleic anhydride with isobutylene [78] or styrene [79].

In one example [79], maleic anhydride (258 g) and styrene (272 g) are mixed with 1752 g of acetone with 1.5 g of peroxypivalate as an initiator. The mixture is heated at 40°C for 20 hours to obtain a solution of the styrene/maleic anhydride copolymer. Diethylene glycol (0.5% based on polymer) was added and the solution was extruded into a water bath to form fibers, which were heated to effect crosslinking. The crosslinked fibers were then neutralized to form an absorbent composition.

In another example [80] ISOBAM 10, a commercial isobutylene/maleic anhydride copolymer for Kuraray Isoprene Chemical company, Ltd., was hydrolyzed and blended with propylene glycol and concentrated slightly to form a spinning solution. Spinning was followed by curing to complete drying and effect crosslinking.

Maleic anhydride copolymers have also been used to produce a superabsorbent film [81]. In one case, a solution of the disodium salt of poly(isobutylene-co-maleic anhydride) was prepared in deionized water. Mixed into 14.7 g of this solution was 0.28 g of 1,3-dichloroisopropanol and ten drops of a 2% solution of sodium lauryl sulfonate. After standing for 40 minutes, the solution was spread on a clean polyethylene sheeting with a 25 mil draw bar. Upon drying, the film readily separated from the polyethylene. The crosslinking reaction is essentially complete after the film is heated to 60°C for 30 minutes followed by 1 hour at 100°C. Utilizing various crosslinking agents such as epibromohydrin higher absorbencies could be achieved.

Another useful superabsorbent film [82] can be produced by dissolving 12.6 g of poly(ethylene-co-maleic anhydride) in water, adding 7.2 g of sodium hydroxide and diluting to a 25% solids solution. This solution was blended with 0.22 g of poly(N-methylol acrylamide) and a film was cast on a polished chrome plate with a 25 mil draw bar. The film was cured for

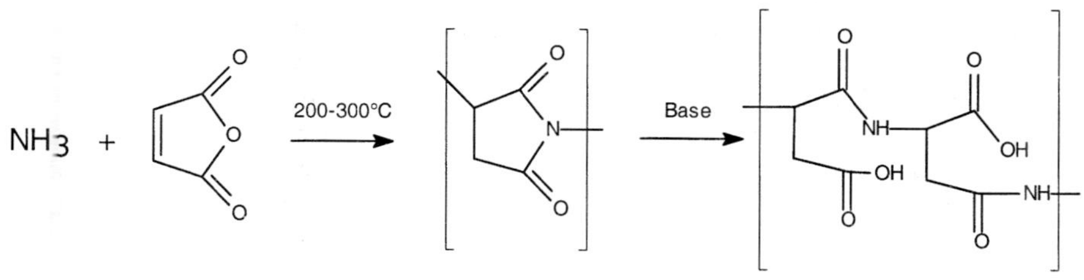

Figure 12: Schematic of the formation of polyaspartic acid

22 hours at 150°C. This superabsorbent absorbed approximately three times as much 0.27 N NaCl solution as did the former. Note that the hydrolyzed version of poly(ethylene-co-maleic anhydride) is structurally identical to a head-to-head polymer of acrylic acid; the normal product of polymerization, however, is head-to-tail.

3.2.3. Polyaspartic Acid. Biodegradable superabsorbents are of interest as landfill disposal becomes more costly [77]. The situation varies considerably in different locations, and generalizations about the market demand for such products are difficult. The challenge of obtaining biodegradability without compromising other properties of the superabsorbent, all while maintaining favorable economics, has limited the commercial success and market penetration of new products. Crosslinked carboxymethyl cellulose is the most familiar chemistry, and that material is discussed in another chapter of this monograph.

A recent candidate for a high volume biodegradable superabsorbent is polyaspartic acid [83]. A naturally occurring amino acid, aspartic acid can be produced both by fermentation and via maleic anhydride and ammonia in a strictly synthetic route. Heating the monomer leads ultimately to a condensation polymer. An even more direct route to the polymer is through heating a maleic anhydride/ammonia mixture. The intermediate polysuccinimide forms and is opened. This last process is shown schematically in Figure 12.

Polyaspartic acid is attractive as a substitute for polyacrylates because its potential charge density, though lower than that of polyacrylic acid, is comparable. Many natural products and derivatives are nonionic or possess only a few carboxyl groups. Perhaps the major shortcoming of the polyaspartic acid materials discussed in the literature is their relatively low backbone molecular weight prior to crosslinking; preparations are commonly around 50,000 g/mol, although 200,000 g/mol has been reported. In order to have efficient high swelling gels, high backbone molecular weight is desirable, and sodium polyacrylate typically has a backbone molecular weight near 1 million or higher.

3.2.4. Nonionic Synthetic Polymer Gels. True *super*absorbents, as noted in the introduction of this chapter, are all polyelectrolytes; the osmotic contribution of the ions leads to much higher swelling in aqueous systems than can be obtained with nonionic gels. Nevertheless, we will very briefly describe the principle polymers of this class and send the interested reader to other sources for additional information.

Poly(vinyl alcohol) is a water soluble polymer formed by the hydrolysis of poly(vinyl acetate). Fluid absorbing polymers can be made from vinyl esters and unsaturated dicarboxylic acids [84]. A polymer of 95/5 vinyl acetate/ monomethyl maleate was saponified in methanolic sodium hydroxide solution to yield a "superabsorbent" product. The saponified monomethyl maleate/vinyl acetate polymer can be blended with an equal part of poly(vinyl alcohol) and converted to water absorbent fiber with high absorption rate [85]. These polymers have been extensively reviewed by Peppas [86].

Poly(ethylene oxide) is a hydrophilic non-ionic polymer that can be crosslinked to form an absorbent gel [87,88]. For example, a mixture of 20 gram of 4% aqueous poly(ethylene oxide) solution and 11 gram of a 2% aqueous poly(4-vinyl-n-butylpyridinium bromide) solution, adjusted to pH 4.9, was crosslinked by irradiation with a 1 MeV van de Graff electron accelerator to a total dosage of 0.7 Mrad, to produce a superabsorbent material. A review of these polymers by Graham [89] can be found in the same monograph cited in reference 86.

Absorptive poly(hydroxymethylene) polymers may be produced that are insoluble but capable of swelling in water [90]. The polymer is prepared by the etherification of poly(hydroxymethylene) in an alkaline solution with an α-halogen carboxylic acid containing 2 to 5 carbon atoms with crosslinking prior to, during, or after etherification with a polyfunctional crosslinking agent. Thus, 6 g of poly(hydroxymethylene) was dissolved in 106 g of 30% aqueous sodium hydroxide solution at 85°C, and 23.6 g of an 80% aqueous monochloroacetic acid solution was added. The mixture was heated for 45 minutes at 85°C and then 0.99 g of bis-acrylamido acetic acid, dissolved in a small amount of hot water, was added. This mixture was heated for another 15 minutes to complete the crosslinking reaction.

3.2.5. Gels from Modified Natural Polymers. Two other chapters in this monograph treat in detail cellulose and starch derivatives, respectively. Although the subject of the present chapter is *synthetic* superabsorbents, we will briefly discuss modified natural gums, which compete in the marketplace to a limited extent.

Most of the industrial gums available today are water soluble or water dispersible. A majority of these gums are used as food thickeners or viscosity controlling agents. In order to use these natural gums as absorbents, they must be crosslinked, which can be accomplished in a variety of ways as noted. The most commonly used gums will be briefly covered here, and their structures are given in Figure 13.

Guar gum is derived from the seed of the guar plant, *Cyamopsis tetragonolobus*. Chemically, guar gum is a galactomannan and consists of a straight chain of (1→4)-β-D-mannopyranosyl units with single (1→6)-α-D-galactopyranosyl attached to every second main chain unit [91]. A dry, flexible, fibrous mass comprised of intertangled fibers has been prepared from guar gum [92]. On contact with water the fibers rapidly absorbed large amounts of water forming a soft gel. Guar gum may be used as the natural gum or in a chemically modified form. The gum may be mixed with other hydrophilic materials such as starch or other gums to increase absorption capacity. Plasticizing agents such as polyvinyl alcohol can be added to improve the adhesion to fibers or to a supporting sheet.

Fibrous guar is prepared by the high speed mixing of a well hydrated mass of swollen gum with a water miscible nonsolvent. A typical procedure was described as follows: Five grams of guar gum were dissolved in 325 mL of water with vigorous stirring until a homogeneous solution was obtained. The mixture was allowed to stand for one hour to fully

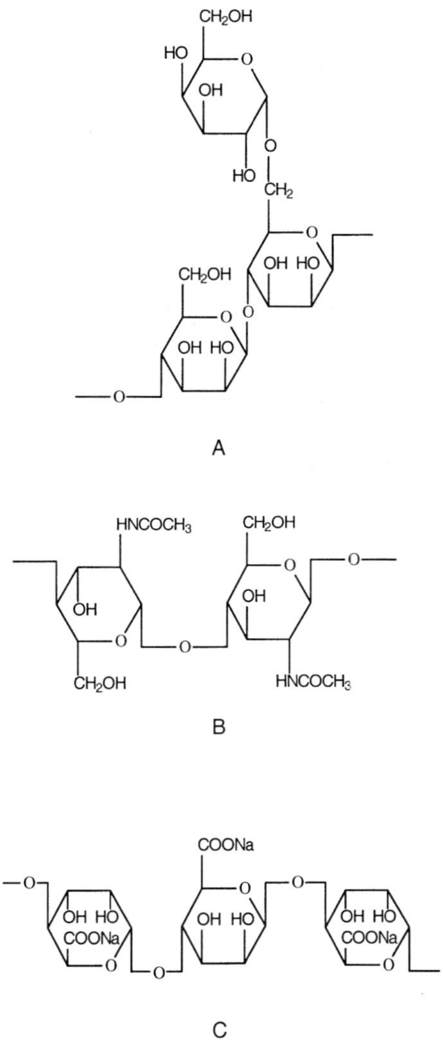

Figure 13: Structures of natural gums. A: Guar gum. B: Chitin. C: Sodium alginate

hydrate and was then stirred vigorously with the addition of one liter of isopropanol. The fibrous precipitate was allowed to settle and the supernatant was decanted.

Addition of borate ions to guar modifies the gelling behavior so that several times its weight of water may be adsorbed to produce a relatively dry, non-sticky and inert gel [93]. A borate salt such as zinc borate in an amount of from 3 to 10% of guar (based on the weight of the borate anion) can be used. The slow rate of release of borate ion is regulated by its solubility coefficient and allows optimal penetration of fluid before a firm gel is produced. The minimum solubility coefficient requires that, upon introduction of liquid, only a small concentration of free borate anion be available to crosslink the almost instantaneously formed guar hydrates. As liquid is introduced and penetrates, additional free borate anion is released from the borate to induce a controlled development of crosslinks.

The production of a crosslinked gel, as well as the stability of the gel, is highly dependent on pH, the optimum pH being between 7.0 and 7.4. This value may be achieved through the addition of an amine such as 2-amino-2-methyl-1,3-propanediol. Carboxymethylhydroxypropyl guar gum is used to prepare a superabsorbent [94]. It can also be crosslinked with aqueous borate solution to produce an effective absorbent material.

Xanthan is a water-soluble extracellular polysaccharide produced by the bacterium *Xanthom campestris*. The polysaccharide is widely utilized as a food thickener and viscosity controlling and suspending agent; it has a molecular weight of over one million [95]. It consists of a cellulose chain derivatized on every second β-D-glucopyranosyl unit at C3 by a trisaccharide of β-D-mannopyranosyl-(1→4)-β-D-glucuronopyranosyl-(1→2)-α-D-mannopyranosyl with a pyruvate ketal at *O*-4,6 of the β-D-mannopyranosyl unit and a 6-*O*-acetyl group on the α-D-mannopyranosyl unit, although not all trisaccharide substituents of the main cellulosic chain contain linked pyruvate and acetate units.

Processes for making a water absorbent from xanthan gum and other carboxylated polysaccharides by crosslinking with an epoxide have been developed [96,97]. Thus, a mixture of 50 g of xanthan gum, 0.1 g of epichlorhydrin, 500 mL of water and 10 mL of 5% potassium hydroxide was left to react for 16 hours, at 30°C, to produce an absorbent product.

Chitin is a linear polymer of N-acetyl-2-amino-2-deoxy-D-glucopyranosyl units linked by β-D-(1→4) bonds. Chitin is the structural polysaccharide in the mantle or exoskeleton of insects and many marine creatures such as shrimp, lobster, and crabs. Chitin is insoluble, as is cellulose, but like cellulose, can be solubilized or modified by derivatization to produce aqueous gels. For example, a water absorbent was prepared from deactylated chitin (chitosan) by hydroxyethylation and crosslinking with formaldehyde [98]. Chitin or chitosan can be modified in a variety of ways, but these generally follow the methods used to produce water-absorbing products from cellulose. Since chitin is more costly than cellulose, its use has not been exploited by industry.

Algin, another natural gum, is a linear polymer of D-mannuronic acid and L-glucuronic acid units. Proportions of these units in various seaweeds differ depending on the source, but D-mannuronic acid is usually 60% of the polysaccharide. D-mannuronic acid units are connected by (1→4)-β-D-linkages and the L-glucuronic acid units are connected by (1→4) linkages, presumably in the α-configuration. Algin or sodium alginate obtained from *Macrocystis perifera*, the giant brown kelp, is widely employed in the food industry as a thickener or suspending agent, but it has special applications where it makes strong gels by complexing with polyvalent metal ions, particularly calcium.

Sodium alginate dissolves rapidly in water, but when mixed with a calcium salt forms a strong gel. The calcium ions form interchain ionic bonds with the carboxyl groups and coordinate with hydroxyl groups between two chains [99]. The powder can be used with fluffed pulp as a fluid absorbent. In this case, the fluid is transported by the fibers mixed with sodium alginate-salt mixture. Ideally, the sodium alginate dissolves quickly and is set to a gel by rapid release of calcium ions. Conveniently, the calcium ions are released from a complexing salt, such as calcium citrate.

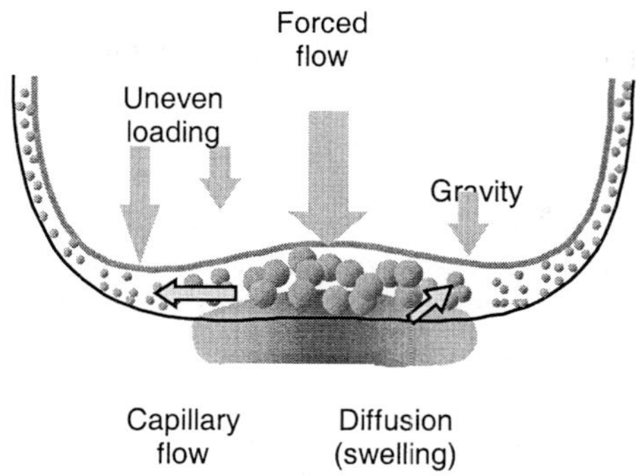

Figure 14: Primary events in fluid absorption and distribution in a diaper.

4. PERFORMANCE AND EVALUATION

A baby diaper represents the primary application for the materials discussed in this chapter. The absorbent portion is a composite pad containing cellulose fluff mixed with granules of superabsorbent polymer. Figure 14 illustrates some of the more significant events that occur in a modern superabsorbent containing diaper during use. Fluid flow in these structures is a complex process; at least three mechanisms operate:

1-An initial insult forces fluid into the porous medium
2-Capillary forces distribute the fluid by wicking
3-Superabsorbent polymer imbibes fluid and swells

These events occur at different time scales but interact with one another. As an example of this type of interaction, consider the swelling superabsorbent. This swelling can open up the structure of the cellulose fluff composite, allowing easier flow; or it can gel block and increase the resistance. Which process dominates determines the performance of the diaper [100].

In the quest for improved performance of personal care products, a variety of evaluation tests for these complicated composite structures have been developed. These methods are important for product design, but outside the scope of this chapter. We will restrict our discussion here to only methods used for evaluation of the superabsorbent resins, particularly those tests cited in the patent art as important for distinguishing one material from another. Two general sources for greater detail on these tests are the chapter by Cutié et al [101] in reference 6 and the methods compilation by the trade group representing European nonwovens producers (EDANA) [102].

Before discussing the tests in detail, however, the nature of the test fluid should be considered. Although this chapter deals only with *aqueous* absorbents, the majority of product is not used to absorb pure water, and the appropriate fluid to use for evaluation will vary depending on the application. Evaluations for baby diapers and adult incontinence products, applications which account for the largest share of superabsorbent production, are often conducted using synthetic urine. Recipes for such these simulants have appeared in patents [103], and commercial quantities can be purchased [104]. However, according to some patents, a 0.9% NaCl solution is often adequate as an approximation for laboratory testing [105]. More complex body fluids are used directly or simulated for evaluation of superabsorbents; selected simulants will be cited in the section on applications.

For agricultural and other outdoor applications, pure water may indeed be the fluid of choice. However, mineral content, especially multivalent ions, will have a profound effect on the swelling of gels, as noted in an earlier section. Performance in sea water (approximately 3% NaCl) is of interest for testing packing of undersea cables.

4.1. Swelling and Modulus

4.1.1. Swelling Capacity. A cube of a polymeric gel, when placed in an excess of swelling fluid, imbibes the fluid and eventually reaches its equilibrium swelling capacity. Its final dimensions can be relatively easily measured and its volume calculated. The ratio of this swollen volume compared to the original volume of the cube is the equilibrium swelling capacity. Even easier, because the density of a highly swollen gel is about that of the fluid itself, the sample can be weighed before and after swelling to obtain an accurate measure of its capacity. Yin [7], for example, made his swelling measurements in this way.

In general, however, the superabsorbent products of commerce are granular, perhaps 400 microns in diameter and often quite irregular in shape. One particle is not a very representative sample, and for a collection of thousands of such particles as comprise even 0.1 gm of sample, accurate measurement is difficult. Simply comparing the swollen weight with the dry weight is complicated by the irreproducible drainage of excess fluid from the swollen sample. Considerable fluid remains trapped in the interstices between the swollen granules.

Several ingenious techniques have been used to circumvent this problem. Simple gravity filtration can yield a useful result [41], but the potential for error is significant, particularly when samples are of different particle size. The interstitial fluid is held more strongly in the smaller pores that form between smaller particles. Vacuum filtration has also been used, but channeling can occur in the gel mass or the air flow can actually dry the sample, all leading to poor reproducibility. Use of a rubber sheet, a common practice in vacuum filtration of crystallized solids, prevents channeling but does apply pressure to the gel mass, and this will attenuate absorption. Another method to account for the interstitial fluid is by adding a large dye molecule (e.g., blue dextrin) to the swelling fluid. Because of its dimensions, this molecule is assumed to be excluded even as the salt and water enter and swell the gel. The change in concentration in the supernatant can be measured optically and the swollen volume calculated [106].

Perhaps the most reliable yet convenient method for determining swelling capacity employs centrifugation [105,107]. A small sample of dry polymer granules is sealed inside a "tea bag". After swelling in an excess of fluid for the appropriate time, typically 30-60 minutes, the bag is placed in a laboratory centrifuge and spun. This could be considered an enhanced gravity filtration, increasing the force of separation. Reliable values are obtained by this method and variations on it are the standard procedure [108].

4.1.2 Modulus of Beds. Utilizing a mechanical spectrometer for measuring the modulus of the monolithic piece of gel illustrated in Figure 6 can give reproducible values if certain precautions are taken [7]. Because the gel is elastic and will not flow, the oscillatory mode must be used, and the amplitude and frequency are important. The amplitude must be maintained within the elastic limit of the material and the frequency must be appropriate for the molecular motion of interest. Slippage between the gel slab and either platen must be avoided, and this can be more difficult if the normal force on the sample causes exudation of fluid to this interface. This exudation occurs when the osmotic pressure retaining the fluid is exceeded, and for this reason gels swollen only up to slightly below their equilibrium capacity should be measured. At equilibrium *any* pressure will cause a liquid slip layer to form, yielding an erroneous value for torque.

Using the method for slabs to measure the modulus of a bed of granular gel is more difficult [106,109]. When the specimen is a collection of swollen particles, the major issue is distinguishing the particle-particle cohesion from the true modulus of the gel that comprises the particles. The fundamental approach is to pay even more attention to the amplitude of the oscillation. If the amplitude is sufficiently small, the particles cannot move past one another, nor even strain the interface. If there is sufficient modulus to produce a discernible signal at this low amplitude (The fundamental signal picked up by this device is that from a strain gauge in the nonmoving platen.), Then the modulus can be determined.

4.1.3. Swelling under Load. Another category of tests has been developed, the results of which depend on *both* swelling capacity and modulus; these are referred to as absorption under load or under pressure. In these evaluations, the gel is constrained by a weight or other application of pressure and its ability to resist this during swelling is determined. This type of demand absorbency test has been used for physical absorbents, *e.g.*, paper towels, for years, and a review of many of these methods can be found in the first edition of this monograph [110]. One such device is the gravimetric absorbency testing system described by McConnell [111], based on principles of a Swedish standard method [112]. Adapting this sort of method to superabsorbents, Kellenberger [113] and later Melius [114] showed correlation between improved absorption under load and favorable performance in diapers. Nagorski [115] corroborated this conclusion in pads. The method recommended by EDANA [116] uses a petri dish to hold an excess of fluid; the methods in the previous references use a supply tube from a constant head reservoir.

To better appreciate these conclusions, it is useful to consider in detail what happens during the swelling process. In a typical test, a monolayer of particles is placed on a screen and a load is applied, for example a weighted piston. Fluid must first wet these particles, be absorbed and develop sufficient osmotic force to raise the imposed load. As the particles continue to swell the path must be maintained for the fluid to move through the bed. When the distortion of the swelling particles leads to gel blocking, the flow stops, limiting the swelling. Sufficient osmotic pressure may still remain in the system to draw in more fluid that would force the piston up, but it can only enter via molecular diffusion through the tightly packed gel bed; this is much too slow a process to be functional in a personal care device. The test duration for superabsorbents used in these products is generally one hour or less. Gel blocking more quickly than this leads to low absorption values. A higher modulus in the particles allows them to resist deformation, maintaining open capillaries and leading to greater value for swelling under load. Of course gel particles that are very stiff because they swell very little yield low values as well. In a series of samples developed with one technology, there tends to be an optimum for absorption under a given load, but which will vary with that load.

4.1.4. Wicking and the Permeability of Gel Beds. As noted in Section 2.1, wicking is the primary method of moving fluids in absorbent structures. The rate of rise of a liquid front is a common measure for physical absorbents [110]. Wicking in superabsorbent beds is also discussed in patents [105]. Complications arise in these experiments because the superabsorbents swell so much that the capillaries become closed off and wicking flow stops. This phenomenon is often referred to as "gel blocking". Important factors affecting wicking in beds of superabsorbents are particle size and distribution and surface hydrophilicity. Surface tension of the fluid is also important and is sometimes affected by materials originally associated with the polymer [117].

Permeability of an already swollen bed of gel particles is a simpler test than wicking with simultaneous swelling. Measuring permeability in porous media in general has a large literature. One description of such a measurement for a superabsorbent system is by Amiya [118].

4.1.5. Kinetics of Swelling. Data from any of the swelling tests can of course be recorded as a function of time; for example, an AUL device fitted with a recorder has been described [101]. References to more elaborate tests are also noted.

A simple test for rapidly screening of multiple samples is the vortex test [119]. In this method, dry polymer is added to a fluid while it is being stirred in a beaker with a magnetic stirrer, such that a vortex is maintained. After addition, the swelling polymer thickens the fluid, causing the vortex to disappear. The time for this to occur is the "vortex time." Primary factors affecting the swelling rate are the equilibrium capacity of the polymer and its surface area. The latter is in turn primarily related to the particle size distribution, but surface roughness and porosity can be quite important, leading to much higher absorption rates than expected [61,62]. This test is not particularly well suited for fibers, which are of interest in part because their small dimensions can lead to quite rapid swelling [77].

4.2. Chemical Analysis

4.2.1. Extractable and Residual Analyses. The content of uncrosslinked material in superabsorbents has been claimed to affect performance directly [106], but in any case can serve as a marker for network connectivity. Measurement methods usually comprise swelling in an excess of solvent and allowing enough time for diffusion of the free polymer into the supernatant. Separation of the excess fluid from the swollen gel and determination of the soluble polymer content in the fluid allows the calculation of the amount of extractable polymer, generally expressed as a weight percent of the original sample.

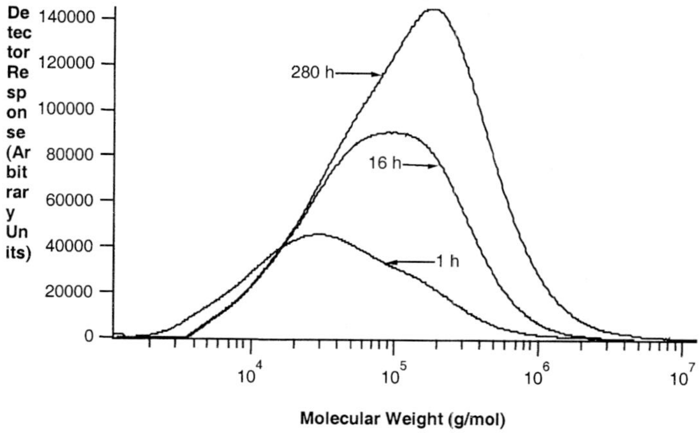

Figure 15: Molecular weight of extracted polymer as a function of extraction time [122].

Polymer in the filtrate or supernatant can be quantified by evaporation and weighing, titration [106], gel permeation chromatography [120], or other method. As commercial products have improved over the years, the extractable polymer content has generally decreased, often below 10%. This leads to the need for rather precise methods to measure such small amounts of material. For example, evaporation is rarely used because the solvent of choice for swelling is 0.9% NaCl, and the salt content dwarfs the polymer content. EDANA recommends a titration method [121].

For very accurate work, chromatography is preferred [120]; this can have the advantage of providing molecular weight data as well. This approach was used to obtain an interesting set of data for a particular network by Allan [122], previously presented in reference 6. Diffusion of the soluble material from the domain of the swollen gel takes significant time, and the shorter chains diffuse more rapidly than the longer chains. At some molecular size the diffusion time is so long as to be indistinguishable from gel.

Unreacted monomers can be extracted into surrounding fluid just as soluble polymer, indeed even more quickly. Analysis of the extract for acrylic acid, the monomer of greatest interest, has been reported with both gas chromatography (GC) [123] and high pressure liquid chromatography (HPLC) [24]. The latter method is preferred because of the lack of thermal stability of acrylic acid makes quantification inaccurate by GC [124,125].

4.2.2. Backbone Molecular Weight. Discussing the "molecular weight" of a crosslinked gel is generally not appropriate. When crosslinked beyond the gel point, the system is formally considered to have an infinite, or at least indefinitely large, molecular weight; the sample becomes a continuum of gel [16]. The more useful parameter for a gel is the molecular weight between crosslinks (see M_c in Eq 10); this is essentially another measure of crosslink density.

Although crosslink density is the parameter that has the greatest impact on the properties of a gel, the molecular weight of the backbone polymer (see M in Eq 10), independent of crosslinks, can often provide useful information as well. During polymerization of the gel, the molecular weight of the developing polymer is largely controlled by the factors that would determine the molecular weight of the analogous non-crosslinked system. Only an occasional di- (or tri-) functional monomer is incorporated into the chain, leading to a covalently crosslinked gel.

As illustrated in Figure 7, these crosslinkers are often esters, *e.g.*, diethylene glycol diacrylate. In these cases, hydrolysis of the ester bonds of the crosslinks will leave the backbone polymer free. Measurement of the molecular weight of the hydrolyzate is then done by techniques noted above for extractable polymers [126].

4.2.3. Chemical Composition. Information about the elemental composition of superabsorbents can be obtained by such methods as atomic absorption, X-ray fluorescence, or mass spectroscopy. Other than which counterion and how much was used, however, little discriminating information about the chemistry of the structure can be obtained this way. Novel chemistry in either the backbone or the crosslinks is usually difficult to ascertain from an elemental approach; C, H, and O dominate and the ratios are not sensitive to monomer choices, considering the requirement of hydrophilicity.

Sodium analysis, however, has been used as a marker for superabsorbents in at least two areas. Soft X-ray analysis of composites has been used to locate and measure the superabsorbent content in unknown structures [127]. Quantification of airborne superabsorbent dust has been reported [128]. When adventitious sodium compromises this method, an alternative ion exchange method has been used [129].

At the level of the chemical groups rather than elements, the classical approach is infrared (IR) spectroscopy. Ultraviolet (UV) spectroscopy is less useful because of the preponderance of carbonyl groups, which absorb strongly in the UV and mask any detail. An IR spectrum of both polyacrylic acid and partially neutralized polyacrylic acid can be found on page 123 of Reference 6. This technique is only useful for gross identification, however, because chemical differences of interest, such as type or amount of crosslinker used, are indistinguishable from the background.

Recently nuclear magnetic resonance (NMR) spectroscopy has been shown useful for following the conversion of monomer to polymer. Cutié et al. [130] found proton (^{1}H) preferable to carbon-13 (^{13}C) nmr because of the greater sensitivity, which led to faster acquisition rates and smaller sample sizes, this in turn allowing more truly isothermal conditions. Chemical enhancement of the ^{13}C content of a sample of TMPTA used as a crosslinker allowed study of its copolymerization with sodium acrylate, as reported by Arriola et al. [37].

4.3. Physical Methods

4.3.1. Thermal Analysis. Cutié et al. [101] have reviewed several conventional thermal techniques in reference 6. Typical objectives have been to obtain heat balance data for the scale-up of polymerization and drying operations and to evaluate hazards. Standard procedures for these analyses generally deliver reliable results, and the details will not be elaborated on here.

A novel thermal analysis method of particular interest, however, is the use of differential scanning calorimetry (DSC) to determine the glass transition temperatures (T_g) of gels. For the partially neutralized commercial products, *i.e.*, near pH 7, moisture content is the major determinant of T_g. Dried polymers show broad and complex transitions up to 140°C and higher, whereas samples with 40% moisture have a rather distinct transition centered at 25°C. The temperature primary transition varies approximately linearly with moisture on a dry weight basis. These techniques have largely been developed by A.J. Pasztor, Jr., of The Dow Chemical Company, but have not yet been published.

4.3.2 Particle Size and Distribution. Average particle size and the distribution of particle sizes can become significant factors in performance. It is not surprising that small particles reach their equilibrium swelling capacity much faster than large particles. The sizes of the dry particles may be important for certain device designs. Fine particles may migrate throughout the composite, or large particles may be palpable. For reasons such as these, measurement of the particle size distribution of a particular sample is often important [109]. Standard sieving techniques are most commonly used, as detailed in the EDANA test methods [131].

4.3.3. Microscopy. Simply visualizing various superabsorbents is frequently informative. Both optical and electron microscopy are commonly used. Cutié et al. [101] have reviewed this area and several images can be found throughout reference 6. Kim [51] illustrated the value of dyes to visualize the core/shell structure of various commercial superabsorbents. Wilson [52] investigated similar structures with photons from a linear accelerator.

5. APPLICATIONS

5.1. Personal Care

By far the majority of global superabsorbent production goes into the personal care market. According to Kuster [132] this share is 96%, of this, 80-85% is used in disposable baby diapers. Most of the balance is used in absorbent products for adults, but this share is increasing as the population ages. External feminine hygiene products consume a smaller fraction.

Disposable baby diapers were first commercialized in the United States by Procter & Gamble during the 1950s, the invention attributed to the late Vic Mills [133]. Early products were generally reserved for use only when travelling or under other non-routine situations, in part due to cost. Common, everyday usage grew in importance as the scale of production increased and unit costs decreased. Relative availability of pulp influenced usage; market penetration occurred sooner in heavily timbered Scandinavian countries than many places. These early products were essentially fluff pads with various configurations of back sheets to hold the diaper on the baby and protect from leakage.

An early patent for the use of superabsorbents in diapers was by Harper [134]. The osmotic storage capability of the gel immobilized urine more effectively than a did a capillary absorbent, such as fluff, alone, and this measurably reduced leakage. The first commercial products containing superabsorbent were sold in Japan in the late 1970s, but were not commercialized in the US until 1984. The most basic construction is a composite pad of fluff and powdered superabsorbent contained in tissue and bonded to an impermeable backsheet. A nonwoven topsheet is usually positioned next to the baby's body.

The original products were functional but, because they still contained considerable cellulose fluff, were bulky. In order to improve the fit, which is both an attractive feature in its own right and also reduces leakage, manufacturers sought to develop thinner diapers. Because the volume of fluid naturally remains constant this tactic requires higher loadings of superabsorbent relative to the fluff. The first observations were that simply adding more polymer was of limited value because of gel blocking, aggravated by crowding of the particles. It was found that gel blocking was mitigated by surface crosslinking and related strategies to develop a "core/shell" particle [42,113,115]. Dramatic improvement in diaper performance was obtained through the introduction of products containing these so-called "second generation" superabsorbents [Some authors refer to this as "third generation"]. These polymer innovations, from many suppliers, directly enabled thinner diapers to be commercialized. More recent innovations in diapers have focused on closures, flaps, backsheets, and fluid distribution, among other features. Although critical to overall performance and of major competitive importance, these aspects do not directly concern superabsorbents and are thus outside the scope of this chapter.

Demographics suggest the elder fraction of the population, regardless of the cutoff age chosen, will increase for the next several decades. At the same time the elderly are staying healthier and living more active lives. These factors combine to lead one to predict significant growth in the market for products that can facilitate normal living for people faced with the incontinence arising from a variety of causes in the senior years. Although personal care products for adult incontinence have may similarities with diapers, there are particular issues, including the following:

Size, each unit is larger both in dimensions and amount of absorbent material. Speed of absorption, volume flow rates for adults can be much larger than for babies. Far more concern with discreetness, both with respect to containment and odor.

These are primarily technical differences, but there are significant marketplace differences as well. For example, there is a substantial institutional market for adult incontinent devices as well as retail, whereas for infants, the institutional portion is a much smaller fraction. Even the retail sales of adult products require different emphases; for example, there is a significant home delivery business for adults, eliminating the need to make the more public purchase at a drug or grocery store. Usage of such products by an individual may extend over many years, as compared to the roughly two years for an infant. A final significant consideration is that these (sold at retail) are primarily purchased by the user, not a surrogate, as is the case for baby diapers. The patent literature rarely cites superabsorbents restricted for use in adult incontinence products, but absorption speed is emphasized in device patents [135].

Immobilizing menses is a far more complicated task than absorbing urine. The composition of this fluid, while always complex, varies among individuals and over time for a given individual. In order to do the extensive laboratory work necessary for product development, synthetic test fluids are useful. Achter et al. [136] describe a synthetic menses fluid that utilizes egg white to mimic the proteinaceous mucin. Superabsorbent developed particularly for feminine hygiene have been occasionally been cited in patents. Modifications include surface treatments [137], specific counterions [138], and hydrogen bonding compounds [139] have been described.

5.2. Other Applications

Maintaining moisture in soil becomes an important goal of agriculture as greater production demands are placed on more marginal soils. Superabsorbent at levels of around 0.1% can be added to improve moisture retention. With heavy soils, the texture is often improved as well. Buchholz [140] covers this and other applications in greater detail and cites Kazanskii and Dubrovskii [141] as a comprehensive review on the behavior of superabsorbents in soil. The ion content of the water in the soil is highly variable, and this can affect moisture retention performance. High concentrations of divalent ions such as calcium can collapse polyelectrolyte gels. For this reason, nonionic polyacrylamides are often used in these applications. Shimomura and Namba [142] report a much increased yield with a sulfonated polymer, presumably because of its lesser sensitivity to multivalent ions. These authors also note the advantage of large particle size absorbents, up to 1-3 mm, for agricultural applications.

Water intrusion into underground and underwater electrical cables leads to current drainage if slight or can create a serious safety hazard if significant. Communication cables, although typically operating at lower current densities, carry more "expensive" information, and even modest degradation of interwire insulation can lead to crosstalk and eventually to total disfunction. Even optical cables are compromised by water. Superabsorbent polymers have been used as a component in the packing of such cables, offering a barrier to rapid water migration when swollen tightly by the initial insult of moisture [143]. A version using a printable superabsorbent recently appeared [144]. In reference 142 systems are described with greatly reduced water intrusion rates, and more detailed information by Hogari and Ashiya [145] can be found in the same monograph. Other applications for superabsorbents in construction to prevent water migration have been reviewed by Buchholz [140].

Sorbents for body fluids in medicine are of growing importance. A patent for a product to immobilize medical waste describes the use of multivalent metals and hydrophobic coating [146]. Systems specific for wound care have been developed [147] and for ostomy bags [148].

Management of both liquid water and water vapor is of interest in food packaging, and Buchholz has recently summarized this area [140]. Controlled release and drug delivery are advanced applications for hydrogels, some of which are ionic and thus superabsorbents. Several specific reviews treat this field in more detail [3,13,149].

6. CONCLUDING REMARKS

Superabsorbent polymers, particularly partially neutralized sodium polyacrylate, have become a high volume chemical since their commercialization less than 20 years ago. They are used in many personal care products, and most people will enjoy the convenience they bring many times during their life. Applications beyond personal care are becoming relatively more important as growing production capacity increases availability.

As ubiquitous as these materials may become, the fundamental understanding of the polyelectrolyte networks that are commercial superabsorbents is still incomplete. The combination of the unique solvent properties of water and the intractable nature of polymer networks will allow these complex systems to remain a challenge for polymer science for some time.

7. GLOSSARY

ΔF	Free energy of the system
ΔF_{elas}	Free energy constituent of elastic restraint
ΔF_{mix}	Free energy associated with mixing in polymer solution
ΔF_{os}	Free energy associated with osmotic effect
ΔH_{mix}	Heat of mixing
k	Boltzman's constant
M	Molecular weight of polymer backbone, i.e. that of the uncrosslinked system
M_C	Molecular weight between crosslinks
n_i	Number of moles of component *i*
q	Swelling volume
ΔS_{mix}	Entropy of mixing
V	Gel volume of swelled crosslinked polymer
V_0	Initial volume of polymer
α^3	Linear deformation factor
v_i	Molar volume fraction of component *i*
v_{sol}	Molar volume of solvent
v_{sp}	Specific volume of polymer
v_e	Number of elastic chains, proportional to crosslink density
x_i	Mole fraction of component *I*
X	Flory-Higgins interaction parameter

8. REFERENCES

1. *Chemical Week*, July 24, 1974, p21.
2. The Dow Chemical Company
3. L. Brandon-Peppas, R. S. Harland, eds., *Absorbent Polymer Technology*, Elsevier Science Publishers, Amsterdam (1990).
4. R. S. Harland, R. K. Prud'homme, eds., *Polyelectrolyte Gels: Properties, Preparation, and Applications, Symposium Series 480*, American Chemical Society, Washington, D.C. (1992).
5. F. L. Buchholz, N. A. Peppas, eds., *Superabsorbent Polymers, Science and Technology, Symposium Series 573*, American Chemical Society, Washington, D.C. (1994).
6. F. L. Buchholz, A. T. Graham, eds., *Modern Superabsorbent Polymer Technology*, Wiley-VCH, New York (1998).
7. Y.-L. Yin, *Swelling Behavior of Polyelectrolyte Gels*, Ph.D. Thesis, Princeton University, Princeton NJ (1993).
8. U. Dayal, S. K. Mehta, M. S. Choudhary, R. C. Jain, *J. Macromol. Sci., Rev. Macromol. Chem. Phys.*, C39, 507 (1999).
9. J. Crank, *The Mathematics of Diffusion, 2nd ed.*, Clarendon Press, Oxford (1975).
10. F. W. Billmeyer, Jr., *Textbook of Polymer Science*, John Wiley and Sons, New York (1965), p 25.
11. G. Gee, *Trans. Faraday Socy.*, 42B, 33 (1946).
12. N. A. Peppas, R. E. Benner, *Biomaterials*, 1, 158 (1980).

13. T. Tanaka in *Polyelectrolyte Gels: Properties, Preparation, and Applications, Symposium Series 480*, R. S. Harland, R. K. Prud'homme, eds., American Chemical Society, Washington, D.C. (1992).
14. e.g., A. A. Collyer, D. W. Clegg, eds., *Rheological Measurement—Second Edition*, Kluwer Academic Publishers, Norwell, Massachusetts (1998).
15. P. W. Atkins, *Physical Chemistry, 3rd ed.*, W. H. Freeman and Company, New York (1986).
16. P. J. Flory, *Principles of Polymer Chemistry*, Cornell University Press, Ithaca (1953)
17. I. Hasa, M. Ilavsiky, K. Dusek, *J. Polym. Sci.*, 13, 253 (1975).
18. W. Oppermann in *Polyelectrolyte Gels: Properties, Preparation, and Applications, Symposium Series 480*, R.S. Harland, R.K. Prud'homme, eds., American Chemical Society, Washington, D.C. (1992).
19. G. Manning, *J. Chem. Phys.*, 51, 924 and 934 (1969).
20. Y.-L. Yin, R. K. Prud'homme, F. Stanley in *Polyelectrolyte Gels: Properties, Preparation, and Applications, Symposium Series 480*, R.S. Harland, R. K. Prud'homme, eds., American Chemical Society, Washington, D.C. (1992).
21. F. L. Buchholz in *Modern Superabsorbent Polymer Technology*, F. L. Buchholz, A. T. Graham, eds., Wiley-VCH, New York (1998).
22. A. T. Graham, L. R. Wilson in *Modern Superabsorbent Polymer Technology*, F. L. Buchholz, A. T. Graham, eds., Wiley-VCH, New York (1998).
23. W. Bauer, Jr., "Acrylic Acid and Derivatives," *Kirk-Othmer Encyclopedia of Chemical Technology, 4th Ed., Vol 1*, J. I. Kroschwitz, M. Howe-Grant, eds. John Wiley & Sons, Inc., New York (1991).
24. S. S. Cutié, G. J.Kallos, P. B. Smith, *J. Chromatogr.*, 408, 349, (1987).
25. F. M. Wampfler, *Formation of Diacrylic Acid During Acrylic Acid Storage*, AIChE Presentation, Houston (March 1987); F. M. Wampfler, *Plant/Operations Prog.*, 7, 183 (1988).
26. T. Tsubakimoto, T. Shimomura, Y. Irie, US 4,666,983 (1987).
27. K. Dahmen, R. Mertens, DE 4020780 (1991).
28. K. Dahmen, H. Redlof, DE 3713601 (1988).
29. S. H. Ganslaw, H. G. Katz, US 4,043,952 (1977).
30. L. T. Uhl, *Rheological Studies of Water–Soluble Polymer Solutions with Interacting Solutes*, Ph.D. Thesis, Princeton University, Princeton NJ (1983).
31. K. Nagasuna, et al., EP 605,150 (1994).
32. H. Ito, S. Shoji, S. Shigeari, *J. Chem. Soc. Jpn. Ind. Chem. Sect.*, 58, 194 (1955).
33. V. A. Kabanov, D. A. Topchiev, T. M. Karaputadze, *J. Polym. Sci. Symp.*, 42, 173 (1973).
34. V. A. Kabanov, D. A. Topchiev, T. M. Karaputadze, L. A. Mkrtchian, *Eur. Polym. J.*, 11, 153 (1975).
35. S. P. Manickam, K. Venkatarao, N. R. Subbaratnam, *Eur. Polym. J.*, 15, 483 (1979).
36. S. S. Cutié, P. B. Smith, D. E. Henton, T. L. Staples, C. Powell, *J. Polym. Sci.: Part B: Polym. Phys.*, 35, 2029 (1997).
37. D. J. Arriola, S. S. Cutié, D. E. Henton, C. Powell, P. B. Smith, *J. Polym. Sci.*, 63, 439 (1997).
38. L. Siddall, T. Johnson, US 4,833,222 (1989).
39. L. Svarz, US 3,732,193 (1973).
40. D. R. Chambers, et al., US 5,145,906 (1992).
41. F. Masuda, K. Nishida, A. Nakamura, US 4,076,663 (1978).
42. T. Tsubakimoto, T. Shimomura, US 4,625,001 (1986). Ref 32
43. Anonymous, *RD 38,514 Res. Disclosures*, 385, 288 (1996).
44. H. Takeda, Y. Taniguchi, US 4,525,527 (1985).
45. W. Alexander, M. Teppo, US 4,820,742 (1989).
46. Y. Irie, K. Iwasaki, T. Hatsuda, K. Kimura, N. Harada, K. Ishikazi, T. Shimomura, T. Fujiwara, US 4,920,202 (1990).
47. A. T. Graham, H. Gartner, US 5,342,899 (1994).
48. Y. Irie, K. Kajikawa, H. Takahashi, T. Fujiwara, EP 454,497 (1991).
49. T. Sutton, K. Jacob, T. Lim, S. Sandor, US 5,064,582 (1991).
50. G. T. Woodrum, T. H. Majette, EP 463,388 (1992).
51. Y.-S. Kim, Conference Proceedings, INDEX 99, EDANA, Geneva.
52. L. R. Wilson, G.E. Mitchell, Conference Proceedings, INSIGHT 99, Marketing/Technology Services, Inc. Kalamazoo, Michigan.
53. T. Tsubakimoto , et al., US 4,734,478 (1988).

54. F. Engelhardt, et al., US 5,331,059 (1994).
55. J. H. Burgert, H. A. Gartner, S. B, Christensen, F. L. Buchholz, T. C. Johnson, A. T. Graham, WO 9420547A (1994).
56. M. Nakamura, S. Obayashi, T. Yamamoto, T. Nakanishi, H. Tanaka, Y. Sakamoto, US 4,683,274 (1987).
57. K. Nagasuna, et al., US 4,973,632 (1990).
58. H. Aoyama, T. Moriya, S. Kondo, US 5,210,159 (1993).
59. K. Masamizu, S. Hirokawa, F. Sanae, US 4,742,086 (1988).
60. S. Hirokawa, et al. DE 3,617,311 (1987).
61. F. W. Stanley, J. C. Lamphere, L.R. Wilson, US 5,744,564 (1998)
62. E. F. Tai, US 5,374,684 (1994).
63. C. J. Long, Z. Amjad, W. F. Masler, W. H. Wingo, US 5,288,814 (1994).
64. T. J. Romack, E. E. Maury, J. M. DeSimone, *Macromolecules* 1995, 912.
65. J. R. Gross, R. T. McFadden, US 3,926,891 (1975); J. R.. Gross, J. L. Potter, F. B. Kassell, US 4,026,932 (1977); R. M. Krajewski, R. E. Erickson, US 4,186,233 (1980); R. E. Erickson, US 4,293,609 (1981); R. E. Erickson, US 4,117,184 (1978).
66. N. D. Burkholder, US 4,076,673 (1978)
67. A. S. Allen, D. Farrar, P. Flesher, EP 269,393 (1988), US 4,861,539 (1989), 4,962,172 (1990), US 5,280,079 (1994).
68. A. S. Allen, US 5,147,956 (1992).
69. See, for example, A. Katchalsky, S. Lifson, H. Eisenberg, *J. Polym. Sci.*, 7, 571 (1951).
70. A. W. Gross, J.C. Dobson, "Methacrylic Acid and Derivatives," *Kirk-Othmer Encyclopedia of Chemical Technology, 4th Ed., Vol 16*, J.I. Kroschwitz, M. Howe-Grant, eds. John Wiley & Sons, Inc., New York (1995).
71. F. L. Buchholz, "Polyacrylamides and Poly(Acrylic Acids)," *Ullmann's Encyclopedia of Industrial Chemistry, Vol. A21*, VCH Verlagsgesellshaft mbH, D-6940 Weinheim (1992).
72. E. Arvanitidou, D. Hoaglund, D. Smisek, in *Polyelectrolyte Gels: Properties, Preparation, and Applications, Symposium Series 480*, R. S. Harland, R. K. Prud'homme, eds., American Chemical Society, Washington, D.C. (1992).
73. A. T. Andrews, *Electrophoresis: Theory, Techniques, and Biochemical and Clinical Applications*, Oxford University Press, New York (1986).
74. J. B. Clarke, EP 101,253B (1987).
75. A. Holst, E. Perplies, G. Buchberger, DE 2,903,267 (1980); Chem Abstr., 93: 169591z; US 4,337,328 (1982).
76. K. Ban, S. Takegami, M. Wakitani, US 4,442,173 (1984).
77. D. S. Allan in *Modern Superabsorbent Polymer Technology*, F. L. Buchholz, A. T. Graham, eds., Wiley-VCH, New York (1998).
78. R. N. Basaw, B. L. Atkins, US 3,983,095 (1976).
79. J. R. Gross, US 3,954,721 (1976).
80. B. Le-Khac, EP 272,074 (1987), US 4,731,067 (1988), US 5,026,784 (1991).
81. J. R. Gross, US 3,966,679 (1976), US 3,980,663 (1976), US 4,017,653 (1977), US 4,018,951 (1977).
82. J. R. Gross, US 3,993,616 (1976).
83. K. C. Low, A. P. Wheeler, L. P. Koskan, *Adv. Chem. Ser.*, 248, 99 (1996).
84. Nippon Synthetic Chemical Industry Co., Ltd., Jpn. Kokai Tokkyo Koho, 55 (1980) 099925; Chem. Abstr., 93: 221470a.
85. Nippon Synthetic Chemical Industry Co., Ltd., Jpn. Kokai Tokkyo Koho, 55 (1980) 132713; Chem. Abstr., 94: 32079q.
86. N. A. Peppas in *Hydrogels in Medicine and Pharmacy, Vol. 2*, N. A. Peppas, ed., CRC Press, Boca Raton FL (1986).
87. P. A. King, US 3,264,202 (1966).
88. P. G. Assarsson, P. A. King, US 3,993,553 (1976).
89. N. B. Graham in *Hydrogels in Medicine and Pharmacy, Vol. 2*, N.A. Peppas, ed., CRC Press, Boca Raton FL (1986).
90. A. Holst, M. Kostrzewa, US 4,061,692 (1977).

91. A. M. Goldstein, E. N. Alter, J. K. Seaman, in R. L. Whistler, (Ed.), *Industrial Gums, 2nd Ed.*, Academic Press, New York, 1973.
92. D. Torr, US 3,645,836 (1972).
93. D. Torr, US 3,903,889 (1975).
94. E. G. Muller, Belg. Patent No. 886,685 (1981); Chem Abstr., 95: 117400h.
95. W. H. McNeely, K. S. Kang, in R. L. Whistler, (Ed.), *Industrial Gums, 2nd ed.*, Academic Press, New York, 1973.
96. Sanyo Chemical Industries, Ltd., Jpn. Kokai Tokkyo Koho, 56 (1981) 005138; Chem. Abstr., 94: 158680x.
97. Sanyo Chemical Industries, Ltd., Jpn. Kokai Tokkyo Koho, 56 (1981) 005137; Chem. Abstr., 94: 141560k.
98. S. Iwase, Jpn. Kokai Tokkyo Koho, 54 (1979) 011955; Chem. Abstr., 90: 170484m.
99. D. A. Rees, *Adv. Carbohydr. Chem. Biochem.*, 24, 267 (1969).
100. T. K. Das, *AICHE Symposum Series No. 319*, 94, 126 (1998).
101. S. S. Cutié, P. B. Smith, R. E. Reim, A. T. Graham in *Modern Superabsorbent Polymer Technology*, F. L. Buchholz, A. T. Graham, eds., Wiley-VCH, New York (1998).
102. EDANA Recommended Test Methods, February 1999.
103. T. A. DesMarais, K. J. Stone, H. A. Thompson, G. A. Young, G. D. Lavon, J. C. Dyer, US 5,260,345 (1993).
104. For example, Jayco Pharmaceuticals, Mechanicsburg PA, as cited in US 5,260,345.
105. S. K. Byerly, D. G. Iwanski, S. R. Kellenberger, J. Qian, W.-H. Shih-Shroeder, K. A. Szymonski, C.-L. Tsai, S. M. Yarbrough, EP 532,002B1 (1997).
106. K. E. Brandt, S. A. Goldman, T. A. Inglin, US Re 32,649 (1988).
107. F. H. Lahrman, C. J. Berg, D. C. Roe, US 5,149,334 (1992).
108. *Centrifuge Retention Capacity in Saline by Gravimetric Determination*, EDANA 441.1-99
109. S. A. Goldman, J. Horney, H. Retzsch, US 5,061,259 (1991).
110. P. K. Chatterjee, H. V. Nguyen, in *Absorbency*, P. K. Chatterjee, ed., Elsevier Science Publishers B.V., Amsterdam (1985), p 60.
111. W. J. McConnell, US 4,357,827 (1982).
112. Svensk Standard SZS 251228, Swedish Textile Research Institute (1971).
113. S. R. Kellenberger, Eur. Patent Appl. 339,461 (1989).
114. M. K. Melius, et al., Eur. Patent Appl. 615,736 (1994).
115. H. Nagorski, in *Superabsorbent Polymers: Science and Technology, ACS Symposium Series 573*, F. L. Buchholz, N. A. Peppas, eds., American Chemical Society, Washington, D.C. (1994).
116. *Absorbency against Pressure by Gravimetric Determination*, EDANA 442.1-99
117. J. D. DeVrieze, T. L. Staples, WO 97-30109 (1997).
118. T. Amiya, et al., UK Appl 2267094A (1993).
119. F. L. Buchholz in *Absorbent Polymer Technology*, L. Brandon-Peppas, R. S. Harland, eds., Elsevier Science Publishers, Amsterdam (1990).
120. S. S. Cutié, W. C. Buzanowski, J. A. Berdasco, *J. Chromatogr.*, 513, 93 (1990).
121. *Extractable Polymer Content by pH Titration*, EDANA 470.1-99
122. D. S. Allan, in *Modern Superabsorbent Polymer Technology*, p 56.
123. A. Vairvanmurthy, M. O. Andreae, J. M. Brooks, *Anal. Chem.*, 58, 2684 (1986).
124. A. R. Blanchette, et al., *Liquid Chromatographic Methods for Determining Extractable Monomeric Acrylate in Polyacrylate Absorbents*, Analytical subcommittee, Institute for Polyacrylate Absorbents, Washington, D.C. (1987).
125. *Determination of the Amount of Residual Monomers*, EDANA 410.1-99
126. S. S. Cutié, S. J. Martin, *J. Appl. Polym. Sci.,* 55, 605, (1995).
127. D. F. Ring, J. Pascente, Conference Proceedings, INSIGHT 94, Marketing/Technology Service, Inc., Kalamazoo, Michigan.
128. P. A. Forshey, T. S. Turban, J. S. Lemmo, S. S. Cutié, D. L. Pytynia, *Anal. Chim. Acta*, 298, 351 (1994).
129. W. L. Rigot, S. S. Cutié, *J. Radioanal. Nucl. Chem.*, 244, 115 (2000).
130. S. S. Cutié, D. E. Henton, C. Powell, R. E. Reim, P. B. Smith, T. L. Staples, *J. Appl. Polym. Sci.,*64, 577 (1997).

131. *Particle Size Distribution—Sieve Fractionation*, EDANA 420.1-99
132. J. B. Kuster, *Nonwovens World*, February-March 2000, 53
133. *Cincinnati Enquirer*, November 7, 1997.
134. B. Harper, R. Bashaw, B. Atkins, US 3,669,103 (1972).
135. J. R. Noel, A. Ahr, US 5,439,458 (1995).
136. A. M. Achter, C. S. Leach, J. N. Lindon, H. A. Sorebo, M.G. Weber, US 5,883,231 (1999).
137. K. Strandqvist, WO 9955393 A1 (2000).
138. R. S. Harland, R. T. Shet, S. M. Yarbrough, US 5,241,009 (1994).
139. D. D. Halabisky, M. R. Hansen, US 5,998,032 (1999).
140. F. L. Buchholz in *Modern Superabsorbent Polymer Technology*, F. L. Buchholz, A. T. Graham, eds., Wiley-VCH, New York (1998).
141. K. S. Kazanskii, S. A. Dubrovskii, *Adv. Polym. Sci.*, 104, 97 (1992).
142. T. Shimomura, T. Namba in *Superabsorbent Polymers, Science and Technology, Symposium Series 573*, F. L. Buchholz, N. A. Peppas, eds., American Chemical Society, Washington, D.C. (1994).
143. R. G. Gravely, S. R. Stokes; T. Tanaka; US 5,642,452 (1997).
144. J. Houben, W. Krug, US 6,043,311 (2000).
145. K. Hogari, F. Ashiya in *Superabsorbent Polymers, Science and Technology, Symposium Series 573*, F. L. Buchholz, N. A. Peppas, eds., American Chemical Society, Washington, D.C. (1994).
146. T. W. Honeycutt, US 5,578,318 (1996).
147. J. A. Gent, Can. Pat. Appl. 2238248 (1998).
148. J. A. Gent, G. E. Steer, GB 2301350 (1996).
149. H. Brønsted, J. Kopeček in *Polyelectrolyte Gels: Properties, Preparation, and Applications, Symposium Series 480*, R. S. Harland, R. K. Prud'homme, eds., American Chemical Society, Washington, D.C. (1992).

Absorbent Technology.
P.K. Chatterjee and B.S. Gupta, editors.

CHAPTER IX

Polymer Grafted Cellulose and Starch

V. T. STANNETT[a], G. F. FANTA[b], W. M. DOANE[b] AND P. K. CHATTERJEE[c]

[a] *Chemical Engineering Dept., North Carolina State University, Raleigh, NC (USA)*

[b] *National Center for Agricultural Utilization Research, USDA-ARS, 1815 N. University St., Peoria, IL 61604 (USA)*

[c] *Nutech International Co., 331 McDowell Drive, East Brunswick, NJ 08816 (USA)*

Contents

1. INTRODUCTION

A graft copolymer consists of a polymeric backbone with covalently linked polymeric side chains. In principle, both the backbone and side chains could be homopolymers or copolymers. Graft copolymers are of great interest in the field of absorbency in a number of aspects. Grafting can be carried out in such a way that the properties of the side chains can be added to those of the substrate polymer without changing the latter. Thus cellulose fibers can be grafted with, say, sodium polyacrylate while still maintaining their fibrous nature and most of their mechanical properties. However, with other types of grafting the crystalline nature of the cellulose, for example, can be largely destroyed. This releases the natural absorbency of cellulose as well as adding that of the hydrophic side chains leading to a very high water absorbency. This can be accomplished by a decrystallization procedure after grafting or, in the case of the hydrolyzed grafted products, by the process itself

The chapter, itself, discusses grafting to cellulose and starch and also the so-called cellulose alloys. The latter consist of regenerated cellulose fibers spun in the presence of an existing hydrophilic polymer. The various methods of grafting, including both chemical and radiation initiation, are first discussed in some detail. There follows an account of the properties and other features of the saponifiable grafts continuing with a similar description of the direct grafting of acrylic and methacrylic acids.

Cellulose alloys are then described followed by a detailed discussion of the synthesis of properties of starch grafts and related absorbents. Finally, an account of the applications of the absorbent cellulose and starch grafted products is presented.

2. METHODS OF POLYMER GRAFTING

In the context of this chapter, the backbone polymers for the grafting are cellulose and starch. The grafted side chains are hydrophilic in nature, either cationic, anionic, or nonionic. These can be prepared by directly grafting such monomers as acrylic acid, or by grafting monomers such as acrylonitrile and hydrolyzing to acrylic acid and its salts.

In principle, there are two general methods for the synthesis of graft copolymers. (1) Side chain polymer A could be linked directed by a suitable chemical reaction to the backbone polymer B; and (2) the backbone polymer B could have active sites such as free radicals or ions formed upon it. These can then be used to polymerize a suitable monomer to produce the side chains of polymer A.

The former method is difficult except in solution and perhaps the most successful has been by treating "living" polymers to a suitably reactive backbone. A good example is the polystyrene- polyvinyl pyridine system where both polymers have been used as backbones and side chains [1,2]. This method has also been used for the grafting of polystyrene to methylated xylans [3,4]. A number of other methods have been developed to link preformed polymers to cellulosic and other carbohydrate polymers. Many of these have been described in the reviews by Krässig and Stannett [5] and by Arthur [6] and in the recent monograph by Hebeish and Guthrie [7]. There are many advantages to this approach which is, in principle, a simple synthetic method. There could be fewer problems of homopolymer formation. More importantly, the length and number of side chains could be readily controlled. This could lead to superior properties, including absorbency because of the higher degrees of substitution and shorter side chains. However, the difficulty of inducing polymer reactions presents a real problem and little promise can be seen with this approach.

The second general method has been much more successful and a large number of techniques have been developed. The techniques related to grafting to cellulose and its derivatives have been described in ref. 5-7 and those related to starch are in ref. 8. Essentially, these are free radical processes although ionic grafting has also been accomplished by more difficult experimental techniques.

A considerable number of free radical grafting methods have been developed. Many of these have not been used in the development of absorbent materials but they are, in principle, applicable and will be briefly outlined. The methods are summarized under a simplified classification technique developed by Stannett and Hopfenberg [9].

2.1. Chain Transfer Methods

In this method radicals are created on the polysaccharide backbone including cellulose and starch by use of the reactions:

$R^{\cdot}$ + Polysaccharide (PS) $\rightarrow PS^{\cdot} + RH$

$PS^{\cdot}$ + Monomer $\rightarrow$ Graft copolymer

$R^{\cdot}$ can be the growing chain of polymers formed by polymerization with a radical initiator in the presence of the polysaccharide, or by the primary radical from the initiator itself. The chain transfer reaction can be enhanced by the deliberate introduction of such groups as halogens or sulfhydryls. The efficiency of this type of grafting reaction is also greatly improved by increasing the ratio of polysaccharide to monomers such as by using a simple swollen system or with the correct choice of swelling agents. With the primary radical approach, the initiator can be absorbed first into the polysaccharide. With redox initiation such as the ferrous ion-hydrogen peroxide system, the iron can be absorbed into the polysaccharide or even ion exchanged with residual carboxylic acid groups, and the hydrogen peroxide added with the monomer. These methods can often be adapted to locate the grafting near the surface or throughout the substrate [10]. An important recent redox system is the use of partial xanthation followed by the addition of hydrogen peroxide and ferrous ion [11].

2.2. Direct Oxidation

A number of oxidizing agents have been found to interact with polysaccharides to form macroradicals which, with monomer, form graft copolymers. The most successful and best studied of these is ceric ion. Briefly the reaction is as follows:

$$PS + Ce^{4+} \rightarrow PS^{\cdot} + Ce^{3+} + H^{+}$$

In fact the reaction is much more complicated and the oxidation-reaction is often preceded by complexing of the ceric ion by the polysaccharides. A number of other oxidizing agents have been used, including pentavalent vanadium, manganese(III) and manganese(IV) and periodate ion. A review of some of these methods, as applied to grafting to wool, has been presented by Nayak [12].

2.3. Initiators for Polysaccharides

Initiators such as peroxides or diazonium salts can be formed directly on the backbone molecules. Hydroperoxides and peroxides of unknown structure can be formed by ozonolysis or by treating with ultraviolet (UV) or high energy radiation in the presence of air. These initiators can then be used to bring about grafting by decomposing in the presence of monomer. The latter can be achieved by heat or by the addition of a reducing agent such as ferrous ammonium sulfate. The use of reducing agents largely eliminates the concurrent formation of homopolymer.

2.4. Polysaccharide Derivatives as Comonomers

A number of vinyl and allyl derivatives of polysaccharides may be synthesised quite readily. Direct free radical polymerization of a suitable monomer in the presence of these derivatives produces a mixture of grafting and cross-linking. With very low degrees of substitution and the proper choice of reactivity ratios and by the controlled addition of chain transfer agents essentially cross-link free grafted products can be prepared.

2.5. Direct Radiation

UV can be used with the addition of suitable photosensitizers. High energy radiation, both isotopic and with accelerated electrons, however, brings about grafting directly. The use of radiation in air to produce peroxides has already been mentioned. In the absence of air two methods are available.

Firstly, direct, mutual, irradiation of the polysaccharide in the presence of the monomer and a suitable swelling agent can be used. This normally produces a considerable amount of homopolymer which can be reduced to a very small proportion by various means, such as increasing the substrate to monomer level, addition of inhibitors, or using vapor phase addition of the monomer.

The second method, often termed the preirradiation method, involves irradiating the polysaccharide and adding the monomer, plus any swelling agent needed, subsequently. This method is very valuable for monomers such as acrylic acid which polymerize rapidly with radiation.

All the free radical methods are subject to inhibition by oxygen, with the preirradiation method being the most susceptible. Extensive references to the above and other methods are given in the various reviews cited earlier.

Grafting to polysaccharides, in general, and to cellulose, in particular, has not been widely practiced on an industrial scale except in a few cases. The reasons are partly due to the balance of the properties imparted by grafting and the economic cost compared with alternative processes. It could well be that the field of highly water-absorbing polymers will prove the turning point in the industrial development of polysaccharide grafts. Technical problems, which will have to be faced, are the concurrent formation of homopolymer and the lack of complete involvement of the polysaccharide molecules in the grafting reaction. Lesser problems are the control of the number and length of the grafted side chains and some concurrent degradation of the backbone polysaccharide molecules. These and other aspects of grafting have been discussed in more detail in a recent paper [13].

3. GRAFTING TO CELLULOSE TO IMPART WATER ABSORBENCY

Cellulose is the key raw material for most commercial absorbent products. Because of the constant demand to increase the absorbency of these products, there has been a concomitant demand for improvement in absorbency of natural and regenerated cellulose fibers. The absorbency of cellulose fibers has been improved by modification of their chemical structure, the known techniques being: (1) by substituting new chemical groups at the site of the original hydroxyl groups of the cellulose fibers; (2) by crosslinking cellulose chains into a network structure; (3) by introducing new groups and crosslinking them together; or (4) by grafting side chains onto the cellulose backbone.

These chemical modifications are generally carried out in liquid (preferably aqueous) slurries, and the resulting modified fibers are then dried into pulpboard, when wood pulp fibers are used as the cellulose source. The pulp board, with modified cellulose, is subsequently ground into pulp fluff. While such methods have produced more absorbent cellulose fibers, the fibers are generally highly brittle and so easily lose their fibrous structure. They reduce to extremely short fibers or powder upon mild mechanical treatment, i.e., when the pulp board is ground into pulp fluff. Additionally, these products tend to have a high degree of bonding between the fibers, so then tend to form agglomerates of hard, knot-like material when dried from water slurries. This process is known as hornification. These knots are no longer fibrous; and when the pulp board is ground to fluff, the knots either break down into powder or remain whole. In either form they are undesirable for absorbent products. While many modified cellulose fibers have greater absorbency then unmodified cellulose fibers, they gain this absorbency at the cost of decreased softness and the loss of other desirable fibrous qualities. Therefore, even though many standard techniques of grafting hydrophilic monomers to cellulose fibers are possible, not all of them result in the most desirable superabsorbent fibers. The ideal superabsorbent fiber would be the one which would exhibit substantially enhanced absorbency, while essentially maintaining the flexibility of the initial fiber substrate. This challenge is being partially met by the introduction of a combination of ionic and non-ionic monomer grafting approach, focusing on meeting the requirements of disposable absorbent products.

The grafting techniques for cellulose superabsorbency are broadly classified under two general categories, saponifiable grafts to cellulose and direct grafting of acrylic and methacrylic acids to cellulose.

3.1. Saponifiable Grafts to Cellulose

In this approach monomers such as acrylonitrile, acrylamide, and various acrylate and methacrylate esters and their mixtures are grafted, followed by saponification to sodium polyacrylate or methacrylate. Nonsaponifiable comonomers are sometimes also used. Most of the work has only been described in patents and these will be briefly summarized later. However, a few papers on the subject have been published. A comprehensive report has been presented by Lepoutre *et. al* in a series of papers [14-17]. Never-dried bleached kraft softwood pulp was used for almost all the experiments described. The monomer used, almost exclusively, was acrylonitrile, followed after grafting by alkaline hydrolysis. A mixture of sodium acrylate with some polyacrylamide groups was obtained by this procedure. This approach had previously been reported by Adams and Hoftiezer with cellulose [18] and by Gugliemelli *et. al* with starch [19]. The water absorbency was found to be enormously increased by the process, as had also been found with similar grafted starches. Three methods of initiating the acrylonitrile grafting were studied: cellulose xanthate plus hydrogen peroxide, the ceric ion method, and the ferrous ion- hydrogen peroxide process. The loss of grafted polymer after hydrolysis was much higher with the xanthate process, less with the ferrous ion process and least with the ceric ion method. The differences were substantial, 50-60% with xanthate, 31-38% with ferrous, and 8-15% with ceric ion. The loss of cellulose was also in the same order, ranging from about 25% with xanthation to only 5% with ceric ion. The former is partly due to concurrent homopolymer

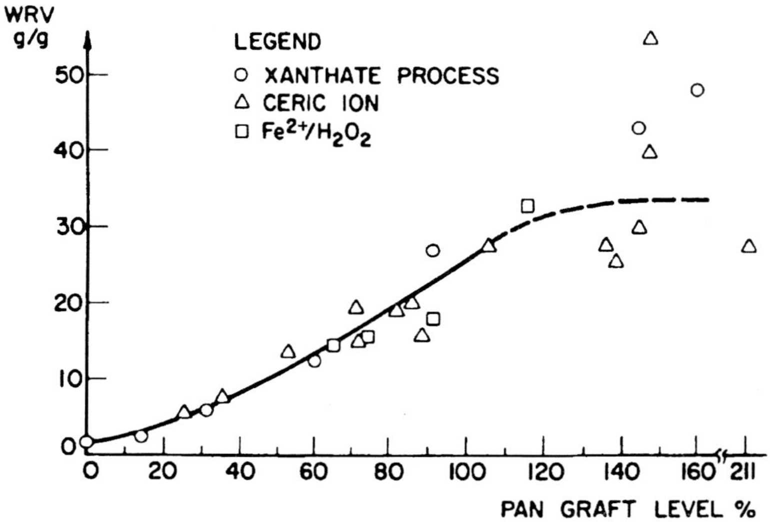

Fig. 1. Water retention value vs. initial (before hydrolysis) PAN graft level [14].

formation and in the case of the xanthate probably by the alkaline hydrolysis of the xanthate linkages as suggested by Krässig [20]. The loss of cellulose was attributed to oxidative degradation. This was presumed to be particularly probable when hydrogen peroxide was involved in the grafting reaction, due partially, perhaps, to the introduction of carboxylic groups. Infrared analysis of the methanol precipitated residues from the hydrolysis solution showed no cellulosic bands suggesting that the loss of polymer during hydrolysis could be mainly homopolymer.

Further details of the ceric ion process were presented in a second paper [15]. It was found that decreasing the pH below 1.7 decreased the yield, and about 2.2 was optimum. This corresponds to the addition of 0.1 % nitric acid to mmolar ceric ammonium nitrate solution. Interestingly it was not found necessary to remove air or to freshly distill the acrylonitrile, purify the ceric salt, or control the temperature closely to get good reproducibility. The method was clearly, therefore, commercially attractive. The type of pulp and the drying history exerted comparatively small effects. Drying the unbeaten pulp decreased the grafting, however, from 70 to 48%. Beating increased the yield back to 68%. This effect appeared to be general. Spruce and pine bleached kraft pulps, commercially dried, gave yields similar to the mixed pulps. The best yields were obtained when the monomer was added first, followed rapidly by the ceric ion. The ceric ion method for grafting acrylonitrile to cellulose has been studied by a number of investigators starting with Schwab *et al.* [21,22]. Other references are given by Lepoutre, together with considerable additional information regarding the kinetics and other details.

The Water Retention Values (WRV) and other properties of the grafted and hydrolysed pulps were also studied by Lepoutre *et al.* [14,17]. The water retention was determined by centrifugation in a metal basket with a stainless steel screen at the bottom for 30 minutes at 900 G. A plot of the water retention versus the initial polyacrylonitrile (PAN), after hydrolysis is shown in Fig. 1. The three methods of grafting gave similar values, the WRV's increasing almost linearly to about 120% grafting. Above this value there was considerable scatter with

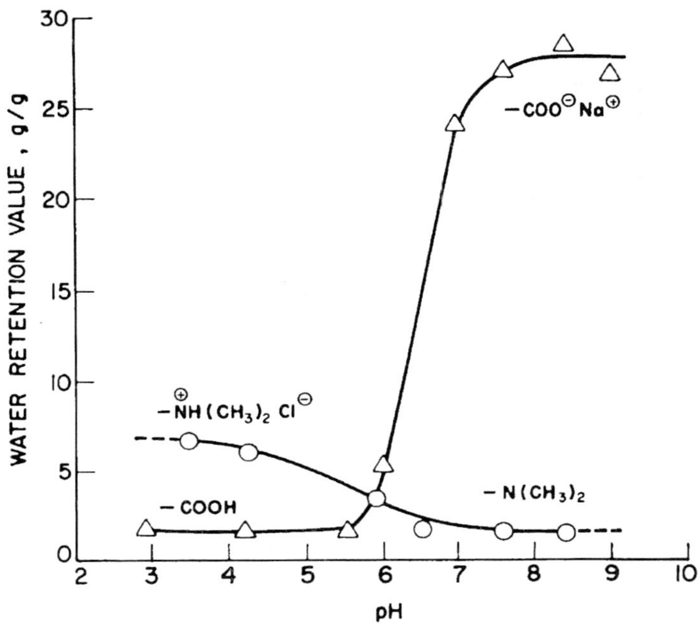

Fig. 2. Water retention value versus pH for cationic and anionic polyelectrolyte grafts [16].

some evidence of leveling off. The scatter was attributed to fragility at very high degrees of grafting plus differences in the packing of the pulp mats during centrifugation. Interestingly, up to WRV's of 30 there was good agreement with the fiber saturation points determined by the method of Stone and Scallan [23]. This indicates that the water is largely present in the fiber walls rather than in the lumen or interfibrilla capillaries. Optical microscopy also indicated the extensive swelling of the grafted hydrolyzed fibers themselves. Drying at 40°C under vacuum decreased the WRV's by as much as 50% with xanthate but not with ceric ion grafting. This could be related to the more numerous, lower molecular weight, side chains, (almost ten fold) produced by the xanthate process causing tighter packing on drying. Extraction of homopolymer caused a reduction in the WRV's. Substituting 1% sodium chloride solution for water reduced the retention values by about 40%. Even at pH 5.0 where only 37% of the carboxylic acid groups were present, the WRV was not changed (Fig. 2).

Cationic grafts using the chloride salt form of dimethylaminoethyl methacrylate and the xanthate initiation method gave much lower WR values both in water and 1% sodium chloride solution [16,17]. A comparison between the WR values of the anionic and cationic pulps at similar grafting levels is presented as a function of pH in Fig. 2. As would be expected, the values are greater at high pH's with the anionic and low pH with the cationic grafts due to ionization effects.

Lepoutre also reported that at high degrees of grafting, the pulps could be dispersed to yield highly viscous colloidal solutions using a Waring Blendor [24]. It was interesting that the dried solutions showed uniform dispersions of short rods with a 35 A width, similar to protofibrils. The drying of these superabsorbents presents a considerable practical challenge. A suitable industrial process has been described [25] involving reducing the pH to about 3 where minimum swelling occurs. The acid groups are then converted to the sodium salt again with sodium hydroxide in aqueous methanol. Considerable data including both batch and continuous processes are presented by Lepoutre.

Ehrnrooth *et al.* [26] have described a vapor phase method of grafting acrylonitrile to bleached kraft pulp using the ferrous ion hydrogen peroxide method of initiation. Less than 5% homopolymer was produced with up to 250% grafting. The products were hydrolysed with sodium hydroxide. TheWRV's were, however, rather low compared with Lepoutre's results, being only about six percent. A number of variables concerning the grafting reaction itself were also investigated.

An intriguing study related to the development of grafted and hydrolysed pulps has been published by Adams and Smith [27]. Acrylonitrile was grafted to bleached kraft with the ceric ion method, followed by alkaline hydrolysis. The products were used to polymerize methyl methacrylate using hydrogen peroxide as the initiator. The resulting fibers had unusual properties and were quite opaque. A number of applications including use as low density opacifying agents and as absorbing and high capacity ion exchange resins for large molecules were suggested.

An interesting discussion of various aspects of highly water absorbing cellulose, including grafted and hydrolyzed products, has been presented by Marchessault *et al.* [28].

In addition to the papers discussed above, a number of key patents have been issued. The first patent specifically concerned with the grafting of a number of monomers to wood pulp and other types of cellulose followed by alkaline hydrolysis appears to be that of Adams and Hoftiezer [18]. A number of wood pulps and methods of grafting including ceric ion were described. Application of such grafted and hydrolyzed cellulose particles in combination with wood pulp have been studied by Chatterjee and Morbey [29].

Chatterjee and Schwenker [30] reported that a proper balance of ionic and non-ionic segments in the grafted chain would be desirable to maintain the flexibility and high liquid absorptivity of the fiber. They developed a fibrous cellulosic product consisting of grafted side chains of polymer molecules providing a novel, highly absorbent, soft, non-flammable fibrous material useful for absorbent hygiene disposable products. The polymer side chains are made up of ionic and non-ionic segments and, on a weight basis, may amount to about 60 to 80% of the total cellulose graft copolymer. The ionic polymer segments could vary from 20% to about 70% by weight of the total copolymer. The cellulose graft copolymer described in the invention had significantly greater absorbency of saline solution than unmodified wood pulp, while still retaining a fibrous form. These cellulose graft copolymer fibers were not as brittle or hornified as those of simply hydrophilic polymer grafted cellulose described earlier. In actual practice, two or more non-ionic monomers having difference in the resistivity of hydrolysis were co-grafted through a free radical process followed by alkaline hydrolysis. By controlling the kinetics of hydrolysis, the grafted chains were converted to produce an optimum balance of ionic and non-ionic moieties.

A different approach is to graft hydrophobic monomers or a combination of hydrophobic and hydrophilic monomers to water-soluble etherified celluloses to produce water-insoluble cellulose derivative graft copolymer. By this approach, a modified cellulose derivative was disclosed [31] which far exceeded the absorption and retention properties of all other modified celluloses. More specifically, an etherified cellulose graft copolymer has been described in this invention, which comprises an alkali metal salt of an etherified cellulose which is soluble in water in the absence of grafting and is insolubilized by having polymer side chains, while still maintaining a high capacity to absorb fluids.

A patent by Hoftiezer and Tilloson [32] describes an improved method of hydrolyzing polyacrylonitrile grafted cellulose fibers. The improvements include hydrolyzing at high solids content with agitation, and subjecting thereafter to a temperature cycling procedure while still damp. These techniques lead to a more coherent and more easily dried superabsorbent product.

A more recent patent by Adams also discusses the effect of aging on the hydrolyzed products. The crumb-like products were found to have increased water absorptive properties after the aging treatment. A high alkaline pressure hydrolysis procedure described by Zimmerer [34] also leads to a more manageable, less gelatinous product.

The rather elaborate drying techniques developed by Lepoutre, and referred to earlier, have also been embodied in a patent [35].

Grafting monomers onto cellulosic materials using a ceric salt initiator to achieve superabsorbency are being are being studied by many others in recent years [36-38]. Gurdag *et al.* [36] found that the maximum grafting yield was obtained when the grafting was carried out at 30°C. It was also observed that the graft copolymer produced at 30°C had the highest water retention capacity. Monomer conversion increased as the reaction temperature increased, and the grafting yield decreased as reaction temperature decreased. However, high temperature favored homopolymers more than the graft polymerization.

Besides ceric salt initiators for grafting, other chemical initiators have also been reported in the recent literature [39-43]. However, among chemical grafting techniques, none of those alternate initiators appear to be more effective than the ceric salt initiator for producing cellulose superabsorbent graft copolymer.

As a completely different approach but using the same grafting technique as illustrated above, a synthetic superabsorbent fiber was developed [44] with polyolefin and polyvinyl alcohol synthetic pulp as the substrate. The product was capable of manifesting high fluid capacity and retention characteristics of grafted cellulose fibers with significantly higher bulk and softer feel than the latter. The method consists of the selection of a synthetic material by choosing a thermoplastic polymer, e.g., polyolefin in admixture with a polymer having sites receptive to accepting hydrophilic polymer grafts, e.g., polyvinyl alcohol. The grafted moiety was selected from the group of polycarboxylic acid type of polymers, preferably hydrolyzed to their alkali salt form. The resultant product had fluff density of less than 0.03 g/cc, which was essentially the density of the starting substrate, the initial fluff, prior to grafting.

3.2. Direct Grafting of Acrylic and Methaerylic Acids to Cellulose

Acrylonitrile and the acrylate and methacrylate esters are rather easy to graft using the ceric ion, the redox and other methods of initiation. Acrylic and methacrytic acids are, on the other hand, rather difficult to graft. The xanthate method and high energy radiation can be used, however. Attempts to use the ceric ion method only led to low degrees of grafting [45-47]. This is presumably due to a preferential reaction of the ceric ions with the monomer acid groups rather than complexation with the cellulose. The latter is widely believed to be the precursor to the grafting initiation reaction. Reasonable grafting yields were, however, obtained with dissolving pulp using carefully controlled procedures and adding the monomer solution first followed by the ceric ion [48]. Pretreating the cellulose dissolving pulp with ceric ion before adding the acrylic acid or methacrylic acid aqueous solutions also gave somewhat lower but good yields. McDowall *et al.* [49] found, however, that the former method gave rather irreproducible yields and low degrees of grafting. Vitta *et al.* [50] found that treating the pulp first with ceric salt did give good yields with reasonable reproducibility and little homopolymer. It seems clear that, with care and specifically modified procedures for each type of cellulose substrate, substantial degrees of grafting and tolerable amounts of concurrent homopolymer formation can be achieved directly with an entirely aqueous system.

A different approach has been made by Gangneux *et al.* [51]. The acrylic acid was dissolved in benzene and added to the cellulose powder (Solka-floc), which had been previously soaked in aqueous ceric ion solution followed by pressing. Since the hydrated ceric ions cannot easily diffuse into the organic layer, large amounts of homopolymer were avoided. A somewhat similar approach using a wide variety of solvents was also studied by Mansour and Nagaty [52]. Good yields but low efficiencies were reported. This method was extended by McDowall *et al.* [53] to grafting to rayon with toluene as the solvent. A number of reaction variables were studied, designed to optimize the procedure.

A more direct method of grafting acrylic and methacrylic acids to cellulose is with initiation by high energy radiation. Since these monomers homopolymerize rapidly with radiation, the preirradiation method is the most convenient. In principle, however, direct irradiation of cellulose in the presence of monomer could be used with the monomer in the vapor phase or in solution containing suitable inhibitors. The only detailed studies which have been reported, however, were with the preirradiation method. This technique involves irradiating the cellulose under high vacuum followed by admitting degassed monomer solution under vacuum and allowing the reaction to proceed. Williams and Stannett [54,55] and Zahran *et al.* [56] have shown that excellent grafting yields can be achieved using this method.

Zahran *et al.* [56] have investigated a number of reaction variables with rayon and cotton grafted with acrylic and methacrylic acids. Typical grafting-time curves are shown in Fig. 3. The grafting yields were found to increase with the percent monomer in aqueous solution, with the total dose and with increasing temperature. The yields presented in Fig. 3 were under close to optimum grafting conditions. Only short reaction times, less than 30 minutes, were needed for yields which are sufficient to achieve superabsorbency. Methacrylic acid grafted much more readily than acrylic acid for reasons that are not clear. In general, rayon gave greater yields than cotton under comparable conditions. This could be ascribed to the greater accessibility of rayon due to its lower crystallinity.

Interestingly, superabsorbency could not be achieved with up to 138% grafted acrylic acid in the acid form, or even after converting to the sodium salt with 3% sodium hydroxide at room temperature. Typical results are presented in Table 1 with rayon and Table 2 with cotton. Similar results were earlier found in the case of cellulose grafted with even larger amounts of ethyl acrylate to produce highly elastic fibers and films [57-60]. This was ascribed to the grafting taking place only in the accessible regions. Since the elongation is highly restricted by thecrystalline morphology, this is understandable. It was found by Williams and Stannet *et al.* [57] that if such ethyl acrylate grafted fibers were treated with a suitable solvent for cellulose, the grafted polymer held the fiber intact. When the solvent was removed, only low degrees of crystallinity were achieved, and high elasticity developed. These results have been published in

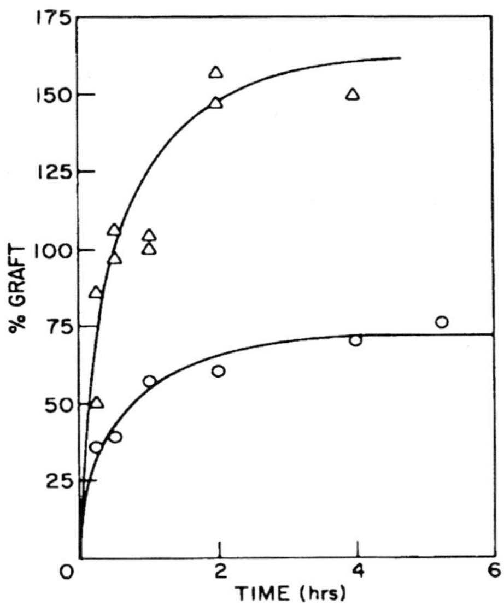

Fig. 3. Preirradiation grafting-time curves for acrylic acid to rayon at 70°C: 75% monomer solution: (0). 1.0 Mrad: (A) 4.0 Mrad [56].

detail and will not be repeated here [57-60].

It is clear that similar effects could be obtained with the acid grafts and water absorbency. This did indeed prove to be the case, as the results included in Table 1 show. However, unlike the elasticity case, it was not necessary to use true cellulose solvents, and hot alkali treatments

Table 1
Water absorbency of acrylic acid grafted rayon before and after various treatments

Graft (%)	Treatment	Water absorbency (cm^3/g)	
		Burette	Centrifuge
0	None	2.86	2.00
138	None	2.39	2.30
138	3% NaOH (room temp)	6.96	8.89
138	3% NaOH (80°C)	-	38.80
144	70% $ZnCl_2$, 40°C, 2h	3.8	-
147	70% $ZnCl_2$, 50°C, 2h plus 3% NaOH (room temp)	-	53.5

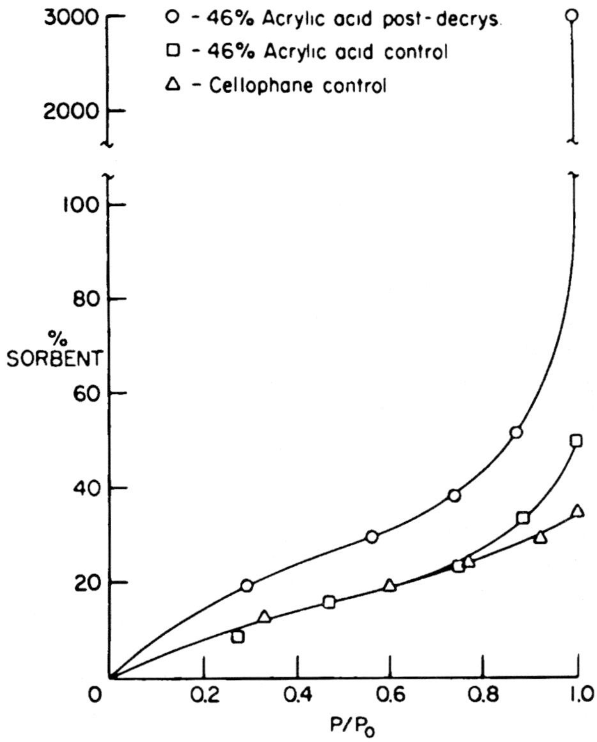

Fig. 4. Water sorption isotherms for acrylic acid grafted rayon (semidull) at 25°C [45].

also led to high water absorbency. Presumably, the swelling and more drastic treatment than with cold sodium hydroxide was sufficient to disrupt the structure. In the case of the super-absorbing grafted celluloses produced by saponification of the grafts, such an effect could presumably be more effective. Even so, the use of cellulose solvents did give even higher absorbencies as the results, also included in Table 1, illustrate. A typical sorption isotherm is shown in Fig. 4. Real superabsorbency only takes place at very high humidities or with liquid water. This is in agreement with the results of Lepoutre *et al.* [14,16]. Higher water absorbency was due to the loss of crystallinity by these post-grafting procedures and was confirmed experimentally in the case of ethyl acrylate grafts but not with the acid grafts.

However, further work is clearly necessary to demonstrate whether this is indeed the case. Some results were also reported with cotton rather than rayon, but only with hot sodium hydroxide after-treatments (see Table 2). Lower absorbencies were attained than with comparable rayon grafts. This could be attributed to the initially higher crystallinity of cotton. It would be interesting to check the effect of treatments with true cellulose solvents, but this has not been studied. Finally, both with rayon and cotton, it was shown that lower absorbencies were

Table 2.
Water absorbency of acrylic and methacrylic acid grafted cotton before and after various treatments.

Grafting	Treatment	Water absorbency (cm^3/g)	
		Burette	Centrifuge
0% acrylic acid	None	4.02	4.03
191 % acrylic acid	None	4.07	3.58
191 % acrylic acid	3% NaOH (room temp)	8.89	12.00
191% acrylic acid	3% NaOH (80°C)	-	26.2
194% methacrylic acid	None	2.89	2.69
194% methacrylic acid	3% NaOH (room temp)	4.40	6.30
194% methacrylic acid	3% NaOH (80°C)	-	16.50

obtained with methacrylic acid compared to acrylic acid at equal levels of grafting. This could be ascribed to the more hydrophobic and stiffer chains of the methacrylic acid side chains.

In a recent study on radiation-induced copolymer grafting, Bilgin and Guthrie [61] concluded that improvements in water retentivity were obtained after decrystallization procedures were carried out on the cellulose copolymers using selected alkali metal salts with methyl alcohol as the continuous medium. Regarding further work on radiation and photo induced grafting, Kubota and Kawabara [62-64] studied gamma ray initiation, ultraviolet light initiation and ceric salt initiation to graft acrylic monomers. Water absorbing function of acrylic acid and methacrylic acid grafted carboxymethyl cellulose (CMC), prepared by ceric salt and radiation initiated systems, were compared to determine the effectivity of a UV initiated grafting technique. It has been concluded that UV grafting using CMC peroxides is useful for preparation of acrylic acid and methacrylic acid grafted CMC samples as compared to ceric and radiation induced grafting. Water absorption of the methacrylic acid grafted samples depends on the grafting conditions of each initiation system.

4. CELLULOSE ALLOYS

These materials are not grafts in the true sense, as it is unlikely that any covalent bonds are formed between the cellulose and the added polymer. A few bonds could be formed, however, but in any case they are normally referred to as cellulose alloy fibers. Methods of formation are, however, remarkably similar. Viscose fibers are spun, but the solution also contains a highly hydrophilic polymer. There has been almost no fundamental studies on these systems, and essentially the entire information is embodied in the patent literature.

According to the early patents [65], it had been known for some years that alloy fibers consisting of sodium carboxymethyl cellulose and regenerated cellulose were useful as absorbent fibers. They were expensive, however, and difficult to dry down to cardable forms without the use of solvents. Their use to absorb body fluids was, however, already described.

More recent patents have also described the process [66,67]. An earlier patent [68] described a modification which involves cross-linking the carboxymethyl cellulose-viscose alloy with formaldehyde after spinning. Improved water absorbencies were obtained and compared with the untreated fibers.

The use of alkali metal or ammonium salts of polyacrylic acid was apparently first described in 1974 [65]. Five to thirty-five parts of the polyacrylate, based on the weight of cellulose, was claimed. The salts were injected into the viscose stream which resulted in other processing refinements. More recently patents [69-72] describe further modifications of the processing conditions with acrylic acid, methacrylic acid and their copolymers.

A commercial absorbent rayon stable fiber has been successfully marketed based on the acrylate salt technology [73]. It is interesting that at 65% RH, the moisture regain is similar to unmodified rayon. In the wet state, however, there is up to twice the absorbency; and the fibers have found use in a number of medical and sanitary applications including nonwoven forms.

In addition to the early use of carboxymethyl cellulose and the acrylic and methacrylic salts, a number of other hydrophilic salts have been studied as additives to viscose. These have included carboxymethyl starch, where good water retention values were obtained [74]. Polyvinyl pyrrolidone has been proposed as the hydrophilic additive in a series of four patents. In the two earlier patents [75,76] cyanoethylated viscose was also incorporated into the viscose. Degree of cyanoethyl substitution was varied from 0.25 to 0.65 and from 5-15% was added plus 5-15% of high molecular weight polyvinyl pyrrolidone. The combination of both additives gave better fluid holding capacities than either of the above. The use of polyvinyl pyrrolidone alone or in combination with carboxymethyl cellulose, and carboxylic acid polymers and copolymers was described in the two additional patents [77,78]. Higher fluid holding capacities were found with the combination of other additives including starch itself [79], cellulose sulfate [80], salts of alginic acid, of 2-acrylamido-2 methylpropanesulfonic acid and its copolymers[81,82] and of copolymers of alkyl vinyl ether and an ethylene dicarboxylic acid [83]. The copolymer of methyl vinyl ether and maleic anhydride was particularly mentioned.

In conclusion, it should be stated that the polymer alloys provide intermediate fluid retention products and are not superabsorbent in the sense that the grafted cellulose and starches are. They have proved practical and useful, however, for a number of medical and sanitary products.

5. GRAFTING TO STARCH TO IMPART WATER ABSORBENCY

Graft polymerizations onto starch are carried out in much the same way as graft polymerizations onto cellulose. Reviews on the synthesis and properties of starch graft copolymers have been compiled [8,8a,8b]. Although hydrolyzed starch-grafted polyacrylonitrile copolymers were not the first polymers recognized as superabsorbents for aqueous fluids, their discovery did much to spark the tremendous interest in this field, and helped create the worldwide superabsorbent industry that we see today.

The synthesis of grafted starch superabsorbent is outlined in Fig. 5. Graft polymerizations may be carried out with either granular, unswollen starch [84] or with starch that has been gelatinized or pasted by heating an aqueous water slurry to about 85-95%°C before room temperature graft polymerization with acrylonitrile [85]. Higher absorbencies are obtained if gelatinized starch is used. Corn starch is most often used as the substrate for graft polymerization; however, starch from other plant sources give copolymers with similar properties.

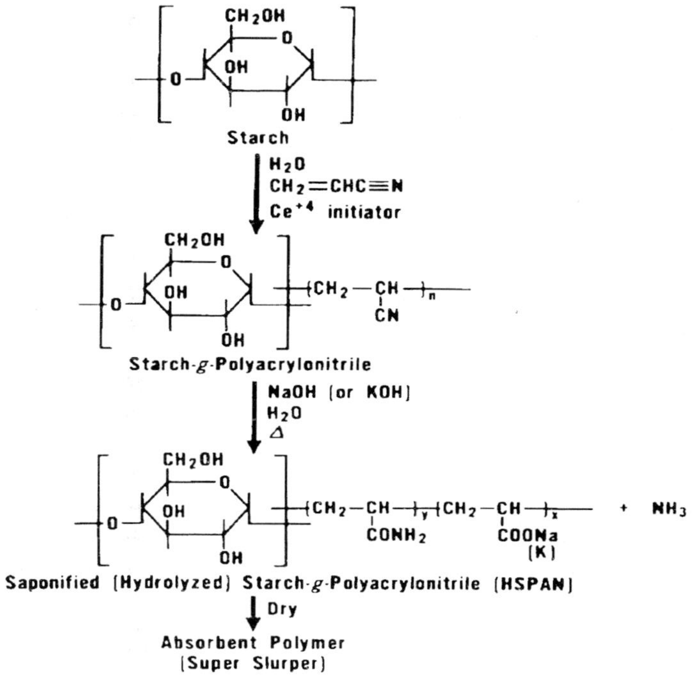

Fig. 5. Preparation of saponified starch-*g*-polyacrylonitrile absorbent polymer.

Starch in the granule state is virtually insoluble in water at ambient temperatures. When an aqueous starch suspension is heated, the granules slowly and reversibly take up water with limited swelling, and then (at about 70°C) they lose their birefringence and undergo irreversible swelling, as hydrogen bonds between individual starch molecules are broken. The temperature at which granules lose their birefringence is known as the gelatinization temperature. Although granules become more swollen and disrupted as the temperature is increased, the starch still remains largely insoluble. For example, only 16% of the starch was dissolved after stirring a 4% water slurry for 1 hr at 85°C [86].

Although a number of methods of initiation, including the Mn^{+3} method of Rånby [87], have been reported for the graft copolymerization of acrylonitrile onto starch, ceric salts are most generally used with this monomer. Graft copolymers containing about 50% grafted polyacrylonitrile (50% add-on) are easily obtained with both granular and gelatinized starch, and copolymers with this approximate weight ratio of starch: polyacrylonitrile are then saponified to yield the superabsorbent copolymer.

Gelatinization of starch prior to ceric-initiated graft polymerization has a profound effect upon the structure of the grafted starch copolymer [73], as seen in Table 3. When starch is graft polymerized as unswollen granules to an add-on of about 50%, the molecular weight of grafted polyacrylonitrile is about 100,000; and the grafting frequency, expressed as the average number of glucopyranosyl units (anhydroglucose units or AGU) per grafted branch, is about 600.

Table 3.
Influence of starch gelatinization on graft copolymer structure

Starch pretreatment (°C)	Graft copolymer[a] PAN content (Wt. %)	M_v of grafted PAN	Grafting frequency (AGU/graft)
25	53	116 000	640
85	56	810 000	3900

[a]After room-temperature extraction with water, dimethylformamide, and dimethylsulfoxide. PAN = polyacrylonitrile. AGU = Anhydroglucose (glucopyranosyl) unit of starch.

However, when starch is gelatinized before graft polymerization, a graft copolymer with about the same percent add-on has polyacrylonitrile grafts with M_v of about 800,000, and the grafting frequency is about 4000 AGU per graft. Graft copolymers thus have only a few high molecular weight polyacrylonitrile grafts per starch molecule. Evidence is not conclusive whether variations in graft copolymer structure are primarily responsible for property differences between superabsorbents prepared from granular and gelatinized starch, or whether these differences are caused by alteration of the starch moiety through gelatinization. One set of experiments has suggested that molecular architecture plays an important role [88].

Saponifications are carried out with either sodium hydroxide or potassium hydroxide at temperatures approaching 100°C [89]. The nitrile substituents of polyacrylonitrile are converted to a mixture of carboxamide and alkali metal carboxylate, and ammonia is generated in the process (see Fig. 5). Carboxamide/carboxylate ratios will vary, depending upon saponification conditions, but are typically on the order of 1:2 [90]. Complete saponification to give poly(sodium acrylate) grafts does not occur. Reaction mixtures initially assume a red-brown color when treated with alkali, due to conversion of polyacrylonitrile into a highly conjugated polymer intermediate [85, 91]. The color gradually fades, however, as the reaction goes to completion; and this color change provides a convenient indicator for the completion of saponification.

Alkaline saponification of polyacrylonitrile grafted starch is usually carried out in water, although aqueous alcohol is also used as a reaction medium [84]. Use of aqueous alcohol affords a superabsorbent product that is easily isolated by filtration; however, the absorbency is lower than that of a comparable product prepared by saponification in water. The hydrolyzed copolymer gives extremely viscous dispersions in water, and the high viscosities of aqueous saponification reaction mixtures thus necessitate the use of a mixer capable of handling a heavy dough-like mass [89].

The hydrolyzed polyacrylonitrile grafted starch remains largely in the form of a highly swollen but insoluble gel after the saponification reaction. Taylor and Bagley confirmed the presence of a substantial insoluble gel fraction by examining a series of plots of reduced viscosity (η_{sp}/c) vs. concentration, which were obtained by diluting an aqueous hydrolyzed copolymer dispersion with sodium chloride solutions of different concentrations [92]. If the hydrolyzed product were truly in solution, the correct sodium chloride concentration for isoionic dilution would give a polyelectrolyte configuration which was constant during dilution and would thus yield a linear plot of reduced viscosity vs. concentration. This was not observed; rather, the isoionic dilution curve dropped off rapidly at low concentrations. Moreover, the presence of insoluble gel was clearly shown by ultracentrifugation experiments.

Taylor and Bagley [92] correctly conclude that the extraordinary thickening action of hydrolyzed polyacrylonitrile grafted starch in water is due to the nearly complete absorption of water by the gel to give a system consisting of highly swollen, deformable gel particles that are closely packed and in intimate contact. Neither the minor amounts of graft copolymer in solution nor the size of the gel particles exerts a large influence on rheological properties. When either the ionic strength of the medium is increased or the swollen gel is diluted to the point where solvent is in excess, the gel particles are no longer tightly packed, and the viscosity of the system thus drops sharply. The condition where get particles are closely packed and no excess solvent is present occurs when the product cQ is greater than 1 [93]. Here, c is equal to the concentration (g polymer/g suspension), while Q is a measure of polymer swelling and is defined as grams of swollen gel per gram of dry polymer in excess solvent of a particular ionic strength.

The conclusions reached through rheological studies are confirmed by scanning electron microscopy. Although graft polymerization occurs predominantly on the granule surface when low add-on (about 20%) polyacrylonitrile grafted starch is prepared, graft copolymers with about 50% add-on are grafted largely throughout the entire granule matrix [94]. Particles of polyacrylonitrile grafted starch have the same outward appearance as ungrafted starch, and the grafted and ungrafted granules are nearly indistinguishable. Moreover, when polyacrylonitrile grafted starch is heated in aqueous alkali to saponify the nitrile substituents, the resulting particles of hydrolyzed copolymer also retain much of the outward appearance of starch granules [95].

Since both starch and saponified polyacrylonitrile are soluble in aqueous alkali, the insolubilty of a graft copolymer of the two components was unexpected; moreover, the fact that the outward appearance of grafted starch granules was still present after saponification was particularly surprising. These observations would, of course, suggest that a cross-linking reaction has occurred, either during the graft polymerization reaction with acrylonitrile or during saponification with alkali. We have found evidence for both of these cross-linking reactions [96]. Cross-linking during graft polymerization probably occurs by way of chain combination of growing polyacrylonitrile macroradicals, and cross-linking also takes place between the starch and polyacrylonitrile components of the graft copolymer during saponifications run in water. Polyacrylonitrile will also cross-link with itself when saponifications are run in solvent systems containing predominantly alcohol.

An important advantage of carrying out saponifications in water is the variety of physical forms of superabsorbent that can be produced from the viscous reaction mass. As mentioned earlier, hydrolyzed copolymer is largely insoluble in water and exists as highly swollen gel particles. When isolated in the dry form, however, these individual gel particles tend to agglomerate together to give macroparticles that will not break up and revert back to the original gel when they are allowed to swell in aqueous fluids. Although the exact mechanism by which gel particles permanently agglomerate is not known, formation of primary chemical bonds between individual gel particles need not be proposed to account for these properties [96]. The behavior of the gel is similar in many respects to that of cross-linked latexes [97,98], and the hydrolyzed grafted starch properties may be similarly explained by assuming interdiffusion of polymer chain ends on the surfaces of water-swollen gel particles, followed by hydrogen bonding between these polymer chains. This permanent agglomeration of gel into macroparticles on simple drying of water dispersions is an extremely valuable property and allows the preparation of the various physical forms of superabsorbent. The superabsorbent copolymer prepared from gelatinized starch exhibits this agglomeration property to a much larger extent than that derived from granular starch.

A commonly used method of isolation is the precipitation of hydrolyzed copolymer from the reaction mixture as a granular solid by addition of a water miscible non-solvent, such as methanol. Since excess alkali and inorganic salts are removed by washing, this method can be used to prepare a purified grade of superabsorbent with a deionized water absorbency of about 1000 g/g, when gelatinized starch is used as the substrate for graft polymerization. The superabsorbent derived from granular starch has roughly a five-fold lower water absorbency. Use of methanol as a precipitant gives higher absorbency products than ethanol, acetone, or isopropanol [99]. Since hydrolyzed polyacrylonitrile grafted starch is a polyelectrolyte, the absorbency of ion-containing fluids is reduced. For example, the amount of 1% sodium chloride solution absorbed is lower than deionized water by roughly a factor of 10. Alcohol-precipitated copolymers, which exhibit faster wicking and reduced gel blocking, are prepared by subjecting the polymers to an ion exchange reaction with a small amount of a high molecular weight quaternary ammonium chloride before before isolation [100]. Aluminum salts of the copolymer also show these desirable properties [101].

If maximum purity and absorbency are not needed (e.g., for agricultural applications), the viscous reaction mass from saponification can be simply dried on heated drums to yield the superabsorbent in the form of coarse flakes. This process is inexpensive and produces no by-products other than steam from the product isolation step. With drum drying, excess alkali must be avoided during saponification to minimize the presence of inorganic salts in the final product. The use of about 0.6-0.8 moles of alkali per mole of acrylonitrile repeating unit allows the saponification to proceed at a reasonable rate without the disadvantages of excess alkali. Although the deionized water absorbency of drum-dried superabsorbent prepared from gelatinized starch is only about 300 g/g, an important advantage of drum-dried flakes is their rapid rate of liquid absorption and the complete absence of gel blocking.

If a viscous water dispersion of the hydrolyzed copolymer (preferably purified) is spread onto a TEFLON-coated tray and allowed to dry, a continuous film is formed. When placed in water, the film will swell as a single entity, producing a continuous sheet of highly swollen gel that can be carefully manipulated without breaking [102]. Moreover, the exact shape of the dry film is retained by the swollen gel sheet. The hydrolyzed copolymer films may be plasticized with polyols, such as glycerol or ethylene glycol, or they may be mixed with low T_g polymer latexes before drying [103] to give films with improved mechanical properties.

Freeze drying a water dispersion of purified product affords the absorbent polymer in the form of a spongy mat [89]. The material in this form tends to gel-block when placed in water; however, this undesirable property can be reduced or eliminated by blending cellulose pulp with the aqueous copolymer dispersion prior to freeze drying. A product containing equal dry weights of hydrolyzed copolymer and cellulose absorbs water rapidly. Because of its expense, freeze drying has not been used to prepare commercial quantities of superabsorbent; however, the freeze-dried absorbent does have unique properties that would make it suitable for some applications, for example, in the medical field.

The hydrolyzed copolymer in any of its physical forms can be reduced back to a smooth gel by applying mechanical shear to a dispersion of the water-swollen polymer. For example, a hydrolyzed copolymer film may be allowed to swell in water, and the swollen film may then be briefly stirred in a Waring Blender to yield a smooth dispersion. If the resulting dispersion is allowed to dry, a new film is obtained whose properties are similar to the original [102]. However, if the amount of mechanical shear is excessive, e.g., that obtained by ultrasonic treatment, the copolymer gel may be degraded to the point where true solubility is achieved [104] In contrast to dispersions of the copolymer gel, viscosities of its solutions are low and are almost Newtonian. Although films cast from these solutions will simply dissolve when placed in water, they can be readily cross-linked again by a number of different techniques. Heating, irradiating with cobalt-60, or aging the soluble films at high relative humidity gives absorbent polymers with properties similar to the original non-degraded polymer.

The hydrolyzed grafted starch copolymer can be converted from the sodium carboxylate form to the free carboxylic acid by adjusting the pH of an aqueous dispersion to about 3 [19]. The carboxylic acid form of the graft copolymer exhibits only limited swelling in water and thus precipitates as a solid, which can be washed with water, dewatered with alcohol, and dried. When dried in the acid form, individual particles of acidified copolymer do not permanently agglomerate together, but tend to revert back to the original micron-sized particles when stirred in water.

The basic starch grafted superabsorbent technology described in this section has been modified to provide new additions to this family of superabsorbents. In one such modification, granular or gelatinized flour is simply substituted for starch in the ceric-initiated graft polymerization with acrylonitrile, and the saponification is carried out in water in the same manner used for starch- based products [92]. Absorbent polymers prepared from both granular and gelatinized yellow corn flour and isolated by three different techniques are described in Table 4. Unpurified absorbent polymers were isolated either by drum drying the saponification reaction mixtures or by diluting the saponificate with water and then allowing it to dry to a film on a TEFLON-coated tray near room temperature. Although differences between drum-dried products prepared from starch and flour were not significant, the tray-dried product from

Table 4.
Water absorbencies of saponified flour-based polymers compared with starch-based polymers

Polymer	Absorbency (g H_2O per g polymer)		
	Drum dried unpurified polymer	Tray dried unpurified polymer	Precipitate with methanol
Gran. starch	180	-	230
Gelat. starch	320	430	1500
Gran. flour	200	-	500
Gelat. flour	270	690	1200

gelatinized flour surpassed the starch product in absorbency. Also, when precipitated with methanol, the granular flour-based polymer had more than twice the absorbency of the corresponding starch-based product.

In a second modification, either granular or gelatinized starch is graft polymerized with a monomer system in which a minor amount (up to 10 mol%) of the acrylonitrile is replaced with a comonomer [105]. Of the various comonomers studied, 2-acrylamido-2-methylpropanesulfonic acid ($AASO_3H$) imparted the highest absorbency to the final saponified polymers (Table 5). Increases in absorbency resulting from addition of $AASO_3H$ are particularly evident in the series of drum-dried and methanol-precipitated polymers where granular starch was used as the substrate for graft polymerization. Large increases are similarly observed for the tray-dried polymers prepared from gelatinized starch. Another benefit realized by incorporation of certain comonomers is a reduction in the length of time needed for saponification. This effect was observed not only for $AASO_3H$, but also for acrylamide, acrylic acid, and methyl acrylate [105].

In other modifications of the starch grafting technology, acrylonitrile has been graft polymerized onto either cross-linked starch [106] or ionic starch ethers, such as carboxymethyl starch or starch sulfopropyl ether [107]. Flour cross-linked with formaldehyde has also been used as a substrate for graft polymerization [108].

Starch graft copolymer absorbents have also been prepared by graft polymerization of acrylic acid and its salts, either alone or in combination with acrylamide. Mooth [109] reports the preparation of absorbent polymers by drum drying mixtures of starch and monomer at elevated temperatures in the presence of an initiator, such as ammonium persulfate.

Table 5.
Water absorbencies of saponified polyacrylonitrile grafted starch prepared from mixtures of acrylonitrile and $AASO_3H$[a]

Polymer[b]	Absorbency (g H_2O per g polymer)		
	Drum dried unpurified polymer	Tray dried unpurified polymer	Precipitate with methanol
0-Granular	180	-	320
0-Gelatinized	320	430	1500
5-Granular	320	-	590
5-Gelatinized	290	1800	1800
10-Granular	350	-	1000
10-Gelatinized	310	1600	2500

[a] $AASO_3H$ = 2-acrylamido-2-methylpropanesulfonic acid, $CH_2 = CHCONHC(CH_3)_2\ CH_2SO_3H$.
[b] Numbers refer to the mol % of $AASO_sH$ in the $AASO_3H$/acrylonitrile monomer mixture as a substrate for graft polymerization [108]. Graft copolymers are then saponified with alkali to yield the absorbent polymers.

Unsaturated derivatives of starch (e.g., acrylamidomethyl starch) have been used as substrates for graft polymerization [110], and cross-linking reactions have also been carried out on the starch moiety after the poly-merization reaction [111].

To help reduce the water solubility of polymers prepared by grafting hydrophilic monomers onto unmodified starch, difunctional monomers may be included in polymerization reaction mixtures to cross-link the synthetic portion of the graft copolymer. A representative example ofthe different procedures that may be found in the patent literature is the ceric-initiated graft polymerization of acrylic acid and sodium acrylate onto corn starch in the presence of N,N-methylenebisacrylamide to yield a product that absorbs about 200 g/g of water [112].

Finally, starch may be etherified in alkali with a number of common reagents, such as chloroacetic acid or ethylene oxide, and the resulting hydrophilic starch derivatives then cross-linked, e.g., with epichlorohydrin [113,114], to yield the superabsorbent. Absorbencies are typically less than 100 g/g and vary with the etherification reagent used, the degree of substitution and the cross-link density. Ionic starch derivatives, for example, carboxymethyl starch or starch sulfate, have also been cross-linked by reaction with polyvalent metal salts, such as aluminum and zirconium [115]. Flexible resilient absorbent composites have been obtained by incorporating starch-based absorbent polymers in reaction mixtures used to prepare polyurethane foams [116].

6. APPLICATIONS

Grafted celluloses and alloys have found applications in absorbent dressings, diapers and tampons. Various catamenial devices, diapers, wound dressings, surgical sponges and incontinence pads have been reported.

Superabsorbents prepared from grafted starch have also been used in disposable soft goods designed to absorb body fluids. Publications dealing with this application have appeared in the United States [117] and in Japan [118,119], and the patent literature also describes methods of incorporating the absorbent polymer into finished products [120,122].

Agriculture perhaps provides the most important end-use application for starch-based super-absorbents. Application of an aqueous slurry of starch-based superabsorbent to the root zone of plants before transplanting prevents roots from drying, reduces wilting and transplant shock, and improves plant survival [123,124]. Hamilton [125] has studied the survival of tobacco plants that were transplanted after dipping the roots into a gel prepared from 4 g of grafted starch super- absorbent and 1 liter of water. Under hot, dry weather conditions at the time of transplanting, 95% of the plants dipped into polymer gel survived, whereas survival of the undipped plants was only 65%.

Use of the superabsorbent as a soil additive has also been studied, and Shrader and Mostejeran [126] reported the effect of superabsorbent addition on water-holding capacity for a variety of different soil types. The superabsorbent not only increased the amount of water held by sandy soils, but the water was held in a form that was readily available to plant roots. The authors concluded that a sandy soil treated with 0.2 weight percent absorbent polymer had about the same water-holding capacity as the best corn belt soil.

Miller [127] investigated the influence of absorbent polymer on water-holding capacities of soils that were packed in columns to simulate the deep drainage encountered under field conditions. As observed by Shrader and Mostejeran [126], the effect of superabsorbent was negligible in loam soils; but the polymer greatly increased the amount of water retained by sand. This water-holding effect was increased by repeated irrigations. The superabsorbent reduced the water infiltration rates in all soils tested. Hemyari and Nofzinger [128] also studied water retention and water infiltration of soils containing varying amounts of superabsorbent, as well as the influence of superabsorbent content on crust strength. Addition of superabsorbent decreased the modulus of rupture of soil crusts, increased water retention of the soils, and decreased the infiltration of water.

Starch-based superabsorbents have also been used to remove suspended water from organic solvents, and successful laboratory tests have been carried out [129] using these polymers as dehydrating agents for ethanol-gasoline mixtures. Although superabsorbent polymers will not efficiently remove water from pure ethanol, addition of the superabsorbent to a cloudy, two-phase mixture of wet ethanol and gasoline yielded a clear solution with a water content of 0.4-0.5%.

Although grafted starch superabsorbents have been known for over two decades, there is still a high level of commercial interest in these materials. We have learned that new starch superabsorbent plants are planned for construction in Puerto Rico and in the United States. These plants will focus on various agricultural applications.

7. REFERENCES

1. J. A. Gervasi, A. B. Gosnell and V. Stannet, J. Polym. Sci. C, 24 (1968) 207.
2. A. B. Gosnell, J. A. Gervasi, D. K. Woods and V. Stannett, J. Polym. Sci. C, 22 (1969) 611.
3. J. J. O'Malley and R. H. Marchessault, J. Polym. Sci., Polym. Letters 3 (1965) 685.
4. J. J. O'Malley and R. H. Marchessault, J. Polym. Sci. C, 24 (1968) 179.
5. H. A. Krassig and V. Stannett, Adv. Polym. Sci., 4 (1965) 111.
6. J. C. Arthur, Adv. Macromol. Chem., 2 (1970) 1.
7. A. Hebeish and J. T. Guthrie, The Chemistry and Technology of Cellulose Copolymers, Publ. Springer-Verlag, New York, 1981.
8. G. F. Fanta and E. B. Bagley, in N. M. Bikeles (Ed.), Encyclopedia of Polymer Science and Technology, Supplement Vol. 2, p. 665, Wiley and Sons, New York, 1977.

8a. G. F. Fanta and W. M. Doane, in Modified Starches: Properties and Uses, O.B. Wurzburg, Ed., CRC Press, Boca Raton, FL, 1986, p. 149.

8b. G. F. Fanta, in Polymeric Materials Encyclopedia, Vol. 10, J.C. Salamone, Ed., CRC Press, Boca Raton, FL, 1996, p. 7901.

9. V. Stannett and H. B. Hopfenberg, in Cellulose and Cellulose Derivatives, Vol. 5, p. 907-936, Wiley and Sons, New York, 1971.
10. D. J. Bridgeford, Ind. Eng. Chem. Prod. Res. and Dev., 1 (1962) 45.
11. W. J. Brickman, Tappi, 56 (1973) 97 and Subsequent papers.
12. P. L. Nayak, J. Macromol. Sci.-Rev. Macromol. Chem. C, 14 (1976) 192.
13. V. Stannett, Am. Chem. Soc. Symposium Series, 187 (1982) 3.
14. P. Lepoutre, S. H. Hui and A. A. Robertson, J. Appl. Polym. Sci., 17 (1973) 3143.
15. P. Lepoutre and S. H. Hui, J. Appl. Polym. Sci., 19 (1975) 1257.
16. S. H. Hui and P. Lepoutre, J. Appl. Polym. Sci., 19 (1975) 1771.
17. P. Lepoutre, S. H. Hui and A. A. Robertson, J. Macromol. Sci., 10 (1976) 681.
18. J. W. Adams and H. W. Hoftiezer, U.S. 3,194,727 (July 1, 1965).
19. L. A. Gugliemelli, M. O. Weaver, C. R. Russell and C. E. Rist, J. Appl. Polym. Sci., 13 (1969) 2007.
20. H. A. Krässig, Svensk. Papperstidn., 75 (1972) 699.
21. E. Schwab, V. Stannett and J. J. Hermans, Tappi, 44 (1961) 251.
22. E. Schwab, V. Stannett, D. H. Rakowitz and J. K. Magrane, Tappi, 45 (1962) 390.

23. J. E. Stone and A. M. Scallan, Tappi, 50 (1967) 496.
24. P. Lepoutre and A. A. Robertson, Tappi, 57 (1974) 87.
25. P. Lepoutre, Am. Chem. Soc. Symposium Series, 58 (1977) 217.
26. E. Ernrooth, R. Jaakkola, L. Koikkalainen, M. L. Weiste and H. Sihtola, Pap. Puu, 59 (1977) 11.
27. J. W. Adams and D. T. Smith, J. Appl. Polym. Sci., Symposium Series, 28 (1976) 831.
28. R. H. Marchessault, M. Dube, J. St. Pierre and J. F. Revol, in Fibre-Water Interactions in Paper Making, Transactions, Oxford, September 1977, Publ. Brit. Paper and Board Industry Federation, London, 1978, p. 795-816.
29. P. K. Chatterjee and G. K. Morbey, U.S. 4,105,033 (August 8, 1978).
30. P. K. Chatterjee and R. F. Schwenker, U.S. 3,889,678 (June 17, 1975).
31. P. K. Chatterjee and R. F. Schwenker, U.S. 4,200,557 (April 29, 1980).
32. H. W. Hoftiezer and A. H. Tilloson, U.S. 3,838,077 (September 24,1974).
33. J. W. Adams, U.S. 4,151,130 (April 24, 1979).
34. R. E. Zimmerer, U.S. 3,793,299 (February 1974).
35. P. Lepoutre, U.S. 4,025,472 (May 24, 1977).
36. G. Gurdag, M. Yasar and M. A. Gurkaynak, J. Appl. Polymer Sci., 66(5) (1997) 929.
37. H. T. Lokhande and P. V. Varadarajan, Bioresour. Tecnol. 45(3) (1993) 161.
38. E. Rezai and R. R. Warner, J. Appl. Polymer Sci., 65(8) (1997) 1463.
39. S-B Park, M. Mitsuhiro and 1. Sakata, Palpu Chongi Gisul 26(2) (1994): CAN 123:201132.
40. Z. Yang and Z. Zhang, Patent CN90-10169 900324; CAN 116:215598.
41. K. Tajiri and H. Tsukamoto, JP 89-158228 890622; CAN 115:10751.
42. B. Rånby and D. Zuchowska, Polymer J. (Tokyo), (1987) 623.
43. P. K. Sahoo, H. S. Samantaray and R. K. Samal, J. Appl. Polymer Sci., 32(7), 5693 (1986).
44. L. Kopolow and P. K. Chatterjee, U.S. 4,423,184, (December 27, 1983).
45. R. J. E. Cumberbirch and J. R. Holker, J. Soc. Dyers and Colourists, 82 (1966) 59.
46. G. N. Richards and E. F. T. White, J. Polym. Sci. C, 4 (1963) 1251.
47. O. Y. Mansour and J. Schurz, Svensk Papperstidn., 76 (1973) 258,288,415.
48. Y. Ogiwara, H. Kubota and M. Tanizaki, Paper presented at 19th High Polymer Conference (Japan), Tokyo, 1970.
49. D. J. McDowall, B. S. Gupta, V. Stannett and J. L. Williams, paper presented at Am. Chem. Soc.-Chem. Soc. of Japan Meeting, Honolulu, Hawaii, April 6, 1979.
50. S. B. Vitta, V. Stannett and E. P. Stahel, J. Makramol. Sci. Chem., to be published.
51. A. Gangneux, D. Wattiez and E. Marechal, Eur. Polym. J., 12 (1976) 535.
52. O. Y. Mansour and A. Nagaty, J. Polym. Sci.-Chem. Ed., 12 (1974) 1887.
53. D. J. McDowall, B. S. Gupta and V. Stannett, Am. Chem. Soc. Symposium Series, in press.
54. J. L. Williams and V. Stannett, J. Appl. Polyin. Sci., 23 (1979) 1265.
55. J. L. Williams and V. Stannett, U.S. 4,036,588 (July 19, 1977).
56. A. H. Zahran, J. L. Williams and V. Stannett, J. Appl. Polym. Sci., 25 (1980) 535.
57. J. L. Williams and V. Stannett, J. Polym. Sci., Polym. Letters, 10 (1972) 665.
58. J. L. Williams, D. K. Woods, V. Stannett, L. G. Roldan, S. B. Sello and C. V. Stevens, Text. Res. J., 43 (1973) 205.
59. J. L. Williams, D. K. Woods, V. Stannett, S. B. Sello and C. V. Stevens, Int. J. Appl. Rad. and Isotopes, 26 (1975) 159.
60. J. L. Williams, V. Stannett, L. G. Roldan, S. B. Sello and C. V. Stevens, Int. J. Appl. Rad. and Isotopes, 26 (1975) 169.
61. V. Bilgin and J. T. Guthrie, Radiat. Phys. Chem., 3 6(4) (1990) 581.
62. H. Kubota and S. Kuabara, J. Appl. Polymer Sci., 64(11) (1997) 2259.
63. S. Kuwabara and H. Kubota, J. Appl. Polymer Sci., 60(11) (1996) 1965.
64. H. Kubota, Y. Fukushima and S. Kuwabara, European Polymer J., 33(l) (1997) 67.
65. F. R. Smith, U.S. 3,844,287 (October 29, 1974).
66. H. Pietsch and W. Horn, U.S. 4,169,121 (September 25, 1979).
67. J. Marini, J. Lenz, J. Schmidtbauer and K. Geppert, Ger. Offen. 3,036,415 (April 30,1981).
68. H. Krässig, F. Gatschy and H. Teichmann, Ger. Offen. 2,809,312 (September 28, 1978).
69. T. C. Allen and D. B. Denning, U.S. 4,066,584 (January 3, 1978).
70. A. W. Meierhofer, U.S. 4,104,214 (August 1, 1978).

71. T. C. Allen, U.S. 4,240,937 (December 23, 1980).
72. T. C. Allen and D. B. Denning, U.S. 4,263,244 (April 21, 1981).
73. Technical Data Sheet Absorbit, Akzona Inc., Enka, NC.
74. F. R. Smith, U.S. 3,847,636 (November 12, 1974).
75. F. R. Smith, U.S. 3,919,385 (November 11, 1975).
76. F. R. Smith, U.S. 3,951,889 (April 20, 1976).
77. F. R. Smith, U.S. 4,041,121 (August 9, 1977).
78. F. R. Smith, U.S. 4,136,697 (January 30, 1979).
79. F. R. Smith, U.S. 4,144,079 (March 13, 1979).
80. Avtex Fibers, Inc., Jap. Patent 80,163,205 (December 19, 1980).
81. F. R. Smith, U.S. 4,063,558 (December 20, 1977).
82. C. Allen, U.S. 4,242,242 (December 30, 1980).
83. D. B. Denning, U.S. 4,165,743 (August 28, 1979).
84. T. Smith, U.S. 3,661,815 (May 9, 1972).
85. M. O. Weaver, E. G. Bagley, G. F. Fanta and W. M. Doane, U.S. 3,997,484 (December 14, 1976)
86. R. C. Burr, G. F. Fanta, C. R. Russell and C. E. Rist, J. Macromol. Sci. Chem. A, 1 (1967) 1381.
87. B. Rånby and C. Rodehed, Polymer Bulletin, 5 (1981) 87.
88. N. W. Taylor, G. F. Fanta, W. M. Doane and C. R. Russell, J. Appl. Polym. Sci., 22 (1978) 1343.
89. M. O. Weaver, R. R. Montgomery, L. D. Miller, V. E. Sohns, G. F. Fanta and W.M. Doane, Staerke, 29 (1977) 413.
90. M. O. Weaver, L. A. Gugliemelli, W. M. Doane and C. R. Russell, J. Appl. Polym. Sci., 15 (1971) 3015.
91. N. S. Batty and J. T. Guthrie, Makromol. Chem., 182 (1981) 71.
92. N. W. Taylor and E. B. Bagley, J. Appl. Polym. Sci., 18 (1974) 2747.
93. N. W. Taylor and E. B. Bagley, J. Appl. Polym. Sci., 21 (1977) 113.
94. G. F. Fanta, F. L. Baker, R. C. Burr, W. M. Doane and C. R. Russell, Staerke, 25 (1973) 15
95. G. F. Fanta, F. L. Baker, R. C. Burr, W. M. Doane and C. R. Russell, Staerke, 29 (1977) 386.
96. G. F. Fanta, R. C. Burr and W. M. Doane, in D.N.-S. Hon (Ed.), Graft Polymerization of Lignocellulosic Fibers, ACS Symposium Series, 187 (1982) 195.
97. S. S. Voyutskii, in H. F. Mark and E. H. Immergut (Eds.), Autohesion and Adhesion of High Polymers, Polymer Reviews, Vol. 4, Wiley, New York, 1963, p. 80.
98. E. B. Bradford and J. W. Vanderhoff, J. Macromol. Chem., 1 (1966) 335.
99. G. F. Fanta, E. B. Bagley, R. C. Burr and W. M. Doane, Staerke, 34 (1982) 95.
100. D. A. Jones and L. F. Elmquist, U.S. 4,159,260, June 26, 1979.
101. L. F. Elmquist, Eur. Pat. Appl. EP 39,542, November 11, 1981.
102. M. O. Weaver, E. B. Bagley, G. F. Fanta and W. M. Doane, Appl. Polym. Symposium, 25 (1974) 97.
103. E. L. Skinner and L. F. Elmquist, U.S. 4,156,664, May 29, 1979.
104. E. B. Bagley and N. W. Taylor, Ind. Eng. Chem., Prod. Res. Dev., 14 (1975) 105.
105. G. F. Fanta, R. C. Burr, W. M. Doane and C. R. Russell, Staerke, 30 (1978) 23.
106. N. Hishiki, H. Hayakawa and T. Ogura, Japan. Kokai 78: 91,994, August 12, 1978.
107. S. Iwase and N. Hishiki, Jpn. Kokai Tokkyo Kohi 78:128,693, November 9, 1978.
108. E. J. Swarthout and P. Antholz, U.S. 4,204,983, May 27, 1980.
109. R. A. Mooth, U.S. 4,155,888, May 22,1979.
110. A. H. Young and F. Verbanac, U.S. 4,115,332, September 19, 1978.
111. F. Masuda, K. Nishida, K. Mita and M. Harada, Japan Kokai 77: 59,690, May 17, 1977.
112. M. Fusayoshi and N. Akira, U.S. 4,076,663, February 28, 1978.
113. L. F. O. Mindt and G. R. Sanderson, S. African 77:00,263, July 18, 1978.
114. A. Holst, Angew. Makromol. Chem. 76/77 (1979) 177.
115. S. H.Ganslaw and H. G. Katz, U.S. 4,090,013, May 16,1978.
116. W. Kelly and A. A. McKinnon, Brit. 1,550,614, August 15, 1979.
117. W. F. Lindsay, Formed Fabrics Industry 8(5) (1977) 20.
118. H. Kubota, Kagaku Kogyo, 30 (1979) 928.
119. S. Aoki and H. Yamasaki, Kagaku Zokan (Kyoto), 81 (1979) 93.
120. E. G. McCabe, Jr., U.S. 4,224,366, September 23, 1980.
121. T. Tamura, Y. Senpuku, N. Hishiki, K. Nishimoto and F. Otsuki, Jpn. Kokai Tokkyo Koho, 79 (1979) 157,788.

122. N. Tochi and A. Ishikawa, Jpn. Kokai Tokkyo Koho, 79 (1979) 139,981.
123. D. Deterling, Progressive Farmer, 96 (1981) 56K.
124. [illegible]. F. Pritchard and J. J. Quinn, Ornamentals South, 4 (1981) 14.
125. [illegible]. L. Hamilton, Tobacco Science, 26 (1982) 17.
126. W. D. Shrader and A. Mostejeran, Coatings and Plastics Prepr. Am. Chem. Soc. Div. Org. Coatings and Plastics Chem., 37 (1977) 683.
127. D. E. Miller, Soil Sci. Soc. Am. J., 43 (1979) 628.
128. P. Hemyari and D. L. Nofzinger, Soil Sci. Soc. Am. J., 45 (1981) 799.
129. G. F. Fanta, R.C. Burr, W. L. Orton and W. M. Doane, Science, 210 (1980) 646.

Absorbent Technology.
P.K. Chatterjee and B.S. Gupta, editors.

CHAPTER X

NONWOVENS IN ABSORBENT MATERIALS

BHUPENDER S. GUPTA
College of Textiles, North Carolina State University, Raleigh, NC 27695-8301, USA

D. K. SMITH
Smith, Johnson & Associates, 2709 Edgewood, Provo, UT 84604, USA

Contents

1. INTRODUCTION

1.1. Nonwoven Definition

In a discussion of absorbency from either a theoretical or practical point of view, the topic of nonwoven materials enters early and frequently. Two characteristics of nonwovens make them particularly suitable for use in an absorbent structure: high bulk for imbibing and holding large amount of fluid per unit mass of material and low cost of converting raw material into final product. Several other properties that have significant impact on the use of nonwovens in an absorbent fabric are disposability, comfort, and ease of fabrication of the absorbent product. The latter is usually composed of a number of components, each supporting different but important functions. These are to receive fluid, imbibe it rapidly, hold it for a period of time, keep the clothing from soiling, keep the skin of the wearer dry, mask odor, be easily worn and removed, and be conveniently disposed off. Accordingly, the intimate and critical role that nonwoven materials have played in the development of commercial absorbent structures is such that it is often difficult to segregate the specific function of one component from that of another in the composite. Indeed, the challenge in the designing of an efficient absorbent product is to organize the structural elements into an integrated whole that benefits from the synergistic interaction between the parts. Where such cooperative functioning occurs, the maximum potential of the structure is realized.

The important role played by nonwoven materials in commercial absorbent structures has resulted in a somewhat parallel growth in the consumption of disposable nonwovens and absorbent products. This is illustrated by the data given in Table 1 [1], which gives the sales figures for disposable nonwovens and absorbent articles in the world. Accordingly, the growth in the sales and, therefore, in consumption, of absorbent products has been phenomenal during the past 20 years, and the trend can be expected to continue, especially as these products become affordable and adopted by the consumers in the developing countries. According to one forecast, the growth in the nonwoven sector, considering all materials, has been between 5–10 % during the past decade and this trend is expected to be continued during the next decade [2].

Table 1.
Approximate sales of disposable nonwoven fabrics and absorbent products in the world. The figures on absorbent products include those for diapers and training pants, feminine hygiene materials, adult incontinence pads and underpads.

	Sales in Billion $					
	1975	1980	1985	1990	1995	2000
Disposable Nonwoven Fabrics	<0.4	2.2	3.7	4.3	6.0	7.9
Absorbent Products	<1.0	6.0	13.0	18.0	25.0	33.0
Diapers/Training Pants	<0.5	5.3	8.7	11.9	15.5	21.0
Adult Incontinence Pads	Small	0.2	0.4	1.5	2.6	3.8

Despite many attempts at defining the term, a satisfactory, universally accepted single definition of nonwovens is still to be developed [3-5]. A major effort, under the auspices of the International Standards Organization (ISO), has proposed the following [6]:

> "A nonwoven is a manufactured sheet of directionally or randomly oriented fibers, bonded by friction, and/or cohesion, and/or adhesion, excluding paper and products which are woven, knitted, tufted, stitch-bonded incorporating binding yarns or filaments, or felted by wet-milling whether additionally needled or not. The fibers may be of natural or man-made origin. They may be staple or continuous filaments or be formed in situ."

This definition suffers from a lack of succinctness and is burdened by additional footnotes, caveats and limitations. This was recognized by Batra et.al. [7] who proposed to classify nonwoven structures into three different groups, namely, the 'fiber web structures', the 'net-like structures', and the 'multiplex structures.' The definition given above closely coincides with that for the 'fiber web structures.' In order to account for the remaining structures, that are not knitted, woven or braided, or falling in the category of paper goods, they suggested the other two terms. However, in spite of the vagueness associated with the single term nonwovens, the latter continues to be used worldwide to represent a broad range of materials that are not knitted, woven or braided, or classified as paper, but are webs based on fibers or fibrils.

1.2. Common Usage of the Term

From a practical standpoint, the characteristic that is unique to nonwoven structures, in contrast to woven and knitted materials, is the absence of a yarn element in the fiber web [8,9]. Thus, nonwovens are an assemblage of fibers, which have not been put into an anisotropic yarn form required by the weaving, knitting or braiding processes.

Also, it should be recognized that nonwoven technology has had its genesis in three originating technologies/industries: textile, paper and polymer processing. Thus, there are segments of nonwoven technology (such as carding in the dry form process) that closely resemble a segment of the conventional textile industry. Other nonwoven structures bear an unmistakable resemblance to products of the paper industry (wet form process), while still other nonwoven materials rely heavily on polymer processing technology and synthetic fiber manufacture as a major sequence in their production (spunbond process).

Thus, there is a frequent tendency for more or less all of the structural elements in absorbent products to be classified as "nonwoven", which in one sense they are, being not of a "woven" nature. The somewhat generic term "disposable" has also heightened the tendency to broaden the scope of the definition. This ready acceptance of incorporating a wide variety of materials under a somewhat imprecise label is sufficiently widespread such that new structural materials initially associated with disposable absorbent products are almost inherently accepted as "nonwovens", even though they do not fit the above definition and may even have no fiber content whatsoever. Thus, perforated films, microporous films and high bulk tissues are frequently so classified.

Notwithstanding these problems of definition and classification, nonwoven materials proliferate, new absorbent products are developed and fundamental principles of absorbency are used and combined in producing unique and innovative articles. All of this has been to the advantage of the industry and the consumer.

In the current discussion, a broad view will be taken of the nonwoven materials, beyond that envisioned by the definition given earlier [6] or that proposed under the term 'fiber web structures' [7].

2. NONWOVEN STRUCTURES

In discussing nonwovens, consideration must be given to the interrelationships of the four major structural elements that influence absorbency:

- Constituent fiber.
- Assembly structure; the web properties resulting from the nature of the fiber assemblage.
- Type of bonding; the chemical or mechanical means of conveying integrity to the fibrous web.
- Finish; chemical and/or mechanical treatments conveyed to the formed and bonded web.

Each of these factors can have an effect on the absorbency of the resulting structure.

2.1. Fiber

Virtually every fiber, both natural and man-made, has been used in absorbent nonwoven structures. Naturally, the hydrophilic fibers have received major attention and utilization. Much of the nonwoven technology in use today was initially developed for the cellulosic fibers. Thus, cotton and rayon gave rise to nonwoven technology coming from the textile sector; wood pulp fiber, on the other hand, was the mainstay of nonwoven technology emerging from the paper industry.

In recent years, synthetic organic fibers have found increasing utilization in absorbent structures. Such synthetic fibers are sometimes utilized because of their inherent hydrophobic properties, whereas other structures utilize their generally superior mechanical properties, especially wet physical properties.

Being the structural element, fibers making up a nonwoven have a major influence on the absorbency characteristics. The major fiber properties exerting such an influence include the following: polymer type, linear density or denier, fiber cross-section shape, crimp, fiber finish and fine structure.

The polymer type, particularly whether hydrophilic or hydrophobic, influences the inherent absorbency properties of the resulting structure. A hydrophilic, swellable fiber gives a structure that is able to absorb liquid via fiber imbibition, giving rise to fiber swelling; such a structure attracts and also holds liquid external to the fiber, in the capillaries and structure voids A hydrophobic fiber, on the other hand, has only the latter mechanism available to it normally.

Fiber linear density or denier, related to the cross-section area, influences void volume, capillary dimensions and the total number of capillaries per unit mass. Fiber cross-section shape can affect many of the same characteristics but also the nature of the capillaries in a structure. Four deep grooved (4DG) polyester, trilobal rayon, L-shaped nylon are examples that are developed to enhance capillary flow [10,11].

Fiber crimp can influence packing efficiency of fibers and the resulting fabric bulk, both of which can influence the absorbency of a nonwoven structure. Also, the nature of the crimp, whether it is two-dimensional or three-dimensional, can have an influence. Fiber finish can also have an impact. The finish usually consists of a small amount (generally 0.1-0.5% by weight) of lubricant and other formulation components applied topically to a fiber. It is used in order to enhance a fiber's processing performance in the mechanical equipment used to open, blend and convert the fiber into a web. Because the finish is on the fiber surface, and the latter is often the only segment of fiber encountered by liquid, the finish can profoundly affect surface wetting, liquid wicking and other properties, directly influencing absorbency.

Further morphological features, such as surface rugosity, core uniformity, etc., can influence absorbency performance somewhat. This possibility has been exploited in a synthetic acrylic fiber produced with a micro-capillary core [12,13]. The micro-capillaries lead to surface holes along the length of the fiber. Although the fiber cannot swell, it can imbibe a significant quantity of liquid (30% by weight and more) by drawing it into the porous core through the surface perforations. A unique feature of such absorbency is the amount of water that can be held without the perception of surface wetness.

Consumption of major fibers by the North American nonwoven industry in total, that includes the absorbent segment of the industry, is provided in Table 2 [1]. The numbers related to the past decade indicate a surge in the use of polyolefins, a somewhat sustained use of polyester, a steady increase in the use of the natural cellulose fiber and a relatively low (as compared to the figure in 1975) but a steady use of the regenerated cellulose fiber. The changes noted in the numbers related to polypropylene and rayon in the table is to a significant extent due to the replacement of the latter by the former in the coverstock application of the absorbent products and the opening of new markets for meltblown

Table 2.
Shipments of synthetic fibers to North American nonwoven producers (in millions of pounds).

	Shipments (Million Pounds)					
	1975	1980	1985	1990	1995	2000
Polypropylene	13	35	139	231	273	458
Polyester	26	167	165	240	282	275
Cotton	*	*	47	50	68	73
Rayon	139	147	117	73	59	68

* Not available

polypropylene fabrics. Chapter VI contains detailed information on the inherent characteristics of fiber types employed in absorbent structures.

2.2. Web Assemblage

The manner in which the fibers are assembled into a web has profound effect on the web properties. The inherent versatility of the nonwoven process is exemplified by the fact that the web structure can range from the extremes of a highly oriented fiber configuration to one of completely random arrangement. As would be expected, the former has considerably different physical properties in different directions. A maximum in tensile strength would be anticipated along the axis of orientation, and minimum at right angles. Such a structure would be expected to have a minimum of tear strength. Such properties exist in the carded fiber webs of the dry form process. On the other hand, a fiber assemblage with a completely random orientation of individual fibers has isotropic properties, with essentially equal physical characteristics in all directions.

Nonwoven structures can also be assembled with fiber arrangement between these two limits, with the physical properties also lying intermediate to the ones given by the two characteristics. Because the fiber configuration in the nonwoven structure will have an effect on such things as packing, pore size, capillary dimensions, capillary orientation, etc., the absorbency properties of the nonwoven structure can also be expected to be sensitive to the nature of fiber arrangement in the web.

Web formation can also be supplemented by localized rearrangement of fibers giving rise to apertured fabrics [8,9]. This has allowed a manufacturer to engineer fabric with precisely controlled perforations and hence obtain modified liquid transfer properties. Further, the bundling of fibers possible by this process can enhance the fabric wicking capabilities, somewhat reminiscent of woven yarn structures.

2.3. Fiber Bonding

In a woven structure, the frictional forces existing at the yarn interlacings is responsible for the strength and integrity of the fabric. In a similar manner, the yarn frictional forces provide the inherent strength and integrity of the interlooped structure in a knitted fabric. In contrast, the nonwoven structure has less possibility for similar forces, as the structural element is a single fiber or group of fibers and the spaces between elements are large and contacts between them few. Consequently, a bonding means is required in a nonwoven web to impart mechanical integrity. In general, the bonding can be characterized as derived from either a chemical basis or a mechanical or physical basis; in some cases, it is a combination of the two systems.

Many nonwovens are bonded by the addition of an external chemical binder. The nature of the latter and, in particular, the method of application of the chemical to the fibrous web has a profound effect on the absorbency of the resultant structure [14]. Because most binder materials are hydrophobic, the net effect usually is some reduction in absorbency [15]. In mechanical bonding, the inherent characteristics of the fibers are usually unaffected; consequently, the effect of mechanical bonding on absorbency of fibers is minimal. However, since mechanical bonding, e.g. needle felting, causes entanglement of fibers, two effects can take place: (1) the entanglement may restrict the inherent ability of structure to swell, and (2) the fabric may become more resilient and resist collapse when subjected to external pressure. These changes in the behavior of a web with mechanical bonding can have a significant impact on the capillary absorption of fluid, as has been shown in Chapter III.

2.4. Binders

Although the usual binder normally detracts from the inherent absorbency properties of a nonwoven structure, it is needed in order to generate strength and integrity in the fibrous web. Indeed, it was only with the introduction of efficient, effective binders, that the full growth of the nonwoven absorbent products has been achieved. The original bonding materials consisted of starch, mucilage and other natural animal and plant adhesives. These binders did convey some integrity, but were water-sensitive, stiff and quite inadequate [16].

The synthetic latex binders, originally based on poly(vinyl acetate), carried the nonwoven materials to a new performance level, permitting widespread use in absorbent, related and many other applications. With the advent of binders based on acrylic ester copolymers, a further enhancement in nonwoven performance was achieved that led to the opening of new and more critical applications for nonwovens [17].

A broad selection of chemical binders has been available to the nonwoven technologist, permitting rather precise engineering of the final product performance. Table 3 lists some of the major binder types, with selected characteristics of interest to absorbent applications [18].

Current commercial practice involves a rather extensive formulation of the binder chemical by the fabric manufacturer, in order to achieve optimum performance from the material. Since the application of chemical involves inevitable compromises between absorbency and mechanical integrity, the preparation used is such that it allows one to achieve best balance of properties. Formulation generally involves water dilution of the binder from the original approximately 50% solids, along with use of an acid or latent acid catalyst (to promote internal cross-linking), viscosity modifier, defoaming agent, and a surfactant. The resin add-on can vary from 5 to 50%. With most absorbent nonwoven structures, a resin content of 10-25% by weight is typical.

Table 3.
Chemical binders for U. S. nonwovens production.

Type (Solids %)	App. Share (%)
Acrylic (50%)	47
Polyvinyl acetate/ copolymers (50%)	22
Styrene butadiene (50%)	12
Polyvinyl chloride copolymers (55%)	4
Nitriles (50%)	1
Polyvinyl alcohol (100%)	1
Ethylene/vinyl acetate (100%)	12
Others	1

Because of the concern with the long term effects of chemicals used in the manufacture of the external binders, there has been a trend in recent years to emphasize "binder-free" products. This interest has also motivated the binder manufacturers to modify their polymerization methods and materials, and led to (1) the introduction of plastic dispersion-type binders, with no reactive capability [19] and (2) the development of "low formaldehyde" and "formaldehyde-free" chemicals [20,21]. Further, the concern with the chemical additives has stimulated the adoption and growth of the thermal techniques for the bonding of nonwovens, as these methods eliminate the need for external chemicals.

2.5. Web Finishing

In nonwoven processes, a web is generally "finished" as soon as the fibers are assembled into an appropriate structure and bonded by the selected chemical or physical means. However, it is possible to give additional chemical or mechanical treatment to the bonded web, which can have an important influence on absorbency and related characteristics. Thus, it is possible to put a chemical treatment on the fabric, which will provide a repellent characteristic [22]. At times, such a treatment can be advantageously employed when it is applied to only a portion of an absorbent structure; thus, a method has been described for the application of a liquid silicone chemical treatment to the sides of a sanitary napkin, to give it a "boat-type" containment capability.

Conversely, a chemical finish can be selected which may enhance the absorbency of the nonwoven structure, by modifying the wetting performance of a fiber surface and, therefore, affecting the wicking behavior.

Mechanical treatments can also be given to the substrate which will affect the absorbency characteristics. Mechanical softening treatments may involve fiber displacement, web buckling or other changes and affect the absorbency of the resulting structure. Embossing, calendering and surface glazing, and other mechanical treatments can be utilized to modify web properties and the related absorbency characteristics.

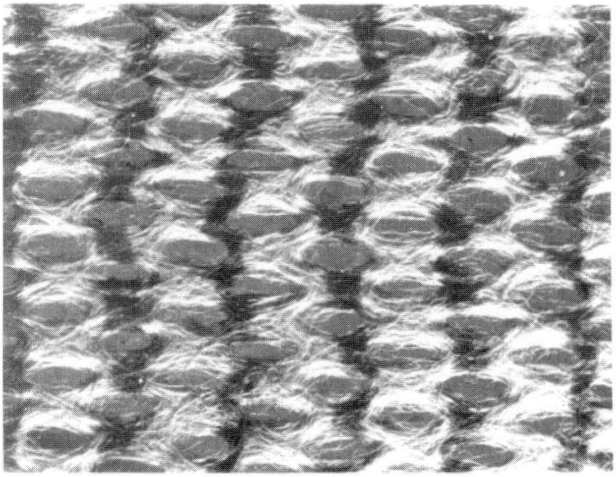

Figure 1. Print-bonded dry form nonwoven.

3. NONWOVEN FABRICATION

Nonwoven structures can be classified by a variety of means, the most common means being the process type or fabrication method. Such classification is usually based on the method of forming the fiber web.

3.1. Uni-Directional Dry Form Process

This process (also called card and bind process or carding process) involves forming a fiber web with fibers predominantly oriented along the longitudinal direction (machine direction). A conventional textile carding machine was originally used for this process; more recently, carding units specifically designed for nonwoven manufacture have been developed and used broadly within the industry [22]. The completeness of fiber parallelization or orientation in the long direction is dependent upon the effectiveness of the carding process, the tension applied to the web during its formation and prior to bonding, and other factors.

A dry form web is characterized by good tensile properties in the direction of the fiber orientation, but by poor properties in the transverse direction. Several methods have been developed to overcome or minimize this deficiency. In print bonding of such a web, careful design (Fig. 1) of the print pattern can help direct the strengthening contribution of the added binder to the weak cross direction [23].

Highly oriented webs can also be laid in plies in the cross direction, conveying fiber strength to that direction. These webs are discussed more thoroughly under the needlepunch process.

3.2. Bonding of Dry Form Web

The bonding method normally employed for the dry form web is chemical, involving a latex binder. The latter can be applied by a variety of methods, including saturation of fabric, spraying a web, or printing binder with a roll. The saturation method is normally not considered for absorbent materials, as the binder covering virtually all fiber surfaces can greatly interfere with the absorbency process. Spray bonding involves applying spherical droplets of binder to the web via airless or air sprays. This leads to bonding where the fine droplets contact the fiber structure at cross-over points of two or more fibers. When a spray droplet fails to migrate or converge to a cross-over point, it serves no bonding purpose, rather adds weight and stiffness without strength. However, with a control on the amount of binder used by spraying, and appropriate binder rheology, most of the binder can be made to migrate to the cross-over points. The result is that a significant portion of the fiber web is left free of the adhesive, meaning that the fiber web retains its inherent absorbency properties, along with other desirable textile-like characteristics, particularly fabric hand.

The placement of the binder resin can be controlled to an even greater extent by means of print bonding. Innovative print pattern designs and technology associated with it have been developed to maximize the effectiveness of the added binder, with a minimum amount of deleterious effects [8,9]. In general, print bonding involves a compromise between fabric tensile properties, that are enhanced, and fabric properties such as absorbency, softness, and handle, that are degraded, by the process. In this method, the binder is positioned on the web in a discrete pattern. The process permits high control of the location of the chemical and its distribution. Processing conditions are selected such that the resin is distributed uniformly through the thickness of the web, without its migrating laterally outside of the pattern. Thus the binder location can he engineered precisely so that there is maximum benefit from the chemical used and minimum interference from it with the primary function of the structure. Consequently, the print bonding method is preferred for most dry form applications. Application of the binder in a specific pattern is normally achieved via a rotogravure process, utilizing engraved rolls [24]. A rotary screen printing unit can also be employed, with the binder resin used in the form of a paste [25]. More recently, the use of binder in the form of foam has been utilized [26]. This technique leads to reduction in the energy required to dry the printed web, as the latex binder is diluted with air rather than water, resulting in lower water load to be evaporated from the web.

A modification of the conventional dry form process has been used in nonwoven facings for absorbent products. This involves the use of thermal techniques for bonding [27,28]. A fabric contains fibers that are thermoplastic and capable of bonding with heat (Fig. 2). The fiber that has found particular utility in this method is polypropylene as its melting temperature range is suitably low and its melt flow viscosity is amenable to appropriate flow for effective bonding. With polypropylene fiber, either alone or in blends containing about 40% or more of the fiber, bonding can be done with hot, smooth, roll calenders. Since this process leads to a fabric with greater stiffness than desired, point embossing of only a portion of the total area is generally used to maintain the needed fabric softness.

Another version of this method involves the use of special binder fibers that have bicomponent structure. These fibers have a higher melting core with a lower melting sheath; the latter is activated by the temperature employed leading to bonding, while the former is unaffected and maintains its fibrous form and provides strength and stability to the web formed. In this case, thermobonding can be done with calender rolls (smooth or embossed)

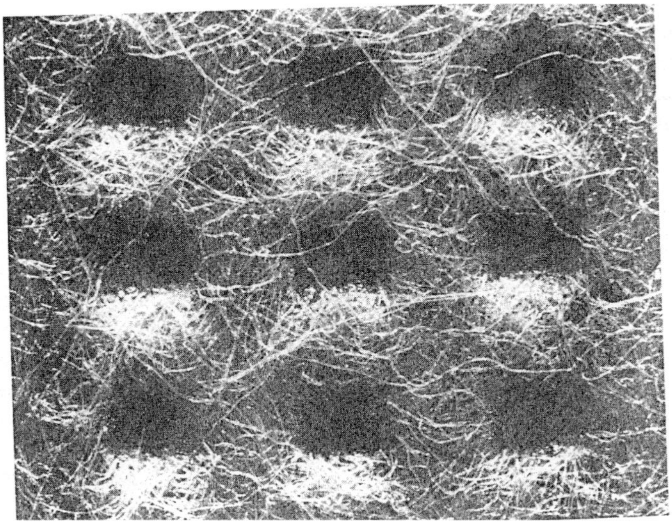

Figure 2. Point emboss thermobonded nonwoven.

or by hot air. Also, ultrasonics, infrared, and other selected sources of controlled heating can be used for the purpose.

3.3. Airlaid Process

A method that is often used to form fiber webs with random or isotropic orientation directly is the airlaid or the air form process. In this, individualized fibers are suspended into an air stream, and allowed to be deposited on a moving pervious belt or drum. By careful control on air flow pattern and receiver speed, the orientation of the fiber onto the forming element can be controlled [29,30]. This ensures that a fiber will have an equal probability of lying in any of the possible directions and lead to an isotropic web. A variety of continuous web formers have been developed that allow air forming of webs of different materials and of a broad range of specifications.

The random orientation of the fibers in the structure with some tendency of fibers to be lifted out of the X-Y plane and become partially oriented in the Z direction results in a web with higher bulk and larger void volume. This situation, of course, can have a significant positive effect on the absorbency properties of the resulting web. Bonding of web made by the air form process can be achieved by the same methods that are utilized for the bonding of the uni-directional dry form structure, i.e. saturation, spray, print or thermal bonding.

3.4. Airlaid Pulp Process

The special adaptation of the air form process has been used extensively in the production of highly absorbent nonwoven structures composed of wood pulp fiber [31,32]. This process, the so-called the air form or the airlaid pulp process, involves generation of an air stream filled with individualized pulp fibers and collecting them on a moving belt. The pulp suspension in air is usually generated by hammermill processing of dry pulp sheets,

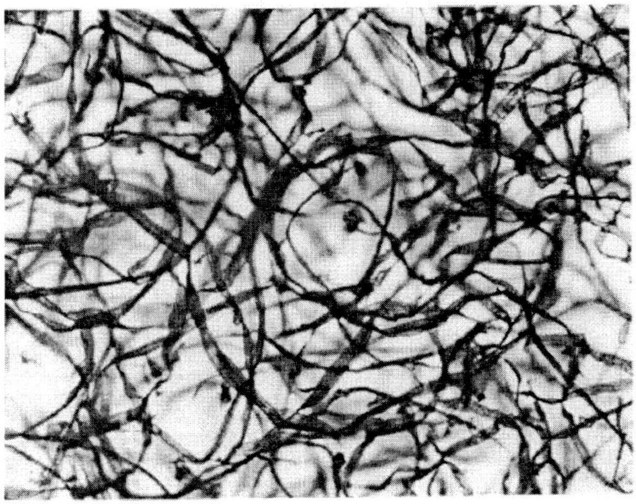

Figure 3. Airlaid pulp web.

followed by a technique that separates clumps. The air stream containing the suspended fibers is then passed through a moving porous belt that gives a random deposition of pulp fibers and leads to a bulky web. Bonding is normally accomplished by spraying of a latex that helps to preserve bulk. This is usually achieved by spraying binder resin first from one side of web, followed by drying and reversing and then spraying from the opposite side. After the second side is dried, the entire web is cured in a hot air oven [33,34]. Typically, the binder content in the final product ranges from 13-25% by weight. Also, thermal fusible fibers are used in some airlaid nonwovens in order to bind the pulp web. With proper processing, such structures can retain a large fraction of the original fabric bulk (Fig. 3), leading to a product that has high potential absorption properties. These materials are particularly suited for use in absorbent cores.

Although not normally recognized as a nonwoven material, the fluff pulp pad in many absorbent products is very similar in structure to that of the air formed pulp nonwoven, except that no bonding agent is employed. In this case, the pulp sheet is comminuted with a hammermill or similar device. In Europe, pin mills have normally been employed for the pulp sheet comminution process [35]. Because such pin mills do not generally have the "grinding" power of a hammermill, sheeted pulp used on such units is generally treated with an agent to facilitate comminution [36]. The pulp fibers are suspended in the air stream generated by the mill and the air-fiber stream is then conveyed to a pad forming station on the converting line [37]. For convenience and ease in manufacture of final products, the pulp pad is often formed onto a carrier tissue or other web substrate. Because the structure generally is not bonded, either by means of an external binder resin or by compressing the web to increase bonding, the fluff pad generally has relatively low integrity. This feature of fluff pad often requires special considerations in designing of the components for assembly in the final product.

There has been a significant effort devoted to improving the absorption performance of fluff pulp pads. The usual low density of the pulp mass generally leads to a significant collapse in the thickness upon wetting. Accordingly, one of the procedures used has been the blending of short hydrophobic, synthetic fibers with pulp in the pad. By adding synthetic fibers to the pulp mass, a less water sensitive matrix is formed which assists in retaining void volume by preventing wet collapse. One of the most successful approaches along this line suggested is the admixture of melt blown micro-fibers with the pulp fiber prior to pad formation [38].

Another idea that has been proposed in the production of materials by the airlaying technology is the multi-bonded airlaid structure [39]. This applies to the preparation of multi-layer composites where multi-forming and multi-bonding stations are organized in one production line that also allows the incorporation of superabsorbent material in the structure. Such system provides a means to design and manufacture finished goods considerably more economically. An example of a product that could be produced on such a system is a diaper core where the complete multi-layer composite, each layer fulfilling a particular task, is produced in one step on an air-laid line.

3.5. Wetlaid Process

The wetlaid or wet form nonwoven process is derived from the paper industry, and bears a considerable resemblance to the normal Fordrinier paper manufacturing process. In this system, short fibers (generally one-quarter inch or less) are suspended in an aqueous slurry along with minor amounts of other constituents. The slurry is subjected to strong agitation in order to cause the fibers to be uniformly distributed. The formation of the web is then achieved by conveying the slurry to a moving wire belt, where the liquid drains through and the fibers are left in the form of a mat in largely random configuration. The web is then removed from the conveyor, taken through a bonding process, and then dried and wound up onto a roll.

Although the method appears largely similar to the papermaking process, there are critical differences [40,41]. Paper-making normally involves a furnish of 100% wood pulp, whereas wet form nonwovens can have compositions that range from a sizeable wood pulp content to a completely synthetic fiber furnish [42]. Wood pulp is frequently used in absorbent, wet form, nonwoven materials because it has an inherent high absorbency, is a low cost fiber, and provides a capability for self bonding through hydrogen linkages. Such bonding between the flat pulp fibers, so critical to paper manufacture, however, results in a product having paper-like properties, such as stiffness, high density, low tear strength, etc. In nonwovens, on the other hand, usually the fibers used are longer and crimped or convoluted, and the frequency of bonding used is lower. These lead to a fabric that is relatively bulkier and softer. To achieve more textile-like properties, in contrast to paper-like, the amount of wood pulp used and, therefore, the degree of hydrogen bonding developed are reduced. An external binder is generally employed to increase strength while maintaining bulk. Such a resin can be applied by spray, print or saturation methods. In addition, there is some use of the technique of adding binder directly to the original fiber slurry. The wet web is then dried and the chemical cured, as necessary. To retain maximum bulk, the web is subjected to minimum pressure during the wet form process. This is achieved by the use of flow through hot air during the drying operation [43,44].

Although not normally considered to be a nonwoven process, the technology for manufacturing high bulk sanitary tissue has a definite relationship to wet form nonwoven

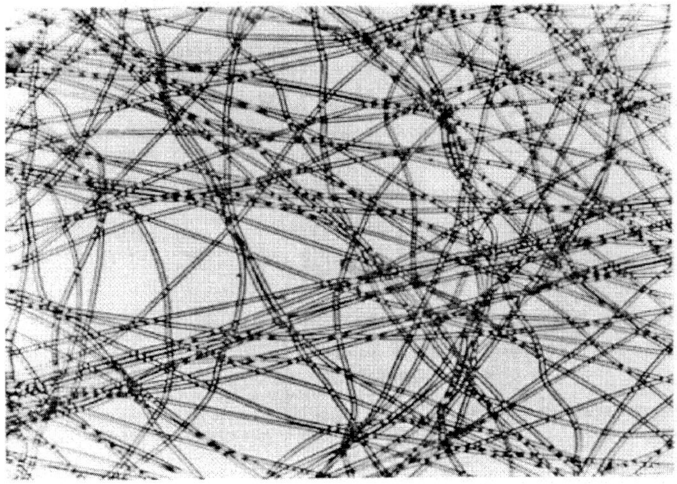

Figure 4. Spunbond nonwoven web.

technology in many respects. Thus, conditions and agents are employed to minimize hydrogen bonding [45]. The formed web is generally dried by a through hot air system to minimize web pressing and resulting web compaction. Also, there is often the use of an external chemical binder that enhances web strength and integrity [46]. In one process, a print bond technique is utilized for producing a bulky tissue. In this, the print bonded tissue is creped on a Yankee dryer; the presence of the bonded regions forces the unprinted regions to crepe, and develop bulk [47]. The resulting structure has the needed physical strength and high bulk, along with excellent softness and absorbency.

3.6. Spunbond Process

The spunbond technology combines the process of extrusion and orientation of thermoplastic synthetic fiber with that of the formation of web and bonding of the nonwoven technology. The result is an integrated process capable of producing a bonded synthetic fiber web of continuous filaments derived from a synthetic polymer in a one-step operation, i.e. it is possible to go from the resin pellet to the finished fabric in one stage [48]. Accordingly, there is significant economic advantage associated with the process. Since the web is derived from continuous filaments and that comparatively strong autogenic bonds are normally formed, the spunbond fabric in general has excellent tensile strength and tear resistance. Also, the web forming step is such that the resulting web is essentially isotropic in structure and properties (Fig. 4).

Inasmuch as this system is based on polymer processing, spunbond structures are normally derived only from man-made filamentous materials, and most usually from thermoplastic resins. Consequently, the spunbond materials are hydrophobic in character. Despite this fact, these materials find extensive use in facings in absorbent structures, especially as the web weight can be reduced to a low level because of the inherent strength of the structure [49].

Spunbond fabrics of particular commercial significance are derived from polyester, polyamide and polypropylene resins. The most frequent method of bonding employed is autogenic bonding, i.e., constituent filaments are made sufficiently fluid so that the filament surfaces touching each other will bond together upon re-solidification of the polymer. However, this often results in excessive bonding points and inadequate filament orientation, with a resulting fabric that is stiff and made up of low strength filaments. Consequently, the more common method used for bonding spunbond webs is to form an isotropic web from the oriented filaments and then pass the web through a thermobonding calender to accomplish bonding. If maximum strength is required, the bonding will be "areal" or over-all of the web. This again will lead to a stiff fabric. To achieve a softer fabric, "point bonding" is used, generally obtained by employing a hot calender roll of elevated points and a smooth back-up roll. Sufficient temperature and pressure are applied to deform and melt weld the filaments within the bond site, but not enough to completely destroy the fiber properties and convert the polymer in the site to an unoriented film.

Principles developed in the design of bond patterns for print bonding dryform nonwovens are also utilized in the design of thermobond print patterns [50]. The amount of bonding area for a typical spunbond fabric ranges from about 10% to 35% or more. Again, by judicious control on the design of the bond site, relative placements, percent bonded area, and other factors, desired balance between strength, softness and other fabric properties can be achieved.

3.7. Needlepunch Process

The needlepunch technology is the most widely used method for mechanical bonding. For this process, the base web employed is normally derived from either the air forming or the carding-cum-cross-lapping process. The cross-laying is done in a continuous manner, often onto another carded uni-directional web. The result, in this case, is a fiber web with one layer having a uni-directional orientation in the machine direction and the other succeeding layers having a uni-directional web oriented at 90° to the machine direction, or more correctly, at an angle of less than 90° to the machine direction because of the continuous nature of the process. By proper selection and juxtapositioning of webs, it is possible to more or less completely control the fiber orientation in the resulting cross-lapped web [51].

Bonding is achieved by the mechanical action resulting from penetration of a barbed needle into the fiber web. The barbs on the needle are arranged such that as the needle penetrates, fibers from the top layer of the web are engaged in the barbs and are driven transversely through the thickness of the web. As the needle retracts, the fibers are released from the one-way barbs. This results in mechanical entrapment and orientation of the portions of fibers in the Z-direction (thickness) of the web. By control of the needle penetration and the repetition of the needling action from the opposite side of the web, a three-dimensional, mechanically entangled network can be achieved [52]. In the resulting structure, often called a needlefelt, the mechanical entanglement is of groups or tufts of fibers rather than of individual fibers. As the extent of needling is increased, a more dense structure results that possesses increased strength, tear resistance and fiber tie-down.

There is a minimum amount of fiber that can be employed in the mechanical entanglement of a needlefelt web. Consequently, most needlefelt structures have a weight of

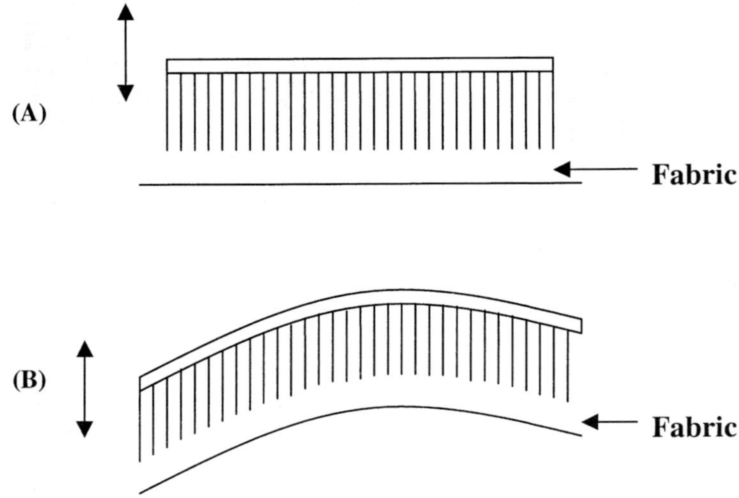

Figure 5. Needle zones. (A) conventional loom, (B) contoured needle zone of H1 needle loom [55].

60 g/m^2 (~ 2 oz/yd^2) or more; a weight of 50 g/m^2 is about the minimum that can be effectively needle bonded.

The absence of added chemical binder gives the needlefelt fabric a structure that possess the inherent absorbency of the structural fibers. The needlepunching process itself results in consolidation of the fibrous web by creating an interconnected network of capillaries in three dimensions. As a result, needlefelt webs are highly resilient and have high capability for absorbing rapidly and holding large amount of fluid.

Advances have taken place in the needlefelt technology in recent years. The feed system has undergone some changes that enhance the precision and the uniformity of the feed and the control of the feed material to the needle loom. One example of these is the micro control feed monitoring system for the needlefelt machine [53]. The greatest advance is in the needle loom design itself [54]. A new loom introduced recently is equipped with a distinctly contoured needle zone (Figure 5) that allows needles to penetrate a web at different angles, both less than and greater than 90° to the X-Y plane (Figure 6) [55]. The result is a better integrated structure, claimed to be obtained at higher production rate [56].

3.8. Spunlace Process

This process utilizes staple fiber web with a specialized type of mechanical bonding [57,58]. The fiber webs can be either oriented or isometric, but in common with other processes, more uniform properties and better overall performance is obtained with the latter. For bonding, the system uses high energy, closely spaced, water jets that emerge from an injector and impinge on a web substrate and entangle loose arrays of fibers (Fig. 7). The process is claimed to have the capability of producing a variety of surface and fabric patterns from many different precursor webs made from essentially any fiber that is not too stiff or

(A)

(B)

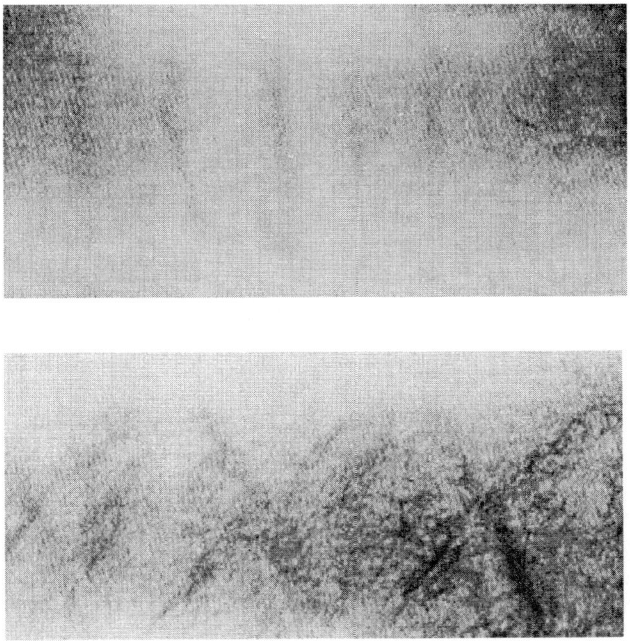

Figure 6. Micrographs of needled fabrics. (A) Conventional needle loom web, (B) H1 needle loom web.

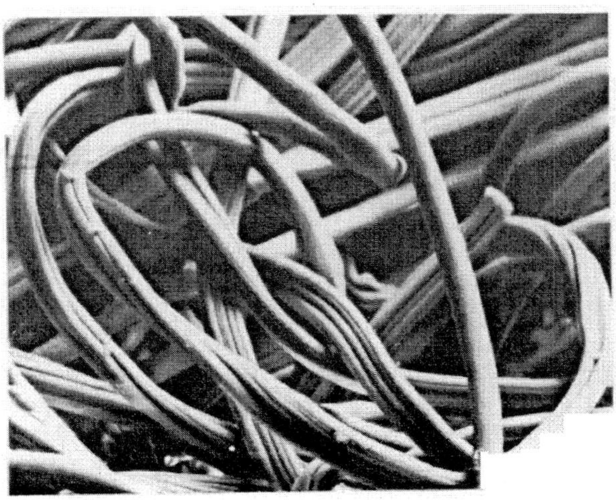

Figure 7. Spunlace web from rayon and polyester.

brittle to bend or move. The injectors are positioned above the moving backing belt or rotating cylinders, which are perforated and carry the unbonded web through the unit.

Two different bonding configurations have been employed widely in the process [59,60]. One is the horizontal system, wherein the fiber web is carried horizontally on a perforated backing belt through the water jet bonding zone; the water jets are also arranged in a horizontal manner, generally with two or more jet manifolds providing the streams. This action accomplishes bonding from one side of the web; if it is desired to entangle from the other side as well, the partially bonded web is led over rolls so that it could be turned and then bonded from the opposite side.

The alternative system utilizes perforated rotating drums for support of the unbonded material. In this case, the water jet manifolds are mounted around the periphery of the drum, and placed close to the drum surface to minimize the distance from the jet orifice to the web. A drum of one-meter can normally accommodate 3 to 8 jet manifolds. Again, when it is desired to entangle the web from both sides, the partially entangled web is removed from one drum, turned or reversed in direction and entangled from the other side.

A system will normally consist of two aprons, one at the feed end and the other at the receiving end, and two cylinders, all four serving to support the web and receiving water streams. The two cylinder system facilitates bonding from both sides of a fabric.

The precursor web can be formed practically by any of the drylaid (airlaid or carded), wetlaid, direct-laid (spunbonded or meltblown) methods for spunlacing with varying degrees of success [60]. For the drylaid precursor, a carded, carded/cross-lapped, or airlaid web of staple fibers is used. For the wetlaid precursor, uniform (isotropic) sheets are made by dispersing fibers in water at very high dilution and depositing on a screen to separate water from fibers.

Wood pulp/polyester fiber blended spunlace fabrics were originally the largest volume of products produced by the process; this resulted from the fact that medical fabrics (for disposable gowns, drapes and packs) were the primary spunlace fabrics used during the early commercialization of the process [61]. In the past few years, however, other spunlace fabric types have become significant commercial successes. Fabrics produced from 100% bleached cotton fibers are included in this category which have been used in a variety of absorbent product applications.

Two processes have been employed for producing the wood pulp/polyester fiber blended spunlace materials. One process involves a special, lightly bonded, low density wood pulp sheet. The latter is formed by conventional papermaking process using a special grade of long fiber wood pulp, usually based on a furnish of western red cedar. This paper sheet is unwound onto the top of a partially bonded isotropic polyester web and then subjected to water jet entangling. The finished sheet has a wood pulp-rich top layer and a polyester-rich bottom layer after entangling from both sides.

The alternative process involves an airlaid web formed by a special air forming equipment [62] fed with a blend of long fiber wood pulp and polyester fiber. This web forming system has the capability of controlling placement of the wood pulp and polyester fibers in the initial web, so that again the final fabric has a wood pulp-rich top layer and polyester-rich bottom layer after entangling.

Most commercial wet-laid machines will generally be limited to fiber lengths less than ½ inch, meaning that the fabrics produced from these webs will usually have lower strengths than those produced from dry-laid precursor webs. However, a resin bonding step can be added if higher strength is required but this is usually done at the expense of flexibility.

In addition to drylaid and wetlaid precursors, a direct-laid precursor can also be used. These are the ones made directly from molten polymer continuously spun into fibers and deposited on a moving apron in a fashion similar to spunbonding and meltblowing processes. These webs do not by themselves entangle well as the continuous filaments lack mobility. However, composite fabrics with excellent three-dimensional stability can be produced by hydroentangling a composite containing a drylaid or wetlaid staple fiber web superimposed on a direct-laid fabric that serves as a reinforcing scrim for the structure [60]. This variant of the spunlace process is especially useful in producing a composite web of a lightweight spunbound fabric around whose filaments a wood pulp layer has been entangled. This can provide a strong yet absorbent fabric from relatively low-cost raw materials.

The water jet employed in the spunlace operation is a means of conveying enough energy to the loose fiber web to cause vigorous movement of individual fibers, sufficient to result in entanglements. The more the entanglements, the greater the strength. To maximize the energy transfer from the jets to the web, the pressure of the water must be relatively high (800 to 2,000 psi or more), the jets must be compact (columnar streams) and not spread out (fan streams). The columnar jets must be parallel and not collide, as this would destroy the entangling capability. These requirements dictate a precision in the design, manufacture and assembly of the manifold components.

The forming belts or supports (patterning belts, backing belts) which support the fiber web as it passes into and through the entangling zone are important elements in the process [63]. In essence, the impaction of water jets forces fibers onto and into the backing belt, giving the fiber web surface the mirror image of the surface of the belt. If the backing belt has larger holes or apertures, an apertured web results; using this system one can actually produce a design or pattern if the latter can be incorporated into the belt. By using a backing belt with very fine apertures, a spunlace fabric approaching a non-apertured appearance and structure can be obtained. This versatility associated with the configuration of the supporting belt provides the capability to produce a wide range of surface and structural features in a spunlace fabric.

The hydroentangling line incorporates a unit for extraction of water by vacuum slots beneath each needling support to remove water penetrating the web. Inefficient water removal can lead to flooding and loss in entangling intensity. Additional water may be removed by high vacuum extraction process to reduce the drying load. The next step in the process is drying, which is often achieved with a through hot air system. A final step in most spunlacing lines is finishing, which is used to impart special characteristics to a product. It may involve a chemical treatment such as a wetting agent or an antimicrobial agent, or a physical treatment such as creping, calendering or heat setting.

Microscopic examination reveals that entanglements are frequent but the fibers are not tightly entangled. The combination of jet orifice spacings and screen patterns produce densely packed loosely entangled areas (see Fig. 11, Chapter III). Unless special care is taken, 100% cellulosic fiber composition is likely to lead to a papery structure; however, a blend containing short cellulose and synthetic fibers can produce a fabric that has excellent drape and bulk, as well as required strength.

The mechanical entanglement of the spunlace process is in contrast to the entanglement of the needlepunch process, which is entanglement of tufts of fibers [64]. A vast majority of spunlace fabrics are produced in basis weights ranging from about 20 g/m^2 to about 100 g/m^2. A web of weight less than 20 g/m^2 does not develop enough integrity unless an auxiliary means of bonding or very low denier fibers providing high number of fibers are used. Webs weighing more than about 100 g/m^2 are usually too heavy to be penetrated by water jets, unless very high energy levels are used.

As normally practiced in commercial operations, the entanglement, especially on fabric surface, is high enough to lead to a strong fabric with good fiber "tie-down." Because chemical binder is not normally used in this system, the spunlace fabric is also characterized by excellent softness and hand. As expected, the absorbency of the structure is excellent, being unhampered by external binder. There can be some restriction to total absorbency, however, by the fact that extensive linkages in densely packed regions can limit the swelling potential of individual fibers, which may result in reduced absorbency compared to that expected in an equivalent non-entangled fiber web. Nevertheless, this nonwoven system constitutes an important process for a number of absorbent products, most specially those used in the health care field.

A modification of the hydroentangling system has also been described and used [65]. This involves fiber entanglement at a relatively low level, which is still sufficient to convey some strength and integrity; this reduced level of cohesion is then supplemented by a limited amount of external latex binder, which conveys additional strength and integrity to the structure. The resulting web has the desired strength and integrity with only a small compromise in absorbency because of the low level of resin used. Also, the modest amount of resin employed (1-5% by weight) does not add much stiffness to the fabric. Consequently, the combination process, involving hydroentangling and latex resin bonding, can provide a fibrous web with a superior balance of absorbency and other physical properties.

A further modification of the spunlace process utilizing no external resin has been described [66]. In this case, the low level of fiber entanglement, achieved as above by using low water jet pressure, is supplemented by adding a small amount (~ 15%) of thermoplastic binder fiber. After drying the fabric, additional heat is applied to melt and thus activate the latter. A polyolefin bicomponent fiber is used for this process, although other binder fibers can also be employed. This same technique, involving addition of a low melt fiber, is now used at times to enhance the physical properties of needle punched fabrics as well [67].

An additional modification described recently is the use of nylon/polyester multi-component splittable fiber. The treatment with water jets splits the material into micro denier nylon and polyester filaments and entangles them to produce a soft, drapable, and strong nonwoven [68].

3.9. Meltblown Process

Meltblowing is one of the more recent technologies that has established itself as an important nonwoven process for producing structures that find applications in a number of products, including absorbent. It is a unique, one-step process that dates back to the work of Manning in 1946 [69], for making filter cartridges, and of Wente in 1954 [70], at the Naval

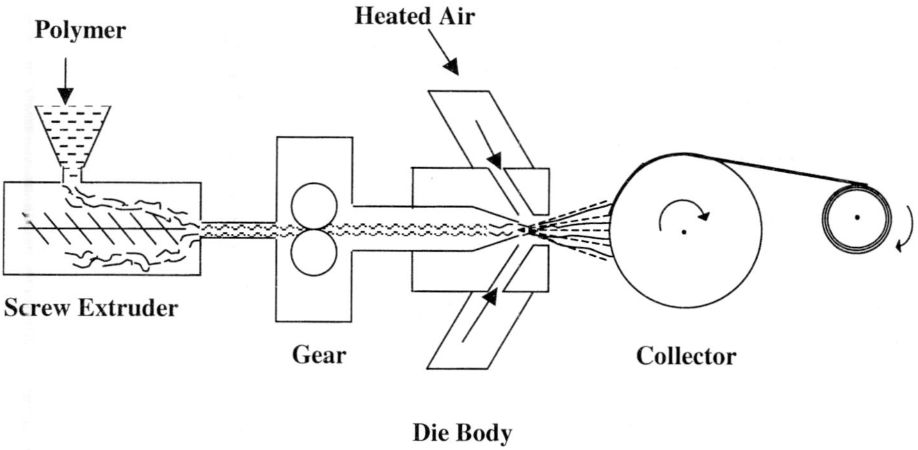

Figure 8. Schematic of a meltblowing process.

Research Laboratory, that showed that the melt of a polymer emerging from orifices could be blown into super fine fibers by hot, high velocity, air.

In this process, a molten polymer is converted into ultra fine fibers and collected on a rotary drum or a forming belt with a vacuum underneath the surface to form a nonwoven web. Two hot air streams at near sonic velocity at the die exit attenuate the extruded stream of polymer. The meltblown process is similar to spunbond in basic concept except it utilizes a die instead of a spinneret and uses air to convert a polymer thread-line into superfine filaments by extreme drawing of the molten polymer (Figure 8). A typical meltblown web contains fibers of a range of sizes (Figure 9), the latter may vary from as low as 0.01 to as high as 20 denier [71]. The fibers are tacky and tend to stick to each other and take the shape

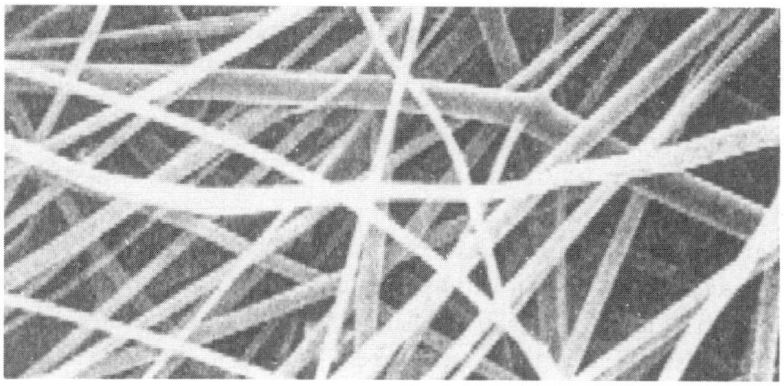

Figure 9. Micrograph of a meltblown fabric showing variable fiber size [72].

of the apron or object on which they are collected. It is recognized that the fibers are weak and lack abrasion resistance. Also, some critical defects such as the presence of micro shots, roping of fibers, and non-uniformity in structure are found to characterize a web [72]. Studies are currently devoted to understanding the causes of such defects and to finding means of alleviating them. Some of the factors that have been suggested to play a role are the air pressure employed at the die exit, the die-collector distance at which fibers are collected, the manner in which the material is quenched, and the nature and the extent of cross flow of air existing at the die [72,73].

Accordingly, the application of meltblown fabrics is currently restricted to those products in which strength and uniformity are not major requirements. Studies have shown that fibers tend to be largely non-crystalline and un-oriented, especially as compared to the fibers obtained from spunbonding of the same polymer. However, the microdenier feature of the fibers in a meltblown fabric make the latter highly suitable for filtration and absorbent applications. The material most widely used currently in the formation of meltblown fabrics is polypropylene, which accounts for 90% or greater of the total production.

Although polypropylene, polyethylene, polyester, and nylon are the major polymers that have been traditionally tried with this technology, studies have shown that successful meltblown materials can also be developed with other materials, such as the biodegradable polyvinyl alcohol [74] and nearly every other thermoplastic material. A great deal of progress has also been made in developing composite structures by combining webs produced by other processes with ones produced by meltblowing. Such products contain fibers of different types and, especially, different deniers, which can lead to unique structures with properties not possible from a single material or process [75].

3.10. Laminate Process

Another important technology for the formation of nonwoven materials is the lamination process. This method involves combining bonded webs into multi-layered structures in order to obtain combination of properties. The individual webs are normally bonded together with latex adhesive, hot melt resin, thermobonding, or by other means. The resulting sheet structure possesses the attributes of the individual webs going into the laminate, modified somewhat by the properties arising from the integrated structure itself. Thus, strength is often enhanced in a laminate structure, generally at the expense of softness. The laminar nature of the composite can affect the total absorbency and particularly the wicking performance. Also, such structures sometimes involve an impervious film as one component of the composite. In this case, it can act as a barrier to absorbency.

3.11. Polymer Web Process

There is another group of miscellaneous web structures that are often classified as nonwoven materials and are used for a variety of functions in absorbent products. These are the sheet materials such as thin foams [76], plastic nets [77-79], perforated films [80,81], and similar materials. Although not truly nonwoven materials, inasmuch as they are not composed of fibers, they often possess a manufacturing or functional relationship to nonwovens. The method of manufacturing these miscellaneous materials are as diverse as their natures. They frequently arise from modified plastic processing techniques and often represent the combination of technologies. In general, they are not hydrophilic, and hence, not inherently absorbent. Because of their structure and characteristics, however, these materials can often substitute for a component of an absorbent product.

3.12. Advanced Composites from Combination of Technologies

During the past several years, advances in equipment, especially those representing meltblowing, spunbonding, hydroentangling, airlaying and carding, have no doubt driven the growth. As an example, turnkey production lines are now available that are capable of producing spunbond polyolefin webs with sizes down to 1.0 denier per filament and finer. Changes in needle loom design and availability of composite lines that combine webs from different processes in one operation are other examples. It appears clear, however, that in the future, product innovations, particularly in areas of composites wherein different materials processed with different technologies are combined, will be the accelerators [75]. Many possibilities exist and have been tried, and some of these have already become commercial realities.

Attractive composites can be made with airlaid/spunlace combination. In the production of many types of products, e.g. wipes, a manufacturer can obtain not only a superior product but also the latter at a fraction of the total cost. For example, infusion of pulp by airlaying into web during spunlacing enables a fabric to be produced that has comparable absorbency to one obtained using staple rayon but at a lower cost. A typical blend for pulp and fiber is 50/50. A good example of an application where airlaid pulp is fused with spunlace on a spunlace line is baby wipe or cleaning cloth.

The greatest emphasis in the development of composite fabrics, however, has been placed on the use of the meltblown and spunbond systems. Examples of structures being widely used include spunbond (SB)/meltblown (MB), known as SM, and SB/MB/SB, or SMS, composites. The production and properties of these are particularly enhanced by the use of bicomponent polypropylene/polyethylene materials in the preparation of MB webs. The spunbond fabric, or spunbond combined with ultra-light weight meltblown fabric, is suited for cover stock for diapers and sanitary products. Other lines are now available that are capable of S, SS, SSS, SMS and SMMS, etc., structures of polypropylene and other synthetic polymers, in speeds exceeding 500 m/min and filament denier lying in the 0.7-10 range.

There has been a continuing interest in the development of polypropylene and polyethylene bicomponent meltblown nonwovens with polypropylene core providing high strength and mechanical stability and polyethylene sheath contributing softness and greater ease of thermal bonding. Recently, core/sheath [82] and side-by-side [83] bicomponent fiber based meltblown fabrics have been developed. Commercial spunbond lines are also providing polypropylene/polyethylene bicomponent material for diaper top sheet and it is noted that the trend to fine denier polypropylene spunbond is continuing with denier lying in the 1.0-1.2 range for diaper use [84].

Most interesting of the composite structures being examined and that relate more directly to absorbent applications are the cotton-surfaced and the cotton-centered nonwovens that are capable of being produced on an integrated line. To make cotton-surfaced products, a cotton/polypropylene web is placed on one or both sides of a polypropylene spunbond web

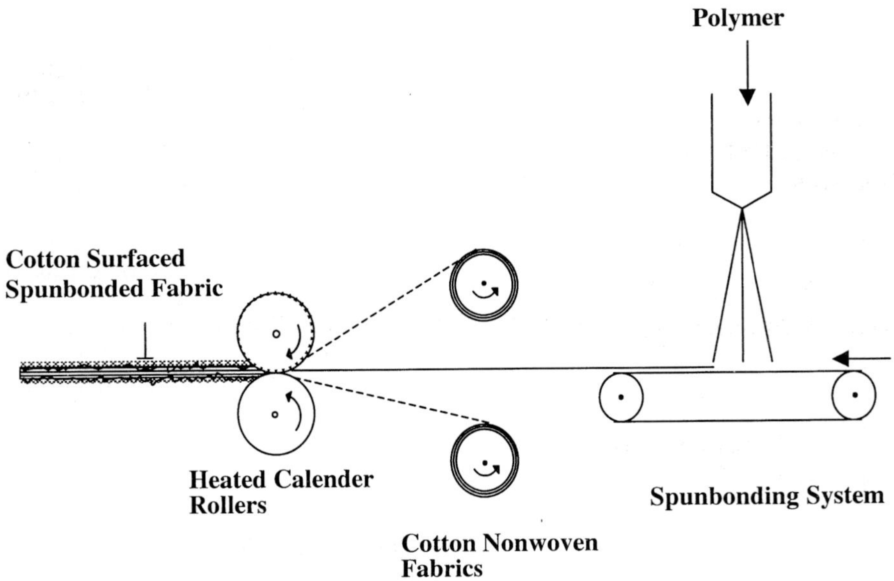

Figure 10. Schematic showing preparation of cotton-surfaced composite on spunbond line.

prior to calender bonding (Figure 10). Recently, publications describing the development of cotton-surfaced materials for personal hygiene applications have appeared in the literature [85,86]. The cotton-centered nonwovens are laminates having cotton sandwiched as a core between two layers of meltblown and/or spunbond webs, with the size and the characteristics of each of the three adjusted to suit the application. For developing such products, the cotton used could be of regular length, fed from a pre-formed roll, or of small or linter size, deposited directly from an airlaying system on the site in coordination with the formation of the spunbond/meltblown webs (Figure 11).

4. ABSORBENT STRUCTURES

In the current context, the major absorbent structures considered are disposable infant diaper, feminine sanitary napkin, underpad, adult incontinence pad, wound dressing (primary and secondary), absorbent wipe and oil absorbent. This list could be lengthened considerably by including derived structures and products, and by considering ancillary potential and applications in a broad range of other materials; however, the examples chosen will serve to illustrate important principles involved.

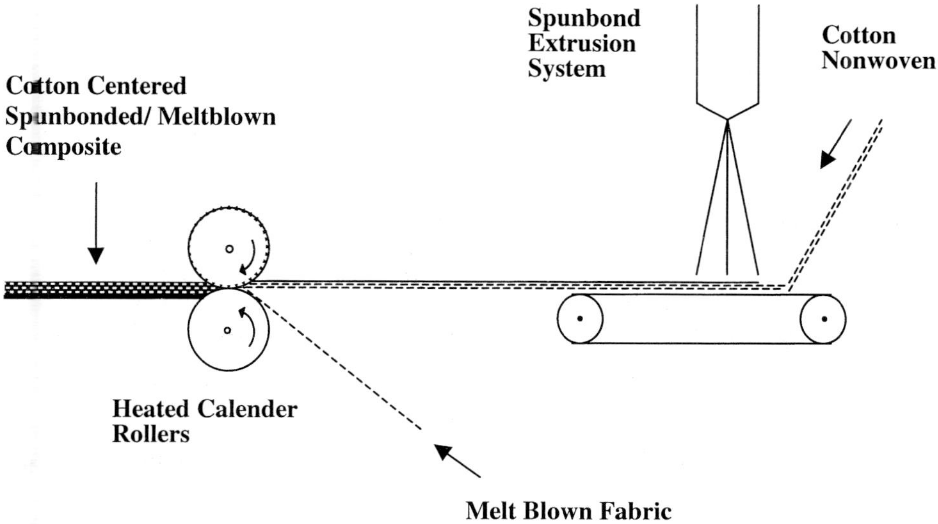

Figure 11. Schematic showing preparation of cotton-centered composite with spunbond and meltblown webs forming the outer surfaces.

4.1. Historical Development of Absorbent Products

Over the past four decades, there has been a rapid shift in the structure of absorbent products, particularly diaper, from the use of traditional woven cotton cloth and cotton wadding to carefully engineered disposable structures employing an increasing amount of specifically designed nonwoven components. During this same period, there has been a pronounced trend toward the use of fluffed wood pulp as the absorbent reservoir in such structures. The universal use of fluff pulp pads, accelerated by the desire to use lower cost materials, has no doubt fostered the increase in the use of nonwoven facings and other components in absorbent products.

Through the years, there has been an evolution in the design of such products to those known today. A variety of materials and configurations have been devised and used; however, in general, the current absorbent structures, such as diaper and incontinence pads, are composed of three or four elements, namely nonwoven facing, absorbent core, with acquisition/distribution layer and superabsorbent polymer, and fluid barrier backing. As an example, the structure of a current "premium" diaper, featuring these and other components, is illustrated in Fig. 12.

4.2. Prototype Structure

In this configuration, an absorbent structure is ready to receive fluid on the upper surface, the nonwoven facing. This surface must be sufficiently wetted by the incoming fluid in order for the former to allow the latter to penetrate into the underlying layer. Ideally, the

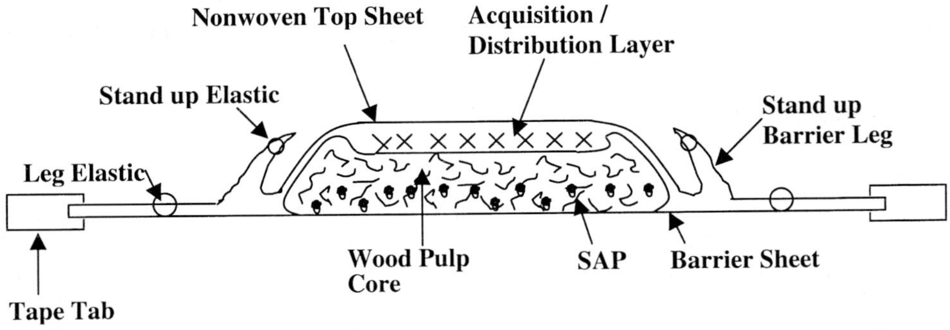

Figure 12. Schematic of a highly engineered diaper product.

wetting (strike-through) is rapid and accompanied by some spreading, to enlarge the area of fluid transfer. These characteristics are usually facilitated by the use of an appropriate surfactant or wetting agent.

The function of the facing in this and similar absorbent products is to present a surface with the following characteristics:

- Softness; pleasing tactile aesthetics, especially in contact with sensitive skin.
- Drapeability; ability to deform and conform to curved surfaces.
- Fluid Permeability; to allow easy passage of fluid into absorbent reservoir.
- Aesthetics; smooth, clean, uniform surface.
- Safety; inert or non-toxic surface for contact with skin.
- Separation; non-adhering material that allows easy removal of product.
- Dryness; to maintain essentially a dry surface in contact with skin.

In some structures, the fluid, after passing through the facing or topsheet, immediately contacts the fluff pulp pad which acts as the absorbent reservoir. In others, an intermediate layer (transition layer) is encountered behind the facing. This tissue is utilized to enhance [87,88] flow of fluid away from the backside of the facing, transport fluid along its plane [89], and increase the volume of the pulp pad involved in absorbing fluid. An additional tissue layer is sometimes used to add superabsorbent polymer to the system. The polymer is frequently inserted at the bottom of the absorbent pad, to minimize gel blockage of the structure by the swollen material.

The backing or back layer of the normal absorbent structure is usually an impervious film, most frequently made of low density polyethylene. It may be a thin multi-layer film that is embossed to provide improved softness and hand and tear resistance [90]. The obvious function of this component is to restrain the fluid within the pad, present a dry surface to the exterior of the product, and provide the user freedom of movement and feeling of comfort. The film may be breathable to add another dimension to the comfort.

While the four-component system is a useful prototype for a variety of absorbent products, many variations and special features have evolved and have been developed for specific articles. The adaptation of the general architecture to specific applications is best illustrated by considering each major absorbent product separately.

4.3. Disposable Infant Diaper

4.3.1 Introduction

A pad of sanitary tissue or paper wadding comprised the first disposable infant diaper product. This hand-shaped pad was combined with a separate square of plastic film by the Swedish mother under the harsh conditions of shortage experienced during World War II [91]. Where the wet strength of the tissue was inadequate, pieces of tissue were frequently left on the moist skin, requiring an unpleasant cleaning up step. Consequently, a cover characterized by sufficient wet integrity was sought. The light weight rayon nonwoven facing was developed for such use, that provided a significant improvement in performance. The product was improved following the war and widely adopted.

In the interests of economy, fluff pulp was introduced as a superior, low cost replacement for tissue as the absorbent reservoir [92]. In Europe, this led to a rectangular shaped, fluff pulp pad, sometimes wrapped with tissue and then over-wrapped with an envelope of nonwoven. This pad could then be used with the separate square of plastic film. For convenience, the plastic film was eventually replaced by a close-fitting plastic panty, with the absorbent pad as an insert. This gave rise to the traditional 2-piece disposable diaper of Europe.

On the U.S. scene, the adoption of the concept centered on an integral, 1-piece structure, the so-called unitized concept. The latter, building on the structure of the flat hospital underpad, resulted in a rectangular pad, which was folded into an appropriate shape, much like a cloth diaper. Closure and fit was accomplished with safety pins, in much the same way as with durable woven cloth diapers. In the mid-60's, a major manufacturer improved upon the idea by folding and glueing, so that a pre-formed product with better fit, easier application, and superior performance resulted. The addition of the self-contained tab closure provided considerable convenience. With many variations, this style led to the current infant diaper of the shaped, one-piece construction [93]. In many parts of the world, this is known as the "American-style" diaper.

This diaper still has four elements. The functions and the structures of the absorbent core and the barrier sheet are well understood and those of the former, in particular, are adequately covered in discussions of the airlaid, wetlaid, spunlace and other technologies in Sec. 3. The functions and the structures of the remaining two elements, namely the topsheet and the acquisition/distribution layer, are not covered. These components of the diaper need detailed discussion, especially the topsheet, since it performs complex and multiple functions. A concern with side leakage has also led to the introduction of leg elastics and cuffs in modern products. These and other aspects of a diaper are attended to in the sections that follow.

4.3.2. Diaper coversheet

The nonwoven fabric that is positioned next to the skin of the wearer of an absorbent sanitary product has been given various names, including: a) cover, b) coversheet, c) coverstock, d) facing, e) topsheet, and others. Despite this variety in descriptive terms, there is no uncertainty about the item involved; it is the soft, nonwoven fabric on the diaper surface that contacts the skin or body of the wearer.

The primary product requirements of the diaper facing have been outlined above. From the standpoint of the diaper technologist, a continual balance, involving inevitable compromises, must be sought among the requirements, along with the minimizing of the cost of this component which is used in such large volume. As might be expected, the nature of the nonwoven facing itself has undergone changes and considerable improvements during this evolution.

Two decades ago, most of the diaper facing was produced from rayon fiber by the dry form and print bond processes, with a small portion derived from the laminate process. At that particular stage of commercial development, there was strong acceptance of the concept that the coverstock used on an infant diaper had to be absorbent. Fiber selection, processing conditions and binder considerations all focused on the requirement for maximum absorbency of the diaper facing.

Within the past two decades, however, there has been a complete turnaround in this concept, to the point that today's basic requirement is for a nonwoven facing to be essentially hydrophobic. This resulted primarily from the desire to provide a dry surface in contact with the skin, which was thought to be more comfortable and less irritating and toxic for the wearer. Such a dry surface is viewed as essential to the maintenance of skin health.

As a consequence, the fiber of choice in sanitary protection facing products has shifted from the hydrophilic rayon to the wholly synthetic hydrophobic fibers. The change that occurred was originally to polyester fiber material, which took over the bulk of the facing market by 1979. This was primarily because a major absorbent product manufacturer in the United States adopted a latex bonded polyester nonwoven fabric as the cover stock of choice [94]. Within a few years, however, another shift took place, this time to thermal bonded polypropylene fiber facing [95].

A major driving force for the change to polypropylene was not only the desire for dry surface properties, but also a recognition of the desirability for having a facing that was binder free. The elimination of formaldehyde and other concommitant chemicals, present in the latex bonded facings, had been viewed as a desirable direction for the future. Consequently, there has been a growing interest in the use of thermal bonded nonwoven technology in order to produce hydrophobic facings without chemical additives. This process involves a fiber web of wholly or partially thermoplastic fiber (generally polypropylene) bonded by means of a hot embossing roll. At the emboss site, partial melting of the binder fiber occurs, resulting in a bond capturing several adjacent fibers (Fig. 2).

The changeover to thermal bonded materials for topsheet was also accelerated by national legislation in Japan that virtually forced the removal of formaldehyde-containing products from applications involving contact with skin. Also, around the same period, substantial interest was shown in the use of thermal bonded facing made with polyolefin bicomponent material as the binder fiber. The most prominent fiber employed consisted of a 3 denier material with a core of polypropylene and eccentric sheath of high density polyethylene [96]. This fiber has been used in 100% form as well as in blend with rayon or conventional polyolefin. When facings are derived from 100% bicomponent material, the bonding has been of the overall type, leading to fabric with strength high enough to allow for the use of very lightweight stock, e.g. 10 g/m^2.

Polypropylene-based fabrics have now become the preferred diaper coverstock material around the world. The use of this fiber provides the desired hydrophobic nature to the facing at a reasonable cost as this resin is one of the lowest cost, high volume, commercial polymers.

The property of hydrophobicity in a diaper facing is necessary but not sufficient for optimum performance. The fabric must also quickly pass liquid through to the interior of the diaper. The perfect facing must function as a "one-way valve", quickly passing liquid through (rapid strike-through), restricting passage of the liquid back through in the reverse direction (wet-back), and presenting a dry and soft fibrous surface to the skin. Consequently, a technology for conveying wetting properties to the surface of the facing to insure rapid strike-through of the voided liquid was necessary. This is most easily achieved by uniformly adding onto the surface of the facing a small amount (0.2 to 0.6 weight %) of an effective rewetting agent. This is generally applied by spraying.

A shortcoming of the topical application of the rewetting agent is that it can be rather quickly removed by multiple discharges, which is not uncommon with infants. Removal of this agent by passing urine through the facing, of course, reverts the fabric surface to its original hydrophobic condition; this can lead to product failure by run-off of the subsequent insults. This situation has led to the development and use of "durable" finishes, which can survive several insults. Such a durable rewetting agent is most easily used by adding the latter to the polymer prior to extrusion of filament in the spunbond process. A similar approach can be taken with polypropylene staple fiber, but with increased production problems.

4.3.3. *Standing leg cuff*

A recent innovation in diaper design has been the introduction of a fold in the facing along the two longitudinal edges of the diaper. This fold is inserted to provide side barriers in the target zone of the diaper. The concept of the design is to provide additional protection against side leakage. This is achieved effectively when the fold is made to stand up from the surface of the diaper, see Figure 12.

This design feature has been given various names, including standing leg cuffs, stand-up cuffs, and others. In order to accommodate this element on the top surface of the diaper, about 30% more facing is required. Further, the facing involved in the fold is more effective in preventing leakage if it is impervious to liquids. Desirably, the cuff should still "breath" or transmit moisture vapor for wearer comfort [97]. The combination of liquid imperviousness and air permeability in a lightweight nonwoven fabric is effectively achieved by a composite of a spunbond web combined with a meltblown web of micro fibers discussed in the next section. The polypropylene and spunbond filament web provides excellent strength and other physical properties, while the meltblown web of extremely fine denier polypropylene fibers functions as a barrier to liquid, still affording reasonable breathability.

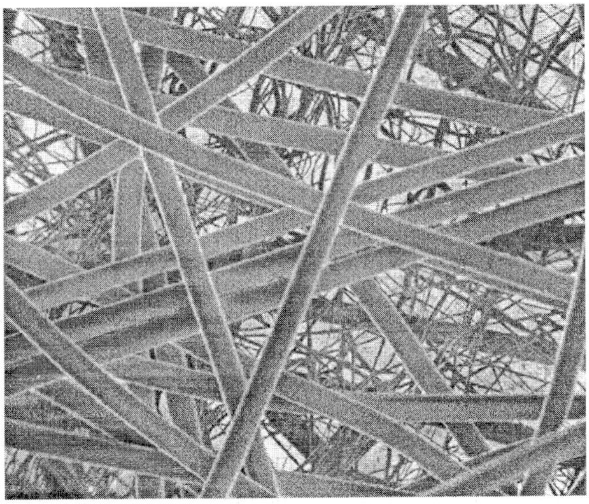

Figure 13. Micrograph of a polypropylene SMMS fabric [99].

4.3.4. Spunbond/meltblown components in coversheet

As pointed out earlier, the meltblown process, licensed by Exxon Chemical, produces a random web derived from micro-denier thermoplastic fiber, usually polypropylene, or a bicomponent fiber based on it [75,98]. This fabric has some self-bonding capability because of modest mechanical entanglement of the fibers as they are formed into a web: the adhesion is often supplemented by thermal bonding in a pattern. Because of lack of strength in meltblown fabrics, a facing consisting of meltblown fibers is usually a composite with the meltblown material sandwiched between layers of spunbonded fabrics and bonding together by calendering.

Such a composite leads to a unique structure, claimed to provide desirable performance at lower basis weight and cost [99]. The most popular combination structure is SMS (or SMMS, SSMMS) in weight ranges from 10-25 g/m^2 comprising 1-5 g/m^2 meltblown microfibers. The spunbond layers provide high tensile strength in both the machine and the cross-directions, whereas the meltblown microfibers greatly improve visual uniformity and liquid barrier properties. The topsheet given proper hydrophilic treatment combines high strike through with effective retention of superabsorbent particles (SAP) smaller than 100 microns. Figure 13 illustrates a low weight (15 g/m^2) SMMS fabric comprized of meltblown fibers accounting for less than 15% of the material by weight.

Table 4 shows the results of one analysis in which the performances of several multilayer spunbond/meltblown fabrics were compared. All structures shown gave acceptable strike-through, against repeated insults, and rewet results. The differences are seen in the values of tensile properties, air permeability and pore size. From the data in the table, it is seen that a 13 g/m^2 SMMS fabric has only slightly lower mechanical properties than found in the 15 g/m^2 SS fabric, but superior particle retention and barrier characteristics. The former is particularly suited for application as a barrier against release of superabsorbent

Table 4.
Comparison of the properties obtained with different spunbond/meltblown configurations [99]

Composite Fabric Structure	SB (dtex)	MB Fiber Diameter (μm)	Basis Weight (g/m^2)	Weight Dist (g/m^2)	Tensile Strength (N/50mm)	Air Perm ($l/m^2/s$)	Pore Size (μm)
SMMS	1.7-1.9	2-6	13	5.5/1/1/5.5	32/13	2200	80-90
SMMS	1.7-1.9	4-8	13	4/2.5/2.5/4	24/10	2000	70-80
SMMS	1.2-1.4	2-6	8	3/1/1/3	28/11	2400	80-90
SS	1.7-1.9	-	15	7.5/7.5	34/16	6000	1000+

through the acquisition layer and the topsheet. Such a composite can also be configured for use as a breathable back sheet having the required liquid barrier properties. Apparently, by adjusting the fraction and nature of the spunbond and meltblown materials in the composite, the nonwoven technologist should be able to engineer a fabric for facing or barrier sheet for numerous applications that presents the best combination of economy and performance.

4.3.5. Secondary facing

With a hydrophobic fabric, the desire for rapid strike-through often requires an absorbent material immediately beneath the nonwoven facing, which not only absorbs rapidly but also has a high rate of wicking away from the wetting site. This situation is generally facilitated by insuring an intimate contact between the topsheet and the pulp pad, such as achieved by inclusion of an absorbent tissue between the back of the sheet and the pad. In some products, this tissue is adhesively bonded to the back side of the facing in order to obtain the desired contact. This may be especially necessary for shaped elastic leg diapers, where the shaping feature may cause the cover material to tend to pull away from the absorbent pad.

During the past few years, there has also been an interest in a "secondary facing" concept, to augment rapid passage of liquid from the back of the cover into the adjacent pulp pad or ply. This has taken the form of a very light weight fiber web with little or no bonding between the primary facing and the absorbent core. In another form, the single facing is fabricated with a rough top (outer surface) and a smooth bottom (inner surface). In this system, liquid tends to move from the rough side to the smooth side, enhancing wicking toward the center.

4.3.6. Acquisition/distribution layer

This mechanism of enhancement of liquid movement has been further advanced by the use of an acquisition/distribution layer (ADL) between the topsheet and the absorbent core. The ADL generally takes the form of a composite fabric with a density gradient through the thickness. The top one-third of the fabric has low density (often higher denier fiber) with relatively large voids and higher void volume; it is especially effective in the acquisition of the presented liquid, even at high discharge rates. The middle one-third of the fabric is at a somewhat higher density with smaller voids, while the lower one-third is of higher density yet, with even finer voids and often smaller denier fibers. The higher density portions of the composite have more and finer capillaries and hence develop greater capillary pressures. As a result, the higher capillary pressures effectively move greater volumes of liquid to the outer regions of the structure, providing greater absorbent capacity through more effective use of the entire absorbent core [100,101].

The ADL layer provides for more rapid liquid acquisition (minimizing flooding in the target zone), and more rapid transport and thorough distribution of the fluid into the remainder of the absorbent region [102].

4.3.7. Diaper Performance Properties

Special fabric characteristics relating to diaper performance have been developed over the years, with assistance from the industry trade associations, such as INDA (USA) [103], TAPPI (USA) [104], EDANA (Europe) [105], ANNA (Japan) [106] and other Standards organizations such as ASTM (USA) [107], as well as from specific industries and commercial organizations [108-112]. These properties include such special tests as 'strike-through' (time in seconds for liquid to penetrate topsheet), 'run-off' (volume of liquid not absorbed when poured onto a fabric maintained at specified angle), and 'wet-back' (volume of liquid squeezed out of an absorbent and back through the facing under standard conditions). Other properties of interest that relate to the performance of the diaper, include 'total Absorption' (saturation to the drip-free point), and 'absorption under load' (total amount absorbed and held under a standard load). For details about testing of absorbent products, see Chapter XI.

4.4. Feminine Sanitary Napkin

A similar evolution to that involved in disposable diaper has occurred with the feminine sanitary napkin. The original manufactured product consisted of paper wadding encased in a woven cotton gauze cover folded into a tubular shape. In some cases and geographical regions, the woven cloth was substituted with a knitted cotton tube. In both instances, excess fabric was employed to provide a tail or tab on each end of the product; the tabs were used to position and retain the napkin in the perineum area by fastening them to a belt worn around the waist. Fluff pulp eventually replaced the wadding as the absorbent core, and a dry form nonwoven facing replaced the woven or knitted cover. Lately, the trend has been towards smaller pads (thin, mini) and elimination of the tab in favor of a pressure sensitive adhesive on the film backing for positioning and affixing the pad to the undergarment.

In common with infant diapers, the nonwoven coverstock formerly used on feminine napkins was required to be as absorbent as possible. Consequently, the typical cover of the second generation product was a printed, dry form nonwoven produced from rayon fiber. This essentially gave in to a more hydrophobic napkin facing. This was first achieved by the use of blends of polyester and rayon and later by the use of wholly synthetic material. Polypropylene spunbond nonwoven has been one of the main structures employed.

Other types of nonwoven facings have been used in limited ways on a variety of feminine absorbent products. These have included a plastic net and a plastic net emboss bonded with a light weight web of polypropylene fiber [113,114]. Another type of facing introduced by a leading manufacturer is a perforated film structure, produced by a thermal forming/perforating process. This gave the film a controlled size apertures and a three-dimensional structure [115]. Although the facing was non-fibrous, the 3-dimensional thermal forming of the film surface and the presence of the perforations conveyed a somewhat cloth-like hand. The facing gave excellent resistance to wet back because of the "apparent" thickness of the structure. Also the strike-through performance was good, provided that the absorbent tissue was in intimate contact with the back side of the cover; this performance leads to a very desirable 'clean and dry' sanitary napkin.

In the fluffed pad itself, steps can be taken to facilitate liquid movement [116]. The so-called "Burgeni Skin" fosters lateral movement along the surface of the pad, hence involving a greater volume of the latter as an absorbent reservoir. This paper-like structure or skin can be formed by spraying a limited amount of water on the pad and then moderately compacting the outermost layer of fibers into a densified "skin" [117]. In essence, this skin provides a somewhat densified layer that facilitates liquid migration by the increased capillary pressure of the compacted fibers. This technique also conveys some structural integrity to the material, facilitating high speed processability.

Because of the requirement for product compactness and the demanding dynamic activities of many consumers of napkin, there has always existed a desire to achieve an anatomical shape for the product. This can be achieved by forming the absorbent pad or core into a 3-dimensional shape rather than into a simple rectangular box configuration. This has been achieved by the use of "pocket formers," or "drum formers," for separate preparation of the core structure [118,119].

The pocket former consists of a shaped mold, fabricated from stainless steel screen, so that a vacuum can be drawn on it. The top of the device is open. Several such units are located on the periphery of a rotary drum. The individualized woodpulp fibers are brought via an airstream and blown into the pervious shaped mold, with the vacuum on the back side to assist in the uniform filling of the latter. Superabsorbent polymer powder can also be positioned in the woodpulp core at the desired position. By providing a short puff of air from the backside of the mold, the shaped core can be released and carried on the manufacturing line for the rest of the processing.

In general, napkin top sheet fabric lighter than the coverstock used on an infant diaper. This results from the fact that the absorbent product is subjected to less severe mechanical action and the recent trend has been toward designing smaller size sanitary napkins. Another consideration that dictates the difference in structures and sizes of the sanitary napkin and the diaper or the incontinence pad is the quantity of fluid discharged at one time and its constitution.

In a dynamic field of product development such as the absorbent sanitary materials, new technical innovations can result in a paradigm shift, giving rise to the use of a different, advanced technology [120,121]. Such is again occurring in the fabrication of the absorbent core for sanitary napkins. In this case, there is a growing use of preformed airlaid pulp fabric to replace the current wood pulp fiber core. The airlaid pulp fabric can be supplied and used on the converting line directly, eliminating the large, noisy, and dusty pulp sheet grinding operation. The fabric can be provided in a range of weights and configurations specifically designed for this use. For example, the web can be provided with the incorporated superabsorbent polymer at the desired position in the structure. Although the raw material in this form may have a higher price, simplified, integrated converting operations make the change attractive.

This same concept is being considered for an integrated absorbent core structure for other absorbent products, such as diapers and adult incontinence pads [122], but the larger core size and greater absorption capacity requirements of some of these can create some economic problems in the use of airlaid pulp fabrics.

A significant use of superabsorbent polymer also occurs in some brands of sanitary napkins. In many cases, the solid polymer in powder form is lightly affixed to a tissue. This composite layer is then inserted between the backing film and the absorbent pulp pad, or in a lower layer of the pad. This location, away from the source of the liquid, is important, as the wetted, swollen, superabsorbent material is a gelled mass that would tend to block the capillaries and retard the progress of fluid into the structure. Hence, the placement of the polymer must be carefully selected to insure that it attracts and holds fluid but does not retard the flow-in of the subsequent additional liquid.

4.5. Underpad

In general, the nonwoven absorbent structures utilized for underpads have been similar in design and configuration to those used for early disposable diapers, the latter being derived from the former. Because the underpad is stationary in use and has less critical performance expectations, lighter weight and lower quality facings have often been used on this product. Lightweight rayon dry form topsheet was the industry standard for many years. Lightweight fabrics produced via the wet form process have also been used. In such cases, the nonwoven fabric has generally been produced from a blend of wood pulp and short cut polyester fiber or short cut rayon fiber in varying proportions. Latex print bonding has normally been employed for these structures.

A successful underpad has been produced by simply combining an impervious polyethylene film with a heavier weight all pulp fabric (85 g/m^2), produced by a modified tissue process. There has been some use of superabsorbent powder in underpads.

An underpad product has also utilized a facing of a plastic net produced by the attenuated extrusion process [123].

4.6. Adult Incontinence Product

A wide variety of absorbent products for adult incontinence have been offered commercially. In most cases, these structures have used conventional absorbent materials and construction details, similar to those of the disposable baby diaper, modified for different body dimensions. There has been a greater use of simple insert nonwoven covered pads utilized with tight-fitting elastic briefs; this product type is reminiscent of the two-piece infant diaper.

Adult incontinence pads similar to the one-piece wing-fold infant diaper design have also been offered commercially. Similarly, elastic leg shaped products have been commercialized in several designs. In a similar fashion, a pull-up or pant style adult pad with extensive elastication of the top, i.e. the waist, has generated considerable interest and commercial success.

In general, nonwoven facings and absorbents similar to those used in infant diapers have been employed. Among these also is a perforated film, similar to the one used in some sanitary napkins [124]. However, the weight of the fabric is sometimes increased to provide for the large volume of fluid involved and the greater mechanical action encountered. Because of the desire to maintain a compact design and to insure user dignity, there has been an increased interest in the use of superabsorbent polymer materials in adult products as well.

Accordingly, the adult pads now also tends to be highly engineered body shaped structures that use components similar to those employed in the modern day diapers and the other personal hygiene items [125].

4.7. Wound Dressing

Nonwoven materials have been used rather extensively as the facing, as well as the entire absorbent pad of a wide variety of wound dressings and surgical pads. Initially, latex bonded uni-directional nonwoven fabrics were utilized as the facing of a limited range of surgical dressings; these fabrics generally replaced the traditional bleached, woven gauze.

The nonwoven topsheet was often combined with an absorbent structure of folded tissue, rayon fiber sliver or similar material. There has been some use of a hydrophobic plastic netting thermally adhered to the surface of pads, or a polyethylene coating applied to the latter, in order to convey easy wound release properties [126,127]. Although such a plastic netting could be visualized as a structure interfering somewhat with absorbency, the netting was porous enough for the fluid to enter into the nonwoven fabric and, subsequently, into the absorbent pad beneath the latter.

During the past many years, there has been a growing interest in the use of a spunlace nonwoven material in a broad range of dressings, surgical sponges, surgical pads and similar important health care products. A widely used fabric for this application has been a spunlace web derived from 70% rayon and 30% polyester in approximately 70 g/m^2 weight. This fabric, folded into a 4-ply sponge, has been able to replace the earlier 16-ply woven gauze sponge with equivalent absorbency, less linting, and softer hand [128].

An absorbent thermally bonded nonwoven web has also been offered as a surgical sponge. This fabric has had the unique point embossed bond pattern, characteristic of the use of twin spirally grooved heated rollers [129]. Although this process could use a variety of fiber compositions, the blend of 50% polypropylene and 50% rayon was given the greatest commercial emphasis.

Needlepunch fabrics from hydrophilic, hydrophobic and blends of fibers have been used for absorbent applications in dressing products. Sufficient structural integrity is often obtained with a modest amount of needling, which provides a fairly bulky structure with good absorbency rate and capacity.

4.8. Absorbent Wipe

One of the most obvious requirements for a wiping product is rapid absorbency. This is true of wipes employed in industrial applications as well as of ones used in institutional and household uses. Disposable and semi-durable nonwoven materials have been extensively employed in these markets.

Wiping performance in the industrial sector is generally concerned either with aqueous or oil-based material pickup. Although some wipes are capable of functioning in both situations, optimum performance is usually obtained when a wipe is designed for either the aqueous or the oil pickup. Superior performance in the oil pickup is generally provided by hydrophobic materials, such as polyester, polyolefin or nylon, whereas the aqueous pickup is enhanced by the use of hydrophilic materials, such as cotton, rayon or wood pulp.

Although substantial use has been made of the waste fabrics, usually woven or knitted, in the industrial wiping sector, there has been growing acceptance and use of clean, specifically engineered, nonwoven structures in this application [130]. A major portion of this market has been served by latex bonded dry form nonwovens with particular use of rayon fiber [131]. Such fabrics have offered acceptable performance in removing oil and superior performance in picking aqueous fluids. In some instances, special latex binder compositions have been required in order to meet the low organic solvent emission specifications of the application. Wet form nonwoven webs have also been offered into this market which have provided performance superior to that given by the disposable paper material that controls a major share of the industrial wipe market. This technology has been especially useful in the development of very low linting products required for clean room application [132].

Spunlace nonwoven fabrics have found ready acceptance in many of the more critical industrial wipe applications. A popular wipe has been spunlace fabric produced from a blend of 40% polyester and 60% wood pulp. The fabric is binder-free and has most of the pulp on one surface. It features good absorbency along with acceptable softness, strength and abrasion resistance.

A major commercial application of meltblown nonwoven materials has been in the industrial wiping sector. These webs are generally composed of 100% polypropylene micro-fibers. They are bonded by emboss calendering and treated with a surfactant to provide easy wetting characteristic for use on aqueous systems [133,134]. Also, these materials are particularly effective in wiping oil-based fluids. For increased durability, the meltblown micro-fiber web is often combined with one or more layers of polypropylene spunbond fabric. This provides an absorbent material with considerably greater strength and surface abrasion resistance.

A range of absorbent wipes produced by the composite process has been used in the institutional wiping sector. The most common structure made by this process has a core of two-ply high bulk tissue and a facing on both sides derived from an apertured rayon, latex bonded dry form nonwoven fabric. The entire composite is adhesively bonded in a distinctive pattern [135].

There has been some use of stitchbonded nonwoven webs derived from various fibers, often secondary or low cost waste fiber. This technology has been particularly utilized in Europe. In a similar manner, several European companies have offered an absorbent nonwoven material for industrial and household markets produced by the needlepunch process. These absorbent materials have usually been made from a blend of 85% rayon and 15% polypropylene fiber. The web is normally given a modest amount of needling on a pre-needle loom, followed by a hot air thermal bonding step in which the polypropylene fiber is melted and converted into binder [136]. Thus, integrity conveyed by the modest needling is further augmented by the thermal bonding action of the polypropylene component.

4.9. Oil Absorbent

There has been a growing interest in recent years in absorbents specifically designed for oil and grease pickup. These are meant to protect the environment by absorbing oil from spills, and avert accidents by absorbing grease and similar fluids from around walkways, plant entrances, machines, kitchen floors, etc. Several commercial absorbent nonwoven materials have been utilized in this application, with very satisfactory results [137].

The polypropylene meltblown micro-fiber webs have been especially useful for the pick-up of these fluids. Such webs float on water, and have a considerable affinity for oil on the surface. The polyolefin materials are particularly effective in this application, as oil and grease will actually displace any water originally present in the pores of the absorbents [138, 139].

Airlaid pulp has also been utilized in this application. In this case, the hydrophilic nature of the wood pulp fiber is masked by the use of a small amount of paraffin wax finish, applied as a post treatment to the fabric. With this finish, the pulp fabric can be used to pick up oil and grease, with the latter again displacing any water (including sea water) originally picked up by the material. Fabrics made by other nonwoven processes, including spunlace, spunbond, and those made by combination of technologies, have also been proposed for use as sorbents of such fluids.

5. POTENTIAL FUTURE DEVELOPMENTS

Because of the advances made in absorbent nonwoven materials and their success in a broad range of commercial products, it is natural to expect that the use of nonwovens in absorbent structures will grow. As the detailed mechanism of absorbency is uncovered and as the critical performance parameters required are delineated, it is obvious that better products and new applications will emerge.

The introduction of specialty fibers in absorbent applications will undoubtedly grow, particularly in the synthetic fiber sector. This will involve the exploitation of the unique capillary structure and potential wicking performance of certain synthetic fibers. Further, the use of micro denier materials and hollow fibers, along with those possessing a porous core, multi-lobal shape and grooved surface, will undoubtedly receive increased attention.

With increasing skill on the part of the fiber technologists to create fibrous precursor webs, including composites with specific orientation and configuration of fibers, the possibility grows for engineering fibrous structures specifically designed to enhance and control absorbency characteristics. As the growth of binder-free systems increase, these structures can utilize the maximum potential offered by the constituent fiber without the interference from a foreign agent. Also, as more bulky, lofty, open structures are devised, the full use of void space can be achieved. Stable structures with as little as 3% fiber volume are currently possible. With appropriate technology, it should be possible to utilize with full advantage this enormous void volume in absorbent structures of the future.

6. ACKNOWLEDGEMENT

The authors gratefully acknowledge the help of Mr. Ian Butler of INDA, the Association of the Nonwoven Fabric Industry, for reviewing the manuscript and making several helpful suggestions. In addition, the authors thank many other individuals from the nonwovens and the absorbent products industry with whom they conferred during the writing of this chapter.

7. REFERENCES

1. I. Butler, INDA, Association of the Nonwovens Fabrics Industry, Private Communication, August 31, 2001.
2. J. Heller, Nonwovens World, August-September, 2001, p. 52.
3. ASTM Method D 123-70, ASTM Standard Methods, ASTM, Philadelphia, PA, 1970.
4. P. Coppin & Co-workers, The Definition of Nonwovens Discussed, EDANA Symposium, Gothenburg, Sweden (June 1974).
5. P. Truelove and G. Nordgren, Index 81 Conference Papers, Vol. 1, EDANA, Brussels, 1981.
6. G. Massenaux, Nonwoven Conference Papers, (ed. G. E. Cusick), UMIST, Manchester (1983), p. 1.
7. S. K. Batra, S. P. Hersh, R. L. Barker, D. R. Buchanan, B. S. Gupta, T. W. George and M. H. Mohamed, Textile Horizons, July 1986, p. 30.
8. A. H. Drelich, Nonwoven Textile Fabrics (Staple Fibers), in Encyclopedia of Chemical Technology, 16, 3rd Edition, p. 104-124, 1981.
9. A. T. Purdy, Developments in Nonwoven Fabrics, Textile Progress 12, The Textile Institute, 1983.
10. Asahi Chemical Industry Co., Ltd., Osaka Japan, Brochure on Ciebet Asahi Kasei Nylon.
11. W. A. Haile and B. M. Phillips, TAPPI Journal, 78 (1995) 139.
12. Anon., Textile Month, p. 26 (January 1977).
13. Bayer Aktiengesellschaft, U.K. Patent 1,532,770 (November 22, 1978).
14. W. F. Schlauch, Nonwoven Fabrics Forum, Clemson University, Clemson, South Carolina (June 1982).
15. J. G. Brodnyan and D. G. Strong, 2nd INDA Technical Symposium, INDA, New York (January 1973), p. 80.
16. R. D. Bakule and J. M. Blickensderfer, TAPPI, 56 (1973) 70.
17. W. G. DeWitt and J. G. Brodnyan, 8th INDA Technical Symposium, INDA, 1980, p. 121.
18. B. J. Obenski, American Dyestuff Reporter 71 (March 1982) 37.
19. D. F. Lunde, Nonwoven Fabrics Forum, Clemson University, Clemson, South Carolina (June 1982).
20. R. D. Gleim, INDEX '93 Congress Papers, Geneva, Switzerland, April 1993.
21. K. H. Schumacher, R. Hummerich, H. Kirsch, R. Rupaner and R. Wüstefeld, INDEX 93 Congress Papers, Geneva, Switzerland, April 1993.
22. W. Wirth, Canada Textile J., 98 (1981) 51; W. Wirth and J. C. Harrison, 9th INDA Technical Symposium, INDA, New York, 1981.
23. A. H. Drelich, 5th, INDA Technical Symposium INDA, New York (March 1976) p. 288; A. H. Drelich, ibid, p. 117-118.
24. A. H. Drelich, In Modern Nonwovens Technology, edited by D. T. Ward, Texpress, Manchester, 1977, p. 61.
25. E. H. Carus and J. W. S. Hearle, Textile Institute Industry, 13 (1975) 147, 226.
26. C. W. Jarvis, P. A. Danforth, B. A. Kolinofsky and E. A. Vaughn, 8th INDA Technical Symposium, INDA, New York March 1979, p. 175.
27. R. G. Mansfield, Advanced Forming/Bonding Conference, INSIGHT 82, Marketing/Technology Service, Kalamazoo, Michigan, 1982, Section II;
28. A. H. Drelich, Advanced Forming/Bonding Conference, INSIGHT 81, Marketing/Technology Service, Kalamazoo, Michigan (September 1981), Section VI.
29. D. E. Wood, In Nonwoven Conference Papers, (edited by G. E. Cusick), UMIST, Manchester, 1980, p. 89.
30. J. Bayer, Fiber Producers Conference, Textile Hall Corporation, Greenville, South Carolina, 1983.
31. H. S. Jensen and S. G. Agren, Advanced Forming / Bonding Conference, INSIGHT 82, Marketing/Technology Service, Kalamazoo, Michigan, 1982, Section VIII.
32. F. Bouda, INDEX 81 Congress Papers, EDANA, Brussels (May 1981), Section 3.
33. J. W. Rohrer, 5th INDA Technical Symposium, INDA, New York (March 1976), p. 107.
34. G. A. M. Butterworth in Nonwoven Conference Papers (edited by G. E. Cusick), UMIST, Manchester, 1980, p. 237.
35. J. H. Field, TAPPI, 25(7) (1982) 93.
36. R. A. Parham and H. I. Hergeth in Proc. Intern. Air-Laid Low-Density Forming Seminar, TAPPI, Atlanta, 1979, p. 12.
37. J. D. Crain, Conference Papers, IDEA 76, INDA, New York 1976, p. 105.
38. R. A. Anderson, R. C. Sokolowski and K. W. Ostermeier (to Kimberly-Clark Corp.), U.S. Patent 4,100,324 (July 11, 1978).
39. D. Feroe, Proc. Fourth International Nonwovens Symposium, Beltwide Cotton Conference, National Cotton Council, Vol. 1 (2001) p. 702.
40. F. Osborne, (to the Dexter Co.) US Patents 2,045,095; 2,045,096 (June 23, 1936).
41. W. T. Heyse in Nonwoven Conference Papers (edited by G. E. Cusick), UMIST, Manchester, 1980. p. 203.
42. W. T. Heyse, Nonwoven Fabrics Forum, Clemson University, Clemson, South Carolina, 1979.

43. J. A. Villalobos, Air-Laid & Advancing Forming Conference, INSIGHT 80, Marketing/Technology Service, Kalamazoo, Michigan, 1980.
44. D. F. Beaumont, Air-Laid Forming in the 80's INSIGHT 80, Marketing/Technology Service, Kalamazoo, Michigan, 1980.
45. J. H. Field, Papers, 1981 TAPPI Annual Meeting, TAPPI, Atlanta, Georgia (March 1981).
46. C. J. Roberts, (to Scott Paper Co.), U.S. Patent 4,000,237 (December 28, 1976).
47. H. E. Becker, A. L. McConnell and R. W. Schutter (to Scott Paper Co.), U.S. Patent 4,158,594 (June 19, 1979).
48. T. M. Holliday, Mod. Textile, 55(1974) 40.
49. T. M. Holliday, Nonwoven Fabrics Forum, Clemson University, Clemson, South Carolina (June 1982).
50. E. E. DeBrunner, 5th Annual TANDEC, Knoxville, TN, (October/November 1995), p. 5.6-1.
51. F. Newmark in Nonwoven Conference Papers (edited by G. E. Cusick), UMIST, Manchester, 1980, p. 123.
52. J. W. S. Hearle, A. T. Purdy and J. J. Jones, J. Textile Institute, 64, (1973) 617, 65 (1974) 6.
53. William Tatham, Ltd., Microfeed 2000, Product Brochure.
54. A. Ollinger, International Textile Bulletin (ITB) Nonwovens/Industrial Textiles, 2000, p. 2.
55. Fehrer AG, High Performance Needle Punching Machines, Product Brochure, Linz, Austria.
56. S. S. Ramkumar,and A.P.S. Sawhney, Proc. 2001 Beltwide Cotton Conference, National Cotton Council, January 9-13, Anaheim, CA, Vol. 1 (2001) p. 674.
57. F. J. Evans (to DuPont), U.S. Patent, 3,485,706 (December 23, 1969).
58. A. Bolton (to Johnson & Johnson), US, Patent 4,144,370 (March 13, 1979).
59. M. Turi, INDA-TEC '93 Technical Symposium, Florida, (September 1993), p. 281.
60. F. J. Medeiros, Proc. INDA-TEC 96, International Nonwovens Conference, Crystal City, VA, 1996, p. 5.1.
61. D. K. Lickfield, TANDEC '93, 3rd Annual Conference, University of TN, November 1993, p. 1-2.1.
62. A. Ruffo and P. Goyal (to Johnson & Johnson), U.S. Patent 3,768,118 (October 30,1973).
63. C. White, F. Zevnik, EDANA Nonwovens Symposium, London, U.K., June 1989.
64. J. C. Shivers, P. Popper and H. W. Saffer, In Modern Nonwovens Technology, (Edited by D. J. Ward), Texpress, Manchester, 1977, p. 141.
65. E. A. Brooks, (to Johnson & Johnson), U.S. Patent 2,045,825 (November 5, 1980).
66. M. Suzuki and T. Kobayashi (to Uni-Charm Corp.), French Patent 2,488,920 (February 26, 1982).
67. S. Tomioka and S. Ejima, Nonwoven Conference Papers, UMIST, Manchester, 1980, p. 63.
68. D. Groitzsch, EDANA Nonwovens Symposium, Prague, Czech (June 2000).
69. F. W. Manning, U.S. Patent 2, 411, 660, (May 22, 1943).
70. V. A. Wente, Tech. Rep. No. PB111437, Naval Research Lab, NRL-4364, 4/15/1954.
71. T. M. Holliday, Proc. INDA-TEC International Nonwovens Conference, Crystal City, VA 1996, p. 8.1.
72. H. Bodaghi, INDA JNR, Vol. 1, No. 1, (1989) 14.
73. M. W. Milligan, F. Lu, L. C. Wadsworth and R. R. Buntin, INDA, JNR, Vol. 5, No. 3 (1993) 8.
74. M. Dever, R. Bensong and N. Fair, INDA JNR, Vol. 5, No. 2 (1993) 27.
75. L. C. Wadsworth, Proc. 2001 Beltwide Cotton Conference, National Cotton Council, January 9-13, Anaheim, CA, Vol. 1 (2001) p. 629.
76. F. Knaus in Nonwoven Conference Papers (edited by G. E. Cusick, UMIST, Manchester, 1980, p. 131.
77. F. Bricklebank, INDEX 81 Conference Papers, EDANA, Brussels (May 1981), Session 3.
78. W. B. Madsen, F. H. Jensen, G. B. Rasmussen, G. Goldstein and Y. Roussin-Moynier (to Beghin-Say). U. K. Patent 1,455,981 and 1,455,982 (August 11,1976).
79. F. Lloyd (to Smith & Nephew Research Ltd.), U.K. Patent 1,496,786 (January 5, 1978).
80. G. E. Raley and J. M. Adams (to Ethyl Corporation), U.S. Patent 4,252,516 (February 24, 1981).
81. E. A. Thompson (to Procter & Gamble Co.), U. S. Patent 3,929,135 (December 30, 1975).
82. F. M. Berger, U.S. Patent 5,633,082, May 27, 1997.
83. D. Zhang, C. Sun, L. Wadsworth, and R. Zhao, Proc. TANDEC 9th Annual Conference, Knoxville, TN, 1999, p. 2.4-1.
84. Y. Morimoto, Proc. TANDEC 9th Annual Nonwovens Conference, Knoxville, TN, November 1999, p. 2.2-1.
85. L. C. Wadsworth, H. Suh, and H. C. Allen, Jr., INDA, INJ, Vol. 9, No. 2 (2000) 13.
86. C. Q. Sun, D. Zhang, L. C. Wadsworth, and E. C. McLean, Jr., Text. Res. J., 70 (2000) 449.
87. L. H. Sanford and J. B. Sisson (to Procter & Gamble Co.), U.S. Patent 3,301,746 (January 31, 1967).
88. J. P. Hanson, Pulp & Paper, 51(1977) 93.
89. J. D. Crain, INDEX 81 Congress Papers, EDANA, Brussels (May 1981), Section 6.
90. G. Pinchard and G. Fous, Proc, Absorbent Product Conference, Insight 93, Kalamazoo, MI, p. VIII-1, 1993.
91. L. Wettergren, First International Absorbents Products Conference, INSIGHT 80, Marketing/Technology Service, Kalamazoo, Michigan (November 1980), Section II.
92. J. L. Ferris, EDANA 1083 Workshops, EDANA, Brussels (May 1983), Section 2.
93. L. Wettergren, Disposable Diapers: Changes in Markets and Materials, Nonwoven Industry 22, (April 1979).
94. E. J. Bouda, 10th INDA Technical Symposium, INDA, New York (November 1982), p. 227.

95. J. R. Starr, Advanced Forming/Bonding Conference, INSIGHT 81, Marketing/Technology Service, Kalamazoo, Michigan (September 1981), Section 1.
96. S. Tomioka and S. Kambe, Advanced Forming Bonding Conference INSIGHT 81, Marketing/Technology Service, Kalamazoo, Michigan, 1981, Section II.
97. L. G. Berglund and F. J. Akin, TAPPI Journal 80 (September 1997), p. 173.
98. R. G. Volkman, 9th INDA Technical Symposium, INDA, New York (March 1981).
99. J. B. Madsen, Nonwovens World, August-September (2001) 69.
100. I. E. Ruzek, INSIGHT '96, Business Conference, p. 88, San Antonio, TX, October 1996.
101. D. L. Proxmire, W. W. Jackson, N. D. Kollin, T. L. Mace, A. L. McCormack, D. R. Schlinz, (to Kimberly-Clark Corporation), U.S. Patent 5,192,606 (March 9, 1993).
102. E. Engle, INSIGHT '92 International Conference, Toronto, Canada, October 1992, p. 74.
103. INDA, Association of the Nonwoven Fabrics Industry, 1300 Crescent Green, Suite 135, P.O. Box 1288, Cary, NC 27512 USA; www.inda.org.
104. TAPPI, Technical Association of the Pulp and Paper Industry, Technology Park/Atlanta, P.O. Box 105113, Atlanta, GA 30348, U.S.A.
105. EDANA, European Disposables and Nonwovens Association, 157 Avenue Eugene Plasky, 1030 Brussels, Belgium; www.edana.org.
106. ANNA, All Nippon Nonwovens Association, Soto-Kanda, 6-Chome, Bldg 3F, 9-2, Soto-Kanda, 6-Chome, Chiyoda-Ku, Tokyo 101, Japan.
107. ASTM, American Society for Testing Materials, West Conshohochen, Pennsylvania, USA.
108. M. J. Welch and J. A. McCombes, 8th INDA Technical Symposium. INDA. New York. 1980.
109. R. L. Shishoo, Adult Incontinence Products Symposium, INSIGHT 82, Marketing/Technology Service, Kalamazoo, Michigan (October 1982), Section III.
110. R. L. Shishoo Absorbent Products Conference, INSIGHT 81, Marketing/Technology Service, Kalamazoo, Michigan, (September 1981), Section VI.
111. J. P. Hanson, New Nonwovens World, Fall (1993) 34.
112. The Marketing / Technology Service Inc (MTS), Kalamazoo, Michigan.
113. A. Wraith, Nonwoven Conference Papers, (edited by G. E. Cusick), UMIST, Manchester, 1983.
114. J. P. Hansen (editor). Facing Materials - New Developments and Applications for Absorbent Products, Marketing/Technology Service, Kalamazoo, Michigan (April 1983), Volume II.
115. J. Alter, Advertising Age, 1; Section 1 (February 7, 1983).
116. A. A. Burgeni and C. Kapur, Textile Res. J., 37 (1967) 356.
117. A. A. Burgeni (to Johnson & Johnson). U.S. Patent 3,017,304 (January 16, 1962).
118. R. W. Collins, INSIGHT 86, (September 1986).
119. G. Richeson, INDA-TEC, Fort Lauderdale, FL (April 1992).
120. H. S. Jensen, INDEX '99, Hygiene 2 Conference, p. 99; Geneva Switzerland, (April 1999), p. 99.
121. J. T. Conley, INSIGHT 98, Absorbent Products Conference, Orlando, FL, (October 14, 1998), p. 188.
122. J. P. Erspamer, S. C. Baer and L. Li, 99 INDEX, Conference Papers, Hygiene I session, Geneva, Switzerland, April 1999.
123. A. G. Patchell, W. D. Murphy and R. Lloyd (to Smith & Nephew Plastics Ltd), U.S. Patent 4,135,021 (January 16, 1979).
124. M. B. Lucas and R. H. Van Coney, (to Procter & Gamble Co.), U.S. Patent 4,151,240 (April 24, 1979).
125. J. Heuchert and P. Mandotti, Proc. Absorbent Product Conference, Insight 94, Kalamazoo, MI, p V-1, 1994.
126. S. I. Varley, S. E. Barnett and J. T. Scales "Wound Release," Conference on Medical Applications and Textiles. University of Leeds (1981), Lecture 19.
127. J. H. Gardner and S. Rizzo. (to Johnson & Johnson). U.S. Patent 3,525,417 (September 15, 1970).
128. M. C. Muller, 11th Technical Symposium, INDA, New York, 1983, p. 42.
129. J. A. Howey, H. A. Patel and F. C. Zevnik, Index 81 Congress Papers, EDANA, Brussels (March 1981), Section 3.
130. J. C. Parrella, Advanced Forming/Bonding Conference—INSIGHT 81, Marketing/Technology Service, Kalamazoo, Michigan (September 1981), Section IV.
131. R. L. Vanderselt and W. G. Isner. IDEA 82 Conference Papers, INDA, New York (March 1982).
132. C. Sheridan. 11th INDA Technical Symposium, INDA, New York (September 1983), p. 35.
133. Anon., Kimberly-Clark KIMTEX, Brochures, W-676-1, W-445-6, CS 2635.
134. W. J. McCulloch and R. A. Van Brederode, Absorbent Products Conference—INSIGHT 81, Marketing/Technology Service, Kalamazoo, Michigan (September 1981), Section VII.
135. R. J. Ludwa, (to Procter & Gamble Co.), U.S. Patent 4,239,792 (December 16. 1980).
136. R. Knaus, Advanced Forming/Bonding Conference— INSIGHT 81. Marketing/Technology Service. Kalamazoo, Michigan (September 1981), Section X.
137. L. Shelton. Absorbent Products Conference—INSIGHT 82, Marketing/Technology Service, Kalamazoo. Michigan (October 1982).
138. E. S. Engle III, TANDEC, 5th Annual Conference, Knoxville, TN (October/November 1995), p. 4-1.1.
139. N. Ecenkova, S. Batchernikova and A. Mikhalkova, EDANA Nonwovens Symposium, Prague, Czech (June 2000), p. 121.

Absorbent Technology.
P.K. Chatterjee and B.S. Gupta, editors.

CHAPTER XI

MEASUREMENT TECHNIQUES FOR ABSORBENT MATERIALS AND PRODUCTS

BHUPENDER S. GUPTA
College of Textiles, North Carolina State University, Raleigh, NC 27695-8301 (USA).

PRONOY K. CHATTERJEE
Nutecа International Co., 331 McDowell Drive, East Brunswick, NJ 08816 (USA).

Contents

1. GENERAL SCOPE AND CONSIDERATIONS

1.1. Challenges in Characterization

Absorbent products are usually highly engineered composites whose in-use performance is ultimately judged by the consumer. Each manufacturer of absorbent articles strives to produce consistently high performing product at a low cost. In order to assure that these materials will stand up to the scrutiny of consumer, the producers must select appropriate fibers and polymeric materials, engineer individual components with desirable characteristics and assemble them into the desired article with optimum performance.

The materials are characterized for properties for a number of reasons [1]. Process control is one. Certain properties are measured to keep the manufacturing equipment functioning efficiently and producing a given product with target specifications. Process optimization is another in which the goal is to design an assembled structure with optimum performance for a given size and weight. There is a continuing effort directed towards producing lighter, thinner and lower cost products that perform as good as or better than the existing product. Quality control could be considered as a third reason. A sound scheme of quality testing is critical since consistency is required for customer to continue to buy the same brand. And finally, the testing is required for research in order to search for superior products that may be radically different, in terms of design and materials, than those available currently.

Research over the past three decades or so have dealt with many different aspects in the field of absorbency, including theoretical modeling of the behavior, development of testing procedures and devices, and statistical evaluation of consumer test data. It is evident that the experimental studies have not kept pace with and reached the same level of sophistication and excellence as has the theoretical work. This could be attributed to some degree to the rate at which the disposable absorbent product market has expanded during the recent years throughout the world. Also, this is partly due to the general lack of clear understanding of what should be measured, and how to measure it. In many instances, the actual use conditions are so complex that the tests conducted using instrumental methods do not exactly meet the situation. A serious problem is that most hygiene and diaper products involve body fluids that change in properties with time and the use extends over several hours. In contrast, the actual tests involve simple saline or a similar fluid and the test is usually completed in a very short time, of the order of a minute.

1.2. Product Structure

A large number of absorbent products exist in the market place. These range from structures as simple as a ball containing randomly collected mass of bleached fibers to as complex as adult incontinent pad and baby diaper that contain multiple components, each performing a specific function. Schematic showing essential layout of such a composite is given in Figure 1. While size and shape, as well as the final end use may differ, they generally share some common structures. Each of the four layers is typified by one of several absorbent characteristics. The fibers that comprise the layers are either hydrophilic or hydrophobic. Hydrophilic is defined as water loving and hydrophobic as water repelling, that is the property of non-wetting by an aqueous fluid. However, vapor may pass through

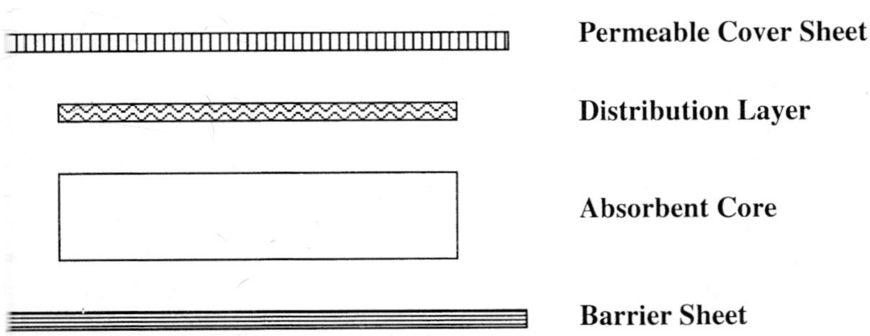

Fig. 1. Composite nature of an absorbent product.

hydrophobic materials, and high pressure can force fluid through them. An impermeable substance will not let fluid pass through even at a reasonably high pressure.

The top sheet is a pervious structure and is the only portion of the product that is in contact with the user's skin. As such, it is extremely important that the body fluid penetrate quickly through this layer. The layer must also be constructed with materials that do not allow the fluid to rewet the skin. Typically, the top sheet is made of hydrophobic fibers. Good transport performance by the top layer ensures that the user will remain dry and comfortable and not develop any skin rashes or chaffing.

The distribution layer, also called wicking or surge layer, is located between the top sheet and the absorbent core. Because the absorbent core cannot absorb the localized insult instantly, the wicking layer imbibes the body fluid away from the point of insult, distributes it laterally, and holds it for the core to absorb it. The wicking layer absorbs the fluid instantaneously by capillary forces. This layer is usually made up of hydrophilic fibers. The pore size of the structure and its distribution are as important as is its wettability.

The absorbent core is the heart of an absorbent hygiene product. In modern diapers, the core consists of hydrophilic wood pulp and superabsorbent polymers. Because the core often must withstand several insults, it must be able to absorb, transport and distribute, and hold body fluid at more than 50% of its own weight.

Finally, the back sheet is an impermeable thin film or barrier fabric that prevents leakage. To improve the handle and aesthetics of the product, particularly diapers and training pants, the back sheet is often embossed for flexibility and for having cloth like appearance.

Thus, different layers perform different functions, some are absorbent, some transport liquid, and some are repellent. This means that test methods should be specific to characterize each of these functions effectively and reproducibly. It has been pointed out earlier (Chapter III) that absorbency, or lack of it, is the result of interactions between fiber (or polymer) and fluid. To optimize the performance of an absorbent article, this interaction needs to be examined and optimized at each layer level. In complex structures, such as an absorbent core of a personal hygiene product, this requires a study and understanding of the roles played by four categories of variables; namely, the fluid properties; the surface, diffusional, swelling and wet mechanical properties of fibers; the pore structure and bonding characteristics of fabrics; and the manner in which, and the conditions under which, the product is tested or used.

1.3. Hierarchy in Testing

1.3.1. Constituent Material

A comprehensive plan of characterization includes testing of materials at many different stages of development. It starts with the characterization of polymers and fibers, characterization of each individual layer or component, and at the end, evaluation of the performance of the final product. The polymers and fibers serve as the raw materials and determine the inherent ability of the building blocks to attract or repel a fluid, or absorb and swell, when a fluid comes into contact with them. The properties of interest are the standard moisture regain, the liquid saturation capacity, water retention value, surface free energy, contact angle, diffusion and swelling characteristics, and wet mechanical properties. Some of these properties are related to each other, but they involve different test methods and often play different roles in different applications.

1.3.2. Two Dimensional Structures

The next stage in a product hierarchy is a structured assembly, usually a two-dimensional web or sheet with some form of bonding for standing mechanical handling and maintaining structural integrity when interacting with fluid. These two-dimensional structures are usually prepared by air laying, carding or wet laying processes, if the fabric is short length or staple fiber based, or meltblowing or spunbonding, if the fabric is continuous filament based. The areal densities of these could vary over a broad range, from as little as 10 g/m^2 or less for coversheet to as much as 200 g/m^2 or more for serving as absorbent core. Since the fibers are loose and could shift during movement of body and/or interaction with the fluid, some form of bonding is essential, and the shorter the fiber, the greater such necessity. The bonding is provided by one of a number of technologies available, namely, needling, hydroentangling, thermal, chemical, or adhesive means. Such two dimensional structures are designed to perform a variety of functions in a product, including repellency, barrier, transport or absorbency of liquid. Accordingly, the properties of interest in characterizing these structures are absorption capacity, rate of absorption, linear wicking, water permeability, air permeability, and repellency.

1.3.3. Assembled Product

The next level is the product itself, which has the required number and types of the two-dimensional fabrics or assemblages, arranged in the order needed and finished in an integral unit form. Each component performs its function in collaboration with the other components and contributes to the overall performance of the article. The single most important property of interest is the customer satisfaction which can, of course, be a very complex and difficult goal to achieve through a single kind of product. Usually, therefore, a line of products, of any given type is marketed that vary in terms of size, configuration, and properties to suit the lifestyle and needs of different customers. The final article is customarily field-tested using selected users. This is time consuming and expensive and is conducted only after the components and the composite have passed the in-house scrutiny of performance. Each major manufacturer, therefore, tends to have its own testing procedure that would simulate to some degree the actual application. The majority of those tests are non-standard, the exact details of which are usually not available in scientific literature. The general criteria in these tests is still to get a measure of the rate of absorption, total absorption capacity, resistance against liquid desorption or release back, and resistance against leakage from the barrier sheet.

1.4. Types of Test Methods

1.4.1. Standard Methods

The broad scope of testing and characterization is accomplished using three different schemes. In this, the first, well-established test methods are used, published by standard organizations, such as the American Society for Testing and Materials (ASTM), the American Association for Textile Chemists and Colorists (AATCC), the Association of the Nonwoven Fabrics Industry (INDA), the Technical Association of Pulp and Paper Industry

(TAPPI), the Scandinavian Pulp and Paper Board Association, the European Disposable and Nonwoven Association (EDANA).

These tests are intended to provide uniform orderly procedures for the determination of certain well defined properties of fibrous materials. The procedures involved are used in assessing properties for the purpose of process control, quality control, ranking of materials from different producers, and for research and development. A major use of the properties measured by these methods is also made in buying and selling of roll goods by manufacturers.

1.4.2. Methods as Research Tools

A second set of methods used are those which do not easily lend to standardization. These are complex in nature and require extensive specimen preparation, delicate controls and calibrations, and they characterize properties that are microscopic in nature. The properties measured are, however, highly valuable as they can provide an insight into the mechanisms governing the performance and a direction for developing new and improved materials and products.

Accordingly, methods such as these are used as research tools by organizations involved in research and development. A typical example of such a method is the fluid intrusion/extrusion procedure used on a structure that provides information on pore size distribution and changes in it during absorption/desorption cycle.

1.4.3. In Vitro and In Vivo Methods

A final set may be considered as one that more or less constitutes *in vitro* or *in vivo* testing, using conditions that more or less simulate actual use environment. As mentioned, the schemes or the methods used are usually specific to a manufacturer and, therefore, vary from one company to another. Syngyna test, for example, would be one such method where the actual equipment and the specific procedure used are likely to vary from one producer of tampons to another.

1.5. Summary and Proposed Plan

In summary, because products must be designed optimally and economically, using carefully selected materials and constructions that would lead to performance acceptable to the customer, manufacturers usually have an elaborate scheme of evaluation for their products. It normally includes a criteria for selecting raw materials, standard test procedures for characterizing the properties of individual components, other methods that lead to a better understanding of the material, in house *in vitro* methods for predicting the performance of assembled units, and field conducted *in vivo* tests and customer surveys for determining the final product performance.

In the following sections, the procedures employed in characterizing the properties of materials and fabrics used in absorbent products, and performance of final products, are illustrated. Only essential information is provided. More specific and detailed information can be obtained by referring to the sources cited. In some instances, where methods are considered as research tools, selected examples of results are also included to provide an insight about the property.

Sections 2 and 3 of the Chapter are devoted to the characterization of some of the properties of fibers that have an impact on the performance of absorbent products. Those covered include the surface energetics, diffusion, swelling and transverse mechanical properties. These are measured by a mixture of standard and non-standard procedures. Section 4 includes description of selected standard test methods available in the literature. The devices and procedures that fall in the category of research tools, and the behaviors characterized by them, are discussed in Section 5. The last section contains a description of some *in vitro* test methods that are used to evaluate the performance of end products.

2. SURFACE ENERGETICS OF FIBERS

2.1. Introduction

Surface wetting force, adhesion and energy are important physical properties that affect the ability of a material to be processed into final products and the performance of the latter in many key applications. Work of adhesion, for example, affects cohesion and bonding between surfaces and would be an important property to consider when selecting materials for applications such as coatings and composites. Likewise, surface contact angle and energy values govern the behavior of fibers in capillary absorption and transport of fluids and would be important quantities to consider when selecting materials for products such as wipes, towels, sanitary pads, tampons and diapers. Another interest in studying surface properties is to know how the surface energy of a fiber, being considered for use in a product, is divided among polar and dispersion components. This can be an important fundamental information that can provide an insight about the affinity the fiber has for the fluid.

2.2. Wettability Parameters

The most common and useful measure of the wettability of a solid is the contact angle, and the most widely used procedure for assessing its value is to place a drop of fluid on a flat surface and measure the angle by some optically enhanced means (Fig. 2). If the angle is less than 90°, the surface is considered wettable since the solid-liquid attraction is more than the liquid-liquid attraction. A value of 90° or more indicates that the drop resists spreading and, therefore, the surface is not wettable. A single value obtained from a static system is, however, not satisfactory for fibers because they are irregular and heterogeneous in characteristics along their lengths. The dynamic contact angle determined by the Wilhelmy technique [2] provides a more satisfactory characterization of fiber-liquid interaction. The technique yields the values of the advancing and the receding contact angle

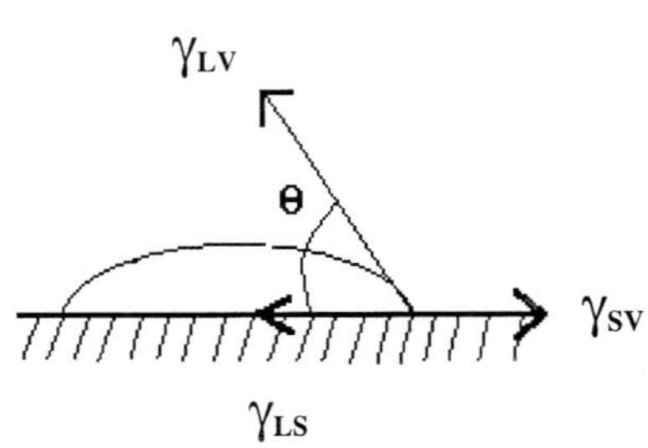

Fig. 2. Interfacial tension and contact angle at equilibrium.

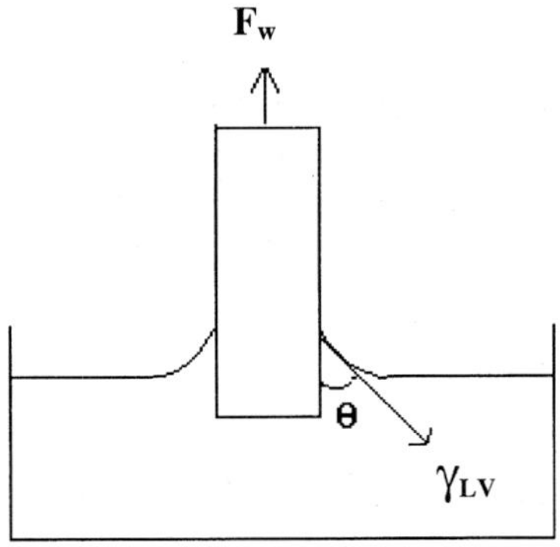

Fig. 3. The Wilhelmy Force.

by evaluating the forces which cause the liquid interline to advance or recede, respectively, over the solid surface.

According to the Wilhelmy principle, the vertical component of the attractive force across the interface between a partially immersed solid and a liquid surface is expressed as follows:

$$F_W = \gamma_{lv} P \cos \theta \quad (1)$$

where F_W = wetting force, γ_{lv} = surface tension of liquid (dyne/cm), P = perimeter of the solid (cm), and θ = contact angle between liquid and solid at the interface (Figure 3). Equations can be written for both the advancing (θ_A) and the receding (θ_R) values of the contact angle as follows:

$$F_W(A) = \gamma_{lv} P \cos \theta_A \quad (2)$$

$$F_W(R) = \gamma_{lv} P \cos \theta_R \quad (3)$$

The surfaces of solids are often characterized for their interaction with liquids in terms of the parameter called the work of adhesion (W_A). It is given by:

$$W_A = \gamma_{sv} + \gamma_{lv} - \gamma_{sl} \quad (4)$$

where the γ_{sv}, γ_{lv}, and γ_{sl} terms represent the interfacial tensions at solid-vapor, liquid-vapor, and solid-liquid interfaces, respectively (Figure 2).
By combining eq. 4 with the Dupre-Young equation (eq. 5), given below, and solving simultaneously for W_A, one gets the relation given by eq. 6.

$$\gamma_{lv} \cos \theta = \gamma_{sv} - \gamma_{sl} \quad (5)$$

$$W_A = \gamma_{lv}(1 + \cos\theta) \tag{6}$$

Equation 6 shows that the work of adhesion is the sum of two quantities, a property of the liquid (γ_{lv}) and a property of the interaction between the liquid and the solid ($\gamma_{lv} \cos\theta$). The latter term, $\gamma_{lv} \cos\theta$, can also be used to define a parameter called Wettability Index [3] which gives a normalized value of the Wilhelmy force, i.e:

$$\text{Wettability Index, } WI = \frac{F_W}{P} = \gamma_{lv}\cos\theta \tag{7}$$

The best method available for determining the surface energy (γ_s), which is the sum of the dispersive component ($\gamma_s^{d'}$) and the polar component ($\gamma_s^{p'}$), is the measurement of the contact angles with two dissimilar liquids whose dispersive and polar contributions to surface tensions are known. Assuming that both the dispersive and the polar interactions across the solid-liquid interface conform to the geometric mean mixing rule, Kaelble [4] expressed the solid-liquid interfacial energies as:

$$\gamma_{sli} = \gamma_s + \gamma_{li} - 2\,(\gamma_s^{d'}\,\gamma_{li}^{d'})^{0.5} - 2\,(\gamma_s^{p'}\,\gamma_{li}^{p'})^{0.5} \tag{8}$$

$$\gamma_{slj} = \gamma_s + \gamma_{lj} - 2\,(\gamma_s^{d'}\,\gamma_{lj}^{d'})^{0.5} - 2\,(\gamma_s^{p'}\,\gamma_{lj}^{p'})^{0.5} \tag{9}$$

Young's equation (eq. 5) is rewritten for a given solid interacting with two fluids as follows:

$$-\gamma_s + \gamma_{sli} = -\gamma_{li}\cos\theta_i \tag{10}$$

$$-\gamma_s + \gamma_{slj} = -\gamma_{lj}\cos\theta_j \tag{11}$$

Contact angles θ_i and θ_j are measured for the liquids, i and j for which the surface tension components, $\gamma_{li}^{d'}$, $\gamma_{li}^{p'}$, $\gamma_{lj}^{d'}$, and $\gamma_{lj}^{p'}$, are known. With these values substituted, equations 8 to 11 are solved simultaneously for determining the dispersive and the polar values of the surface free energy of the material.

2.3. Experimental Procedure

The dynamic wetting force is measured by a continuous immersion/emersion technique using a highly sensitive electrobalance, with the sensitivity of the order of 10^{-2} dyne. The general arrangement [3] is shown in Figure 4. The wetting liquid container resting on a movable platform is traversed up and down at a predetermined speed, usually very small (~ 500 μm/min), via a computer programmed motor-controlled drive. Since the surface properties are sensitive to the state of the surface, the test material is appropriately cleaned and conditioned prior to test. For analysis in which a specimen is tested for properties against two different fluids, the tested specimen/foil assembly from the first fluid is dried and conditioned in a dessicator before it is retested in the second fluid.

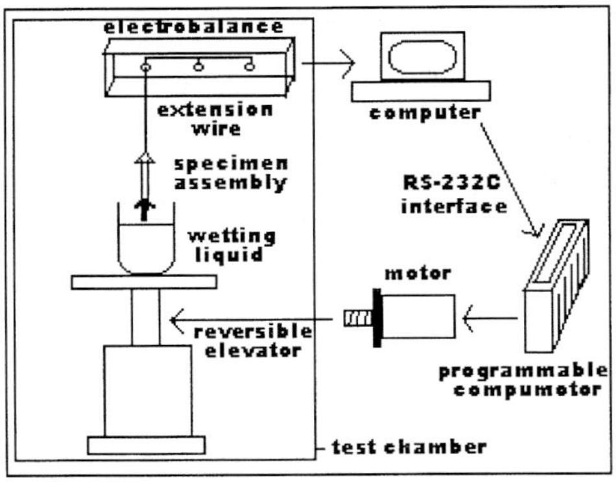

Fig. 4. Wilhelmy Device for Measuring Wettability [3].

It is obvious that a fiber specimen should remain vertical and straight, i.e. should be free of crimp and not curl or float when contacting the fluid. The technique can work well without involving any special specimen preparation if the fiber is of large cross-sectional area and free of crimp, so that it is straight and inherently stiff, and its surface is hydrophilic, so that it is attracted to water. However, in order to use the method on fine diameter fibers of textile dimensions (1-15 denier) or on fibers or films that are non-wettable, a special procedure, involving the use of a high energy metal that loads the lower end, is required. The metal wire used is made of platinum and acts as a sinker. It is attached to the lower end of the specimen. High force of attraction between the fluid and the metal causes the end of the specimen to immerse and remain vertical even if the latter should repel against the fluid. The length and the diameter of the wire being known the buoyancy of the submerged sinker could be calculated.

The prepared specimens involving a fine denier fiber and a rectangular polymeric film are shown in Figure 5. The two ends of the test fiber are glued to a triangular piece of aluminum foil. The sinker appropriately configured is hooked to the loop so formed at the lower end, which produces two vertical sections of the specimen. The mounting tab is suspended from the hangdown wire of the electrobalance by means of a hole punched near

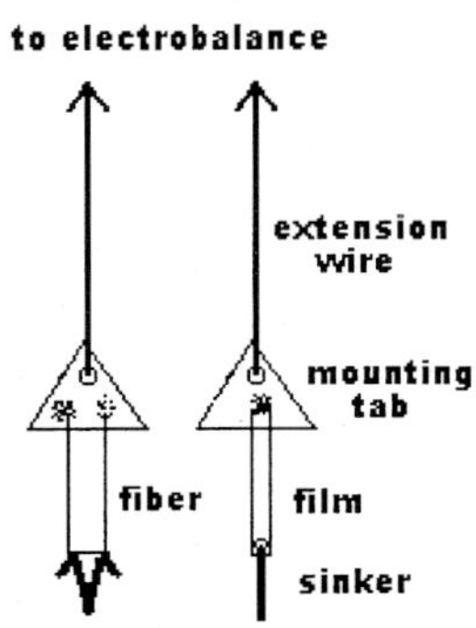

Fig. 5. Specimen preparation for wetting force measurements [3].

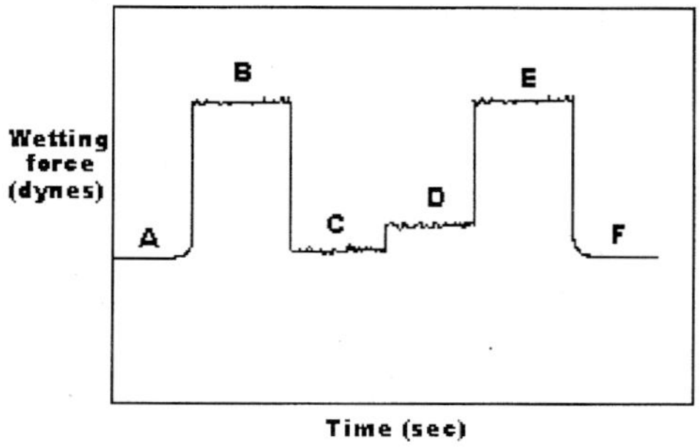

Fig. 6. Hypothetical trace obtained in dynamic contact angle experiment on specimen with a sinker.

the vertex of the foil. For studies on film, a thin rectangular shaped specimen (~10mm x 1mm) cut carefully with a sharp razor blade may be used. One end of the film is glued to the foil, the other end has a small hole punched that allows a platinum wire with the top end bent sharply to be suspend from it.

The basic experimental run consists of the following cycle: starting with the liquid surface just clear of the specimen, the drive is activated so that the liquid moves up and wets the solid. After the liquid surface has traversed about half the specimen length, the drive is reversed and the liquid moves down until it breaks free of the specimen assembly.

The force detected on the specimen is given by the following equation:

$$F = mg + F_W - F_B - F_S \pm F_f \tag{12}$$

In this, F is the measured force on the sample, F_W is the actual wetting or Wilhelmy force, F_B is the buoyancy force of the sample, F_S is the buoyancy force of the submerged sinker, mg is the weight of the fiber assembly, and F_f is the frictional drag of the sinker/fiber (mostly sinker) against the fluid. The sign of F_f is (-) for advancing and (+) for receding. The first term on the right is usually tared out at the beginning and the third and the fifth are negligible in comparison with the magnitude of the wetting force. Therefore, the equation simplifies to:

$$F = F_W - F_S = \gamma_{lv} P \cos\theta - F_S \tag{13}$$

Since the measured force is the result of the pull of liquid surface in the direction determined by the angle of contact, the observed effect should be an upward push when this angle is greater than 90°. This produces an apparent loss in weight, but this "negative" force value can be used in eq. 1 without any discontinuity. When the angle is 90°, the measured force would be zero but the result would be as significant as any other; that is, it would still indicate a finite attraction, though with no vertical component.

2.4. Interpretation of the Wetting Force Data

A somewhat idealized but largely representative trace of the force, F, obtained in Wilhelmy test, involving a sinker, is shown in Figure 6. Regions A, B, C, D, E and F in the profile represent, respectively, the force signal of the assembly before the contract is made with the fluid, during immersion of the sinker, during immersion of the test fiber, during withdrawal of the same, during withdrawal of the sinker, and after the assembly is out of the fluid. Because of the large buoyancy effect of the sinker, regions C and D do not reveal slopes. In actuality, region C involves a small negative slope due to increasing buoyancy and region D involves a small positive slope due to decreasing buoyancy of the test material. The transition between C and D is not as abrupt but gradual as it represents transition of the meniscus from a larger advancing angle, θ_A, to smaller receding angle, θ_R. This is due to hysteresis, discussed later. When a fiber material is being scanned, there is always some noise in the force profile and that is due to chemical heterogeneity and localized surface variations that affect the wetting force.

2.5. Perimeter of Test Specimen

Determination of contact angle (eq. 1) by the Wilhelmy method requires that the perimeter of the specimen be known. On a round fiber of reasonably large size, perimeter can be determined accurately by measuring diameter with an optical microscope. On fibers, which are fine and irregular in cross-sectional shape, this method can lead to inaccurate results. One suitable option is to determine the value by image analysis of SEM micrograph of the cross-section. This procedure may be laborious but it should provide accurate results. A second option would be to assume, as suggested in the literature [5], that the receding value of the contact angle θ_R, will be zero if the surface is interacted with a liquid of low surface tension such as hexadecane. If this assumption is true and adopted, then the value of the perimeter can be obtained by simply measuring the magnitude of the receding wetting force and dividing it by the surface tension of the fluid:

$$F_R = \gamma_{lv} P \cos \theta_R, \quad or \quad P = F_R / \gamma_{lv} \tag{14}$$

However, in a study conducted on cellulose acetate fibers in which both procedures were used, it was noted that the perimeter values obtained by the method in which hexadecane was used and θ_R was assumed to be zero, were significantly lower (~ 19%) than those obtained using image analysis of fiber cross-sections [3]. This indicates that θ_R for hexadecane on cellulose acetate was not zero, but greater. On cellulosic fibers, namely cotton and rayon, as well, hexadecane did not yield a θ_R of zero, but a somewhat higher value.

Miller discussed a hypothetical situation under which perimeter can be obtained without either the image analysis of a cross-section or the use of low surface tension fluid [6]. In Miller's test, a fiber specimen is cut squarely and tested without any sinker attached to the end. The specimen is taken through the full cycle with the end immersing in the advancing mode and then withdrawing totally from the fluid in the receding mode. As the meniscus moves to the bottom of the specimen and moves around the edge, at some point in this transition the direction of the pull is vertically down. At this point, the contribution of the surface tension to the pull out force is maximum (Fig. 7) since the effective angle with the vertical wall is presumably zero. Under this situation, the maximum force $F(M)$ registered is given as:

$$F(M) = \gamma_{lv} P \tag{15}$$

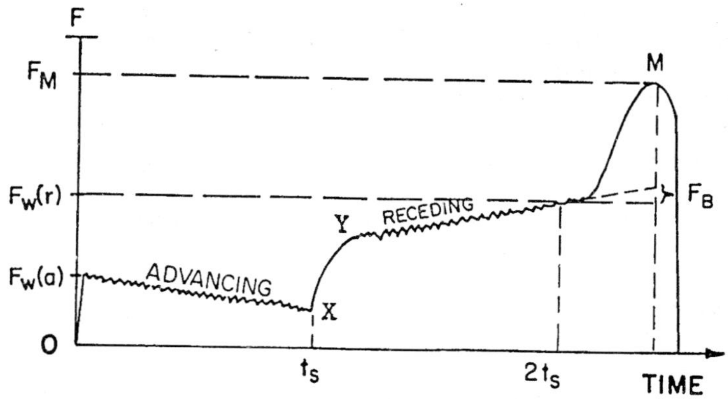

Fig. 7. Maximum pullout force M in Miller test [6].

Accordingly, the value of the perimeter, P, can be obtained simply by dividing the maximum pull out force with the fluid surface tension. The value of P so obtained could then be used in eqs. 2 and 3 and the values of the advancing contact angle, θ_A, and the receding contact angle, ϑ_R, can be determined.

This procedure, although theoretically feasible, has serious practical difficulties when applied to fibers of textile dimensions and properties. Two most obvious problems are that (1) fibers being highly flexible and deformable do not lend to a sharp 90° cut perpendicular to fiber axis and (2) the presence of texture and crimp, typical of all fibers used in absorbent and textile applications, require the use of bottom end loading with a sinker. The latter system means that the end of the specimen is not free but a part of an assembly. Also, the procedure will not work on materials that have receding contact angles greater than 90°, which some materials such as the polyolefins have [3].

2.6. Contact Angle and Wettability Index

Average values of contact angle and the wettability index are given for selected textile fibers in Table 1. The values of the advancing contact angle, which is of relatively greater importance, vary substantially among the fibers, with the trilobal rayon having the lowest (22°) and the olefin fiber having the highest (96°) value. Accordingly, all fibers listed in the table, except polypropylene, have wettable surfaces. The differences in angles are given by the differences in the chemical and the morphological structures. The receding value is smaller and is considered to be governed largely by the chemical structure of the fiber [3]. The three cellulosic fibers (cotton and two types of rayon) being chemically similar have nearly the same values of θ_R. Cellulose acetate, with 66 or greater percentage of the hydrophilic hydroxyl groups replaced by largely hydrophobic acetyl groups, has significantly larger value of θ_R. The differences in θ_A among the three cellulosic fibers is due largely to

Table 1.
Average values of contact angle and wettability index of selected textile fibers measured by the Wilhelmy technique [3].

Fiber	Contact Angle (°) θ_A	Contact Angle (°) θ_R	Wettability Index (dynes/cm) WI_A	Wettability Index (dynes/cm) WI_R
Cotton	34	20	60	68
Trilobal Rayon	22	18	68	69
Regular Rayon	55	17	41	69
Cellulose Acetate	56	44	40	52
Polyester	79	68	14	27
Polypropylene	96	93	-8	-4

the differences in morphological structures. The lowest value noted in the trilobal fiber is most likely due to the ridges formed by the trilobal shape on fiber surface that imbibed fluid and made surface more readily wettable. And the highest value seen in the round or regular rayon was attributed to the relatively lower value of molecular orientation found in this fiber [7]. Polyester has low attraction for fluid but still has a wettable surface whereas the polyolefin, being strictly hydrocarbon, has a non-wettable surface.

Wettability index is given by the product of surface tension of water (a constant) and the cosine of the contact angle (eq. 7). It gives a measure of the normalized attractive force exerted by a solid on a fluid and is directly proportional to the cosine of the contact angle. Accordingly, the fibers that have the lowest value of the contact angle also have the highest value of the attractive force for the fluid, and this value, as expected, is greater for the receding than for the advancing mode.

2.7. Work of Adhesion and Hysteresis

Since the only variable in the equation for the work of adhesion is the cosine of the contact angle (eq. 6), the differences in the contact angle values among the fibers and between the modes also reflect the differences in the WA values, given in Table 2. This table also lists the values of hysteresis given by the ratio of the work of adhesion in the receding to that in the advancing mode [8]. Among the factors that have been mentioned as being responsible for hysteresis are the surface roughness and contamination and the time-dependent interaction of the solid with the liquid [9,10]. A material that interacts with fluid and reaches equilibrium rapidly or does not interact with it much even after a period of contact will usually have a low value of hysteresis. The interaction of fluid with cotton and trilobal rayon support the former and that with polyolefin the latter behaviors. On the other hand, a material which has the potential to interact on account of its chemical composition but requires contact with fluid over a length of time to reach equilibrium, due either to roughness or lack of molecular order or orientation, will usually have a high value of hysteresis. Among the fibers listed in the table, regular or round rayon, cellulose acetate and polyester fall in this group.

Table 2.
Average values of the works of adhesion and hysteresis. [3,7]

Fiber	Work of Adhesion (dynes/cm) WA_A	WA_R	Hysteresis (WA_R / WA_A)
Cotton	133	141	1.06
Trilobal Rayon	140	142	1.01
Regular Rayon	114	142	1.25
Cellulose acetate	113	125	1.11
Polyester	86	100	1.16
Polypropylene	65	69	1.06

2.8. Solid Surface Energy

Average values of the total as well as the polar and the dispersion energies are given in Table 3. The determination of the energy value is based on the measurements of the advancing contact angles. The total energy is highest for trilobal rayon (68 dynes/cm) and cotton (61 dynes/cm) and lowest for polypropylene (23 dynes/cm). The three cellulosic fibers, i.e. the two rayons and cotton, have approximately the same value of the dispersion energy (~ 19 dynes/cm); the differences in the total, therefore, arise from the differences that exist in their polar values. These three fibers, being largely hydrophilic, have relatively higher polar components. Among the other three fibers, cellulose acetate has the highest value of surface energy (52 dynes/cm), followed first by polyester (42 dynes/cm) and then by polypropylene (23 dynes/cm). The polar fractions of the energies are 33% in cellulose acetate, 10% in polyester, and 8% in polypropylene. Among these materials, therefore, only cellulose acetate has a sizable polar value, which must be due to the presence of hydroxyl groups in the chains. Accordingly, the surface energies of these materials are largely dispersion, which in polyolefin accounts for 92 % or greater of the total.

Table 3.
Average values of the dispersion, polar, and total surface energies of the fiber [3,7].

Fiber	Surface Energy (dynes/cm) Polar	Dispersion	Total
Cotton	42	19	61
Trilobal Rayon	52	16	68
Round Rayon	24	22	46
Cellulose Acetate	17	35	52
Polyester	4	38	42
Polypropylene	2	21	23

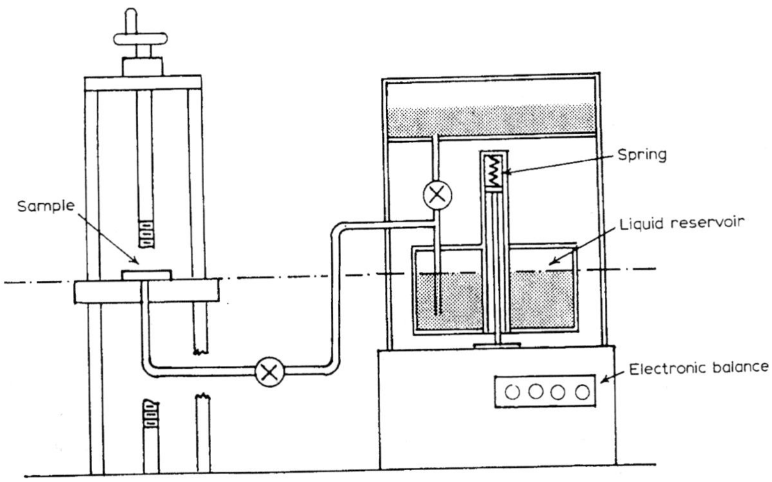

Fig. 23. Gravimetric absorbency testing system [60].

tubing (Fig. 20) or a burette with an air bleeder (Fig. 19), the liquid source is a relatively large vessel resting on the top of an electronic balance via a coil spring. The material of the latter has a predetermined spring constant and is capable of compensating the weight loss (due to absorption) or weight gain (due to exsorption) of liquid in the vessel such that the liquid level can be maintained constant.

The amount of liquid absorbed is measured continuously by the electronic balance, the amount absorbed as a function of time can be recorded continuously via a strip chart recorder (or digitally with a microcomputer). A modified version that incorporates thickness measuring sensors and is fully computer controlled and operated has been illustrated and discussed in sec. 2, Chapter III [21].

The advantage of this design, other than its ability to record infinitesimal changes, is its flexibility in removing or returning the fluid and the application of a large volume of liquid. A wide variety of test cells, including the porous plate, point source, wicking, etc., allowing different modes of contact between the absorbent sample and the fluid, for assessing different behaviors can be used with this equipment.

A siphon-type attachment used for measuring “flow conductivity”, based on Darcy’s law, has been used with this testing system [63]. This tester made use of the constant level of source liquid, which was maintained by the device. A moisture vapor transmission test cell was also developed and patented by Gillespie and Farrington [64] that could be used in place of the normal cell and measure the moisture vapor transmission value of a fabric. The cell included a built in fan that drew outside air, circulated it over the sample and then exhausted it out to the atmosphere.

With a porous plate attachment, and an electro-mechanical system, which can raise or lower the porous plate in a programmable fashion, the equilibrium sorption isotherms [33] can be obtained conveniently. The system incorporates a sample thickness-measuring device

[21] which allows monitoring of bulk volume changes under constant load; the load can be preprogrammed for carrying out cycle loading tests, as discussed in the following section.

5.6. Sorption Equilibria and Bulk Volume Changes

In problems related to absorbency, very often absorption and exsorption can occur simultaneously in the same porous structure. The study of the capillary pressure versus equilibrium saturation relationships is important for the characterization of porous medium in terms of its relative ability to attract, retain and distribute liquid.

The measurement of capillary pressure and equilibrium saturation relationships for fiber masses has been carried out by a number of workers [33,56-59,65,66]. The apparatus for measuring the isotherms is based on porous plate support as shown in Fig. 20 and described in the previous section.

The hydrostatic tension represented by the column of water of height *h*, functions as the withdrawing pressure, which opposes the absorption pressure of the sample C located on top of the porous plate B (Fig. 20). This pressure is changed by raising or lowering the funnel assembly. The change is made in small increments and the sample is allowed to reach new equilibrium saturation at each position. The equilibrium saturation is determined by reading the relative displacement of the meniscus in the horizontal glass capillary tube.

The sample is always kept under a small confining pressure during the test. For absorption equilibria, *h*, starts out at maximum and decreases in predetermined increments. At each position, when equilibrium is reached, the amount of liquid absorbed is recorded. The procedure is repeated until *h* is almost zero and the sample is completely saturated. For exsorption, the process is reversed: starting from complete saturation *h* is increased incrementally and the equilibrium measurements are made until maximum *h* is reached. The whole process can be repeated again for the second and subsequent cycle of isotherms. The first cycle of sorption isotherm for an uncompressed rayon is shown in Fig. 24.

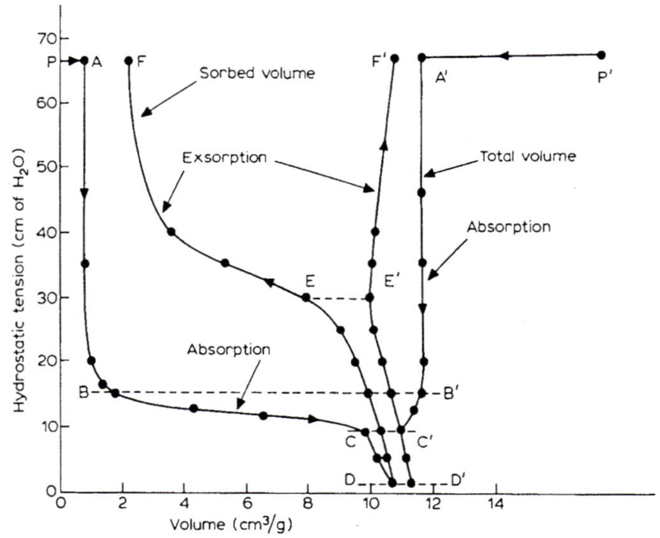

Fig. 24. Sorption isotherms for uncompressed rayon web [33].

contains a small fraction of unopened board chips which introduces heterogeneity into the system effecting the fluid flow mechanism. Those chips will have finer capillaries than the rest of the fluff system. Therefore, the fluid flow, equilibrium capacity, and retention in those discrete elements would be different from that of the bulk of the pad. Also, they will introduce additional interfaces within a layer, which would indirectly influence the transfer of fluid to adjacent layers. The interface problem discussed with respect to unopened board chips in a fluff pad is also applicable to the case where other additives are distributed within the fluff system if their physical structures are different from that of the individualized wood pulp fibers. Unless the flow mechanism in those interfaces is clearly understood, no theory can fully define the fluid flow mechanism in the entire system to an extent where one can precisely predict the overall performance of the composite.

1.4. Characteristics of Absorbent Components

The use of fiber masses in the manufacture of absorbent products provides a broad scope for the development of constructions that are specifically designed to meet a given set of functional requirements. There has been studies reported in the literature concerning the rate at which liquid is absorbed [4-7] as well as to the total volume of liquid that can be stored in the interior of a porous medium [8,9]. In order to understand these functions, it is important to characterize the porous medium in terms of its ability to attract, retain, and distribute fluid in the pores between fibers. A detailed analysis of this type of phenomena, using a porous plate absorbency tester [4-7,10] has been reported by Burgeni and Kapur [11].

The technique described by Burgeni and Kapur provides a complete absorption and exsorption isotherm and the volume changes occurring to the sample during this period as discussed earlier. This type of investigation can be carried out with samples at different densities and under different environmental pressures. The area under the absorption and exsorption branches represents the energy output and input on the capillary sorption cycle. The difference between absorption and exsorption energy is the hysteresis of the sorption process. The sorption hysteresis is a useful criterion to determine the efficiency of a sorption process. Its magnitude is of particular significance in the qualitative interpretation of phenomena related to the wicking of fluid in a fiber web as well as to the transfer of fluid from one web structure to another. Wide hysteresis loops indicate that a large portion of the surface-free energy fails to be converted into work done in absorption. Further studies at different web density provided the detail explanation concerning the fluid flow characteristic of absorbent products containing compressed elements either as an integral part of the uncompressed system known as fluff skin [12] or as an insert such as tissue or nonwoven.

The data obtained through the equilibrium absorption measurement under variable hydrostatic tension can be converted to so-called equivalent pore size distribution of the structure. Such an analysis provides the evaluation of different materials in terms of the structure property relationships.

As described in the subsequent chapters, porosity is not the only determining factor of absorbency. The wetting and contact angle of fibers play important roles toward understanding the structure-property relationship. Aberson [12a] has proposed a method to estimate the effective contact angle of relatively small fibers and fiber fragments of wood pulp. Low contact angles were shown to be associated with high rates of absorption in pads. All other factors remaining equal, the highest rates of absorption were associated with the lowest liquid holding capacity. The fiber length of wood pulp was shown to have only a small effect on the absorption rate but a relatively large effect on equilibrium capacity. Pads prepared from thick fibers were found to have large initial absorption rates compared with thin fibers although the liquid holding capacity was reduced. Burgeni and Kapur [11] also reported that the size and shape of the fibers, as well as their alignment, influenced the capillary sorption cycles. They claimed that the evaluation of capillary sorption equilibrium in fiber masses could be used as a performance criterion in the construction of absorbent products for specific end uses.

Among many factors, the modulus of single fibers is an important criterion, which has a significant influence on the pore structure of an absorbent product during its use. The initial pore structure of an absorbent pad under a definite pressure during the use of the pad will depend upon the load-bearing capacity of fibers individually as well as collectively. Fibers with high curl factor, and high bending modulus would resist external pressure and thus maintain their original pore structure. Fibers that offer less resistance would collapse more, resulting in lower porosity and thus low absorption capacity. For a high absorbency product, it is desirable to have a higher porosity material and, therefore, one would select high modulus fibers as long as they are aesthetically pleasing.

Compressional and absorptive behavior of bulk fiber systems has been studied by Gottlieb et al. [8] with a variety of synthetic fibers as well as natural fibers. It was shown that the amount of water absorbed increased linearly with the reciprocal of the square root of the product of the true density and denier of the fiber, which was consistent with the theoretical concept originally, proposed by Preston and Nimkar [12b]. It has been also demonstrated that there is an optimum dry bulk volume and correspondingly an optimum bulk density at which maximum absorption of water by a system of fibers will take place. Pads with glass, Fortisan and Dacron, which had high elastic moduli, showed high absorptive capacity. These authors also suggested that a successful way of improving the capacity of a fiber system to take up water could be achieved by increasing its stiffness by chemical or physical means.

Application of hydrophobic synthetic fibers, e.g., polyester or polypropylene fiber in combination with hydrophilic fibers, either as separate layers or blended together, has been reported in patent literature [13] as a means to reduce the wet collapse and thus maintain the original pore structure of the pad. When the fibers are all hydrophilic they become soft and readily compressible on absorbing liquid. Thus when the wet pad is placed under loads during use, the pad collapses and the corresponding inter fiber spaces in the pad are reduced. On the other hand, hydrophobic fibers provide a high degree of resilience even when they are wet. In the case of a layered structure, if a layer of hydrophobic synthetic fibers is placed on the top of a hydrophilic fiber layer and the pad under a confining pressure is exposed to liquid, the bottom layer will collapse to a higher degree than the top one. This difference in collapsibility will create a difference in capillary dimension, larger capillaries being at the top and smaller ones at the bottom. Consequently, the body fluid absorbed at the top layer would be readily transmitted to the bottom layer and will be retained therein [14]. Thus, relatively dry surface can be maintained in the composite while it is being subjected to a liquid source.

A critical problem occurs if the initial structure of an absorbent product at the dry stage is desired to have a low bulk volume. A high modulus fiber will maintain its high bulk at the dry state also and, therefore, a low volume product would be difficult to make without introducing a chemical bonding agent or by thermal bonding. Steiger [15] has applied the wet cross-linking technology [16] to achieve a structure, which could be compressed easily to a low bulk volume, or low porosity but expand and increase the porosity as it absorbs the liquid. This phenomena can be attributed to the wet resiliency inherent in wet cross-linked cellulose. The dry cross-linked cellulose [17] on the other hand provides high porosity or high bulk both under wet and dry conditions. Using the technique of Burgeni and Kapur [11], Steiger and Kapur [18] demonstrated that for maximum liquid-holding capacity, an uncompressed fibrous assembly requires wettable fibers of high wet modulus. Tests with rayons of various denier, state of finishing, and modulus indicate that, for compressed absorbents, it is desirable to minimize the required compacting forces by reducing the dry modulus and resilience of the fibers.

1.5. Water Swellable Fibers

Water swellable absorbent fibers have been used in modifying the absorbent structures. However, these fibers would modify different structures in different ways depending upon the chemistry of those materials and physical characteristics of the structures.

As discussed earlier, when a fluff pad consisting of short fiber systems interacts with water, the water goes mainly into the interfiber capillaries. Only a small portion (5 to 6%) of the total absorbed fluid is held inside the fiber or adhered on the fiber surface. The total amount of absorption is called the maximum capacity, whereas, the amount retained after squeezing or applying pressure may be called as the fluid retention capacity. The maximum capacity is influenced by the following factors: the apparent density of the pad, the wet and dry compressional resilience of the fluff, and the wettability of individual fibers.

If a given swellable fiber is used in compressed and uncompressed forms, the absorbency will be different in those cases because of the difference in the apparent densities. The mechanism of absorption will also be different, because a highly compressed fiber system will expand on absorbing water, whereas an uncompressed or lightly compressed fiber system will collapse on absorbing water [11]. Hence, the absorbency in the two systems (compressed and uncompressed) should be treated separately.

However, the swellable type of fibers become soft as they absorb water. The softness at equilibrium swelling is dependent upon the degree of swelling. If the degree of swelling is very high, the fiber loses its structural integrity and tends to dissolve in water.

In a compressed system (density: approximately 0.5 g/cm^3), if the fibers swell and maintain their structural integrity at wet stage, the fibers system will absorb more liquid due to the swelling and increased capillary size. The capillary size increases since the swelled fibers push one another due to the lack of space for free expansion. Therefore, swellable fibers, when used in compressed form, improve the maximum capacity as well as the fluid retention capacity of the system.

If the degree of swelling is too high, the fibers become soft (or, the structural integrity is reduced), the fibers will flatten out, fill up the capillaries and reduce the absorption. In an extreme case, they will completely block the pores, and a gel layer may form [19].

In an uncompressed fiber system (density: approximately 0.03 g/cm^3), if the fibers swell and maintain their structural integrity, the fluid retention capacity increases, but the total capacity (or the maximum capacity) may or may not increase. As the fibers swell, they readily fill up the pores, since, in this case, the free expansion space is available within the system itself. The retention capacity increases at the cost of the capacity of the capillaries. The total capacity may not alter, unless the fibers become too large on swelling. Again, if the degree of swelling is too high, gel formation may occur in this case also.

From the above discussion it is evident that the absorption capacity of a compressed swellable fiber system increases on increasing the swelling capacity, and decreases on increasing the softness of the fibers at wet stage. Thus, any modification to fibers that will increase the swelling characteristics and decrease the softness of the fibers at wet stage should improve the absorption capacity of compressed fiber systems.

In actual case, if the swelling capacity of a fiber is increased, the structural integrity at wet stage is decreased or vice versa. The softness and swelling capacity both change during a chemical reaction. In general, the absorption capacity is proportional to the swelling factor of the fiber, and inversely proportional to the softness of the wet fiber.

Because of these opposing effects, an optimum balance between swelling and softness should result in the maximum improvement in the absorption capacity. At a very high degree of swelling, softness becomes infinitely large and the absorption capacity of the fibers will be zero. However, before reaching this limiting condition, the gel stage occurs and the fibers become practically ineffective as an absorbent material.

This generalized concept is applicable to an absorbent structure composed of hundred percent swellable fiber. In actual practice, the swellable fiber is usually blended with conventional fibers such as rayon, cotton, pulp, or synthetic fibers. Such a blend could help to overcome the gelling problem and may, in fact, show synergism in absorbency. The extent of synergism, however, depends upon the physical structure of the absorbent system and the chemistry of the absorbent fiber.

2. ABSORBING FLUID CHARACTERISTICS

Most of the absorbent products available in the present day market deal with the absorption of water or body fluids. The theories available in the literature are based on the absorption of pure water or water containing simple electrolytes such as sodium chloride.

Although it has been long recognized that the mechanism of water absorption is quite different from body fluid absorption, no systematic study has been reported in the literature attempting to elucidate the latter mechanism. Perhaps the complexities of different body fluids, viz., urine, perspiration, blood, menstrual fluid, are so great and their chemistry is so poorly defined that it would be a futile effort to tackle the problem with the present state of knowledge. More biochemical and rheological information on different body fluids are required to define their flow characteristics in capillary system and diffusion through polymers present in the system.

Additionally, because of the existence of particulate matters (platelets) in the blood, a filtering effect occurs during the absorbing process, which gradually changes the pore structure of the absorbing medium. Figure 1 shows the absorption capacity versus time curves for radial wicking of water and of blood in spun-laced nonwoven fabrics [20]. The expected linearity was observed with the water while some deviation from this linearity and a generally lower absorption rate are observed with blood as the test fluid.

Unlike water, the characteristic properties of all body fluids change with time and environmental conditions. The changes may be accelerated or inhibited as soon as the fluid will interact with a fiber and it may continue to change throughout the absorption period. This continual change of chemical nature of the fluid may result in a constant change of the flow mechanism. Therefore, to deal with the absorbency of body fluid, one would have to

It is still true today that there has not yet been invented a single, simple test which can be used to objectively evaluate "absorbency", either as a practical performance measure of a material or as a scientific phenomenon [60]. Since absorbency is vital in many commercial and health maintenance functions, industry has had to develop a set of "standards" despite the inability to find incontrovertible terminology [61].

"Absorbency" can have many definitions, of which we offer a few examples:
initial uptake of fluid, equilibrium saturation, horizontal and vertical wicking, fluid migration, re-wetting and fluid perviousness.

The reader is invited to peruse the previous chapters of this book in order to find other synonyms which may be considered useful. Efforts continue to be made by various trade organizations (Scan, TAPPI, INDA, Zallcheming, etc.) and individuals to produce methods and equipment [62] which may improve the situation. It appears that a totally new data base must be acquired if we are to progress into the future [63].

Today's research has revealed that few, if any, of the existing methods for characterization of fibrous matrices can be applied to combination or composite structures of "normal" plus "super" absorbents. The need and opportunity to invent useful combination materials and to devise methods to demonstrate some unique property thereof were never greater than at present.

6.3. Absorbent Products

Over the centuries, man has devised a variety of useful materials based on fibers assembled so as to provide interstices of sufficient number and volume under ambient conditions, such that the "absorbency" fulfilled a needed function to imbibe and store fluid.

In certain parts of the world, there have arisen centers of knowledge (in terms of mechanical expertise) which have led to the development of rather unsophisticated methods for "forming" these absorbent structures. Whether built by the product producers themselves or by an equipment supplier, the machinery for absorbent product manufacture uses early technology including: cardin, knitting, and air-laying. All of these methods transport individual fibers from a compact source (shipping economics) to a final form which will meet the desired performance [64].

"Normal" absorbents take up and immobilize the absorbate via interconnecting "cells" which have the desired total liquid-holding volume under the conditions prescribed for use of the product. Picture, then, the effect of placing "super" absorbents within some of these cells. The total absorbent capacity of the structure or product may be increased or, alternatively, by maintaining the original capacity, the cost may be reduced since some of the less efficient "normal" absorbent material can now be removed. Thus, these new superabsorbent materials offer flexibility in design to make less bulky, more comfortable products. However, capacity may be reduced by (a) reduction of the compressive modulus of the dry matrix containing the superabsorbent or (b) gel-blocking of the capillaries and interconnecting passageways in the structure.

Absorbent consumer products are designed to hold a specific quantity of absorbed fluid, using a minimum of absorbent, and with design features to provide the required combination of capacity, rate of absorption, comfort and esthetics. This is true of diapers, sanitary protection products, underpads and other absorbent products.

The lack of appreciation of simple hydrodynamic engineering principles may have resulted in some of the early product failures. Superabsorbents merely immobilize the fluid by "entrapment" rather than by holding it in interstitial locations. This principle is vital in realizing that the hydrogel does not necessarily add to the fluid containment volume of the product, but may "steal" from the pore volume because of its physical requirements. Thus, the total product volume has probably not been changed at full saturation, whether using simple or super or mixed absorbents. The most effective products need to provide these features:

1. A special and stable zone for superabsorbent action.
2. Fixing this zone physically.
3. Mechanisms for immediate acceptance of exudates.
4. Means to prevent squeeze out.
5. Reduction of strikeback when reducing absorbent bulk.
6. Maintenance of product integrity during packaging, shipping, and consumer application.

It is apparent that engineering in both products and materials rather than substitution alone is the requirement for realizing the potential of superabsorbents. In the future, we will be exposed to engineered composites, incorporating some of the major features listed earlier.

Superabsorbents could be available now in millions of pounds annually but there is limited production because the market has not yet developed to its full potential. In addition to needed applications technology, there is the problem of cost; producers offer the current brands of fibrous, powder and laminated superabsorbents at much higher than the wood pulp.

The cost plus an understanding of hydrodynamic requirements provides opportunity for future development. The current best products in the market indicate the trend of innovation required in product engineering together with concomitant equipment development to assemble these products at a rate to be economically viable. The challenge then is to devise geometric structures of sufficient hydraulic volume to meet the need, by use of the most economic combination of materials to transport, immobilize and retain the fluid to be absorbed [65].

In addition, there is incipient activity in the application of superabsorbents in agriculture[66]. This is based upon the inherent properties of these hydrogel materials to attract and retain large amounts of water. Thus, a means is available to hold water in dry soil or sand until the plant needs it. Some reports have surfaced concerning the use of superabsorbents in medicinals. The materials per se in specific particle form provide a soft "scrubber" for initial attention of traumatic wounds. Any action in other fields is tempered by the availability of natural materials, costs and lack of application technology. Potential suppliers need to develop the economics of scope [plus production facilities] which will provide materials with a unique function at an economically viable level of cost performance.

6.4. Application of Biotechnology

The newest frontier for the development of next generation absorbent materials is biotechnology [67-70]. Certain microorganisms also exist for the biosynthesis of hydrocarbon polymers, called "biopolymers" from which synthetic-like fibers and plastics can be made [71]. The techniques of recombinant DNA/genetic engineering can be expected to provide these new biomaterials at low, stable prices and without dependence upon petrochemicals, timber, animals or crops as is the case today.